CAMBRIDGE MATHEMATICAL TEXTBOOKS

Cambridge Mathematical Textbooks is a program of undergraduate and beginning graduate level textbooks for core courses, new courses, and interdisciplinary courses in pure and applied mathematics. These texts provide motivation with plenty of exercises of varying difficulty, interesting examples, modern applications, and unique approaches to the material.

A complete list of books in the series can be found at www.cambridge.org/mathematics Recent titles include the following:

"This is an excellent text for a first course in analysis in one and several variables for students who know some linear algebra. The book starts with the real numbers, does differentiation and integration first in one variable, then in several, and finally covers differential forms and Stokes' theorem. The style is friendly and conversational, and hews to the principal of going from the specific to the general, making it a pleasure to read."

– John McCarthy, Washington University in St. Louis

"Conway's previous texts are all considered classics. "A First Course in Analysis" is destined to be another. It is written in the same friendly, yet rigorous, style that his readers know and love. Instructors seeking the breadth and depth of Rudin, but in a less austere and more accessible form, have found their book."

– Stephan Ramon Garcia, Pomona College

"This is a beautiful yet practical introduction to rigorous analysis at the senior under-graduate level, written by a master expositor. Conway understands how students learn, from the particular to the general, and this informs every aspect of his text. Highly recommended."

– Douglas Lind, University of Washington

"A First Course in Analysis charts a lively path through a perennially tough subject. Conway writes as if he's coaching his reader, leavening the technicalities with advice on how to think about them, and with anecdotes about the subject's heroes. His enjoyment of the material shines through on page after page."

– Bruce Solomon, Indiana University, Bloomington

"This year-long undergraduate book carefully covers real analysis "from sets to Stokes" and is done in a friendly style by an experienced teacher and masterful expositor. There are plenty of examples, exercises, and historical vignettes that both give the student the opportunity to gain technical mastery of the material and to whet their appetites for further study."

– William T. Ross, University of Richmond

"A First Course in Analysis is a beautifully written and very accessible treatment of a subject that every math major is required to learn. It will join Conway's other textbooks as a classic in Advanced Calculus. Those who teach and learn analysis through Conway's book will appreciate his cheerful and easy-to-understand style."

– Wing Suet Li, Georgia Institute of Technology

A First Course in Analysis

John B. Conway

The George Washington University, Washington, DC, USA

CAMBRIDGE
UNIVERSITY PRESS

CAMBRIDGE
UNIVERSITY PRESS

University Printing House, Cambridge CB2 8BS, United Kingdom

One Liberty Plaza, 20th Floor, New York, NY 10006, USA

477 Williamstown Road, Port Melbourne, VIC 3207, Australia

4843/24, 2nd Floor, Ansari Road, Daryaganj, Delhi - 110002, India

79 Anson Road, #06-04/06, Singapore 079906

Cambridge University Press is part of the University of Cambridge.

It furthers the University's mission by disseminating knowledge in the pursuit of education, learning and research at the highest international levels of excellence.

www.cambridge.org
Information on this title: www.cambridge.org/9781107173149
DOI: 10.1017/9781316779811

First published 2018

Printed in the United States of America by Sheridan Books, Inc.

A catalog record for this publication is available from the British Library.

ISBN 978-1-107-17314-9 Hardback

For Ann
As always

Contents

Preface

This book is intended for the year-long course on analysis taught to undergraduates near the end of their education and graduate students at the beginning of theirs. It is recommended for students in the physical sciences, economics, and statistics. Sometimes this course is titled *Advanced Calculus* and sometimes just *Analysis*. The subject matter comprises the basics of differentiation and integration of functions of one and several variables, topped off with some form of the theorems of Green, Gauss, and Stokes.

How is the material presented in this book? A guiding principle I have followed for a long time when I teach or when I write a textbook, as opposed to a monograph, is to go from the particular to the general. This practice came about from a combination of what I observed in the classroom and the historical way mathematics developed. The present book contains many instances of this approach, but a dramatic illustration is what happens in the last two chapters on surfaces and the Green, Gauss, and Stokes Theorems. I begin (Chapter 8) with an exposition of what happens in $\mathbb{R}^2$ and $\mathbb{R}^3$ including proofs of these just mentioned theorems in this setting. Unlike the rest of this book, however, there are places here where I relax the rigor. This is especially so when I discuss the orientation of surfaces in $\mathbb{R}^3$. In the following chapter (Chapter 9) I introduce differential forms on $\mathbb{R}^p$, constantly illustrating everything with reference to what was seen in the lower dimensional spaces. Here rigor is enforced. After we establish the Generalized Stokes Theorem, we go back and prove the particular cases of the three big theorems in $\mathbb{R}^2$ and $\mathbb{R}^3$ as a consequence. I think this is a better approach than going directly to a treatment of differential forms and surfaces in $\mathbb{R}^p$. My experience is that most students at this level are not ready for such a direct route without some preparation.

Another philosophy of mine in writing is not to introduce a concept until it's needed; I want the reader to quickly see the concept in action. For example I wanted to give a rather detailed diagnosis of the nature of a critical point for a function from $\mathbb{R}^p$ into $\mathbb{R}$. For me this entails introducing symmetric matrices and the Spectral Theorem. This could have been done in a special chapter on linear algebra. Instead I waited until we were ready to discuss critical points. In the first course in Linear Algebra the usual practice, due to time constraints, is never to talk about the Spectral Theorem. I therefore prove it in $\mathbb{R}^p$.

Speaking of linear algebra, that's a major difficulty for anyone teaching a version of this course. It was certainly a problem when I was writing this book. Linear algebra is a stated prerequisite, but I know most of the students whom I've taught over the years have forgotten some of the linear algebra they learned. Nevertheless when we teach a course on multivariable analysis we cannot reteach linear algebra. The path I take is to recall some of the high points, usually without proofs. On the other hand I need and use more on determinants than I think students know or have even seen. As a consequence I define determinants (6.4.18) and derive their properties, though the proofs tend more toward the sketchy than elsewhere in this book.

Another of my beliefs about teaching and writing a text is that just because I know some topic that extends the basic material doesn't mean I should teach it in a class or include it in the text. Besides, I want to keep the book lean. I also want to have confidence that anyone using this book could make it to the multivariable versions of the Fundamental Theorem of Calculus. I suspect many instructors will regret the absence of some topic they like. If this is your case, I'm sorry; but I think you'll find other interesting things here.

Universities with graduate programs frequently have graduate students enrolled in this course. When I teach a mixed class I focus on the undergraduates; this is what I did while writing this book. This has several consequences. I assume a minimum background: three semesters of calculus and a semester of linear algebra. I have to also assume that the students have some level of comfort reading and writing proofs. I certainly am aware that someone with my stated minimum background may not be comfortable with proofs. As a partial bridge, the proofs that are in the first half of the book are more detailed.

Finally, three additional statements about this book. When I write a text I always imagine myself presenting the material in front of a classroom. Some have commented on my "chatty" style and this is the source. Sections marked with an asterisk (*) are optional and are not needed for the rest of the book. You might also observe that to some extent computational exercises are sparse (but not absent). It seems to me that students have done a lot of such things in the three semesters of calculus that is assumed of the readers of this text. If they survived that, I didn't see a reason to put more of the same here.

Synopsis of the Chapters

The book begins with a chapter on the real numbers and some other basic material. It may be that some students could skip part of this, but I suspect it will be rare that there is a student in a class using this book who could skip the entire chapter. The starting point is a development of the real numbers assuming that you understand the rationals. I give two approaches to this: an abbreviated one and a more thorough one. (This thorough one still has some gaps as this is not the focus of this book.) The chapter also includes material on sequences, open and closed sets in $\mathbb{R}$, continuity, as well as countable and uncountable sets. The point of this chapter is to give all readers a chance to fill in gaps and expose them to proofs. (The proofs here are straightforward and presented in more detail than elsewhere in the book.)

The next three chapters contain the core of the basic material in one dimension. Chapter 2 is on differentiation in one variable and concludes with l'Hôpital's Rule and Taylor's Theorem. Chapter 3 gives the theory of the Riemann integral over subsets of $\mathbb{R}$. It also contains starred sections on Lebesgue's Theorem giving a necessary and sufficient condition for integrability and the Riemann–Stieltjes integral. Chapter 4 is entitled "Sequences of Functions." It covers the standard material on uniform convergence as well as power series.

Chapter 5 is entitled "Metric and Euclidean spaces." I am aware that many books at this level avoid metric spaces, but I decided not to. Less you think this violates my principle of going from the particular to the general, don't forget that almost everything here has first appeared as a result for $\mathbb{R}$ in Chapter 1. When this is combined with the fact that the exposition is replete with examples in $\mathbb{R}^p$ that illustrate the results, I feel comfortable that my principle is intact. It also seems to me that proofs in the abstract spaces are easier. In fact if you state a result in Euclidean space and try to fashion a proof of it, you are strongly tempted to get involved with coordinates and such while the proof in metric spaces is clean and shows what is really going on. The chapter ends with a section on spaces of continuous functions. Included here are the Stone–Weierstrass Theorem and the Arzela–Ascoli Theorem, two results I consider basic but are frequently omitted at this level. (In fact the Stone–Weierstrass Theorem is used to prove Fubini's Theorem in Chapter 7.)

Chapter 6, "Differentiation in Higher Dimensions," covers the standard topics in this subject. An introduction and recollection of most of the linear algebra needed for the rest of the book is presented and used here. The treatment of critical points may differ from most books. Here the observation that the second derivative of a $C^{(2)}$-function from $\mathbb{R}^p$ into $\mathbb{R}$ is a hermitian linear transformation is central. This permits the application of the Spectral Theorem and allows us to carefully analyze the behavior of the function at a critical point, even when it's a saddle point.

Chapter 7 is titled "Integration in Higher Dimensions" and covers the Riemann integral in an abbreviated way. It seems to me that some books spend too much time on this integral, time that I think can be better spent on other things. The treatment of Riemann integration given in Chapter 7 achieves simplicity and rigor by only integrating continuous functions. Students who continue their study of mathematics will see the Lebesgue integral, which is not only more general but a lot easier to understand than an in-depth treatment of the Riemann integral in $\mathbb{R}^p$. For example, the issue of integrability is largely dormant in the Lebesgue case but complicated in the Riemann case. Students who don't continue in mathematics are not hurt, since when they encounter integration in higher dimensions in their future life it is almost always an iterated integral and what we do here is ample.

Chapter 8, "Curves and Surfaces," focuses on these objects in $\mathbb{R}^2$ and $\mathbb{R}^3$. Chapter 9, "Differential Forms," extends this to $\mathbb{R}^p$. I've already discussed these chapters and the approach taken.

References and Sources

As I indicate at various places in the body of this book, I have used [7], [11], and [15] while I was writing this book. I have certainly used the last two as sources of exercises. Also I've made use of the Internet more heavily than with any of my previous books. Sometimes I used articles from the Web that, as far as I know, are not available in print. When I did this I usually corresponded with the author if (s)he was identifiable. In the Bibliography I list this reference with the site *as I found it during one of my final readings of this book*. I also found a lot of my exercises online. Sometimes these were found on the Web where no author was designated; in such a case I did not reference the site.

Biographical notes

As in my last two books I have added a short biographical note for any mathematician whose work is quoted. There is no scholarship on my part in this, as all the material is from secondary sources, principally what I could find on the Web. In particular, I made heavy use of the St Andrews University site www-history.mcs .st-andrews.ac.uk/history/BiogIndex.html and some use of Wikipedia.

I emphasize the personal aspects of the mathematicians we encounter along the way, rather than recite their achievements. This is especially so when I discover something unusual or endearing in their lives. I figure many students will see their achievements if they stick with the subject and most, at this stage of their education, won't know enough mathematics to fully appreciate the accomplishments. In addition I think the students will enjoy learning that these famous people were human beings.

For Students

From a simplistic point of view, this course repeats calculus. That's grossly misleading. True, we'll talk about differentiation and integration in one and several variables and I am assuming that all students have completed the standard three-semester sequence in calculus. On the other hand everything done here will be presented with complete mathematical rigor. The emphasis is not on computing but on understanding the concepts; part of that understanding is proving the results as well as working out examples and exercises. In addition, we'll see new material in the second semester when we discuss integration of functions defined on surfaces.

I've long thought this course is the most difficult in the undergraduate curriculum. It's also one of the most important for your future in mathematics, whether you go on to graduate school or begin a career just after graduation. So work hard at this.

My advice to all students is to read the book with paper and a writing implement of your choice. Draw pictures, if this is needed. For goodness sake, read the material between the proofs; these paragraphs contain a lot that will help your understanding. I leave a lot of detail checking to the reader and frequently insert such things as (Why?) or (Verify!) in the text. I want you to delve into the details and answer these questions. It will check your understanding and give some perspective on the proof.

I am also convinced there are some details that should not appear in books and that professors should not present in the classroom. Good examples of such material are detailed calculations and long proofs that contain a jungle of subscripts. You just can't absorb such things by watching and listening. One place where I leave a lot of details to the reader is the definition and properties of the determinant of a square matrix. This starts at (6.4.18). You aren't going to understand this course by watching someone else do the work. You need to go to your study desk and push on ahead. As I have often said, learning mathematics is not a spectator sport.

I also strongly advise you to at least read all the exercises. With your schedule and taking other courses, you might not have the time to try to solve them all, but at least read them. They contain additional information.

Thanks

First I thank all the students who have sat in my classes over the years. I think you learned something from me, but I also learned something from you. I've had conversations during my career with several professional mathematicians on subjects that appear here. These are too numerous to remember let alone list, but I thank you all. I specifically want to thank Waclaw Szymanski who showed me a nice improvement of the proof of the Stone–Weierstrass Theorem. Undoubtedly my greatest thanks go to Professor Mark Hunacek who read many of the chapters, gave me several exercises, and provided valuable feedback.

Thanks for your attention.
Live long and prosper

1 The Real Numbers

In this chapter we'll develop the properties of the real numbers, the foundation of this book. But first we start with an introduction to set theory.

1.1. Sets and Functions

The concept of a set is the basis for all of mathematics. The reader is likely to have come across many of the notions associated with set theory, and if this experience has been extensive, please skip to the next section.

A set X is a collection of things called *elements* or *points*. We'll treat the idea of a set and an element of a set as a fundamental notion and not try to give a more formal definition. The notation $x \in X$ is used to denote the fact that x is an element that is contained in the set X. You might also sometimes see $x, y \in X$; this means that x and y are both elements of X. (Note that the statement $x, y \in X$ does not imply that these two points are distinct.) When x fails to be an element of the set X, we write $x \notin X$.

A basic concept of set theory is that of a *subset*. We say that A is a subset of X if A is a set and every element of A is also an element of X. Using symbols we can write this relation by saying that if $x \in A$, then $x \in X$. We introduce the notation $A \subseteq X$ to denote this; this is read as "A is contained in X," or "A is a subset of X." We could also write $X \supseteq A$, read as "X contains A." Another notation for $A \subseteq X$ that the reader will encounter in the literature is $A \subset X$, but we won't use this in this book. Note that if $A = X$ it is also the case that $A \subseteq X$. We say that A is a *proper subset* of X provided $A \subseteq X$ but $A \neq X$.

There are many ways to define sets and subsets, depending on the situation. For example, consider the set of *natural numbers*; that is, the set of positive integers denoted by $\mathbb{N}$. To define this set we could also write $\mathbb{N} = \{1, 2, 3, \ldots\}$. If we want to denote the subset of $\mathbb{N}$ consisting of even integers, we can write $\{2n : n \in \mathbb{N}\}$. The odd integers can be expressed by $\{n - 1 : n \in \mathbb{N} \text{ and n is even}\}$, or we could write $\{2n - 1 : n \in \mathbb{N}\}$. We also have the set $\mathbb{Z}$ of all integers, positive and negative as well as zero. So

$$\mathbb{Z} = \{n - m : n, m \in \mathbb{N}\}$$

Hence $\mathbb{N} \subseteq \mathbb{Z}$. (We are assuming that the reader is familar with the natural numbers, $\mathbb{N}$, the integers, $\mathbb{Z}$, the rational numbers, $\mathbb{Q}$, and with their properties. In the next section we'll introduce the real numbers, $\mathbb{R}$.)

If we are given a set X and A and B are both subsets of X, then in a similar way as above we say that A is a *subset* of B if every element of A is also an element of B; that is, if $x \in A$, then $x \in B$. In symbols we write this as $A \subseteq B$ or $B \supseteq A$. We can also say that A is *contained in* B or B *contains* A. The two sets A and B are *equal* if we have both $A \subseteq B$ and $B \subseteq A$.

There are two special subsets of any set X: the set X itself and the *empty set* $\emptyset$ consisting of no elements. The empty set may take a moment to digest, but it plays an important role in mathematics. A distinction of $\emptyset$ is that it is a subset of every subset of X, the only such subset of X. If $A \subseteq X$ and $x \in X$, there is also the notation $x \notin A$ to indicate that the point x does not belong to A. So if $x \in X$, then $x \notin \emptyset$. There is another special subset – actually a collection of them. If $x \in X$, then $\{x\}$ denotes the set consisting of the single element x. This is called a *singleton* set. So $x \in \{x\}$ and $\{x\} \subseteq X$.

We want to define some operations between subsets. So we consider X as the universe and examine its various subsets $A, B, C, \ldots$

1.1.1. Definition. If $A, B \subseteq X$, the *intersection* of A and B is the set $A \cap B$ defined by

$$A \cap B = \{x \in X : x \in A \text{ and } x \in B\}$$

1.1.2. Example. (a) If $A \subseteq B$, then $A \cap B = A$.

(b) Consider the set $\mathbb{Q}$ of all rational numbers, positive, negative, and 0. That is

$$\mathbb{Q} = \left\{ \frac{n}{m} : n, m \in \mathbb{Z} \text{ and } m \neq 0 \right\}$$

If $A = \{x \in \mathbb{Q} : -7 \leq x \leq 4\}$, then $A \cap \mathbb{N} = \{1, 2, 3, 4\}$.

(c) If $A = \{x \in \mathbb{Q} : x \leq 0\}$ and $B = \mathbb{N}$, then $A \cap B = \emptyset$.

(d) Note that if $A \subseteq B$, then $A \cap B = A$.

If $A \cap B = \emptyset$, we say that the sets A and B are *disjoint*.

1.1.3. Definition. If $A, B \subseteq X$, then the *union* of A and B is the set $A \cup B$ defined by

$$A \cup B = \{x \in X : x \in A \text{ or } x \in B \text{ or both}\}$$

It's worth emphasizing that the use of the word "or" in the preceding definition is not the exclusive "or". In other words, when we say in the definition that $x \in A$ or $x \in B$ we do not exclude the possibility that x belongs to both A and B. That is, we do not insist that $A \cap B = \emptyset$.

1.1.4. Example. (a) If $A \subseteq B$, then $A \cup B = B$.

(b) $A \cap B \subseteq A \cup B$.

(c) For any subset A of X, $A \cup \emptyset = A$.

(d) If $A = \{x \in \mathbb{Q} : -7 < x \leq 1\}$ and $B = \{x \in \mathbb{Q} : -1 < x < 5\}$, then $A \cup B = \{x \in \mathbb{Q} : -7 < x < 5\}$.

(e) If $A = \{x \in \mathbb{Q} : -7 < x \leq 1\}$ and $B = \{x \in \mathbb{Q} : 2 \leq x \leq 5\}$, then $A \cup B = \{x \in \mathbb{Q} : either -7 < x \leq 1 \text{ or } 2 \leq x \leq 5\}$.

1.1.5. Proposition. *The distributive laws hold for union and intersection. That is:*

(a) *if $A, B, C \subseteq X$, then $A \cap (B \cup C) = (A \cap B) \cup (A \cap C)$; and*
(b) *if $A, B, C \subseteq X$, then $A \cup (B \cap C) = (A \cup B) \cap (A \cup C)$.*

Proof. This proof is the prototype for establishing that two sets are equal: we take an element of the left-hand side and show it's an element of the right-hand side; then take an element of the right-hand side and show it's an element of the left-hand side.

(a) If $x \in A \cap (B \cup C)$, then $x \in A$ and $x \in B \cup C$. Thus $x \in B$ and $x \in C$. But this says that either "$x \in A$ and $x \in B$" or "$x \in A$ and $x \in C$." Therefore $x \in (A \cap B) \cup (A \cap C)$. Conversely, assume that $x \in (A \cap B) \cup (A \cap C)$. So either $x \in (A \cap B)$ or $x \in (A \cap C)$. In the first case, $x \in A$ and $x \in B$; in the second case, $x \in A$ and $x \in C$. Thus in either case, $x \in A$; also, depending on the case, either $x \in B$ or $x \in C$. Therefore $x \in A \cap (B \cup C)$.

(b) The proof of this part has a slightly different flavor than the proof of the first part. If $x \in (A \cup B) \cap (A \cup C)$, then $x \in A \cup B$ and $x \in A \cup C$. So either $x \in A$ or $x \in B$. If $x \in A$ then we have that $x \in A \cup (B \cap C)$. If $x \notin A$, then the fact that $x \in A \cup B$ and $x \in A \cup C$ implies that $x \in B$ and $x \in C$; hence $x \in B \cap C$. Therefore $x \in A \cup (B \cap C)$. The proof of the other half of (b) is Exercise 2. ∎

We also define the *difference* of the two sets as

$$A \backslash B = \{x \in X : x \in A \text{ but } x \notin B\}$$

Some mathematicians use the notation $A - B$ instead of $A \backslash B$. I prefer the notation $A \backslash B$ because in some situations $A - B$ is ambiguous. For example, if A and B are subsets of $\mathbb{Q}$, we will use the definition $A - B = \{a - b : a \in A, b \in B\}$. The same applies when A and B are subsets of a vector space. So throughout this book the difference of two sets will be denoted using the backslash.

1.1.6. Example. (a) If X is any set and $A \subseteq X$, then $A \backslash \emptyset = A$ and $A \backslash X = \emptyset$.

(b) If $A = \{x \in \mathbb{Q} : 0 < x < 1\}$ and $B = \{x \in \mathbb{Q} : \frac{1}{2} < x \leq 3\}$, then $A \backslash B = \{x \in \mathbb{Q} : 0 < x \leq \frac{1}{2}\}$.

(c) If $A = \{x \in \mathbb{Q} : 0 < x < 1\}$ and $B = \{x \in \mathbb{Q} : 3 \leq x < \infty\}$, then $A \backslash B = A$.

For any subset A of X, the difference $X \backslash A$ is called the *complement* of A. Elsewhere the reader might encounter the notation A^c or $\tilde{A}$ to denote the complement of A. Note that $X \backslash (X \backslash A) = A$ (Exercise 4).

1.1.7. Proposition (De Morgan's[1] Laws). *If X is any set and A and B are subsets of X, then:*

(a) $X\backslash(A \cup B) = (X\backslash A) \cap (X\backslash B)$;
(b) $X\backslash(A \cap B) = (X\backslash A) \cup (X\backslash B)$.

Proof. We prove (a) and leave the proof of (b) as Exercise 5. Again we use the standard approach to proving that two sets are equal. If $x \in X\backslash(A \cup B)$, then $x \notin A \cup B$. The only way this can happen is that both of the following two statements are true: $x \notin A$ and $x \notin B$. That is, $x \in X\backslash A$ and $x \in X\backslash B$; equivalently, $x \in (X\backslash A) \cap (X\backslash B)$.

Now assume that $x \in (X\backslash A) \cap (X\backslash B)$. This says that $x \in X\backslash A$ and $x \in X\backslash B$; that is, $x \notin A$ and $x \notin B$. But combining these two statements means $x \notin A \cup B$, or that $x \in X\backslash(A \cup B)$. ∎

We now extend the concepts of intersection and union to multiple sets. Indeed, we'll extend these to infinite collections of sets. Namely, assume that $A_1, A_2, \ldots$ are subsets of X and define

$$\bigcap_{n=1}^{\infty} A_n = \{x \in X : x \in A_n \text{ for all } n \geq 1\}$$

$$\bigcup_{n=1}^{\infty} A_n = \{x \in X : x \in A_n \text{ for some } n \geq 1\}$$

There is a version of De Morgan's Laws for this as well.

1.1.8. Theorem (De Morgan's Laws). *If X is a set and $\{A_n : n \geq 1\}$ is a collection of subsets, then:*

(a) $X\backslash[\bigcup_{n=1}^{\infty} A_n] = \bigcap_{n=1}^{\infty}(X\backslash A_n)$;
(b) $X\backslash[\bigcap_{n=1}^{\infty} A_n] = \bigcup_{n=1}^{\infty}(X\backslash A_n)$.

The proof of this last theorem is Exercise 6 and proceeds like the proof of Proposition 1.1.7.

We conclude this section with a discussion of functions. If X and Y are two sets, then a *function* from X into Y is a rule, denoted by $f : X \rightarrow Y$, that assigns to each x

[1] Augustus De Morgan was born in 1806 at Madura, India (now Madurai). His father was an officer in the British Army stationed there. The young Augustus lost his sight in one eye shortly after birth and the family returned to England when he was seven months old. He entered Trinity College, Cambridge in 1823. He received his BA but refused to take a theology exam, which was required for an advanced degree. He returned to London in 1826 to study for the Bar, but instead he became the first professor of mathematics at the newly established University College London in spite of the fact that he had never published in the subject, a fact he soon remedied. On a matter of principle he resigned in 1831. He was once again appointed in 1836, but resigned again in 1866. In 1838 he introduced the term "mathematical induction." The process had been in use before, but without clarity; De Morgan managed to give it a rigorous basis. Through his life he was a prolific author with hundreds of articles and many books. He introduced the laws of the present proposition and was a great reformer of mathematical logic. In addition he founded the London Mathematical Society and became its first president. He was quite dogmatic, as his two resignations might indicate. He never voted and never visited the House of Commons, the Tower of London, or Westminster Abbey. He died in 1871 in London.

in X a unique point y in Y. Synonyms for function are the terms *map* and *mapping*. The set X is called the *domain* of f and the set Y is called the *range* of f. The set $f(X) = \{f(x) \in Y : x \in X\}$ is called the *image* of f. Note the distinction between range and image. Now that you have noted the distinction you should be aware that some mathematicians define the range of a function to be what we have called the image and vice versa. When they do this they sometimes use the term codomain for what we call the range. Confused? Don't worry too much about it except when you consult other sources; we will consistently use the terms as we defined them above. Frankly, the distinction will affect very little that is said.

1.1.9. Example. (a) $f : \mathbb{Q} \to \mathbb{Q}$ defined by $f(x) = x^2$ is a function. Its domain is $\mathbb{Q}$ and its image is the set of rational numbers in $[0, \infty)$. What is its range? You could say it's the same as its image or you could say it's $\mathbb{Q}$. Perhaps this vagueness is unappealing, but there is no "correct" answer. What you call its range might depend on your purpose or the context of the discussion.

(b) If for each x in $\mathbb{Q}$ we let $f(x) = +1$ when $x \geq 0$ and $f(x) = -1$ when $x \leq 0$, then this is not a function since the value of $f(0)$ is not uniquely defined. If we were to redefine f by stating that $f(x) = -1$ when $x < 0$, then it is a function.

(c) If X and Y are sets, $y_0 \in Y$, and $f(x) = y_0$ for every x in X, then $f : X \to Y$ is a function – called a *constant function*.

(d) If X is any set and $A \subseteq X$, define $\chi_A : X \to \mathbb{R}$ by

$$\chi_A(x) = \begin{cases} 1 & \text{if } x \in A \\ 0 & \text{if } x \notin A \end{cases}$$

This function is called the *characteristic function* of A. Some call this the *indicator function*. Observe that for all x in X, $\chi_\emptyset(x) = 0$ and $\chi_X(x) = 1$.

If $f : X \to Y$ and $g : Y \to Z$, then the *composition* of f and g is the function $g \circ f : X \to Z$ defined by

$$g \circ f(x) = g(f(x))$$

for all x in X. So, for example, if $f(x) = x^2$ and $g(x) = x^3$, then $g \circ f(x) = (x^2)^3 = x^6$. Similarly $f \circ g(x) = x^6$, so that in this case $g \circ f = f \circ g$. If again $f(x) = x^2$ and $g(x) = \sin x$, then $g \circ f(x) = \sin(x^2)$ while $f \circ g(x) = (\sin x)^2$. So it is not true that composition is commutative.

1.1.10. Definition. A function $f : X \to Y$ is called *surjective* if for each y in Y there is at least one x in X such that $f(x) = y$. f is *injective* if for x_1, x_2 in X, the relation $f(x_1) = f(x_2)$ implies that $x_1 = x_2$. f is *bijective* if it is both injective and surjective.

In the literature the reader will often see the term "onto" instead of surjective and one-to-one instead of injective. I have no problem with the term one-to-one; I often use it and the reader might see it in this book. I have a language problem, however, with using onto as an adjective when it is a preposition. While I might use it that way in a casual conversation, when I am being a bit more formal I won't.

We'll encounter the terms surjective, injective, and bijective frequently as we progress.

1.1.11. Example. (a) The function $f : \mathbb{Q} \to \mathbb{Q}$ defined by $f(x) = x + 1$ is bijective.

(b) The function $f : \mathbb{N} \to \mathbb{N}$ defined by $f(x) = x^2$ is injective but not bijective.

(c) The function $f : \mathbb{Z}\backslash\{0\} \to \{n^2 : n \in \mathbb{N}\}$ defined by $f(x) = x^2$ is surjective but not injective.

Exercises

In these exercises X is a given set and $A, B, C, \ldots$ are subsets of X.

(1) Let $A = \{2n : n \in \mathbb{N}\}$, $B = \{2n + 1 : n \in \mathbb{N}\}$, $C = \{3n : n \in \mathbb{N}\}$, $D = \{3n + 1 : n \in \mathbb{N}\}$, $E = \{3n + 2 : n \in \mathbb{N}\}$. Determine each of the following sets: (a) $A \cap B$; (b) $A \cap C$; (c) $A \cup D$; (d) $D \cup E$; (e) $(A \cap E) \cup (B \cap E)$.
(2) Complete the proof of Proposition 1.1.5(b).
(3) Prove that the associative laws apply to unions and intersections: $A \cap (B \cap C) = (A \cap B) \cap C$ and $A \cup (B \cup C) = (A \cup B) \cup C$.
(4) Give a detailed proof that $X\backslash(X\backslash A) = A$.
(5) Prove part (b) of Proposition 1.1.7.
(6) Prove Theorem 1.1.8.
(7) If $f : X \to Y$ is a function, prove that the following statements are equivalent. (a) f is injective. (b) $f(A\backslash B) = f(A)\backslash f(B)$ for all subsets A and B of X. (c) $f(a \cap B) = f(A) \cap f(B)$ for all subsets A and B of X.

1.2. The Real Numbers

I am aware that the reader has been working with the real numbers $\mathbb{R}$ for the entirety of his/her mathematical life. However I suspect that many students studying the present material for the first time may be unaware of some of the properties of $\mathbb{R}$ that are crucial for making calculus work. Indeed one of these properties, The Completeness Property, is not shared by $\mathbb{Q}$ and makes it impossible for calculus to survive as a theory only involving rational numbers.

For the teacher there are two ways to handle this. One is to just state the needed properties of $\mathbb{R}$ and proceed to develop calculus. Another is to start with the properties of $\mathbb{Q}$ and carefully develop the definition of the real numbers and derive the needed properties. This second approach gives a solid understanding of the material and gives a grounding in writing proofs in analysis. The disadvantage is that going through this material takes time that an instructor may not have. The first approach has the advantage of quickly getting to the more advanced topics. If I were the instructor, I can easily imagine different circumstances under which I would be led to use either of these approaches.

So which approach will be adopted in this book? In some sense both. We will begin with a quick survey of the first approach and then go through a more thorough grounding. Why? I assure you that doing it this way is not chosen because of intellectual indecision or cowardice. As I said before, there are reasons that support taking either route. I'm going to leave it to individual instructors and readers to figure out if they have the time to do the second or are comfortable only taking the first.

Quick Approach

In this approach we assume the reader is familiar with the arithmetic properties of the set of real numbers, $\mathbb{R}$, as well as its usual order relations, $<$ and $\leq$. We also assume the reader knows the distinction between rational numbers and irrational numbers. The first important fact about $\mathbb{R}$ that the reader may not be fully conscious of is the following.

1.2.1. Axiom (Density Property). *If a and b are rational numbers and a $<$ b, then there is an irrational number x with a $<$ x $<$ b. Similarly, if a, b are irrational numbers and a $<$ b, then there is a rational number x with a $<$ x $<$ b.*

The Density Property will be used frequently as we develop the theory of differentiation and integration. The other important property we need involves the ordering on $\mathbb{R}$.

If $E \subseteq \mathbb{R}$ we say that E is *bounded above* if there is a number a such that $x \leq a$ for all x in E. Such a number a is called an *upper bound* of E. Similarly E is *bounded below* if there is a number b with $b \leq x$ for all x in E; b is called a *lower bound* of E. It is easy to see that E is bounded above with an upper bound a if and only if the set $-E = \{-x : x \in E\}$ is bounded below with a lower bound $-a$. For this reason any statement about the upper bound of a set has its analogue for the lower bound and we will frequently only do the upper version. A set E is *bounded* if it is both bounded above and bounded below.

1.2.2. Definition. If $E \subseteq \mathbb{R}$ and E is bounded above, then a *least upper bound* or *supremum* of E is a number α that satisfies: (i) α is an upper bound for E; (ii) $\alpha \leq a$ for any other upper bound a for E. Similarly, if E is bounded below, then the *greatest lower bound* or *infimum* of E is a number β that satisfies: (i) β is a lower bound for E; (ii) $b \leq \beta$ for any other lower bound b for E. In symbols we write $\alpha = \sup E$ and $\beta = \inf E$. (The reader may have seen the notation $\alpha = \text{lub } E$ and $\beta = \text{glb } E$, but we will use the sup and inf notation.)

1.2.3. Axiom (Completeness Property). *If a non-empty subset E of $\mathbb{R}$ has an upper bound, it has a supremum. If a non-empty subset E of $\mathbb{R}$ has a lower bound, it has an infimum.*

We also have uniqueness for the supremum.

1.2.4. Proposition. *If the subset E of* $\mathbb{R}$ *is bounded above, its supremum is unique. That is, if* α *and* α' *are both the supremum of E, then* $\alpha = \alpha'$. *If E is bounded below, its infimum is unique.*

Proof. Let α and α' be as in the statement of the proposition. Since α' is an upper bound for E and α is a supremum, the definition of a least upper bound implies that $\alpha \leq \alpha'$. Similarly, since α' is also a supremum, we have that $\alpha' \leq \alpha$. Thus $\alpha = \alpha'$.

That the infimum is unique when it exists can be proved in a manner analogous to the preceding proof or you can use Exercise 2. ∎

Also see Exercise 3.

The density and completeness properties may seem obvious to you, but that is probably because you have always thought of $\mathbb{R}$ as having them. Nevertheless, unless you are in possession of an exact definition of the real numbers, as will be carried out shortly, you cannot give a rigorous proof of their existence. Let's also remark that the set $\mathbb{Q}$ does not have the Completeness Property. For example $\{a \in \mathbb{Q} : a^2 < 2\}$ does not have a supremum within the set $\mathbb{Q}$. (See Proposition 1.2.9 below.) Of course it has a supremum in $\mathbb{R}$, namely $\sqrt{2}$, but this is not a rational number as is established in the proof of Proposition 1.2.9.

More Thorough Approach

Here we will define the real numbers. From the student's point of view this may seem unnecessary. After all, you have been working with real numbers since high school, solving equations and doing calculus. Most of the time calculus is presented in what might be called a "naive" way: the presentation glosses over some intrinsic properties of the real numbers that make calculus work. Here we want to present the theory in a precise mathematical way. To do this we need a precise definition of the real numbers.

Caveat. I have entitled this subsection with the words "More Thorough" rather than just "Thorough." I am not going to present every detail of a thorough approach. To prove every detail and totally explore the definition of the real numbers would make it impossible to complete our study of functions of a single variable in one semester. I will present more than enough of the material to establish the density and completeness properties. But some topics will not be encountered. In addition the proofs of many facts will be left as exercises. If the reader is interested in seeing a complete development of the real numbers, the books [6] and [13] will provide them.

How do we define the set of real numbers? There has to be a starting point. In some treatments the beginning is a development of the properties of the natural numbers. In others it is the properties of the rational numbers. We are going to start somewhere in between these two. We will definitely assume you know the natural numbers. There are, however, some properties of $\mathbb{N}$ that I think you will readily accept but may not have seen explicitly stated. Here is one that is a first cousin of the fact that there is no largest integer. No proof is given.

1.2.5. Lemma. *If $m, n \in \mathbb{N}$, then there is a natural number N such that $Nn > m$*

The definition of $\mathbb{Q}$ has already been given in (1.1.2). It's an algebraically defined entity and we are certainly going to assume the reader is knowledgeable of all its algebraic properties. There are some non-algebraic properties of $\mathbb{Q}$ that we need and that some readers may not have been exposed to. These will involve the order structure of $\mathbb{Q}$. Here is one that is the version of the preceding lemma for $\mathbb{Q}$.

1.2.6. Proposition. *If x and ϵ are positive rational numbers, then there is an n in $\mathbb{N}$ with $n\epsilon > x$.*

Proof. Put $x = a/b$ and $\epsilon = c/d$ with a, b, c, d in $\mathbb{N}$. If $n \in \mathbb{N}$, then

$$\epsilon - \frac{x}{n} = \frac{c}{d} - \frac{a}{nb} = \frac{nbc - ad}{nbd}$$

By the preceding lemma we can choose n such that $nbc > ad$. For that value of n, $n\epsilon > x$. ∎

In the Quick Approach above we defined the concept of a set of real numbers that is bounded above or below. The same concept applies to subsets of $\mathbb{Q}$, which are, of course, subsets of $\mathbb{R}$. But here we want to underline that we must choose the upper bound to be a rational number. This is not so important until we also discuss the concept of the supremum and infimum of subsets of $\mathbb{Q}$. Here for a bounded subset of $\mathbb{Q}$ to have a supremum it must be rational. There are subsets of $\mathbb{Q}$ that are bounded above but do not have a supremum in $\mathbb{Q}$. To do this we present two lemmas.

1.2.7. Lemma. *There is no rational number x with $x^2 = 2$.*

Proof. In fact if there were such a rational number, we could write it as $x = n/m$, where $n, m \in \mathbb{Z}$ and n and m have no common divisor other than ± 1. If $2 = (n/m)^2$, then we get $n^2 = 2m^2$; so n^2 is even. This implies that n is even; in fact if it weren't, we would have that it's odd. That is, we would have that $n = 2k + 1$. But then $n^2 = (2k+1)^2 = 4k^2 + 4k + 1 = 2(2k^2 + 2k) + 1$, which is odd. This is in direct contradiction to the fact that n^2 is even. But if n is even we can write $n = 2p$, so that $4p^2 = n^2 = 2m^2$. Dividing by 2 we get that m^2 is even; as before we get that m is even. That is, we have shown that 2 is a common divisor of both n and m, contradicting the fact that they were chosen with ± 1 as the only common divisor. ∎

1.2.8. Lemma. *If a, b are positive rational numbers with $a^2 < b^2$, then $a < b$.*

Proof. By the hypothesis we have that $0 < b^2 - a^2 = (b - a)(b + a)$. But since a and b are positive, $a + b > 0$. Thus $b - a > 0$. ∎

1.2.9. Proposition. *The set $A = \{a \in \mathbb{Q} : a^2 < 2\}$ is bounded above but has no supremum in $\mathbb{Q}$.*

Proof. It is immediate that 2 is an upper bound for A so it remains to show that A has no supremum in $\mathbb{Q}$. The proof is by contradiction; so assume there is an x in $\mathbb{Q}$ with $x = \sup A$. We will show that $x^2 = 2$, thus contradicting Lemma 1.2.7 and finishing the proof. We show that $x^2 = 2$ by showing that x^2 can be neither larger nor smaller than 2. First assume that $x^2 > 2$. We will show that there must be a number w in $\mathbb{Q}$ with $0 < w < x$ and $2 < w^2$. In fact suppose such a w exists. It then follows that if $a \in A$, then $a^2 < 2 < w^2$. But if $a < 0$, then $a < w$; if $a > 0$, then Lemma 1.2.8 shows $a < w$. This establishes that w is an upper bound of the set A. Since $w < x$, we have our desired contradiction.

In $n \in \mathbb{N}$, then

$$\left(x - \frac{1}{n}\right)^2 = x^2 - \frac{2x}{n} + \frac{1}{n^2} > 2 - \left[\frac{2x}{n} - \frac{1}{n^2}\right]$$

Using (1.2.6) we can choose n in $\mathbb{N}$ such that $2x > \frac{1}{n}$. Thus $2x/n - 1/n^2 > 0$ and the above inequality shows that $w = x - \frac{1}{n}$ works.

Now assume that $x^2 < 2$. Again let $n \in \mathbb{N}$ and examine $x + \frac{1}{n}$. We have that

$$2 - \left[x + \frac{1}{n}\right]^2 = [2 - x^2] - \left[\frac{2x}{n} + \frac{1}{n^2}\right]$$

As we did before, choose n sufficiently large that $w = x + \frac{1}{n}$ satisfies $2 - w^2 > 0$. This says that $w \in A$. But $w > x$, contradicting the fact that x is an upper bound of A. So again we arrive at a contradiction. ∎

The last proposition inspires the definition of the real numbers.

1.2.10. Definition. A *Dedekind*[2] *cut*, or simply a *cut*, is a non-empty subset A of $\mathbb{Q}$ satisfying the following three conditions: (a) A is a proper subset of $\mathbb{Q}$; (b) if $a \in A$ and $b < a$, then $b \in A$; (c) if $a \in A$, there is a b in A with $a < b$.

An example of a cut can be obtained from Proposition 1.2.9 except we have to adjoin to the set A in that proposition all the negative numbers. That is, an example of a Dedekind cut is

1.2.11 $\qquad\qquad A = \{a \in \mathbb{Q} : a^2 < 2 \text{ or } a < 0\}$

This can be shown to be a cut by using the techniques of the preceding proof (Exercise 10). Note that if $q \in \mathbb{Q}$, then the set $B = \{x \in \mathbb{Q} : x \leq q\}$ is not a cut (Why?) while $C = \{x \in \mathbb{Q} : x < q\}$ is. (See (1.2.13).)

....

[2] Richard Dedekind was born in 1831 in Braunschweig, Germany also referred to as Brunswick. His father was a professor at the Collegium Carolinum in Brunswick, an institution between a high school and a university. He received his doctorate in 1852 from the University of Göttingen and was Gauss's last student. In 1854 he began to teach at Göttingen and was joined by Riemann and then Dirichlet, who influenced him greatly. In 1858 he accepted a position in Zurich and it was during his first year there while teaching calculus that the idea of a cut came to him. In 1862 he accepted a position at the Brunswick Polytechnikum in his hometown. He made many contributions to number theory and algebra, where he introduced the idea of an ideal in a ring. He never married and he had no PhD students. He died in 1916 in the same town where he was born.

A useful observation about cuts is the following. If A is a cut and $x \in \mathbb{Q} \backslash A$, then $a < x$ for every a in A. In fact otherwise there is an a in A with $x \leq a$. Since $x \notin A$, it must be that $x < a$. But then (b) in the definition says $x \in A$.

In the literature there is another definition of a Dedekind cut as a pair of non-empty subsets (A, B) of $\mathbb{Q}$ having certain properties. See [8].

1.2.12. Definition. The set of *real numbers* is the collection of Dedekind cuts of $\mathbb{Q}$. The real numbers are denoted by $\mathbb{R}$.

Does this definition strike you as strange?

I thought it strange the first time I saw this. But think of the cut given in (1.2.11). Isn't it natural to call this $\sqrt{2}$? Nevertheless, you have a right to pause when we define a number to be a set.

We have a rather involved task before us. We want to show that $\mathbb{R}$ as defined above has all the properties we are used to thinking the set of real numbers has: the ability to add and multiply as well as a notion of order. We'll also see that it has the Density and Completeness properties, (1.2.1) and (1.2.3). A caveat is that many details will be left to the reader to prove.

We start by showing that $\mathbb{Q} \subseteq \mathbb{R}$ in a natural way. But first let's make a temporary agreement about notation. For elements of $\mathbb{R}$ we'll use small Greek letters, while rational numbers will be denoted by small roman letters.

1.2.13. Proposition. *If $a \in \mathbb{Q}$, then $\alpha = \{b \in \mathbb{Q} : b < a\}$ is a real number; that is, α is a cut.*

Proof. Clearly α is a proper subset of $\mathbb{Q}$, and if $b \in \alpha$ and $c < b$, then $c \in \alpha$. Finally if $b \in \alpha$, then $c = (b+a)/2 \in \mathbb{Q}$ and $b < c < a$. Thus α is a cut. ∎

What is happening in the last proposition is that we have embedded $\mathbb{Q}$ into $\mathbb{R}$ by using the map $a \mapsto \alpha$. We identify each rational number a with its image α in $\mathbb{R}$. Unless there is a good reason to make a distinction between a rational number a and its image α under this map, no distinction will be made. Thus we write

$$a = \{b \in \mathbb{Q} : b < a\}$$

In particular, $0 = \{b \in \mathbb{Q} : b < 0\}$. So we use the symbol 0 to denote both the rational number 0 and the corresponding cut.

We introduce the concept of an order on $\mathbb{R}$.

1.2.14. Definition. If $\alpha, \beta \in \mathbb{R}$, say that $\alpha < \beta$ if $\alpha \subseteq \beta$ and $\alpha \neq \beta$. Say that $\alpha \leq \beta$ if $\alpha \subseteq \beta$ with the possibility they are equal. We define $\alpha > \beta$ and $\alpha \geq \beta$ similarly.

As usual, if $\alpha > 0$, we say α is positive; if $0 > \alpha$, we say α is negative. The proof of the next result is Exercise 11.

1.2.15. Proposition. (a) *If $a \in \mathbb{Q}$ and $\beta \in \mathbb{R}$ such that $a < \beta$, then $a \in \beta$.*
(b) *If $\alpha, \beta \in \mathbb{R}$, then either $\alpha \leq \beta$ or $\beta \leq \alpha$.*
(c) *(Trichotomy Law) If $\gamma \in \mathbb{R}$, then either $\gamma < 0$, $0 < \gamma$, or $\gamma = 0$.*

1.2.16. Theorem. *If E is a non-empty subset of $\mathbb{R}$ that is bounded above, then* $\sup E$ *exists. That is, $\mathbb{R}$ has the Completeness Property.*

Proof. So E is a set of cuts; that is, a set of subsets of $\mathbb{Q}$. Take a moment to absorb this and the notation and reasoning that follow will seem more natural. Let ζ be an upper bound for E and let $\alpha = \bigcup\{\epsilon : \epsilon \in E\}$. The crux of the proof is to first show that α is a cut, and so $\alpha \in \mathbb{R}$; then we'll show that $\alpha = \sup E$. To see that α is a cut, observe that it is non-empty since $E \neq \emptyset$. Since $\epsilon \subseteq \zeta$ for every ϵ in E, $\alpha \subseteq \zeta$ and so $\alpha \neq \mathbb{Q}$. To show that α satisfies the other two parts of the definition of a cut, fix b in α and let $\epsilon \in E$ such that $b \in \epsilon$. If $y \in \mathbb{Q}$ with $y < b$; it follows that $y \in \epsilon$ because ϵ is a cut; hence $y \in \alpha$. Finally, since ϵ is a cut there is a rational number x in ϵ with $b < x$; since $x \in \epsilon$, $x \in \alpha$. Therefore α is a cut.

Now that we have that $\alpha \in \mathbb{R}$, we show that $\alpha = \sup E$. Since $\epsilon \subseteq \alpha$ for every ϵ in E, α is an upper bound for E. On the other hand, if ζ is any upper bound for E, then $\epsilon \leq \zeta$ for all ϵ in E. That is, $\epsilon \subseteq \zeta$ for all ϵ in E. Therefore $\alpha \leq \zeta$. ∎

Now that we have the Completeness Property, we can prove half of the Density Property. In fact it is a direct consequence of the definition of $\mathbb{R}$ that if $\alpha, \beta \in \mathbb{R}$ and $\alpha < \beta$, then there is a rational number γ with $\alpha < \gamma < \beta$. Indeed, since $\alpha \neq \beta$, there is a b in β such that $b \notin \alpha$. If we let γ be the cut defined by b, then $\alpha < \gamma < \beta$. The proof of the other half of the Density Property must wait until we have established the arithmetic properties of the real numbers. We start with the definition of addition in $\mathbb{R}$. To justify this we need a proposition.

1.2.17. Proposition. *If $\alpha, \beta \in \mathbb{R}$, then*

$$\gamma = \{a + b : a \in \alpha, b \in \beta\}$$

is a cut.

Proof. (a) To show that γ is a proper set we first note that it is clear that it is not empty. Now take positive rational numbers p, q such that $a < p$ for all a in α and $b < q$ for all b in β. (Why do p and q exist?) It follows that $p + q > a + b$ for all a in α and b in β. Thus $p + q \notin \gamma$, and so $\gamma \neq \mathbb{Q}$.

(b) Let $x \in \gamma$ and suppose $y < x$. Put $x = a + b$, where $a \in \alpha$ and $b \in \beta$. So $y - a < b$; hence $y - a \in \beta$. But then $y = a + (y - a) \in \gamma$.

(c) Take $a + b$ in γ and let $a' \in \alpha$ such that $a < a'$. So $a' + b \in \gamma$ and is strictly bigger that $a + b$. ∎

1.2.18. Definition. If $\alpha, \beta \in \mathbb{R}$, then $\alpha + \beta$ is the cut γ defined in the preceding proposition.

1.2.19. Proposition. *If $\alpha, \beta, \gamma \in \mathbb{R}$, the following hold.*

(a) $\alpha + \beta = \beta + \alpha$.
(b) $\alpha + (\beta + \gamma) = (\alpha + \beta) + \gamma$.
(c) $\alpha + 0 = \alpha$.

The proof is an exercise.

Now we start the path to the definition of the negative of a real number α. It is convenient to introduce the notation $\mathbb{Q}_+ = \{a \in \mathbb{Q} : a > 0\}$. The method for defining $-\alpha$ may seem a bit opaque at first, so let's take a minute to motivate it. Examine the cut $A = \alpha$ introduced in Proposition 1.2.9 to show that $\sqrt{2}$ is not rational; suppose we want to find the cut associated with $-\sqrt{2}$. This would be $\beta = \{b \in \mathbb{Q} : b < -\sqrt{2}\}$. (We have departed somewhat from the notation and spirit we have been following up to this point, but this is an intuitive discussion and so we take some liberties.) Note that $b \in \beta$ if and only if $\sqrt{2} < -b$. Now for any b in β there is an r in $\mathbb{Q}_+$ such that $b + r \in \beta$. Thus $b \in \beta$ if and only if there is an r in $\mathbb{Q}_+$ such that $-b - r \notin \alpha$. Note that this necessary and sufficient condition for membership in $-\alpha$ is consistent with the rigorous approach we are following for defining $\mathbb{R}$; remember this intuitive discussion when we define the negative of a real number in the proposition below.

1.2.20. Proposition. *For every α in $\mathbb{R}$ there is a β in $\mathbb{R}$ such that $\alpha + \beta = 0$.*

Proof. Fix α in $\mathbb{R}$ and let

$$\beta = \{b \in \mathbb{Q} : \text{there is an } r \text{ in } \mathbb{Q}_+ \text{ such that } -b - r \notin \alpha\}$$

The first task is to show that β is a cut. To see that $\beta \neq \emptyset$, assume $x \notin \alpha$ and put $b = -x - 1$. Since $-b - 1 = x$, $b \in \beta$. On the other hand, if $a \in \alpha$ and $r \in \mathbb{Q}_+$, $a - r < a$ and so $-(-a) - r \in \alpha$ for every $r > 0$. That is, $-a \notin \beta$ and $\beta \neq \mathbb{Q}$. The remainder of the proof that β is a cut is left to the reader.

Now observe that if $b \in \beta$, then $-b \notin \alpha$. In fact if $-b \in \alpha$, then for every r in $\mathbb{Q}_+$, $-b - r \in \alpha$ since it is smaller than $-b$; this violates the definition of membership in β. Hence for any a in α, $a < -b$ and so $a + b < 0$. Therefore $\alpha + \beta \subseteq 0$. To show that $0 \subseteq \alpha + \beta$, let $z \in 0$ and put $x = -z/2$; so $x > 0$. By Lemma 1.2.6 there is an n in $\mathbb{Z}$ such that $nx \in \alpha$ but $(n+1)x \notin \alpha$. (Details?) So $b = -(n+2)x \in \beta$ and $-b - x \notin \alpha$. Therefore $z = nx + b \in \alpha + \beta$ and we have shown that $0 \subseteq \alpha + \beta$. ∎

1.2.21. Definition. For α in $\mathbb{R}$, we define $-\alpha$ to be the cut β in Proposition 1.2.20.

The proof of the next proposition is Exercise 12.

1.2.22. Proposition. *Let $\alpha, \beta, \gamma \in \mathbb{R}$.*

(a) *The element $-\alpha$ is unique. That is, if $\alpha + \beta = \alpha + \gamma = 0$, then $\beta = \gamma$.*
(b) $-(-\alpha) = \alpha$.
(c) *If $\alpha \geq 0$, then $-\alpha \leq 0$.*

Defining multiplication is more complicated than defining addition due to the fact that the product of two negative rational numbers is positive. Consequently we begin with the definition of the product of two elements of $\mathbb{R}_+ = \{\alpha \in \mathbb{R} : \alpha > 0\}$ and later we will see how to extend this to the definition of the product of arbitrary real numbers. The proofs of the various properties of the definition are very similar and will be left as exercises. The proof of the next proposition is Exercise 13.

1.2.23. Proposition. *If $\alpha, \beta \in \mathbb{R}_+$, then*

$$\{ab : a \in \alpha, b \in \beta \text{ and } a > 0\}$$

is a cut.

1.2.24. Definition. If $\alpha, \beta \in \mathbb{R}_+$, then $\alpha\beta$ is the cut defined in the preceding proposition.

The proof of the next proposition is Exercise 14.

1.2.25. Proposition. *If $\alpha, \beta, \gamma \in \mathbb{R}_+$, then the following hold.*

(a) $\alpha\beta = \beta\alpha$.
(b) $(\alpha\beta)\gamma = \alpha(\beta\gamma)$.
(c) *If 1 is the cut $1 = \{x \in \mathbb{Q} : x < 1\}$, then $\alpha = 1 \cdot \alpha$.*
(d) $\alpha(\beta + \gamma) = \alpha\beta + \alpha\gamma$.
(e) $\alpha^{-1} = \{c : c < \frac{1}{a} \text{ when } a \notin \alpha\}$ *is a cut and* $\alpha(\alpha^{-1}) = 1$.

Now we extend the definition of the product to all real numbers.

1.2.26. Definition. Let $\alpha, \beta \in \mathbb{R}$. (a) Define $\alpha 0 = 0\alpha = 0$. (b) When neither α nor β is 0, define

$$\alpha\beta = \begin{cases} (-\alpha)(-\beta) & \text{if } \alpha < 0, \beta < 0 \\ -[(-\alpha)\beta] & \text{if } \alpha < 0, \beta > 0 \\ -[\alpha(-\beta)] & \text{if } \alpha > 0, \beta < 0 \end{cases}$$

I am going to leave it to the interested reader to formulate and prove a version of Proposition 1.2.25 for products of arbitrary real numbers. There are also properties relating the order structure and the arithmetic structure of $\mathbb{R}$, such as $\alpha\beta < 0$ when $\alpha > 0$ and $\beta < 0$, that will be left to the reader.

Define the *absolute value* of α, $|\alpha|$, by

$$|\alpha| = \begin{cases} \alpha & \text{if } \alpha \geq 0 \\ -\alpha & \text{if } \alpha \leq 0 \end{cases}$$

1.2.27. Proposition. *If $\alpha, \beta \in \mathbb{R}$, the following hold.*

(a) $|\alpha| = |-\alpha|$.
(b) $|\alpha\beta| = |\alpha||\beta|$.
(c) (Triangle Inequality) $|\alpha + \beta| \leq |\alpha| + |\beta|$.
(d) (Reverse Triangle Inequality) $||\alpha| - |\beta|| \leq |\alpha - \beta|$.

Proof. The proof of the first three parts is left to the reader in Exercise 14. The proof of (d) is easy if we apply the triangle inequality and use a little trick that will be used frequently in the future, namely adding and subtracting a quantity. Observe that $|\alpha| = |(\alpha - \beta) + \beta| \leq |\alpha - \beta| + |\beta|$. Subtracting we get that $|\alpha| - |\beta| \leq |\alpha - \beta|$. Now reverse the roles of α and β in this inequality to get $|\beta| - |\alpha| \leq |\alpha - \beta|$. Using the definition of absolute value we get (d). ∎

The triangle inequality will be used frequently as we progress in this book and we will not cite it by name; it's that fundamental. Also be aware that adding and subtracting a term and then applying the triangle inequality is a method of proof you will also see used often in the future.

The reader knows that the definition of an irrational number is an element in $\mathbb{R}\backslash\mathbb{Q}$. In the language of cuts, it is a cut that is not of the form given in Proposition 1.2.13.

1.2.28. Theorem. (a) *If $\alpha, \beta \in \mathbb{Q}$, then there is an irrational number γ with $\alpha < \gamma < \beta$.*
(b) *If $\alpha, \beta \in \mathbb{R}\backslash\mathbb{Q}$ and $\alpha < \beta$, then there is a rational number x with $\alpha < x < \beta$.*

Proof. The proof of (b) was given just after Theorem 1.2.16, so it remains to prove (a). To start we show there is an irrational number t with $0 < t < 1$. Indeed, $t = \sqrt{2}/2$ works, where $\sqrt{2}$ denotes the irrational number corresponding to the cut $\{a \in \mathbb{Q} : a^2 < 2 \text{ or } a < 0\}$. If $\alpha, \beta \in \mathbb{Q}$ and $\alpha < \beta$, then it is easy to see that $\alpha < \gamma = (1-t)\alpha + t\beta < \beta$. Also $t = (\beta - \alpha)^{-1}(\gamma - \alpha)$, so if it were the case that γ is rational, we would have that t is rational. ∎

Before we conclude the section, we introduce the concept of infinity as related to $\mathbb{R}$ as well as the set of *extended real numbers* $\mathbb{R}_\infty$. We first define $\mathbb{R}_\infty$ as the set $\mathbb{R} \cup \{\infty, -\infty\}$, where $\pm\infty$ are two abstract points, and we define an order on this set by declaring that for any α in $\mathbb{R}$ we have that $-\infty < \alpha < \infty$. We assume that the reader has some familiarity with the idea of infinity and understands that $\pm\infty$ are not real numbers.

This concludes the present section. The reader interested in going deeper into these matters can consult [6] and [13]. We now abandon the formalism of this section and starting in the next section we return to using $\mathbb{R}$ as we have all known it.

Exercises

(1) For real numbers a and b write the quantities $a + b + |a - b|$ and $a + b - |a - b|$ without using absolute values.

(2) If $\alpha = \sup E$, show that $-\alpha = \inf\{-E\}$.

(3) Let E be a subset of $\mathbb{R}$ that is bounded above and set $A = \{a \in \mathbb{R} : a \text{ is an upper bound for } E\}$. Show that $\sup E = \inf A$.

(4) Give a proof that $\sqrt{3}$ is not a rational number.

(5) Fill in the missing details in the proof of Proposition 1.2.9.

(6) For each of the following sets X find $\inf X$ and $\sup X$ if they exist and say if they belong to X. (a) $X = \{x : |3x - 9| < 5\}$. (b) $X = \{x : |3x - 1| > \frac{1}{3}\}$. (c) $X = \{x \in \mathbb{Q} : x^2 > 7\}$.

(7) Find the infimum and supremum of the set $X = \{\frac{1}{2}, \frac{2}{3}, \frac{3}{4}, \frac{4}{5}, \ldots\}$.

(8) Find the infimum and supremum of the set $X = \{3 - \sqrt{\frac{1}{2+n}} : n \in \mathbb{N}\}$.

(9) If E and F are bounded subsets of real numbers, show that $\sup\{x - y : x \in E, y \in F\} = \sup E - \inf F$.

(10) Show that the set $A = \{a \in \mathbb{Q} : a^2 < 2 \text{ or } a < 0\}$ is a cut.

(11) Prove Proposition 1.2.15.

(12) Prove Proposition 1.2.22.

(13) Prove Proposition 1.2.23.

(14) Prove Proposition 1.2.27.

(15) If A and B are two non-empty subsets of $\mathbb{R}$ that are bounded, give a necessary and sufficient condition that $\sup A = \inf B$.

(16) If E is a non-empty bounded subset of $\mathbb{R}$ and $a, b \in \mathbb{R}$, find formulas for $\sup\{ax + b : x \in E\}$ and $\inf\{ax + b : x \in E\}$ in terms of $\sup E$ and $\inf E$.

1.3. Convergence

We divorce ourselves from the notation convention of the preceding section where we denoted real numbers as lower case Greek letters. We will often use the idea of an interval in $\mathbb{R}$, of which there are several types. An *open interval* I is a subset of the form $I = \{x \in \mathbb{R} : c < x < d\}$, where $-\infty \leq c < d \leq \infty$; it is denoted by (c, d). A *closed interval* is a set of the form $[c, d] = \{x \in \mathbb{R} : c \leq x \leq d\}$, where $-\infty < c \leq d < \infty$. A *half-open* interval is a set having one of two forms: $[c, d) = \{x \in \mathbb{R} : c \leq x < d\}$ or $(c, d] = \{x \in \mathbb{R} : c < x \leq d\}$, where there are appropriate restrictions on c and d. (What restrictions?) An *interval* in $\mathbb{R}$ is a set I of any one of these types.

In this section we will examine the convergence of sequences of real numbers. Recall that a *sequence* in $\mathbb{R}$ is an enumeration of some real numbers: $a_1, a_2, \ldots$ It is usually denoted by $\{a_n\}$.

1.3.1. Definition. A sequence $\{a_n\}$ in $\mathbb{R}$ *converges* to a if for any open interval containing a, I contains all but a finite number of terms of the sequence. This is denoted by writing $a_n \to a$ or $a = \lim_n a_n = \lim a_n$.

Observe that if $a_n \to a$, what happens with the first finite number of elements of a sequence is irrelevant to whether it converges. That is, for any integer n_0, $a_n \to a$ if and only if the sequence $\{a_n : n \geq n_0\} \to a$. The first result is that when a sequence converges, the limit is unique.

1.3.2. Proposition. *A convergent sequence can only converge to one point. That is, if $a_n \to a$ and $a_n \to b$, then $a = b$.*

Proof. Adopting the notation in the statement of the proposition, assume that $a \neq b$. Thus we may assume that $a < b$. It follows that

$$a \in \left(a - \frac{b-a}{2}, a + \frac{b-a}{2}\right) \text{ while } b \in \left(b - \frac{b-a}{2}, b + \frac{b-a}{2}\right)$$

Since these two intervals are disjoint, it cannot be that they each contain all but finitely many members of the sequence $\{a_n\}$. This contradicts the hypothesis that the sequence converges to both a and b.   ■

Before presenting some examples, let's establish the next result; this proposition quantifies the definition of convergence.

1.3.3. Proposition. *Let $\{a_n\}$ be a sequence in $\mathbb{R}$ and let $a \in \mathbb{R}$.*

(a) $a_n \to a$ *if and only if for every $\epsilon > 0$ there is an N such that $|a_n - a| < \epsilon$ for all $n \geq N$.*
(b) $a_n \to a$ *if and only if $(a_n - a) \to 0$.*

Proof. (a) Suppose $a_n \to a$ and $\epsilon > 0$. Since $I = (a - \epsilon, a + \epsilon)$ is an open interval that contains A, I contains all but a finite number of the terms of the sequence. This means there is an integer N such that $a_n \subset I$ when $n \geq N$; that is, $|a - a_n| < \epsilon$ when $n \geq N$. Conversely, assume the condition holds and $I = (c, d)$ is an open interval containing a. Since $c < a < d$, there is an $\epsilon > 0$ such that $c < a - \epsilon < a < a + \epsilon < d$. Thus there is an integer N such that $|a_n - a| < \epsilon$ when $n \geq N$. That is $\{a_N, a_{N+1}, \ldots\} \subseteq I$.
  (b) Exercise 1.   ■

1.3.4. Example. For the sequence $\{a_n\}$ defined by $a_n = n^{-2}$, we have that $a_n \to 0$. Note that if $\epsilon = 1/100$ and $N = 11$, then $|a_n| < \epsilon$ when $n \geq N$. We could also take $N = 20$ and get that $|a_n| < \epsilon$ when $n \geq N$. In other words, in the preceding proposition that value of N obtained for a given ϵ is not unique. If $\epsilon = 1/10,000$, then we could take $N = 101$. In general, for any ϵ we could take for N any integer larger than $1/\sqrt{\epsilon}$.

It is worth observing that in part (a) of the preceding proposition the condition $n \geq N$ can be replaced by $n > N$ and the conclusion $|a_n - a| < \epsilon$ can be replaced by $|a_n - a| \leq \epsilon$. (Verify!) The value of part (a) above is that it quantifies the idea of convergence as the preceding example shows. See Exercise 2.

1.3.5. Example. (a) If $a_n = a$ for all n, then $a_n \to a$.
  (b) $\frac{1}{n} \to 0$. In fact if $\epsilon > 0$, then by Proposition 1.2.6 there is an N in $\mathbb{N}$ with $N\epsilon > 1$. It follows that for $n \geq N$, $\frac{1}{n} < \epsilon$.
  (c) If $a_n \to 0$ and $\{b_n\}$ is a bounded sequence, then $a_n b_n \to 0$. In fact let $|b_n| \leq B$ for all $n \geq 1$ and let $\epsilon > 0$. Choose N such that $|a_n| < \epsilon/B$ when $n \geq N$. If $n \geq N$, then $|a_n b_n| \leq B|a_n| < \epsilon$.
  (d) If $a_n \to a$, then $\{a_n\}$ is bounded. In fact choose an integer N such that $|a - a_n| < 1$ for $n \geq N$. Since $\{a_1, \ldots, a_N\}$ is finite, there is a number A with $|a_n| \leq A$ for $1 \leq n \leq N$. If $n \geq N$, then $|a_n| \leq |a_n - a| + |a| \leq 1 + |a|$. Therefore for every $n \geq 1$, $|a_n| \leq M = \max\{A, 1 + |a|\}$.

1.3.6. Proposition. *If $a_n \to a$ and $b_n \to b$, the following hold.*

(a) $a_n + b_n \to a + b$.
(b) $a_n b_n \to ab$.
(c) *If $b \neq 0$, then $b_n \neq 0$ except for possibly a finite number of integers n. If $b_n \neq 0$ for all n, then $a_n/b_n \to a/b$.*

Proof. The proof of (a) is left to the reader. To prove (b) note that

$$
\begin{aligned}
|a_n b_n - ab| &\leq |a_n b_n - a_n b + a_n b - ab| \\
&\leq |a_n b_n - a_n b| + |a_n b - ab| \\
&\leq |a_n||b_n - b| + |b||a_n - a|
\end{aligned}
$$

By Example 1.3.5(d) there is a constant C such that $|a_n| \leq C$ for all n. Choose $M > \max\{C, |b|\}$. If $\epsilon > 0$, let N be such that $|a_n - a| < \epsilon/2M$ and $|b_n - b| < \epsilon/2M$ whenever $n \geq N$. From the above inequality we have that when $n \geq N$, $|a_n b_n - ab| < \epsilon$ and so (b) holds.

To start the proof of (c), choose n_0 such that $|b_n - b| < |b|/2$ when $n \geq n_0$. Thus for $n \geq n_0$, $|b| = |b - b_n + b_n| \leq |b - b_n| + |b_n| < |b|/2 + |b_n|$, and it follows that $|b_n| > |b|/2$. In particular we have the first part of the statement of (c). Now observe that we actually proved that we have a constant $c > 0$ such that $|b_n| \geq c$ for all but a finite number of integers n. As we observed after Proposition 1.3.3, we can assume that $|b_n| \geq c > 0$ for all n. Note that if we prove that $b_n^{-1} \to b^{-1}$, an application of part (b) shows that (c) is valid. But

$$
\left| \frac{1}{b_n} - \frac{1}{b} \right| = \left| \frac{b - b_n}{bb_n} \right| \leq \frac{1}{c^2}|b - b_n|
$$

From here the reader can easily supply the details to demonstrate that $b_n^{-1} \to b^{-1}$. ∎

A sequence $\{a_n\}$ is said to be *increasing* if $a_n \leq a_{n+1}$ for all n; it is *strictly increasing* if $a_n < a_{n+1}$ for all n. It should be noted that some call a sequence satisfying $a_n \leq a_{n+1}$ for all n a *non-decreasing sequence* and reserve the term "increasing" for what we have called strictly increasing. Similarly we define $\{a_n\}$ to be *decreasing* if $a_{n+1} \leq a_n$ for all n, and we can define *strictly decreasing* if the $\leq$ sign is replaced by $<$. Observe that if $\{a_n\}$ is a decreasing sequence, then $\{-a_n\}$ is increasing. So every time we prove a result for increasing sequences we have an analogous result for decreasing sequences.

1.3.7. Proposition. *If $\{a_n\}$ is an increasing sequence that is bounded above and $a = \sup_n a_n$, then $a_n \to a$. Similarly, if $\{a_n\}$ is a decreasing sequence that is bounded below and $a = \inf_n a_n$, then $a_n \to a$.*

Proof. As we observed before the statement of the proposition, we need only prove the statement concerning increasing sequences. If $\epsilon > 0$, then by definition of the supremum, $a - \epsilon$ is not an upper bound of $\{a_n\}$. Thus there is an integer N

such that $a - \epsilon < a_N$. But since $\{a_n\}$ is increasing we have that $a - \epsilon < a_n \leq a$ whenever $n \geq N$. Thus $a_n \to a$. ∎

1.3.8. Definition. If $\{a_n\}$ is a sequence and $\{n_k\}$ is a sequence of positive integers with $n_1 < n_2 < \cdots$, then $\{a_{n_k}\}$ is called a *subsequence* of $\{a_n\}$.

Note that a subsequence is a new sequence, so it makes sense to discuss the convergence of a subsequence.

1.3.9. Example. (a) $\{a_{2n}\}$ is a subsequence of $\{a_n\}$.
(b) $\{a_1, a_5, a_3, a_9, a_7, \ldots\}$ is not a subsequence of $\{a_n\}$. (Why?)
(c) $\{a_1, a_1, a_1, \ldots\}$ is not a subsequence of $\{a_n\}$. (Why?)

The proof of the next proposition is Exercise 8. Also see Exercise 9.

1.3.10. Proposition. *If $a_n \to a$ and $\{a_{n_k}\}$ is a subsequence of $\{a_n\}$, then $\{a_{n_k}\} \to a$.*

The next result is extremely important, our crucial result on convergence. As might be expected, its proof is rather complex.

1.3.11. Theorem (Bolzano[3]–Weierstrass[4] Theorem). *A bounded sequence has a convergent subsequence.*

....

[3] Bernard Bolzano was born in Prague in 1848. His father was from northern Italy and migrated to Prague where he was an art dealer. Bernard was the fourth of twelve children, but only he and a brother made it to adulthood. Bernard had delicate health his entire life. In 1796 he entered the Charles University in Prague, where he studied philosophy, physics, and mathematics. In 1800 he began a course in theological studies while simultaneously writing a doctoral thesis in geometry; he earned his doctorate in 1804. Two days after receiving that degree he was ordained as a Catholic priest. He soon realised he wanted to teach rather than minister to people. Nevertheless he was given a chair in philosophy and religion at the university, though his views differed from those of the Habsburgh rulers. Between 1821 and 1825 he was tried for heresy but refused to recant. He resigned his chair and spent most of his time working on mathematics. It was earlier, in 1817, that Bolzano proved the present theorem, which he used to prove the Intermediate Value Theorem. (See Theorem 1.7.7 below.) It was proved 50 years later by Weierstrass and it was recognized as a fundamental result. Bolzano died in 1848 in Prague. It seems a bit mysterious why Weierstrass shares the credit for this result, given the amount of time that elapsed before he found it. I suspect it has something to do with the great prestige Weierstrass has in mathematics.

[4] Karl Weierstrass was born in 1815 in Ostenfelde, Germany. His early schooling was rocky, though he exhibited greater than usual mathematical ability. The difficulty was that his father wanted him to pursue a career in finance, but his passion was for mathematics. At first he reacted to this with a rebellious approach, neglecting all his studies and focusing on fencing and drinking. Nevertheless he studied mathematics on his own with extensive readings, even though he was supposed to follow a course in finance. After having left the university of Bonn without taking any examinations, he seems to have reached an understanding with his father and attended the academy at Münster with the intention of becoming a high school teacher. Here he came under the influence of Christoph Gudermann, a mathematician of note, and impressed his mentor with a paper on elliptic functions. In 1841 Weierstrass passed the exam to become a teacher, a career he followed for some years. In 1854 he published a paper on Abelian functions and attracted considerable attention from the research world – sufficient for the University of Königsberg to give him an honorary doctorate, enabling him to launch his university career at Braunsberg. He obtained a chair at the University of Berlin in 1856, where he remained for the rest of his life. He had a profound influence on mathematics, setting new standards of rigor and fostering the careers of numerous mathematicians including many whose contributions were profound. He became known as the father of modern analysis. He was plagued by health problems that periodically surfaced and then ebbed. Starting in the early 1860s he lectured while seated and while a student assistant wrote on the board. During his last three years he was confined to a wheelchair and died of pneumonia in Berlin in 1897. He never married.

Proof. As we mentioned the proof is involved so be careful. Start by letting $\{a_n\}$ be the bounded sequence and assume that x and y are real numbers such that $x \le a_n \le y$ for all n. The proof is trivial if the sequence has only a finite number of distinct elements since in that case there is at least one point that is repeated infinitely often; hence we can take a subsequence all of whose entries are equal to that point. So assume $\{a_n\}$ has an infinite number of distinct points. Divide the interval $I = [x, y]$ into two equal parts and look at which points in the sequence $\{a_n\}$ belong to each half. Since $\{a_n : n \in \mathbb{N}\}$ is an infinite set, one of the intervals has an infinite number of these points; call it J_1 and let $\mathbb{N}_1 = \{n \in \mathbb{N} : a_n \in J_1\}$. Observe that the length of J_1, $|J_1|$, equals $2^{-1}(y - x)$. Put $b_1 = \inf\{a_n : n \in \mathbb{N}_1\}$, and let $n_1 = \min \mathbb{N}_1$. We have the first element in our desired subsequence, a_{n_1}, and it satisfies $|b_1 - a_{n_1}| \le |J_1| = 2^{-1}(y - x)$.

To find the second element in the desired subsequence, we proceed in a similar way. Divide the interval J_1 into two equal parts, and let J_2 be a half interval that contains infinitely many of the points $\{a_n : n \in \mathbb{N}_1\}$. Put $\mathbb{N}_2 = \{n \in \mathbb{N}_1 : a_n \in J_2\}$, and let $b_2 = \inf\{a_n : n \in \mathbb{N}_2\}$. We are tempted to put $n_2 = \min \mathbb{N}_2$. It might happen, however, that $n_1 \in \mathbb{N}_2$, in which case $n_1 = \min \mathbb{N}_2$ and we would have that $n_2 = n_1$. Since we want to define the second element a_{n_2} in a subsequence, we need that $n_2 > n_1$. Thus put $n_2 = \min\{n \in \mathbb{N}_2 \setminus \{n_1\}\}$. From the construction we have that

$$|J_2| = 2^{-1}|J_1| = 2^{-2}(y - x) \text{ and } |b_2 - a_{n_2}| \le 2^{-2}(y - x)$$

Also because $\mathbb{N}_2 \subseteq \mathbb{N}_1$, we have that

$$b_1 \le b_2$$

Continue this process. We obtain decreasing intervals $J_1, J_2, \ldots$; infinite sets of integers $\mathbb{N}_1, \mathbb{N}_2, \ldots$ with $\mathbb{N}_1 \subseteq \mathbb{N}_2 \subseteq \cdots$; and integers $n_1 < n_2 < \cdots$. We define $b_k = \inf\{a_{n_k} : n_k \in \mathbb{N}_k\}$. We have that these satisfy:

$$|J_k| = \frac{1}{2^k}(y - x)$$

$$b_k \le b_{k+1}$$

$$|b_k - a_{n_k}| \le \frac{1}{2^k}(y - x)$$

Now these conditions imply that $\{b_k\}$ is an increasing sequence and it is bounded above by y. By Proposition 1.3.7 there is a number a such that $b_k \to a$. Therefore

$$\begin{aligned}
|a_{n_k} - a| &= |a_{n_k} - b_k + b_k - a| \\
&\le |a_{n_k} - b_k| + |b_k - a| \\
&\le \frac{1}{2^k}(y - x) + |b_k - a| \\
&\to 0
\end{aligned}$$ ■

The reader would do well to study the preceding proof. To begin, be conscious of the fact that we have used the Completeness Property of the real numbers. (Where?)

There are many bounded sequences in $\mathbb{Q}$ that do not have a subsequence that converges to another rational number. For example, take a sequence in $\mathbb{Q}$ that converges to $\sqrt{2}$. Also not only does the proof establish an important theorem, but it uses techniques that can and will be employed in the future.

1.3.12. Definition. A sequence $\{a_n\}$ is called a *Cauchy*[5] *sequence* if for every $\epsilon > 0$ there is an integer N such that $|a_n - a_m| < \epsilon$ when $m, n \geq N$.

1.3.13. Theorem. *A sequence converges if and only if it is a Cauchy sequence.*

Proof. Half of the proof is straightforward. Assume that $a_n \to a$ and $\epsilon > 0$. There is an integer N such that when $n \geq N$, $|a_n - a| < \epsilon/2$. Hence when $m, n \geq N$, $|a_m - a_n| \leq |a_m - a| + |a - a_n| < \epsilon$.

Now assume that $\{a_n\}$ is a Cauchy sequence.

Claim. If $\{a_{n_k}\}$ is a subsequence of $\{a_n\}$ that converges to a, then $a_n \to a$.

In fact let $\epsilon > 0$ and choose an integer N_1 such that $|a_{n_k} - a| < \epsilon/2$ when $n_k \geq N_1$. Choose N_2 such that $|a_n - a_m| < \epsilon/2$ when $n, m \geq N_2$. Put $N = \max\{N_1, N_2\}$, and suppose $n \geq N$. Pick any integer $n_k \geq N$. Since we also have that $n_k \geq N_1$, it follows that $|a_{n_k} - a| < \epsilon/2$. Since $n, n_k \geq N_2$, we have that $|a_n - a_{n_k}| < \epsilon/2$. Hence $|a_n - a| \leq |a_n - a_{n_k}| + |a_{n_k} - a| < \epsilon$ and we have established the claim.

From here the proof is an easy consequence of Theorem 1.3.11. In fact there is an integer N such that $|a_n - a_m| < 1$ when $m, n \geq N$. Thus when $n \geq N$, $|a_n| \leq |a_n - a_N| + |a_N| \leq 1 + |a_N|$, and so $\{a_n : n \geq N\}$ is bounded. Since what remains of the sequence is finite, $\{a_n\}$ is a bounded sequence. By the Bolzano–Weierstrass Theorem $\{a_n\}$ has a convergent subsequence. Hence the claim proves the theorem. ∎

In light of this theorem, we see the importance of the concept of a Cauchy sequence – if we are given a Cauchy sequence, we don't have to produce a limit in order to know that the sequence converges. We will see the importance of this as we proceed.

We conclude this section with two important concepts for sequences. First, however, we need to introduce the idea of a sequence converging to $\pm\infty$. Say that a sequence $\{x_n\}$ *converges to* ∞ if for each real number R there is an integer N such that $x_n > R$ for all $n \geq N$. Similarly, $x_n \to -\infty$ if for each real number R there is

....

[5] Augustin Louis Cauchy was born in Paris in August 1789, a month after the storming of the Bastille. He was educated in engineering and his first job was in 1810 working on the port facilities at Cherbourg in preparation for Napoleon's contemplated invasion of England. In 1812 he returned to Paris and his energies shifted toward mathematics. His contributions were monumental, with a plethora of results bearing his name. His collected works fill 27 published volumes. As a human being he left much to be desired. He was highly religious with a totally dogmatic personality, often treating others with dismissive rudeness. Two famous examples were his treatment of Abel and Galois, where he refused to consider their monumental works, which they had submitted to him. Both suffered an early death. Perhaps better treatment by Cauchy would have given them some recognition that would have resulted in a longer life and a productive career to the betterment of mathematics; we'll never know. He had two doctoral students, one of which was Bunyakowsky. Cauchy died in 1857 in Sceaux near Paris.

an integer N such that $x_n < R$ for all $n \geq N$. It is easy to see that $x_n \to -\infty$ if and only if $-x_n \to \infty$. Also see Exercise 13.

If $\{a_n\}$ is a given sequence, let $x_n = \sup\{a_n, a_{n+1}, \ldots\}$ and consider the sequence $\{x_n\}$. Note that as n increases, we are taking the supremum over a smaller collection of terms. Thus $x_n \geq x_{n+1}$ for all n; that is, $\{x_n\}$ is a decreasing sequence. As such it has a limit, though that limit might be $\pm\infty$. (Note that it could be that $x_n = \infty$ for all n if $\{a_n\}$ is not bounded above; in this case $x_n \to +\infty$. Though it is impossible that any $x_n = -\infty$, it might still happen that $x_n \to -\infty$.) Similarly if we set $y_n = \inf\{a_n, a_{n+1}, \ldots\}$, then $\{y_n\}$ is an increasing sequence and so it converges though it may be that the limit is $\pm\infty$. We formalize this discussion in the following.

1.3.14. Definition. If $\{a_n\}$ is a sequence of real numbers, the *limit superior* or *upper limit* of $\{a_n\}$ is defined as

$$\limsup_n a_n = \limsup a_n = \lim_n [\sup\{a_n, a_{n+1}, \ldots\}]$$

The *limit inferior* or *lower limit* of $\{a_n\}$ is defined as

$$\liminf_n a_n = \liminf a_n = \lim_n [\inf\{a_n, a_{n+1}, \ldots\}]$$

We will frequently just say the *limsup* or *liminf* of the sequence $\{a_n\}$.

Notice that because we allow $\pm\infty$ as the limits, the limsup and liminf of a sequence always exists as an element of $\mathbb{R}_\infty = \mathbb{R} \cup \{-\infty, \infty\}$.

1.3.15. Example. (a) If $a_n = (-1)^n$, then $\limsup a_n = 1$ and $\liminf a_n = -1$.

(b) If $a_n = 1$ when n is odd and $a_n = \frac{1}{n}$ when n is even, then $\limsup a_n = 1$ and $\liminf a_n = 0$.

(c) If $\{a_n\}$ is the sequence

$$\{1/2, 1/4, 2/4, 3/4, 1/8, 2/8, \ldots, 7/8, 1/16, \ldots, 15/16, \ldots\}$$

then $\limsup a_n = 1$ and $\liminf a_n = 0$.

(d) If $a_n = -n$ when n is odd and $a_n = n$ when n is even, then $\limsup a_n = +\infty$ and $\liminf a_n = -\infty$.

The proof of the next proposition is Exercise 14.

1.3.16. Proposition. *If $\{a_n\}$ is a sequence of real numbers, then the following hold.*

(a) $\liminf a_n \leq \limsup a_n$.
(b) $\{a_n\}$ *is bounded above if and only if* $\limsup a_n < \infty$ *and it is bounded below if and only if* $\liminf a_n > -\infty$.
(c) $a_n \to a$ *if and only if* $\limsup a_n = a = \liminf a_n$.
(d) *If $\{b_n\}$ is another sequence and $a_n \leq b_n$ for all $n \geq 1$, then* $\liminf a_n \leq \liminf b_n$ *and* $\limsup a_n \leq \limsup b_n$.

1.3.17. Proposition. *If $\{a_n\}$ is any sequence in $\mathbb{R}$, then*

$$\liminf a_n = -[\limsup(-a_n)]$$

This last result allows us to reduce the proof of an assertion about lim inf to proving an analogous assertion about lim sup. The reader is asked to prove this as Exercise 15. In fact anyone who is not thoroughly familiar with lim sup and lim inf should complete this exercise to solidify the concepts in their brain.

1.3.18. Proposition. *If $\{a_n\}$ is a sequence in $\mathbb{R}$, $\alpha = \limsup a_n$, and $\beta = \liminf a_n$, then there is a subsequence of $\{a_n\}$ that converges to α and another that converges to β.*

Proof. We only prove the statement about $\alpha = \limsup a_n$ as the other half of the proposition follows by the preceding proposition. In addition we will assume $\alpha \in \mathbb{R}$; when $\alpha = \pm\infty$, we leave the proof to the reader. Let $s_n = \sup\{a_n, a_{n+1}, \ldots\}$; so $s_n \searrow \alpha$. By the definition of the limit superior, for each $k \geq 1$ there is an m_k with $\alpha \leq s_{m_k} < \alpha + k^{-1}$. By the definition of supremum there is an $n_k \geq m_k$ such that $\alpha - k^{-1} < a_{n_k} \leq s_{m_k}$. Thus $|\alpha - a_{n_k}| < k^{-1}$ and so the subsequence $\{a_{n_k}\}$ converges to α. ∎

1.3.19. Corollary. *Let $\{a_n\}$ be a sequence in $\mathbb{R}$.*

(a) *If $a \in \mathbb{R}$ such that $\limsup a_n > a$, then there are infinitely many values of n such that $a_n > a$.*
(b) *If $a \in \mathbb{R}$ such that $\limsup a_n < a$, then there is an integer N such that $a_n < a$ for all $n \geq N$.*

Proof. (a) By the preceding proposition there is a subsequence $\{a_{n_k}\}$ such that $a_{n_k} \to \limsup a_n > a$. Thus there is an N such that $a_{n_k} > a$ for all $n_k \geq N$.

(b) If $\limsup a_n < a$, then the definition of the lim sup implies there is an N with $\sup\{a_N, a_{N+1}, \ldots\} < a$. Hence part (b). ∎

Exercises

(1) Prove Proposition 1.3.3(b).
(2) For each of the following sequences $\{a_n\}$ find the value of the limit and for each stipulated value of ϵ, find a value of N such that $|a - a_n| < \epsilon$ when $n \geq N$. (a) $a_n = n^{-1}, \epsilon = .0001$. (b) $a_n = 2^{-n}, \epsilon = .0001$. (Are the values for N you found the smallest possible? This has no bearing on the convergence, but it's a bit more challenging to find the smallest possible N.)
(3) If $a_n \to a$ and $\{a_{n_k}\}$ is a renumbering of the original sequence, does a_{n_k} converge to a?
(4) Prove that the sequence $\{n^2\}$ does not converge.
(5) (a) Show that $\{\frac{2^{\frac{1}{n}}}{n}\}$ converges and find its limit. (b) What is $\lim_n \frac{n+5}{2n+6}$? (c) Show that $\lim_n \frac{5n^2+2}{8n+10n^2} = \frac{1}{2}$.
(6) Show the following. (a) $\sqrt{n+1} - \sqrt{n} \to 0$. (b) $\sqrt{n^2+2n} - n \to 1$. (c) $\frac{\sqrt{n+1}}{n} \to 0$.

(7) Show that the following sequences converge. (a) $\{x_n\} = \{\frac{n}{2^n}\}$. (b) $\{x_n\} = \{\frac{1 \cdot 3 \cdots (2n-1)}{2 \cdot 4 \cdots (2n)}\}$.

(8) Prove Proposition 1.3.10.

(9) If $\{x_n\}$ is a sequence in $[a, b]$ with the property that every convergent subsequence of $\{x_n\}$ converges to the same point x, show that $x_n \to x$.

(10) Prove the converse of Proposition 1.3.7. That is, show that if $\{a_n\}$ is an increasing sequence and $a_n \to a$, then $\{a_n\}$ is bounded above and $a = \sup_n a_n$.

(11) (a) If $p > 0$, show that $\lim_n n^{-p} = 0$. (b) If $|x| < 1$, show that $\lim_n x^n = 0$.

(12) If $\{x_n\} \to 0$, show that

$$\frac{x_1 + \cdots + x_n}{n} \to 0.$$

(13) Let $\{x_n\}$ be a sequence in $\mathbb{R}$. (a) If $x_n \to \infty$ and $\{x_{n_k}\}$ is a subsequence, show that $x_{n_k} \to \infty$. (b) Find an example of a sequence $\{x_n\}$ then converges to ∞ and a sequence $\{y_n\}$ such that $y_n \to y$, but $\{x_n y_n\}$ does not converge to ∞. (c) Give a condition on a sequence $\{y_n\}$ such that if $x_n \to \infty$ and $y_n \to y$, then $x_n y_n \to \infty$. (d) If $\{x_n\}$ is a sequence and $x_n \to \infty$, show that $x_n^{-1} \to 0$. (e) If $x_n \to 0$, is it true that $x_n^{-1} \to \infty$?

(14) Prove Proposition 1.3.16.

(15) Prove Proposition 1.3.17.

(16) If $\{a_n\}$ and $\{b_n\}$ are sequences of positive numbers such that $b_n \to b > 0$, show that $\limsup a_n b_n = b \limsup a_n$. Are the requirements that $a_n, b_n, b > 0$ needed?

(17) Show that there are sequences of real numbers $\{a_n\}$ and $\{b_n\}$ such that $\limsup_n (a_n + b_n) \neq \limsup_n a_n + \limsup_n b_n$.

1.4. Series

If $\{a_n\}$ is a sequence in $\mathbb{R}$, and $m \geq 1$ we define the m-th *partial sum* of $\{a_n\}$ as $s_m = \sum_{n=1}^m a_n = a_1 + \cdots + a_m$. The notation $\sum_{n=1}^\infty a_n$ is called the *infinite series* or just the *series* and stands for the sequence $\{s_m\}$ of partial sums. Sometimes we may want to start the summing process at a different number than $n = 1$. In particular we will often examine infinite series of the form $\sum_{n=0}^\infty a_n$ as we do in Proposition 1.4.4 below.

1.4.1. Definition. If $\{a_n\}$ is a given sequence of numbers in $\mathbb{R}$ and $s \in \mathbb{R}$, we say that the infinite series $\sum_{n=1}^\infty a_n$ *converges* to s if the sequence of partial sums $\{s_m\}$, $s_m = \sum_{n=1}^m a_n$, converges to s. When this happens we call the series a *convergent series* and we write $\sum_{n=1}^\infty a_n = s$. If the series $\sum_{n=1}^\infty a_n$ fails to converge to any real number s, then we say that the series *diverges* or that it is a *divergent series*.

1.4.2. Proposition. *If the infinite series $\sum_{n=1}^\infty a_n$ converges, then $a_n \to 0$.*

Proof. Suppose $\sum_{n=1}^\infty a_n = s$ and let s_m be the m-th partial sum. So $s_m \to s$. Thus $a_n = s_n - s_{n-1} \to s - s = 0$. ∎

Most of you learned the last result in calculus as well as the fact that the condition that $a_n \to 0$ does not imply that the series converges. Here is the standard example.

1.4.3. Example. Consider the series $\sum_{n=1}^{\infty} n^{-1}$. This is called the *harmonic series*. The origin of this name has to do with the concept of overtones in music, a topic we won't go into. To see that this diverges we rewrite the series grouping together the terms as follows:

$$\sum_{n=1}^{\infty} \frac{1}{n} = 1 + \frac{1}{2} + \left(\frac{1}{3} + \frac{1}{4}\right) + \left(\frac{1}{5} + \cdots + \frac{1}{8}\right) + \left(\frac{1}{9} + \cdots + \frac{1}{16}\right) + \cdots$$

$$> 1 + \frac{1}{2} + \left(\frac{1}{4} + \frac{1}{4}\right) + \left(\frac{1}{8} + \cdots + \frac{1}{8}\right) + \left(\frac{1}{16} + \cdots + \frac{1}{16}\right) + \cdots$$

In other words we group the terms together in such a way that each sum within parentheses is larger that a half. So the next parenthetical expression would contain 16 terms with the last term being $\frac{1}{32}$. We therefore see that the partial sums grow infinitely large. (This example was important to present at this moment in the text, but what we just did was to use Corollary 1.4.8 below. Nevertheless in this specific instance the application is evidently valid.)

The next result is an example, but its importance elevates it to a proposition.

1.4.4. Proposition. *When $|x| < 1$, the series $\sum_{n=0}^{\infty} x^n$ converges to $(1-x)^{-1}$. When $|x| \geq 1$ the series $\sum_{n=0}^{\infty} x^n$ diverges.*

Proof. We begin with the following.

Claim. If $x \neq 1$ and $n \geq 1$, then

$$1 + x + \cdots + x^n = \frac{1 - x^{n+1}}{1 - x}$$

In fact this is demonstrated by verifying that $(1 - x)(1 + x + \cdots + x^n) = 1 - x^{n+1}$.

Now that the claim is established we invoke Exercise 1.3.11(b) to get that the series converges to $(1 - x)^{-1}$ when $|x| < 1$. Notice that when $|x| \geq 1$, $\lim_n x^n \neq 0$ so the series diverges by Proposition 1.4.2. ∎

The series in the preceding proposition is called the *geometric series*.

We can use Theorem 1.3.13 to obtain a necessary and sufficient condition for an infinite series to converge.

1.4.5. Proposition. *An infinite series $\sum_{n=1}^{\infty} a_n$ converges if and only if for every $\epsilon > 0$ there is an integer N such that when $N \leq n < m$, $|\sum_{k=n}^{m} a_k| < \epsilon$.*

Proof. If s_n denotes the n-th partial sum of the series and $n < m$, then $\sum_{k=n}^{m} a_k = s_m - s_{n-1}$. Hence the condition stated in this proposition is equivalent to the condition that $\{s_n\}$ is a Cauchy sequence. By Theorem 1.3.13 this is equivalent to the condition that $\{s_n\}$ converges. ∎

If all the terms of a series are positive, it becomes a bit easier to discuss convergence. In fact note that if $a_n \geq 0$ for all n, then $s_n = a_1 + \cdots + a_n$ defines an increasing sequence. Hence the next result is immediate from Proposition 1.3.7.

1.4.6. Proposition. *If $a_n \geq 0$ for all n, then $\sum_{n=1}^{\infty} a_n$ converges if and only if the sequence of partial sums is bounded.*

Also see Exercise 3.

1.4.7. Theorem (Comparison Test). *If $\sum_{n=1}^{\infty} a_n$ and $\sum_{n=1}^{\infty} b_n$ are two infinite series such that there is an integer N with $|a_n| \leq b_n$ when $n \geq N$ and if $\sum_{n=1}^{\infty} b_n$ converges, then $\sum_{n=1}^{\infty} a_n$ converges.*

Proof. First note that the hypothesis implies that $b_m \geq 0$ for all n. Let s_n denote the n-th partial sum of the series $\sum_{n=1}^{\infty} a_n$. If $\epsilon > 0$, the fact that $\sum_{n=1}^{\infty} b_n$ converges implies that we can choose $M \geq N$ such that when $m > n \geq M$, $\sum_{k=n+1}^{m} b_k < \epsilon$. Thus when $m > n \geq M$, $|s_m - s_n| = |\sum_{k=n+1}^{m} a_n| \leq \sum_{k=n+1}^{m} |a_n| \leq \sum_{k=n+1}^{m} b_n < \epsilon$. That is the sequence $\{s_n\}$ is a Cauchy sequence and hence must converge. ∎

1.4.8. Corollary. *If $\sum_{n=1}^{\infty} a_n$ and $\sum_{n=1}^{\infty} b_n$ are two infinite series such that there is an integer N with $0 \leq a_n \leq b_n$ when $n \geq N$ and if $\sum_{n=1}^{\infty} a_n$ diverges, then $\sum_{n=1}^{\infty} b_n$ diverges.*

Proof. If it were the case that $\sum_{n=1}^{\infty} b_n$ converges, then the theorem would imply that $\sum_{n=1}^{\infty} a_n$ converges. ∎

It is worth underlining that unlike in the theorem, in the corollary we are assuming that each term of the series $\sum_n a_n$ is non-negative.

1.4.9. Definition. An infinite series $\sum_{n=1}^{\infty} a_n$ *converges absolutely* if the series of positive terms $\sum_{n=1}^{\infty} |a_n|$ converges.

1.4.10. Proposition. *If a series converges absolutely, then it converges.*

Proof. Suppose $\sum_{n=1}^{\infty} a_n$ is an absolutely convergent series. Let $\epsilon > 0$ and choose an integer N such that when $n \geq N$, $\sum_{k=n}^{\infty} |a_n| < \epsilon$ (Exercise 3). If s_n is the n-th partial sum of the series $\sum_{n=1}^{\infty} a_n$, then for $n > m > N$ we have that $|s_n - s_m| = |\sum_{k=m+1}^{n} a_k| \leq \sum_{k=m+1}^{n} |a_k| < \epsilon$. That is, $\{s_n\}$ is a Cauchy sequence and hence it converges. ∎

1.4.11. Example. Consider the *alternating harmonic series* $\sum_{n=1}^{\infty} (-1)^n \frac{1}{n}$. By Example 1.4.3 we know this series is not absolutely convergent. On the other hand for $n > m$,

$$\left| \sum_{k=m+1}^{n} (-1)^k \frac{1}{k} \right| = \left| \frac{1}{m+1} - \frac{1}{m+2} + \cdots + (-1)^{n-m-1} \frac{1}{n} \right|$$

Now observe that in this sum, if we group successive pairs of terms after $(m+1)^{-1}$, each of those pairs is positive and they are all subtracted from $(m+1)^{-1}$. Hence we have that

$$\left| \sum_{k=m+1}^{n} (-1)^k \frac{1}{k} \right| < \frac{1}{m+1}$$

By Proposition 1.4.5, the alternating harmonic series converges. So the converse of Proposition 1.4.10 is false: there are convergent series that are not absolutely convergent.

1.4.12. Theorem (Root Test). *Let $\{a_n\}$ be a sequence in $\mathbb{R}$ and put*

$$r = \limsup |a_n|^{\frac{1}{n}}$$

(a) *If $r < 1$, then the series $\sum_{n=1}^{\infty} a_n$ converges absolutely.*
(b) *If $r > 1$, then $\{a_n\}$ does not converge to 0 and the series $\sum_{n=1}^{\infty} a_n$ diverges.*

Proof. (a) Let $r < x < 1$. By Corollary 1.3.19(b) there is an integer N such that $|a_n|^{\frac{1}{n}} < x$ for all $n \geq N$. Now the geometric series $\sum_{n=1}^{\infty} x^n$ converges since $0 \leq x < 1$. Thus the comparison test implies that $\sum_{n=1}^{\infty} |a_n|$ converges.

(b) Since $r > 1$, Corollary 1.3.19(a) implies that $|a_n|^{\frac{1}{n}} > 1$ for infinitely many values of n. Hence it cannot be that $a_n \to 0$ and so $\sum_{n=1}^{\infty} a_n$ diverges. ∎

1.4.13. Example. When $\limsup |a_n|^{\frac{1}{n}} = 1$ there is no conclusion. Consider the harmonic series $\sum_{n=1}^{\infty} n^{-1}$, which diverges. We claim that $(\frac{1}{n})^{\frac{1}{n}} \to 1$. To see this we need two results from later in this book: L'Hôpital's Rule (2.5.1) and properties of the natural logarithm (§3.3). Since the reader has encountered both of these in calculus and we are only concerned with an example here, we will use these results from the future to explore the present example. By L'Hôpital's Rule, $(\log x)/x \to 0$ as $x \to \infty$. (See Exercise 3.3.14.) Hence $\log[(\frac{1}{n})^{\frac{1}{n}}] = -\frac{1}{n}\log n \to 0$, and so $(\frac{1}{n})^{\frac{1}{n}} = \exp[\log(\frac{1}{n})^{\frac{1}{n}}] \to 1$. On the other hand the alternating harmonic series converges and satisfies $\limsup |a_n|^{\frac{1}{n}} = 1$.

We should also underscore an additional disparity between the two parts in the Root Test. Part (a) concludes that the series converges absolutely, while (b) says that the series diverges, not merely that it fails to converge absolutely. The same disparity pertains to the next result.

1.4.14. Theorem (Ratio Test). *Let $\{a_n\}$ be a sequence in $\mathbb{R}$ and assume that $r = \lim_{n \to \infty} |a_{n+1}/a_n|$ exists.*

(a) *If $r < 1$, then the series $\sum_{n=1}^{\infty} a_n$ converges absolutely.*
(b) *If $r > 1$, then the series $\sum_{n=1}^{\infty} a_n$ diverges.*

Proof. The proof is similar to that of the Root Test. If $r < x < 1$, then there is an N such that $|a_{n+1}/a_n| \leq x$ for all $n \geq N$. Thus $|a_{n+1}| \leq x|a_n|$ whenever $n \geq N$. Therefore for $k \geq 1$, $|a_{N+k}| \leq x|a_{N+k-1}| \leq x^2|a_{N+k-2}| \leq \cdots \leq x^k|a_N|$. By the Comparison Test with the geometric series $\sum_{n=1}^{\infty} x^n$, $\sum_{n=1}^{\infty} |a_n|$ converges. The proof of part (b) is similar to the proof of (1.4.12(b)) and is left to the reader. ∎

1.4.15. Example. The series

$$\sum_{n=0}^{\infty} \frac{x^n}{n!}$$

converges absolutely for every real number x. (Recall that $0! = 1$.) In fact we can apply the Ratio Test to find that

$$\frac{x^{n+1}/(n+1)!}{x^n/n!} = \left[\frac{x^{n+1}}{(n+1)!}\right]\left[\frac{n!}{x^n}\right] = \frac{x}{n+1} \to 0$$

The preceding series is very important and we'll see it again later. In fact the reader may remember from calculus that this series converges to e^x. We'll establish this and more when we discuss convergence of functions in Chapter 4.

Absolutely convergent series have many additional properties. For example it can be shown that if $\sum_{n=1}^{\infty} a_n$ converges absolutely and $\sigma : \mathbb{N} \to \mathbb{N}$ is a bijection, then $\sum_{n=1}^{\infty} a_{\sigma(n)}$ converges. On the other hand, if the series $\sum_{n=1}^{\infty} a_n$ is *conditionally convergent*, that is, it converges but not absolutely, then for any real number x there is a bijection σ on $\mathbb{N}$ such that $\sum_{n=1}^{\infty} a_{\sigma(n)}$ converges to x. This amazing result is called the Riemann[6] series theorem. A proof can be found at http://en.wikipedia.org/wiki/Riemann_series_theorem

Exercises

(1) If the series $\sum_{n=1}^{\infty} a_n$ converges, then prove that for any integer $m \geq 1$ the series $\sum_{n=m}^{\infty} a_n$ converges.

(2) (a) Show that if $\sum_{n=1}^{\infty} a_n$ converges to A and $\sum_{n=1}^{\infty} b_n$ converges to B, then $\sum_{n=1}^{\infty}(a_n + b_n)$ converges to $A + B$. (b) If $\sum_{n=1}^{\infty} a_n$ converges to A and $x \in \mathbb{R}$, show that $x \sum_{n=1}^{\infty} a_n$ converges to xA.

(3) Show that a series $\sum_{n=1}^{\infty} a_n$, where each $a_n \geq 0$, converges if and only if for every $\epsilon > 0$ there is an integer N such that $|\sum_{k=N}^{\infty} a_k| < \epsilon$.

(4) Prove Theorem 1.4.14(b).

(5) Prove the following. If $\{a_n\}$ and $\{b_n\}$ are two sequences of strictly positive numbers such that $a_{n+1}/a_n \leq b_{n+1}/b_n$ for all $n \geq 1$, then the following hold: (a) If $\sum_n b_n$ converges, then $\sum_n a_n$ converges; (b) if $\sum_n b_n$ diverges, then $\sum_n a_n$ diverges.

---

[6] See the definition of Riemann integrable in §3.1 below for a biographical note.

(6) Does

$$\sum_{n=1}^{\infty}(-1)^n\frac{n^2 2^n}{n!}$$

converge absolutely?

(7) Suppose that the series $\sum_{k=0}^{\infty} a_k$ converges absolutely. (a) If $\{b_k\}$ is a bounded sequence of numbers, show that $\sum_{k=0}^{\infty} a_k b_k$ converges absolutely. (b) By giving an example, show that if it is only assumed that the series $\sum_{k=0}^{\infty} a_k$ converges (not absolutely), then $\sum_{k=0}^{\infty} a_k b_k$ may diverge.

(8) If $\sum_n a_n$ converges absolutely, does $\sum_n a_n^2$ converge absolutely?

(9) If $\sum_n a_n$ converges and $b_n \to 0$, does $\sum_n a_n b_n$ converge?

1.5. Countable and Uncountable Sets

Here we will explore the notions of countable and uncountable sets. The fact that not all infinite sets are equivalent is one that comes as a surprise to many people. Indeed historically this was a shock to the world of mathematics when Cantor[7] first revealed it.

1.5.1. Definition. A set X is *countable* if there exists a subset A of the natural numbers $\mathbb{N}$ and a bijective function $f : A \to X$. It is called *uncountable* if it fails to be countable.

1.5.2. Example. (a) Any finite set is countable. For infinite sets that are countable, we will say they are *countably infinite*. Some say that such an infinite set is *denumerable*. Below we show the existence of sets that are uncountable.

(b) The set of all integers, $\mathbb{Z}$, is countable. In fact, $0, 1, -1, 2, -2, 3, \ldots$ describes a bijective function from $\mathbb{N}$ onto $\mathbb{Z}$. For convenience we will often show a set is countable by describing how to exhaust the set by writing it as a sequence as we just did. Such an undertaking tells us how to define a bijective function even though finding a formula for that function may be unclear. In the present case it is not difficult to write a formula for this function. Indeed if we define $f : \mathbb{N} \to \mathbb{Z}$ by

$$f(n) = \begin{cases} \frac{n}{2} & \text{if } n \text{ is even} \\ -\frac{n-1}{2} & \text{if } n \text{ is odd} \end{cases}$$

then this is the function that gives the correspondence described above. In other situations writing a formula for the function may range from challenging to impossible. Understand, however, that proving the existence of such a function does not mean we have to write its formula. If we describe a process or algorithm for determining

---
[7] See the biographical note in Theorem 1.6.8.

which element of the set corresponds to each integer and if this process exhausts the set, then we have described the required function.

(c) Any subset of a countable set is countable. In fact this is immediate from the definition of a countable set.

The next two propositions are useful in showing that a given set is countable. See, for example, Corollary 1.5.5 below.

1.5.3. Proposition. (a) *If X is any set such that there is a subset A of $\mathbb{N}$ and a surjective function $f : A \to X$, then X is countable.*
(b) *If X is a countable set, Y is another set, and there is a surjection $f : X \to Y$, then Y is countable.*

Proof. To prove (a), let f and A be as in the statement. For each x in X let n_x be the first integer n in A with $f(n) = x$; that is, $n_x = \min f^{-1}(x)$. So $B = \{n_x : x \in X\}$ is another subset of $\mathbb{N}$ and $g : B \to X$ defined by $g(n_x) = x$ is a bijection. Part (b) is immediate from (a) and the definition of a countable set. ∎

We take a moment to interrupt the discussion of different types of infinite sets to return to the discussion of set theory. We specifically want to define the *cartesian product* of two sets, $X \times Y$, something many readers may have already seen. This is the set

$$X \times Y = \{(x, y) : x \in X, y \in Y\}$$

Similarly if $X_1, \ldots, X_n$ are a finite number of sets we could define the cartesian product

$$X_1 \times \cdots \times X_n = \{(x_1, \ldots, x_n) : x_k \in X_k \text{ for } 1 \le k \le n\}$$

1.5.4. Proposition. *If X and Y are countable sets, then so is $X \times Y \equiv \{(x, y) : x \in X, y \in Y\}$.*

Proof. We only consider the case where the two sets are infinite. To prove the proposition it is equivalent to show that $\mathbb{N} \times \mathbb{N}$ is countable. (Why?) Here we want to define a bijection $f : \mathbb{N} \to \mathbb{N} \times \mathbb{N}$. Again we need only show how to arrange the elements (m, n) in $\mathbb{N} \times \mathbb{N}$ in a sequence. So imagine $\mathbb{N} \times \mathbb{N}$ as an infinite square array of pairs of positive integers. On the first row are all the pairs $\{(1, n) : n \in \mathbb{N}\}$; on the second $\{(2, n) : n \in \mathbb{N}\}$; etc. We write down the following sequence of entries.

$$(1, 1), (2, 1), (1, 2), (3, 1), (2, 2), (1, 3), (4, 1), (3, 2), (2, 3), (1, 4), \ldots$$

If you write the array on paper and draw northeast diagonal lines connecting these pairs, you should be able to discern the pattern. (Many other patterns are possible.) This describes the bijection. ∎

1.5.5. Corollary. *The set of rational numbers is countable.*

Proof. Writing each rational number as a fraction in reduced terms we see that there is a bijection between $\mathbb{Q}$ and a subset of $\mathbb{Z} \times \mathbb{Z}$, which is countable by the proposition. ∎

Using induction and the preceding proposition we can obtain the following corollary.

1.5.6. Corollary. *If $X_1, \ldots, X_n$ are countable sets, then so is $X_1 \times \cdots \times X_n$.*

1.5.7. Proposition. *If $X = \bigcup_{n=1}^{\infty} X_n$ and each of the sets X_n is countable, then X is countable.*

Proof. We write $X_n = \{x_n^1, x_n^2, \ldots\}$. If X_n is infinite, we can do this with $x_n^k \neq x_n^j$ for all $k \neq j$; if X_n is finite, repeat one of the points an infinite number of times. Thus $f : \mathbb{N} \times \mathbb{N} \to X$ defined by $f(n, k) = x_n^k$ is surjective. It follows by Proposition 1.5.3(b) that X is countable. ∎

1.5.8. Corollary. *The set of all finite subsets of $\mathbb{N}$ is countable.*

Proof. If F denotes the set of all finite subsets of $\mathbb{N}$, then note that $F = \bigcup_{n=1}^{\infty} S_n$, where S_n is the set of all subsets of $\{1, 2, \ldots, n\}$. But S_n is a finite set. (In fact from combinatorics we know that S_n has 2^n elements.) By the preceding proposition, F is countable. ∎

Now we turn to some results showing the existence of uncountable sets.

1.5.9. Proposition. *The set of all sequences of zeros and ones is not countable.*

Proof. Let X be the set of all sequences of zeros and ones, and suppose it is countable; so we can write $X = \{x_1, x_2, \ldots\}$. We manufacture an element a in X such that $a \neq x_n$ for any $n \geq 1$. This will furnish a contradiction to the assumption that we have an exhaustive list and thus prove the proposition. Suppose that for each $n \geq 1$, $x_n = x_n^1 x_n^2 \cdots$ is a sequence of zeros and ones; in other words, $x_n = \{x_n^k : k = 1, 2, \ldots\}$. If $n \geq 1$ and $x_n^n = 0$, let $a^n = 1$; if $x_n^n = 1$, let $a^n = 0$. This defines an $a = a^1 a^2 \cdots$ in X. Since $a^n \neq x_n^n$, $a \neq x_n$ for any $n \geq 1$. This gives our desired contradiction. ∎

1.5.10. Corollary. *The collection of all subsets of $\mathbb{N}$, $2^{\mathbb{N}}$, is not countable.*

Proof. In fact by looking at the characteristic functions of subsets of $\mathbb{N}$, we see that the set of all sequences of zeros and ones is in bijective correspondence with $2^{\mathbb{N}}$. ∎

To prove the next proposition we have to consider dyadic expansions of numbers in the unit interval. For $0 \leq x \leq 1$ we can write

$$x = \sum_{n=1}^{\infty} \frac{x_n}{2^n}$$

where each $x_n = 0$ or 1. (This series always converges since it is dominated by $\sum_{n=1}^{\infty} 2^{-n} = 1$.) The proof that each x in the unit interval can be so expanded is not too complicated and proceeds as follows. Consider x and divide the interval into its equal halves: $[0, \frac{1}{2}]$ and $[\frac{1}{2}, 1]$. If x belongs to the first half, let $x_1 = 0$; if $x \in [\frac{1}{2}, 1]$, let $x_1 = 1$. (We note an ambiguity here if $x = \frac{1}{2}$ and we will address this shortly.) Note that in either case we have that $|x - x_1/2| < \frac{1}{2}$. Now consider whichever half interval contains x and divide it into two equal halves; let $x_2 = 0$ if x belongs to the first half and $x_2 = 1$ if it belongs to the second half. Now we have that

$$\left| x - \left(\frac{x_1}{2} + \frac{x_2}{2^2} \right) \right| < \frac{1}{2^2}$$

Continue this process and we see that the series so defined will converge to x. (The reader who wants to write out the details can formulate an induction statement based on what we just did and prove it. See Exercise 2.)

What about the ambiguity? If $x = a/2^n$ for some $n \geq 1$ and $0 < a < 2^n$, then the choice of x_n can be either 0 or 1. In fact this is the only way such an ambiguity arises. In fact using the summation for a geometric series,

$$\sum_{k=n}^{\infty} \frac{1}{2^k} = \frac{1}{2^n} \sum_{k=0}^{\infty} \frac{1}{2^k} = \frac{1}{2^n} \frac{1}{1 - \frac{1}{2}} = \frac{1}{2^{n-1}}$$

It follows that if $\{x_n\}, \{y_n\}$ are two sequences of zeros and ones, then the only way that we can have that $\sum_{n=1}^{\infty} x_n/2^n = \sum_{n=1}^{\infty} y_n/2^n$ is that either there exists an integer N such that $x_n = y_n$ for all $n \geq N$, or one sequence ends in all zeros and the other ends in all ones. See Exercise 3.

1.5.11. Proposition. *The interval* $(0, 1)$ *is not countable.*

Proof. In a sense, this proposition is a corollary of Proposition 1.5.9, but its proof is a bit more involved than you usually associate with a corollary. Let X be the set of all sequences of zeros and ones that are not constantly one from some point on. Let Y be the set of all sequences of zeros and ones that are constantly one from some point on. Note that $X \cap Y = \emptyset$ and $X \cup Y$ is the set of all sequences of zeros and ones. Now by considering the characteristic functions of subsets of $\mathbb{N}$, there is a bijective mapping between $X \cup Y$ and $2^{\mathbb{N}}$. Hence $X \cup Y$ is uncountable by Corollary 1.5.10.

Let Y_1 be the singleton consisting of the identically 1 sequence and for each $n > 1$ let Y_n be the set of all sequences $\{x_k\}$ in Y with $x_k = 1$ whenever $k \geq n$. We note that there is a bijection between Y_n and the set of all subsets of $\{1, \ldots, n-1\}$. Hence Y_n is finite. By Proposition 1.5.7, $Y = \bigcup_{n=1}^{\infty} Y_n$ is countable. Again using Proposition 1.5.7 we have that the only way for $X \cup Y$ to be uncountable is for X to be uncountable. Therefore $(0, 1)$ is uncountable. ∎

(1) Show that if A is an infinite subset of $\mathbb{N}$, X is a countably infinite set, and $f : A \to X$ is a bijection, then there is a bijection $g : \mathbb{N} \to X$. (Hint: First show that if A is an infinite subset of $\mathbb{N}$, then there is a bijection $h : \mathbb{N} \to A$.)
(2) Write out a detailed proof that each x in the unit interval has a dyadic expansion.
(3) If $\{x_n\}$, $\{y_n\}$ are two sequences of zeros and ones, show that

$$\sum_{n=1}^{\infty} x_n / 2^n = \sum_{n=1}^{\infty} y_n / 2^n$$

if and only if there is an integer n such that $x_k = 0$ and $y_k = 1$ for all $k \geq n$.

1.6. Open Sets and Closed Sets

Here we examine certain special subsets of $\mathbb{R}$ that together with their extensions to higher dimensional space underpin analysis.

1.6.1. Definition. A subset F of $\mathbb{R}$ is said to be *closed* if whenever $\{x_n\}$ is a sequence of points in F and $x_n \to x$ we have that $x \in F$. A subset G is said to be *open* if $\mathbb{R} \backslash G$ is a closed set.

1.6.2. Example. (a) $\mathbb{R}$ and $\emptyset$ are simultaneously closed and open.
   (b) Finite subsets of $\mathbb{R}$ are closed.
   (c) A closed interval is a closed set. (Why?) Similarly intervals of the form $(-\infty, a]$ and $[b, \infty)$ are closed. A closed interval is not, however, an open set.
   (d) The union of two closed intervals is a closed set. In fact if $F = [a, b] \cup [c, d]$ and $\{x_n\}$ is a sequence in F such that $x_n \to x$, then there is a subsequence $\{x_{n_k}\}$ contained in either $[a, b]$ or $[c, d]$. Since $x_{n_k} \to x$, we have that $x \in F$ and F must be closed.
   (e) An open interval (a, b) is an open set since its complement is the union of two closed intervals. An open interval is not, however, a closed set unless it is the interval $(-\infty, \infty)$.

1.6.3. Proposition. (a) *If $F_1, \ldots, F_n$ are closed sets, then $\bigcup_{k=1}^{n} F_k$ is closed.*
(b) *If $F_1, F_2, \ldots$ are closed sets, then $\bigcap_{k=1}^{\infty} F_k$ is closed.*
(c) *If $G_1, \ldots, G_n$ are open sets, then $\bigcap_{k=1}^{n} G_k$ is open.*
(d) *If $G_1, G_2, \ldots$ are open sets, then $\bigcup_{k=1}^{\infty} G_k$ is open.*

Proof. (a) Let $\{x_n\}$ be a sequence in $\bigcup_{k=1}^{n} F_k$ such that $x_n \to x$. It follows that at least one of the sets $F_1, \ldots, F_n$ contains infinitely many of the points in $\{x_n\}$. That is, there is a set F_j, $1 \leq j \leq n$ and a subsequence $\{x_{n_k}\}$ contained in F_j. Since $x_{n_k} \to x$, $x \in F_j \subseteq \bigcup_{k=1}^{n} F_k$, so this union is closed.

(c) By De Morgan's Law, $\mathbb{R}\backslash\bigcap_{k=1}^{n} G_k = \bigcup_{k=1}^{n}(\mathbb{R}\backslash G_k)$, which is closed by (a).

(b) If $\{x_n\}$ is a sequence in $\bigcap_{k=1}^{\infty} F_k$ that converges to x, then for $1 \le k < \infty$ $\{x_n\}$ is a sequence in F_k. Thus $x \in F_k$ since F_k is closed. Thus $x \in \bigcap_{k=1}^{\infty} F_k$, which is therefore closed by definition.

(d) As in the proof of (c), $\mathbb{R}\backslash\bigcup_{k=1}^{\infty} G_k = \bigcap_{k=1}^{\infty}(\mathbb{R}\backslash F_k)$ so that $\bigcup_{k=1}^{\infty} G_k$ must be open by (b). ∎

Also see Exercises 4 and 3. Now we give an equivalent formulation of open sets that will be used more often than the definition.

1.6.4. Proposition. *A subset G of $\mathbb{R}$ is open if and only if for each x in G there is an $\epsilon > 0$ such that $\{y \in \mathbb{R} : |y - x| < \epsilon\} = (x - \epsilon, x + \epsilon) \subseteq G$.*

Proof. Suppose G is open and $x \in G$. If the stated condition in the proposition is false, then for every $n \ge 1$ there is a point x_n in $\mathbb{R}\backslash G$ with $|x_n - x| < \frac{1}{n}$. Thus $x_n \to x$. Since $\mathbb{R}\backslash G$ is closed it follows that $x \in \mathbb{R}\backslash G$, a contradiction. Now assume that G satisfies the stated condition; we want to show that $\mathbb{R}\backslash G$ is closed. Let $\{x_n\}$ be a sequence in $\mathbb{R}\backslash G$ that converges to a point x. If it were the case that $x \notin \mathbb{R}\backslash G$, then $x \in G$. By assumption there is an $\epsilon > 0$ with $(x - \epsilon, x + \epsilon) \subseteq G$. But since $x_n \to x$ there is an n such that $|x_n - x| < \epsilon$ and so $x_n \in G$, a contradiction. ∎

1.6.5. Example. Let $a, b \in \mathbb{R}$ with $a < b$. If $E = \{x : a < x < b \text{ and } x \in \mathbb{Q}\}$, then E is not open. In fact if $x \in E$, we can choose $\epsilon > 0$ such that $(x - \epsilon, x + \epsilon) \subseteq (a, b)$, but the Density Property of $\mathbb{R}$ implies there is an irrational number y in $(x - \epsilon, x + \epsilon)$. Hence $(x - \epsilon, x + \epsilon)$ is not a subset of E. Similarly, $\{x : a \le x \le b \text{ and } x \in \mathbb{Q}\}$ is not a closed set.

We need a lemma that characterizes intervals in $\mathbb{R}$ – whether they are bounded or not, closed, open, or neither. To be clear an interval is any set of the form $[a, b]$, (a, b), $[a, b)$, or $(a, b]$, where a and b are any real numbers and there is the possibility that $a = -\infty$, or $b = \infty$, or both.

1.6.6. Lemma. *If $I \subseteq \mathbb{R}$, then I is an interval if and only if whenever $a, b \in I$ with $a < b$ it follows that $[a, b] \subseteq I$.*

Proof. It is clear that every interval has the stated property, so we need only assume that I has the property and show that it is an interval. To do this let $\alpha = \inf I$ and $\beta = \sup I$; it may be that $\alpha = -\infty$ or $\beta = \infty$ or both. If $\alpha < x < \beta$, then the definition of supremum and infimum implies there are points a and b in I with $\alpha \le a < x < b \le \beta$. By hypothesis, $[a, b] \subseteq I$; in particular, $x \in I$. Hence $(\alpha, \beta) \subseteq I$. Therefore it must be that I is one of the four possible intervals with endpoints α and β. ∎

1.6.7. Proposition. *A subset G of $\mathbb{R}$ is open if and only if it is the union of a countable number of pairwise disjoint open intervals.*

Proof. Clearly if G is the union of a sequence of open intervals, then it is open by Proposition 1.6.3(d). Now assume that G is an open set. By the preceding proposition whenever $x \in G$, there is an $\epsilon > 0$ such that $(x - \epsilon, x + \epsilon) \subseteq G$. Let

$\mathcal{I}_x$ denote the collection of all open intervals that contain the point x and are contained in G. By what we just said, $\mathcal{I}_x \neq \emptyset$; put $J_x = \bigcup\{(a,b) : (a,b) \in \mathcal{I}_x\}$. We'll use the preceding lemma to show that J_x is an interval. Let $a, b \in J_x$ and assume that $a < b$. By definition there are intervals I_a, I_b in $\mathcal{I}_x$ such that $a \in I_a$ and $b \in I_b$. Since I_a is an open interval containing a as well as x, $I_a = (c,d)$ and $c < a, x < d$. Similarly $I_b = (c', d')$ and $c' < b, x < d'$. Let's assume that $a < b \leq x$. It follows that $c < a < b \leq x < d$, so $[a,b] \subseteq I_a \subseteq J_x$. Now let's assume that $a \leq x \leq b$; so $[a,x] \subseteq I_a \subseteq J_x$ and $[x,b] \subseteq I_b \subseteq J_x$ and thus $[a,b] \subseteq J_x$. The remaining case that $x \leq a < b$ is similar to the first. By the lemma, J_x is an interval and it contains x by its definition.

We claim that if x and y are distinct points in G it follows that either $J_x = J_y$ or $J_x \cap J_y = \emptyset$. In fact if $J_x \cap J_y \neq \emptyset$, then Exercise 5 implies that $J_x \cup J_y$ is an open interval. As such it must belong to both $\mathcal{I}_x$ and $\mathcal{I}_y$. By definition, $J_x = J_y$.

Therefore $\{J_x : x \in G\}$ is a collection of pairwise disjoint open intervals. On the other hand the Density Property of $\mathbb{R}$ implies each J_x contains a rational number. Since $\mathbb{Q}$ is countable, $\{J_x : x \in G\}$ must be a countable collection of open intervals whose union is G. ∎

For any set E contained in $\mathbb{R}$ define its *diameter* as

$$\operatorname{diam} E = \sup\{|x - y| : x, y \in E\}$$

Note that the set E is bounded if and only if it has finite diameter.

1.6.8. Theorem (Cantor's[8] Theorem). *Let $\{F_n\}$ be a sequence of non-empty subsets of $\mathbb{R}$ satisfying:* (i) *each F_n is closed, and* (ii) $F_1 \supseteq F_2 \supseteq \cdots$.

(a) *If one of the sets F_n is bounded, then $\bigcap_{n=1}^{\infty} F_n \neq \emptyset$.*
(b) *If the sequence of sets $\{F_n\}$ also satisfies* (iii) $\operatorname{diam} F_n \to 0$, *then $\bigcap_{n=1}^{\infty} F_n$ is a single point.*

Proof. Let $\{F_n\}$ be as in the statement of the theorem and put $\bigcap_{n=1}^{\infty} F_n = F$. By (i) and Proposition 1.6.3, F is closed. (a) If F_n is bounded, then condition (ii) implies that F_k is bounded for all $k \geq n$. Without loss of generality we may assume that F_1 is bounded. For each $n \geq 1$ pick a point x_n in F_n; $\{x_n\}$ is a bounded sequence. By the Bolzano–Weierstrass Theorem there is a subsequence $\{x_{n_k}\}$ and a point x such that $x_{n_k} \to x$. Fix any positive integer m; by (ii), $x_{n_k} \in F_m$ for all $n_k \geq m$. By (i), $x \in F_m$. Since m was arbitrary, $x \in F$ and so $F = \bigcap_{n=1}^{\infty} F_n \neq \emptyset$.

....

[8] Georg Cantor was the child of an international family. His father was born in Denmark and his mother was Russian; he himself was born in 1845 in St. Petersburg where his father was a successful merchant and stock broker. He is recognized as the father of set theory, having invented cardinal and ordinal numbers and proved that the irrational numbers are uncountable. He received his doctorate from the University of Berlin in 1867 and spent most of his career at the University of Halle. His work was a watershed event in mathematics, but it was condemned by many prominent contemporary mathematicians. The work was simply too radical, with counterintuitive results such as $\mathbb{R}$ and $\mathbb{R}^p$ having the same number of points. He began to suffer from depression around 1884. This progressed and plagued him the rest of his life. He died in a sanatorium in Halle in 1918.

(b) Note that by condition (iii), from some point on the sets F_n are all bounded. So by part (a), $F \neq \emptyset$. On the other hand, $F \subseteq F_n$ for all n, so diam $F \leq$ diam F_n for all n. Therefore (iii) implies diam $F = 0$ and so it can have only one point. ∎

In most of the literature you will find that Cantor's Theorem is part (b) alone. Part (a) is separated because we will need it later in this book. At this point Exercise 6 is compulsory.

1.6.9. Theorem. *If X is a closed and bounded subset of $\mathbb{R}$ and $\{G_n\}$ is a sequence of open sets such that $X \subseteq \bigcup_{n=1}^{\infty} G_n$, then there is an integer N such that $X \subseteq \bigcup_{n=1}^{N} G_n$.*

Proof. Let $F_n = X \setminus [\bigcup_{k=1}^{n} G_k]$. So F_n is closed and bounded and $F_1 \supseteq F_2 \supseteq \cdots$. If $F_n \neq \emptyset$ for all n, then Cantor's Theorem implies $\emptyset \neq \bigcap_{n=1}^{\infty} F_n = X \setminus [\bigcup_{n=1}^{\infty} G_n]$; but this contradicts the assumption that $X \subseteq \bigcup_{n=1}^{\infty} G_n$. Therefore it must be that there is an N such that $F_N = \emptyset$, proving the theorem. ∎

For any non-empty subset E of $\mathbb{R}$ define the *distance* from a point x to E by

$$\text{dist}\,(x, E) = \inf\{|x - y| : y \in E\}$$

1.6.10. Definition. For a non-empty subset E of $\mathbb{R}$ define the closure of E to be the set cl $E = \{x \in \mathbb{R} : \text{dist}\,(x, E) = 0\}$.

1.6.11. Proposition. *Let E be any non-empty subset of $\mathbb{R}$.*

(a) *The closure of E is a closed set.*
(b) *If F is any subset of $\mathbb{R}$ such that $E \subseteq F$, then cl $E \subseteq$ cl F.*
(c) *If E is a closed set, then $E =$ cl E.*

Proof. Suppose E is not empty and $\{x_n\}$ is a sequence in cl E that converges to x; we want to show that $x \in$ cl E. Since $x_n \in$ cl E, there is a y_n in E with $|y_n - x_n| < \frac{1}{n}$. Thus $|x - y_n| \leq |x - x_n| + |x_n - y_n| \to 0$. Hence $x \in$ cl E (Why?), proving (a). If F is a set that contains E and $x \in$ cl E, then $0 = \text{dist}\,(x, X) = \inf\{|x - y| : y \in E\} \geq \inf\{|x - y| : y \in F\} = \text{dist}\,(x, F)$. Hence $x \in$ cl F. The proof of (c) is Exercise 7. ∎

The next result is a corollary of the proof rather than the statement of the preceding proposition.

1.6.12. Corollary. *If $E \subseteq \mathbb{R}$, then*

$$\text{cl}\, E = \{x \in \mathbb{R} : \text{there is a sequence } \{x_n\} \text{ in } E \text{ such that } x_n \to x\}$$

The preceding proposition says that the closure of E is the smallest closed set that contains E.

Exercises

(1) Prove that a closed interval is a closed set.
(2) Prove part (d) of Example 1.6.2.

(3) Show that parts (a) and (c) of Proposition 1.6.3 are false if the finite collections of sets are replaced by an infinite sequence of sets.

(4) If G is an open set and F is a closed set, prove that $G \backslash F$ is open and $F \backslash G$ is closed.

(5) Show that if I and J are two interval in $\mathbb{R}$ and $I \cap J \neq \emptyset$, then $I \cup J$ is an interval. Note that I and J are not assumed to be open or closed intervals. (Hint: Lemma 1.6.6 may be useful.)

(6) Consider the three restrictions (i), (ii), and (iii) placed on the sets $\{F_n\}$ in Cantor's Theorem. (a) Find a sequence of sets $\{F_n\}$ that satisfies (i) and (ii), but $\bigcap_{n=1}^{\infty} F_n = \emptyset$. (b) Find a sequence of sets $\{F_n\}$ that satisfies (i) and (iii), but $\bigcap_{n=1}^{\infty} F_n = \emptyset$. (c) Find a sequence of sets $\{F_n\}$ that satisfies (ii) and (iii), but $\bigcap_{n=1}^{\infty} F_n = \emptyset$.

(7) Prove part (c) of Proposition 1.6.11.

(8) Prove Corollary 1.6.12.

(9) If $\{a_n\}$ is a sequence in $\mathbb{R}$ and

$$E = \{a \in \mathbb{R} : \text{ there is a subsequence } \{a_{n_k}\} \text{ that converges to } x\}$$

show that E is a closed set.

(10) Find the closure of the set E in Example 1.6.5.

(11) If $E \subseteq \mathbb{R}$, say that x is a *limit point* of E if for every $\epsilon > 0$, there is a point e in E with $0 < |x - e| < \epsilon$. (Note that we insist that $e \neq x$ though there is nothing to preclude x being in the set E.) A point x is called an *isolated point* of E if $x \in E$ but x is not a limit point. (a) Show that E is a closed set if and only if it contains all its limit points. (b) Show that x is a limit point of E if and only if there is a sequence of distinct points in E that converges to x.

1.7. Continuous Functions

In this section we begin to focus on functions defined on subsets of $\mathbb{R}$. This focus and the related one of functions defined on subsets of higher dimensional Euclidean space will remain for the rest of the book. We begin by studying properties that many functions have. Here is an important and desirable elementary property.

1.7.1. Definition. If $X \subseteq \mathbb{R}$ and $a \in X$, a function $f : X \to \mathbb{R}$ is *continuous at a* if whenever $x \in X$ and $\{x_n\}$ is a sequence in X that converges to x, $f(x_n) \to f(x)$. It is said to be *continuous on X* if it is continuous at each point of X.

Easy examples of continuous functions are the *constant function*, $f(x) = a$ for all x, and the *identity function*, defined by $f(x) = x$ for all x in X. Usually we will be looking at functions defined on intervals and we may want to restrict the range to something less than the entirety of $\mathbb{R}$. This, however, does not affect the definition.

Suppose X is any set, not just a subset of $\mathbb{R}$, and $f, g : X \to \mathbb{R}$. (We are taking X to not necessarily be a subset of $\mathbb{R}$ because we will need this in the future.) Define the following functions from X into $\mathbb{R}$. $f \pm g : X \to \mathbb{R}$ is defined by $(f \pm g)(x) = f(x) \pm g(x)$ for all x in X. $fg : X \to \mathbb{R}$ is defined by $(fg)(x) = f(x)g(x)$ for all x in X. If $g(x) \neq 0$ for all x in X, then $(f/g)(x) = f(x)/g(x)$.

The proof of the next proposition follows immediately from Proposition 1.2.6.

1.7.2. Proposition. *If* $X \subseteq \mathbb{R}$ *and* $f, g : X \to \mathbb{R}$ *are functions continuous at a point a in X, then the following hold.*

(a) $f + g : X \to \mathbb{R}$ *is continuous at a.*
(b) $fg : X \to \mathbb{R}$ *is continuous at a.*
(c) *If* $g(x) \neq 0$ *for all x in X, then* $f/g : X \to \mathbb{R}$ *is continuous at a.*

1.7.3. Corollary. (a) *Every polynomial is a continuous function on all of* $\mathbb{R}$.
(b) *A rational function is continuous at every point where its denominator does not vanish.*

Proof. The identity function and the constant functions are continuous from $\mathbb{R}$ into itself. Now repeatedly apply the preceding proposition. ∎

We will see many more examples of continuous functions as we progress. In particular, the trig functions introduced in the next section will be seen to be continuous. Recall the definition of the composition of two functions.

1.7.4. Proposition. *Let* $X, Y \subseteq \mathbb{R}$, $f : X \to \mathbb{R}$ *such that* $f(X) \subseteq Y$, *and let* $g : Y \to \mathbb{R}$. *If* f *is continuous at a and* g *is continuous at* $\alpha = f(a)$, *then* $g \circ f : X \to \mathbb{R}$ *is continuous at a.*

Proof. If $\{a_n\} \subseteq X$ and $a_n \to a$, then $f(a_n) \to f(a)$ by the continuity of f. Thus $g \circ f(a_n) = g(f(a_n)) \to g(f(a))$ by the continuity of g. ∎

Here is an equivalent formulation of continuity.

1.7.5. Theorem. *If* $X \subseteq \mathbb{R}$, $a \in X$, *and* $f : X \to \mathbb{R}$, *then* f *is continuous at a if and only if for every* $\epsilon > 0$ *there is a* $\delta > 0$ *such that* $|f(x) - f(a)| < \epsilon$ *whenever* $x \in X$ *and* $|x - a| < \delta$.

Before beginning the proof of the theorem, let's take a moment to understand this equivalent formulation of continuity at the point a. (Let me add that many books take the statement of this theorem as the definition of continuity.) If you want to phrase the condition that is equivalent to continuity in words (with the equivalent symbolic statement inserted in brackets) you would say: Tell me how close you want $f(x)$ to come to $f(a)$ [for every $\epsilon > 0$] and no matter how close you want to get I can always tell you how close you must take x to a [there is a $\delta > 0$] so that it always works.

Proof. Assume f is continuous at a and let $\epsilon > 0$. Suppose no $\delta > 0$ can be found such that the condition is satisfied; that is, for any $\delta > 0$ there is at least one x with $|x - a| < \delta$ and $|f(x) - f(a)| \geq \epsilon$. It follows, by taking $\delta = \frac{1}{n}$, that for every n in $\mathbb{N}$ there is a point a_n in X with $|a_n - a| < \frac{1}{n}$ but $|f(a_n) - f(a)| \geq \epsilon$. But then we have that $a_n \to a$ and $\{f(a_n)\}$ does not converge to $f(a)$, violating the definition of continuity.

Now assume that f satisfies the stated condition and let $\{a_n\}$ be a sequence in X that converges to a. If $\epsilon > 0$, then we know we can find δ as in the condition. But since $a_n \to a$, there is an N such that $|a_n - a| < \delta$ when $n \geq N$. Thus $|f(a) - f(a_n)| < \epsilon$ when $n \geq N$. By definition, $f(a_n) \to f(a)$ so we have shown that f is continuous at a. ∎

At this point we will stop discussing functions that are continuous at a point and focus on functions continuous on a subset of $\mathbb{R}$, usually an interval.

1.7.6. Theorem (Extreme Value Theorem). *If $[a, b]$ is a bounded closed interval in $\mathbb{R}$ and $f : [a, b] \to \mathbb{R}$ is a continuous function, then f is a bounded function and there are points x_0 and y_0 in $[a, b]$ such that $f(x_0) \leq f(x) \leq f(y_0)$ for every x in $[a, b]$.*

Proof. Let's prove that the point x_0 exists. To do this put $\alpha = \inf\{f(x) : x \in [a, b]\}$, with the full realization at this stage in the argument that it could be that $\alpha = -\infty$. That is, it could be that f is not bounded below on the interval. The fact that this cannot happen will follow when we show the existence of the point x_0 such that $f(x_0) = \alpha$. By the definition of an infimum there is a sequence $\{x_n\}$ in $[a, b]$ such that $f(x_n) \to \alpha$. But $[a, b]$ is a bounded interval and so $\{x_n\}$ is a bounded sequence. By the Bolzano–Weierstrass Theorem there is a subsequence $\{x_{n_k}\}$ and a point x_0 such that $x_{n_k} \to x_0$; of necessity, $x_0 \in [a, b]$ since $[a, b]$ is a closed interval. But then $f(x_{n_k}) \to f(x_0)$, so $f(x_0) = \alpha$.

The proof of the existence of y_0 follows by using the first half of the proof applied to the function $-f$. ∎

The theorem states that the function attains its maximum and minimum values, hence its designation as the Extreme Value Theorem. We will refer to this theorem as the EVT. If the interval is not closed and bounded the theorem is no longer valid; just consider the the identity function on any interval that is either not closed or not bounded.

1.7.7. Theorem (Intermediate Value Theorem). *If I is an interval in $\mathbb{R}$, $f : I \to \mathbb{R}$ is a continuous function, $a, b \in I$, and $\tau \in \mathbb{R}$ such that $f(a) < \tau < f(b)$, then there is a t in I that lies between a and b such that $f(t) = \tau$.*

Proof. First note that since I is an interval, $[a, b] \subseteq I$. Let $A = \{x \in [a, b] : f(x) < \tau\}$; so $A \neq \emptyset$ since $a \in A$ and $x < b$ for all x in A. Let $t = \sup A$. By the definition of supremum there is a sequence $\{a_n\}$ in A that converges to t. Since f is continuous, $f(a_n) \to f(t)$; thus $f(t) \leq \tau$. We'll show that $f(t) = \tau$. In fact if $f(t) < \tau$, there is an $\epsilon > 0$ such that $f(t) + \epsilon < \tau$. By Theorem 1.7.5 there is a $\delta > 0$ such that $|f(x) - f(t)| < \epsilon$ when $|x - t| < \delta$. In particular when $t < x < t + \delta$, $f(x) < f(t) + \epsilon < \tau$ and so $x \in A$. Since $t = \sup A$, this contradiction shows that it must be that $f(t) = \tau$. ∎

We will refer to the Intermediate Value Theorem as the IVT.

The preceding proof shows the importance of Theorem 1.7.5. In the future we will use this equivalent formulation of continuity without citing the reference (1.7.5). Now we state a consequence of the Intermediate Value Theorem together with the Extreme Value Theorem.

1.7.8. Corollary. *The image of a closed and bounded interval under a continuous function is a closed and bounded interval.*

Proof. Given a bounded interval $[a, b]$ and a continuous function $f : [a, b] \to \mathbb{R}$, put $\alpha = \inf f([a, b])$, $\beta = \sup f([a, b])$. Thus $f([a, b]) \subseteq [\alpha, \beta]$. The Extreme Value Theorem implies $\alpha, \beta \in f([a, b])$. The preceding theorem implies that if $\alpha < \tau < \beta$, then $\tau \in f([a, b])$. Thus $f([a, b]) = [\alpha, \beta]$. ∎

1.7.9. Definition. If f, g are two functions defined on a set X and taking values in $\mathbb{R}$, define the two functions $f \vee g : X \to \mathbb{R}$ and $f \wedge g : X \to \mathbb{R}$ as follows:

$$f \vee g(x) = \max\{f(x), g(x)\}$$
$$f \wedge g(x) = \min\{f(x), g(x)\}$$

For obvious reasons the function $f \vee g$ is called the *maximum of f and g* and $f \wedge g$ is called the *minimum of f and g*.

If $|f|$ is the function defined as $|f|(x) = |f(x)|$, let's point out that $|f| = f \vee 0$ and $-|f| = f \wedge 0$. The proof of the next lemma is Exercise 7.

1.7.10. Lemma. *If $X \subseteq \mathbb{R}$ and $f : X \to \mathbb{R}$ is continuous, then $|f|$ is a continuous function on X.*

1.7.11. Proposition. *If $X \subseteq \mathbb{R}$ and $f, g : X \to \mathbb{R}$ are continuous functions, then $f \vee g : X \to \mathbb{R}$ and $f \wedge g : X \to \mathbb{R}$ are both continuous.*

Proof. In light of Exercise 6, we need only show that $f \vee g$ is continuous. Begin by showing that for any two real numbers a and b, $\max\{a, b\} = \frac{1}{2}[a + b + |a - b|]$. Hence for the functions f and g,

$$f \vee g = \frac{1}{2}[f + g + |f - g|]$$

The proposition now follows from the preceding lemma and other previous results. ∎

The preceding proposition allows us to construct new continuous functions from old ones. *For your cultural edification*, what's going on here is that we have shown that if we define $C(X)$ to be the collection of all continuous functions from X into $\mathbb{R}$, then, with the operations $\vee$ and $\wedge$ defined above, $C(X)$ becomes what is called a *lattice*. The interested reader can look up the definition on the Web. (Actually it must be checked that the axioms defining a lattice are all satisfied.)

We will close this section by introducing and exploring the following concept.

1.7.12. Definition. If $X \subseteq \mathbb{R}$, then a function $f : X \to \mathbb{R}$ is *uniformly continuous* on X if for every $\epsilon > 0$ there is a $\delta > 0$ such that whenever $x, y \in X$ and $|x - y| < \delta$ we have $|f(x) - f(y)| < \epsilon$.

Note that by Theorem 1.7.5 every uniformly continuous function is continuous. The difference is in that theorem when we are given the ϵ, the choice of δ is allowed to depend on a point. For a uniformly continuous function we can obtain the δ independent of the point. Keep this in mind as you consider the following example.

1.7.13. Example. The function $x \mapsto x^2$ from $\mathbb{R}$ into itself is not uniformly continuous even though it is continuous. In fact $|x^2 - y^2| = |x - y||x + y|$. So if we are given an $\epsilon > 0$, no matter how small we make δ if we take $x > y > \epsilon/\delta$, then when $|x - y| = \delta/2$ we have $|x^2 - y^2| = (\delta/2)(x + y) > (\delta/2)(2\epsilon/\delta) = \epsilon$.

If $E \subseteq \mathbb{R}$, $f : E \to \mathbb{R}$ is called a *Lipschitz*[9] *function* if there is a positive constant M such that $|f(x) - f(y)| \le M|x - y|$ for all x and y in E. As we progress in this book, we'll see many examples of Lipschitz functions, including one below. First, however, we establish one of their fundamental properties, a property that is easy to prove.

1.7.14. Proposition. *A Lipschitz function is uniformly continuous.*

Proof. If $\epsilon > 0$, let $0 < \delta < \epsilon/M$ and the definition of uniform continuity is satisfied. ∎

Recall the definition of the distance from a point to a set E, dist (x, E), given in §1.6.

1.7.15. Proposition. *If $E \subseteq \mathbb{R}$ and $x, y \in \mathbb{R}$, then*

$$|\text{dist}\,(x, E) - \text{dist}\,(y, E)| \le |x - y|$$

Consequently the function $f : \mathbb{R} \to \mathbb{R}$ defined by $f(x) = \text{dist}\,(x, E)$ is a Lipschitz function.

Proof. If $e \in E$, then $|x - e| \le |x - y| + |y - e|$; so taking the infimum over all e in E we get $\text{dist}\,(x, E) \le \inf\{|x - y| + |e - y| : e \in E\} = |x - y| + \text{dist}\,(y, E)$.

....

[9] Rudolf Lipschitz was born in Königsberg, Germany (now Kaliningrad, Russia) in 1832 to well-to-do parents. (As you read these biographical notes, you might note that most of the mathematicians discussed here came from professional families. Long ago higher education was not readily available to the children of those who did physical labor.) He began his studies at the University of Königsberg but moved to Berlin. With a year off for health reasons he received his doctorate from Berlin in 1853. After four years teaching in a gymnasium he became a privatdozent at the University of Berlin. In 1862 he became an extraordinary professor at Breslau. During this time he married a woman who had lived near him in Königsberg. Then in 1864 he left Breslau for Bonn where he spent the rest of his distinguished career, making substantial contributions in a variety of fields including number theory, Fourier series, differential equations, mechanics, and potential theory. He died in Bonn in 1903.

Reversing the roles of x and y we have $\text{dist}\,(y, E) \leq |x - y| + \text{dist}\,(x, E)$, whence we get the desired inequality. ∎

Here is the main result on uniformly continuous functions.

1.7.16. Theorem. *If X is a bounded closed subset of $\mathbb{R}$ and $f : X \to \mathbb{R}$ is continuous, then f is uniformly continuous.*

Proof. Assume f is not uniformly continuous. So there is an $\epsilon > 0$ such that for any $\delta > 0$ there are points x, y in $[a, b]$ with $|x - y| < \delta$ but $|f(x) - f(y)| \geq \epsilon$. Letting $\delta = \frac{1}{n}$ we obtain sequences $\{x_n\}$ and $\{y_n\}$ in X such that $|x_n - y_n| < \frac{1}{n}$ and $|f(x_n) - f(y_n)| \geq \epsilon$. Since X is bounded there is a subsequence $\{x_{n_k}\}$ and a point x in X such that $x_{n_k} \to x$. Because $|x_{n_k} - y_{n_k}| < n_k^{-1}$, $y_{n_k} \to x$. Since f is continuous, $f(x_{n_k}) \to f(x)$ and $f(y_{n_k}) \to f(x)$. This gives a contradiction to the fact that $|f(x_{n_k}) - f(y_{n_k})| \geq \epsilon$ for all integers n_k. ∎

Before presenting the final result of this important section, we start with a discussion. Suppose we are given a closed and bounded interval $[a, b]$ and a closed subset X of $[a, b]$. What is the nature of the set $[a, b] \backslash X$? Such a set in mathematics is called a *relatively open* subset of $[a, b]$. In other words, it wants to be open in relation to $[a, b]$. In fact Exercise 1.6.4 implies that $(a, b) \backslash X$ is open. So the only way that $[a, b] \backslash X$ fails to be an open subset of $[a, b]$ is if $a \notin X$ or $b \notin X$ or both. Remember this when we prove the next result.

Now suppose $a < c < d < b$ and we have a continuous function $f : [c, d] \to \mathbb{R}$. We want to get an *extension* of f to a continuous function $\tilde{f} : [a, b] \to \mathbb{R}$. That is we want $\tilde{f}$ to be a continuous function on $[a, b]$ such that $\tilde{f}(x) = f(x)$ when $x \in [c, d]$. There is a simple way to do this. Let α and β be any real numbers and let $\tilde{f}$ be the function defined on $[a, b]$ whose graph looks as follows: (i) between a and c the graph of $\tilde{f}$ is the straight line connecting the points (a, α) and $(c, f(c))$; (ii) between c and d it has the same graph as f; and (iii) between d and b the graph is the straight line between $(d, f(d))$ and (b, β). (The interested reader can write down the equations that define $\tilde{f}$. In fact, in the proof of the theorem below we'll need this.) The choice of α and β was arbitrary, but if we want $\tilde{f}$ to have an additional property enjoyed by f, a restriction on the choice of α and β must be made. For example, if $m \leq f(x) \leq M$ when $x \in [c, d]$ and we choose $n \leq \alpha, \beta \leq M$, then we will have $m \leq \tilde{f}(x) \leq M$ for all x in $[a, b]$. In particular, if f is a positive function, we can make $\tilde{f}$ positive by stipulating that $\alpha, \beta \geq 0$. The next result generalizes this discussion and the proof uses the method just described.

1.7.17. Theorem. *If X is a closed subset of $[a, b]$ and $f : X \to \mathbb{R}$ is a continuous function with $m \leq f(x) \leq M$ for all x in X, then there is a continuous function $\tilde{f} : [a, b] \to \mathbb{R}$ that is an extension of f such that $m \leq \tilde{f}(x) \leq M$ for all x in $[a, b]$.*

Proof. Consider $G = [a, b] \backslash X$. For the moment assume that $a, b \in X$ so that $G = (a, b) \backslash X$ is open (Exercise 1.6.4). By Proposition 1.6.7, $G = \bigcup_{n=1}^{\infty} (a_n, b_n)$, where the intervals $\{(a_n, b_n)\}$ are pairwise disjoint. Define $\tilde{f}(x)$ to be $f(x)$ when $x \in X$; and, when $x \in (a_n, b_n)$, define it so that the portion of its graph lying above this interval is the straight line connecting the points $(a_n, f(a_n))$ and $(b_n, f(b_n))$ as discussed prior to the statement of this theorem. It is clear that $\tilde{f}$ is an extension of f and $m \le \tilde{f}(x) \le M$ for all x in $[a, b]$. It remains to show that it is continuous. Clearly $\tilde{f}$ is continuous at the points in G. Assume $x \in X$ and that $\{x_k\}$ is a sequence in $[a, b]$ that converges to x; we want to show that $\tilde{f}(x_k) \to \tilde{f}(x) = f(x)$. The sequence $\{x_k\}$ can be partitioned into two subsequences: one that consists of the points that belong to X and the other the points that belong to G. In the first case we have that $\tilde{f}$ of that subsequence converges to $f(x)$ since $\tilde{f}$ is an extension of f. Therefore the proof will be finished if we assume $\{x_k\}$ is a sequence entirely lying in G. Thus for every $k \ge 1$ there is a integer n_k such that (a_{n_k}, b_{n_k}) contains x_k.

Claim. $[f(b_n) - f(a_n)] \to 0$ and $f(b_{n_k}) \to f(x)$.

By Theorem 1.7.16, f is uniformly continuous. So if $\epsilon > 0$ there is a $\delta > 0$ such that $|f(z) - f(y)| < \epsilon$ when $z, y \in X$ and $|z - y| < \delta$. Since G is a subset of the bounded interval $[a, b]$, G is bounded and so $\sum_{n=1}^{\infty}(b_n - a_n) < \infty$; hence $(b_n - a_n) \to 0$ and there is an integer N with $b_n - a_n < \delta/2$ for $n \ge N$. (The reason for using $\delta/2$ rather than δ will become clear a little later.) It follows that $|f(b_n) - f(a_n)| < \epsilon$ when $n \ge N$, establishing the first part of the claim. Now when $n_k \ge N$, since $a_{n_k} < x_k < b_{n_k}$ it follows that $b_{n_k} - x_k < \delta/2$. Since $x_k \to x$ and $n_k \to \infty$, there is an integer K such that $|x_k - x| < \delta/2$ and $n_k \ge N$ when $k \ge K$. Therefore when $k \ge K$, $|b_{n_k} - x| \le |b_{n_k} - x_k| + |x_k - x| < \delta$. Therefore $|f(x) - f(b_{n_k})| < \epsilon$ when $k \ge K$, establishing the second part of the claim.

From the definition of $\tilde{f}(x_k)$ we have that

$$\tilde{f}(x_k) = f(a_{n_k}) + \frac{f(b_{n_k}) - f(a_{n_k})}{b_{n_k} - a_{n_k}}(x_k - a_{n_k})$$

Keeping in mind that $x_k - a_{n_k} < b_{n_k} - a_{n_k}$, we have that

$$\tilde{f}(x_k) - f(x) = f(a_{n_k}) + [f(b_{n_k}) - f(a_{n_k})]\frac{x_k - a_{n_k}}{b_{n_k} - a_{n_k}} - f(x)$$
$$< f(a_{n_k}) + [f(b_{n_k}) - f(a_{n_k})] - f(x)$$
$$= f(b_{n_k}) - f(x) \to 0$$

by the claim. Therefore $\tilde{f}$ is continuous at x.

What happens when one or both of a and b do not belong to X? For example, assume $a \notin X$ and put $c = \inf X$. Since X is closed, $c \in X$; by definition $[a, c) \subseteq G$. If in addition $b \in X$, then $G = [a, c) \cup \bigcup_{n=1}^{\infty}(a_n, b_n)$, where the intervals $\{(a_n, b_n)\}$ are pairwise disjoint. If neither a nor b belong to X, then there is a number d such that $G = [a, c) \cup (d, b] \cup \bigcup_{n=1}^{\infty}(a_n, b_n)$. If $a \in X$ but $b \notin X$, then there is a number d such that $G = (d, b] \cup \bigcup_{n=1}^{\infty}(a_n, b_n)$. In any of these three cases the proof proceeds as in the case that both endpoints are in X. The details are left to the reader. ∎

The preceding theorem can be generalized and is called the Tietze Extension Theorem, something the reader will see if (s)he continues the study of mathematics.

A cursory examination of the proof above shows that it uses properties particular to $\mathbb{R}$ and any generalization necessitates a completely different proof.

Exercises

(1) Using Proposition 1.7.2, perform an induction argument to show that every polynomial is a continuous function on $\mathbb{R}$.

(2) If f is a continuous function defined on $[a, b]$ and $|f(x)| = 1$ for all x in $[a, b]$, show that f is a constant function.

(3) Let $f : \mathbb{R} \to \mathbb{R}$ be defined by $f(x) = 0$ when x is irrational and $f(a/b) = 1/b$ when $a \in \mathbb{Z}, b \in \mathbb{N}$, and a and b have no common divisor except 1. Where is f continuous?

(4) Use the IVT to show that if $f : [0, 1] \to [0, 1]$ is a continuous function, then it has a fixed point. That is, there is a point x with $f(x) = x$. Equivalently there is a point x with $x - f(x) = 0$.

(5) If $f : [0, 1] \to \mathbb{R}$ is a continuous function such that $f(0) < 0$ and $f(1) > 1$, show there is a number c in $[0, 1]$ such that $f(c) = c^3$.

(6) If f, g are two functions defined on a set X and taking values in $\mathbb{R}$, show that $(-f) \vee (-g) = -[f \wedge g]$. Similarly $(-f) \wedge (-g) = -[f \vee g]$.

(7) Prove Lemma 1.7.10.

(8) Complete the proof of Proposition 1.7.11.

(9) Let $f : [a, b] \to \mathbb{R}$ be an increasing function and assume that f has the intermediate value property. That is, assume that if $f(a) \le c \le f(b)$, then there is a number x_0 in $[a, b]$ such that $f(x_0) = c$. Prove that f is continuous.

(10) If $f : \mathbb{R} \to \mathbb{R}$ is continuous and $f(x) \in \mathbb{Q}$ for every x in $\mathbb{R}$, show that f is constant.

(11) Give an example of a bounded function that is continuous on $(0, 1)$ but not uniformly continuous.

(12) Let $a > 0$ and show that $f(x) = (x + 1)^{-1}$ is uniformly continuous on $[a, \infty)$.

(13) Let $X_1, \ldots, X_n$ be pairwise disjoint closed and bounded subsets of $\mathbb{R}$. If $f : X \to \mathbb{R}$ and f is uniformly continuous on X_k for $1 \le k \le n$, show that f is uniformly continuous on X.

(14) For a function $f : (a, b) \to \mathbb{R}$ and for each c in (a, b) and $\delta > 0$, define $\omega(c, \delta) = \sup\{|f(x) - f(c)| : |x - c| < \delta\}$. (a) Show that f is continuous at c if and only if $\lim_{\delta \to 0} \omega(c, \delta) = 0$. (b) Show that f is uniformly continuous if and only if

$$\lim_{\delta \to 0}[\sup\{\omega(c, \delta) : c \in I\}] = 0.$$

(15) Is the composition of two uniformly continuous functions a uniformly continuous function?

(16) If f and g are uniformly continuous functions on X, are $f \vee g$ and $f \wedge g$ uniformly continuous?

(17) If $E \subseteq \mathbb{R}$ and $f : E \to \mathbb{R}$ is uniformly continuous, show that when $\{a_n\}$ is a Cauchy sequence in E it follows that $\{f(a_n)\}$ is a Cauchy sequence in $\mathbb{R}$. Show that the

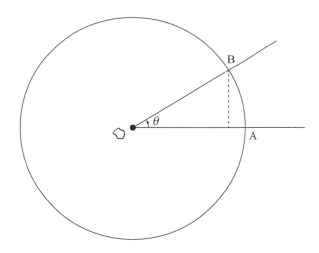

Figure 1.8.1

function $x \mapsto x^2$ on $\mathbb{R}$ maps Cauchy sequences into Cauchy sequences even though it is not uniformly continuous (1.7.13).

(18)   In reference to the preceding exercise, show that if f is only assumed to be continuous it does not follow that $\{f(a_n)\}$ is a Cauchy sequence in $\mathbb{R}$ when $\{a_n\}$ is a Cauchy sequence in E.

(19)   If $f : X \to \mathbb{R}$, show that f is uniformly continuous if and only if whenever $\{x_n\}$ and $\{y_n\}$ are sequences in X such that $x_n \to x \in X$ an $y_n \to y \in X$ we have that $|f(x_n) - f(y_n)| \to 0$.

(20)   If E is a closed subset of $\mathbb{R}$, $f : E \to \mathbb{R}$ is a continuous function, and $\{a_n\}$ is a sequence in E, is $f(\limsup a_n) = \limsup f(a_n)$? (If the answer is yes, prove it; if the answer is no, give a counterexample.) (b) If the answer in (a) was no, can you add a condition on f that makes the statement true?

1.8. Trigonometric Functions

We are going to assume that the reader knows the definition of all the trig functions as well as the various trig identities. What we want to concentrate on here is showing that they are continuous. We start with the sine function and we show it is continuous at $\theta = 0$. (*Note*: all angles in this book are measured in radians. They are the natural measurement of angles and are the only measurement we can use for calculus.)

Consider the circle of radius 1 centered at the origin; this is referred to as the *unit circle*. Let $\theta > 0$ and draw a line starting at the origin O and making an angle having θ radians with the positive x-axis. The point B where this line meets the unit circle has coordinates $B = (\cos\theta, \sin\theta)$. Let $A = (1, 0) \in \mathbb{R}^2$. (See Figure 1.8.1.) Note that the length of the base of the triangle $\triangle AOB$ is $|OA| = 1$, while its height is $\sin\theta$. We get that Area $\triangle AOB = \frac{1}{2}\sin\theta$. On the other hand we know that the area of the sector of the unit circle subtended by the arc of length θ is $\frac{1}{2}\theta$. (This is where we need to have θ in radians.) Since $\triangle AOB$ is contained in this sector, we get that $0 < \sin\theta < \theta$ when $\theta > 0$. On the other hand we also know that $\sin(-\theta) = -\sin\theta$. So when $\theta > 0$, $|\sin(-\theta)| = \sin\theta < \theta$. Thus we arrive at the fact that $|\sin\theta| < |\theta|$ when $\theta \neq 0$. So if $\theta_n \to 0$, we have that $\sin\theta_n \to 0$. This proves half the following.

1.8.1. Lemma. *The functions* sin *and* cos *are continuous at* 0.

Proof. We have the statement involving the sine function. Now for θ close to 0 we know that $\cos\theta = \sqrt{1 - \sin^2\theta}$. So if $\theta_n \to 0$, $\cos\theta_n \to 1 = \cos 0$. ∎

1.8.2. Proposition. *The functions* sin *and* cos *are continuous everywhere.*

Proof. If $a \in \mathbb{R}$, then we know that $\sin(\theta + a) = \sin\theta\cos a + \cos\theta\sin a$. From the lemma we see that as $\theta_n \to 0$, $\sin(\theta_n + a) \to \sin a$, so that the sine function is continuous at a. The proof for the cosine function is similar. ∎

Now using Proposition 1.7.2 we can discover where the other trig functions are continuous. For example, $\tan\theta = \sin\theta/\cos\theta$ is continuous everywhere $\cos\theta \neq 0$. That is, $\tan\theta$ is continuous everywhere except the points $\{\frac{\pi}{2} \pm n\pi : n \in \mathbb{N} \cup \{0\}\}$.

The next example is not only revealing but will be important as we progress.

1.8.3. Example. If $a \in \mathbb{R}$, let $f : \mathbb{R} \to \mathbb{R}$ be defined by

$$f(x) = \begin{cases} \sin\frac{1}{x} & \text{if } x \neq 0 \\ a & \text{if } x = 0 \end{cases}$$

It follows that no matter how a is chosen, f is not continuous at 0 but it is continuous at each point $x \neq 0$. In fact the continuity away from $x = 0$ follows by Proposition 1.7.4. On the other hand if $a_n = (2\pi n)^{-1}$, then $a_n \to 0$, but $f(a_n) = 0$ for all $n \geq 1$; if $b_n = (\pi/2 + 2\pi n)^{-1}$, then $b_n \to 0$, but $f(b_n) = 1$ for all $n \geq 1$. Therefore no matter how we define $f(0)$ it cannot be that f is continuous at 0.

Exercises

(1) Prove that $\{\sin n\}$ has a convergent subsequence.
(2) Where is the function $x \mapsto \tan x$ continuous?
(3) Where is the function $x \mapsto \tan x - \sec x$ continuous?

2 Differentiation

2.1. Limits

In this section we want to introduce and explore the concept of a limit of a function. We require the reader to consult Exercise 1.6.11 for the concepts of an isolated point and a limit point of a set. These will be used in the next definition.

2.1.1. Definition. If $X \subseteq \mathbb{R}$, a is a limit point of X, and $f : X \to \mathbb{R}$, say that the *limit* of f on X at a is the number L provided that for every $\epsilon > 0$ there is a $\delta > 0$ such that $|f(x) - L| < \epsilon$ whenever $x \in X$ and $0 < |a - x| < \delta$. In symbols we denote this by

$$\lim_{x \to a} f(x) = L$$

or

$$f(x) \to L \text{ as } x \to a$$

If a is an isolated point of X, then we define $\lim_{x \to a} f(x) = f(a)$.

There are a few additional things worth emphasizing. First, it is not necessary that a belong to the set X if it is a limit point; the restriction that a is a limit point, however, insures that there are points in X that are arbitrarily close to a. In particular we do not require that the function f be defined at a. Second, look at the part of the definition "$x \in X$." It may well be that there are other points, not in X, where f is defined; these do not influence whether $f(x) \to L$ as $x \to a$ with x in X. Third, look at the part of the definition that states that $|f(x) - L| < \epsilon$ holds when "$0 < |x - a| < \delta$." Even if $a \in X$ so that $f(a)$ is defined, we do not insist that $|f(a) - L| < \epsilon$. Now consider the case that a is an isolated point of X. (For example it might be that $a = 1$ and $X = [-1, 0] \cup \{1\}$.) In this case there are sufficiently small values of δ such that there are no points x in X satisfying $0 < |x - a| < \delta$. Technically or logically, if we used the first part of this definition, we would have the sorry state of affairs that no matter what we choose for L the conclusion $|f(x) - L| < \epsilon$ holds whenever $x \in X$ and $0 < |a - x| < \delta$. In other words the value of $\lim_{x \to a} f(x)$ could be anything. Of course this is intolerable, so we make a separate definition of the limit of $f(x)$ for isolated points. Finally note that the case when a is a limit point or an isolated point can be covered with the statement that $a \in \operatorname{cl} X$, though the definition of the limit still differs from one case to the other.

As we progress we won't make an issue of the comments in the last paragraph, though the reader should keep them in mind. In fact we see that these issues do not appear in the notation used to denote the limit. There are certain situations, however, where we want to emphasize how the variable x approaches a. When X is an interval and a is one of the endpoints of the interval, we introduce special notation for the limit. (Note that in this case a is not an isolated point of X.) For example if a is the left-hand endpoint of the interval X we will use the notation

$$\lim_{x \to a+} f(x) = L = f(a+)$$

This is called the *right-hand limit* of f as x approaches a. If $b < a < c$ and $f : (b, c) \to \mathbb{R}$ we might also discuss $\lim_{x \to a+} f(x)$, where we apply the original definition to the restriction of f to the set $X = (a, c)$. Similarly if a is the right-hand endpoint of an interval X we use

$$\lim_{x \to a-} f(x) = L = f(a-)$$

This is called the *left-hand limit* of f as x approaches a. Again when $b < a < c$ and $f : (b, c) \to \mathbb{R}$ we might also discuss $\lim_{x \to a-} f(x)$, where we apply the original definition to the restriction of f to the set $X = (b, a)$.

Finally note that in the definition above both a and L are numbers. Later in this section we will explore the notion of a limit being equal to $\pm\infty$ as well as taking a limit as $x \to \pm\infty$. The reader will note a similarity between the proof of this next proposition and that of Theorem 1.7.5.

2.1.2. Proposition. *If $X \subseteq \mathbb{R}$, a is a limit point of X, and $f : X \to \mathbb{R}$, then $\lim_{x \to a} f(x) = L$ if and only if for every sequence $\{a_n\}$ in $X \backslash \{a\}$ that converges to a, $f(a_n) \to L$.*

Proof. Suppose $\lim_{x \to a} f(x) = L$ and let $\{a_n\}$ be a sequence in $X \backslash \{a\}$ that converges to a. If $\epsilon > 0$, let $\delta > 0$ such that $|f(x) - L| < \epsilon$ when $0 < |x - a| < \delta$. By the definition of convergence there is an N such that $0 < |a_n - a| < \delta$ when $n \geq N$. It follows that $|f(a_n) - L| < \epsilon$ when $n \geq N$, and so $f(a_n) \to L$. Now assume that $f(a_n) \to L$ whenever $\{a_n\}$ is a sequence in $X \backslash \{a\}$ that converges to a, and let $\epsilon > 0$. Suppose no $\delta > 0$ can be found to satisfy the definition. Thus for every $n \geq 1$ there is a point a_n in X such that $0 < |a_n - a| < \frac{1}{n}$ but $|f(a_n) - L| \geq \epsilon$. It follows that $\{a_n\}$ is a sequence in $X \backslash \{a\}$ that converges to a, but $\{f(a_n)\}$ does not converge to L, thus furnishing a contradiction. ∎

2.1.3. Corollary. *A function $f : X \to \mathbb{R}$ is continuous at a point a in X if and only if $\lim_{x \to a} f(x) = f(a)$.*

The next result follows from the fact that $\lim_{x \to c} f(x) = L$ if and only if $\lim_{x \to c+} f(x) = L = \lim_{x \to c-} f(x)$. (Verify!)

2.1.4. Proposition. *If $a < c < b$ and $f : (a, b) \to \mathbb{R}$, then f is continuous at c if and only if $f(c-)$ and $f(c+)$ exist and are equal to $f(c)$.*

The next result is a consequence of Proposition 1.3.6 applied to Proposition 2.1.2.

2.1.5. Proposition. *Suppose $X \subseteq \mathbb{R}$, $a \in \operatorname{cl} X$, $f : X \to \mathbb{R}$, and $g : X \to \mathbb{R}$. If $\lim_{x \to a} f(x) = L$ and $\lim_{x \to a} g(x) = K$, then the following hold.*

(a) $\lim_{x \to a}[f(x) + g(x)] = L + K$.
(b) $\lim_{x \to a}[f(x)g(x)] = LK$.
(c) *If $g(x) \neq 0$ for all x in X and $K \neq 0$, then $\lim_{x \to a}[f(x)/g(x)] = L/K$.*

We return to Proposition 2.1.4 above to make a closer examination of how a function can be discontinuous at a point.

2.1.6. Definition. If $a < c < b$ and $f : (a, b) \to \mathbb{R}$, then f has a *simple discontinuity* or a *jump discontinuity* at c provided $\lim_{x \to c+} f(x) = f(c+)$ and $\lim_{x \to c-} f(x) = f(c-)$ exist but $f(c+) \neq f(c-)$.

2.1.7. Example. (a) If $f(x) = -2$ when $x \leq 0$ and $f(x) = 1$ when $x > 0$, then $f : \mathbb{R} \to \mathbb{R}$ has a simple discontinuity at 0.
    (b) If $f : [-1, 1] \to \mathbb{R}$ is defined by $f(x) = \sin(x^{-1})$ when $x \neq 0$ and $f(0) = 0$, then the discontinuity of f at 0 is not simple. See Example 1.8.3.

If $X \subseteq \mathbb{R}$, a function $f : X \to \mathbb{R}$ is *increasing* if $f(x) \leq f(y)$ whenever $x, y \in X$ and $x \leq y$. The reader should look once again at the definition of an increasing sequence in §1.3 where there is a discussion of different terminology. The same discussion applies here. When discussing functions we will also use terms like *strictly increasing*, *decreasing*, and *strictly decreasing*; the definitions should be apparent from their use with sequences. The function is *monotonic* if it is either increasing or decreasing and *strictly monotonic* if the monotonicity is strict. Note that $f : X \to \mathbb{R}$ is increasing if and only if $-f$ is decreasing. So any result established for an increasing function has a companion result that holds for decreasing functions.

2.1.8. Proposition. *If $f : (a, b) \to \mathbb{R}$ is a bounded increasing function, then $f(x-)$ and $f(x+)$ exist for every x in (a, b). Moreover*

$$\sup_{y<x} f(y) = f(x-) \leq f(x) \leq f(x+) = \inf_{y>x} f(y)$$

and every discontinuity of f is simple. In addition $f(a+)$ and $f(b-)$ exist with

$$f(a+) = \inf\{f(y) : a < y\} \text{ and } f(b-) = \sup\{f(y) : y < b\}$$

Proof. Fix x in (a, b) and put $L = \sup_{y<x} f(y)$. If $\epsilon > 0$ let $y_0 < x$ such that $L - \epsilon < f(y_0) \leq L$. Because f is increasing we have that $L - \epsilon < f(y_0) \leq f(y) \leq L$ whenever $y_0 \leq y \leq x$. Letting $\delta = x - y_0$ shows that when $x - \delta < y < x$ we have $y_0 < y < x$ and so $L - \epsilon < f(y) \leq L$. Thus $f(x-)$ exists and equals L. Also note that since f is increasing, $f(x)$ is an upper bound for $\{f(y) : y < x\}$ so that $f(x-) \leq f(x)$. The proof of the statement for $f(x+)$ is similar and left to the reader as are the statements for $f(a+)$ and $f(b-)$.

Finally, the only way that f can fail to be continuous at x is for the limit $\lim_{y \to x} f(y)$ not to exist. By what we have just shown this means that $f(x-) < f(x+)$ and so there is a jump discontinuity at x. $\blacksquare$

Of course there is an analogue of the preceding proposition for decreasing functions. See Exercise 3. Note that in light of this proposition when f is a bounded monotonic function on (a, b) we can extend f to a function $f : [a, b] \to \mathbb{R}$ by letting $f(a) = f(a+)$, $f(b) = f(b-)$. The extended function remains monotonic, and, moreover, is continuous at the endpoints of $[a, b]$.

2.1.9. Proposition. *If $f : (a, b) \to \mathbb{R}$ is a monotonic function, then the number of discontinuities of f is countable.*

Proof. Assume f is increasing and for the moment assume f is bounded. As remarked just before the statement of this proposition, we can assume that $f : [a, b] \to \mathbb{R}$ is increasing and continuous at the endpoints. Let $D_n = \{x \in (a, b) : f(x+) - f(x-) \geq \frac{1}{n}\}$. Assume that D_n has an infinite number of points; thus there is a sequence $\{x_k\}$ of distinct points in D_n. If we think geometrically, each $f(x_k+) - f(x_k-)$ measures the vertical "gap" in the graph of f at the discontinuity x_k. The sum of all the gaps at the points x_k must be less than the gap $f(b) - f(a)$. That is, $\sum_{k=1}^{\infty}[f(x_k+) - f(x_k-)] \leq f(b) - f(a) < \infty$. But by the definition of D_n, $\sum_{k=1}^{\infty}[f(x_k+) - f(x_k-)] = \infty$, a contradiction. So D_n is finite. By Proposition 1.5.7, $D = \bigcup_{n=1}^{\infty} D_n$ is countable; by the preceding proposition D is the set of discontinuities of f. When f is not bounded, consider the restriction of f to $[a + \frac{1}{n}, b - \frac{1}{n}]$; this restriction is bounded. (See Exercise 6.) If E_n denotes the set of discontinuities of f inside this closed interval, then E_n is countable by the first part of the proof. Since $\bigcup_{n=1}^{\infty} E_n$ is the set of discontinuities of f on (a, b), again Proposition 1.5.7 implies that the set of discontinuities of f is countable. $\blacksquare$

2.1.10. Definition. If $X \subseteq \mathbb{R}$, a is a limit point of X, and $f : X \to \mathbb{R}$, say that the *limit* of f on X at a is ∞ provided that for every positive number P there is a $\delta > 0$ such that $f(x) > P$ whenever $x \in X$ and $0 < |a - x| < \delta$. In symbols we denote this by

$$\lim_{x \to a} f(x) = \infty$$

An alternate phrasing of the same thing is to say $f(x) \to \infty$ as $x \to a$. Say that

$$\lim_{x \to a} f(x) = -\infty$$

if for every negative number N there is a $\delta > 0$ such that $f(x) < N$ whenever $x \in X$ and $0 < |a - x| < \delta$.

We leave it to the reader to formulate the definition of $\lim_{x \to a+} f(x) = \infty$. Also we are only going to state results for the case that $f(x) \to \infty$ and leave the statements for the case that $f(x) \to -\infty$ to the reader. The analogue for Proposition 2.1.2 seems sufficiently clear that we leave it as Exercise 5. The only result we pay close attention to here is the analogue of Proposition 2.1.5 as it contains a few dangerous curves.

2.1.11. Proposition. *Suppose $X \subseteq \mathbb{R}$, $a \in \mathrm{cl}\, X$, $f : X \to \mathbb{R}$, and $g : X \to \mathbb{R}$. If $\lim_{x \to a} f(x) = \infty$ and $\lim_{x \to a} g(x) = K$ where $0 < K \leq \infty$, then the following hold.*

(a) $\lim_{x \to a}[f(x) + g(x)] = \infty$.
(b) $\lim_{x \to a}[f(x)g(x)] = \infty$.
(c) *If $g(x) \neq 0$ for all x in X and $K \neq \infty$, then $\lim_{x \to a}[f(x)/g(x)] = \infty$.*

The proof is Exercise 7. If in the preceding proposition we were to allow K to take on negative values then all hell can break loose. See Exercise 8.

Recall that a subset X of $\mathbb{R}$ is not bounded above if for every positive number P there is an x in X with $x > P$.

2.1.12. Definition. If $X \subseteq \mathbb{R}$ such that X is not bounded above, $f : X \to \mathbb{R}$, and $L \in \mathbb{R}$, say that the *limit* of f on X as x approaches ∞ is L provided that for every $\epsilon > 0$ there is a positive number P such that $|f(x) - L| < \epsilon$ when $x \in X$ and $x > P$. In symbols we denote this by

$$\lim_{x \to \infty} f(x) = L$$

An alternate phrasing of the same thing is to say $f(x) \to L$ as $x \to \infty$.

Needless to say there is a definition of the statement $\lim_{x \to -\infty} f(x) = L$ as well as the statements $\lim_{x \to \infty} f(x) = \infty$, $\lim_{x \to \infty} f(x) = -\infty$, etc. It doesn't seem worthwhile to explicitly state all these.

Exercises

(1) If $b < a < c$ and $f : (b, c) \to \mathbb{R}$, show that $\lim_{x \to a} f(x) = L$ if and only if $\lim_{x \to a-} f(x) = L = \lim_{x \to a+} f(x)$.

(2) Let $b < a < c$ and $f : (b, c) \to \mathbb{R}$. (a) Show that $\lim_{x \to a+} f(x) = L$ if and only if for every decreasing sequence $\{a_n\}$ in (a, c) that converges to a, $f(a_n) \to L$. (b) Show that $\lim_{x \to a-} f(x) = L$ if and only if for every increasing sequence $\{a_n\}$ in (b, c) that converges to a, $f(a_n) \to L$.

(3) State and prove the version of Proposition 2.1.8 for decreasing functions.

(4) The following result is called the *squeezing principle*. Suppose f, g, h are functions on X and $a \in \mathrm{cl}\, X$. Show that if $g(x) \leq f(x) \leq h(x)$ for all x in X and $\lim_{x \to a} g(x) = \lim_{x \to a} h(x) = L$, then $\lim_{x \to a} f(x) = L$.

(5) State and prove a version of Proposition 2.1.2 for $f(x) \to \infty$ as $x \to a$.

(6) Why in the last half of the proof of Proposition 2.1.9 can we assume that $a + \frac{1}{n} \leq b - \frac{1}{n}$?

(7) Prove Proposition 2.1.11.

(8) Find examples of functions f and g defined on all of $\mathbb{R}$ such that the following happen. (a) $\lim_{x\to 0} f(x) = \infty$ and $\lim_{x\to 0} g(x) = -\infty$, but

$$\lim_{x\to 0} f(x)g(x) = 0$$

(b) $\lim_{x\to 0} f(x) = \infty$ and $\lim_{x\to 0} g(x) = -\infty$, but

$$\lim_{x\to 0} f(x)g(x) = -\infty$$

(c) $\lim_{x\to 0} f(x) = \infty$, $\lim_{x\to 0} g(x) = -\infty$, but

$$\lim_{x\to 0} f(x)g(x)$$

does not exist.

(d) Is it possible that $\lim_{x\to 0} f(x) = \infty$ and $\lim_{x\to 0} g(x) = -\infty$, but

$$\lim_{x\to 0} f(x)g(x) = \infty$$

2.2. The Derivative

2.2.1. Definition. Suppose (a, b) is a given open interval, $x \in (a, b)$, and $t_0 > 0$ such that $(x - t_0, x + t_0) \subseteq (a, b)$. Say that a function $f : (a, b) \to \mathbb{R}$ is *differentiable* at x if the function defined on $(x - t_0, x + t_0)$ by

$$t \mapsto \frac{f(x+t) - f(x)}{t}$$

has a finite limit as $t \to 0$. When this happens we call the value of this limit the *derivative* of f at x and denote it by

2.2.2 $$f'(x) = \lim_{t\to 0} \frac{f(x+t) - f(x)}{t}$$

Of course another notation for the derivative of f is

$$\frac{df}{dx}$$

There are equivalent ways to define these notions. Using the same set-up as in the preceding definition but not making specific the restrictions to make everything precise, we can say that f is differentiable at x if and only if

$$f'(x) = \lim_{y\to x} \frac{f(y) - f(x)}{y - x}$$

exists. As we just did, we will henceforth drop specifying the eligible values of y where this quotient is defined.

The reader can consult other books to see discussions of the right and left derivatives of f at x. These are defined just as above except that in (2.2.2) we take the left-hand and right-hand limits. These have a use, especially when f is defined on a closed interval and we want to define the derivative at an endpoint as we do now.

2.2.3. Definition. If $f : [a, b] \to \mathbb{R}$ we say that f is *differentiable* at a if

$$f'(a) = \lim_{t \to 0+} \frac{f(x+t) - f(x)}{t}$$

exists. Similarly we can define f to be *differentiable* at b. Say that f is *differentiable* on $[a, b]$ or (a, b) if it is differentiable at every point of the interval.

In what follows we often state results about functions differentiable at a point of $[a, b]$, but we will only prove them for the behavior of the function at a point x in the open interval (a, b). The proofs in the case where $x = a$ or $x = b$ are left to the reader.

2.2.4. Proposition. *If $f : [a, b] \to \mathbb{R}$, $x \in [a, b]$, and f is differentiable at x, then f is continuous at x.*

Proof. Assume that $x \in (a, b)$. (The case that x is an endpoint is similar.) Note that

$$\lim_{y \to x} |f(y) - f(x)| = \lim_{y \to x} \left| \frac{f(y) - f(x)}{y - x} \right| |y - x|$$

$$= \lim_{y \to x} \left| \frac{f(y) - f(x)}{y - x} \right| \cdot \lim_{y \to x} |y - x|$$

$$= |f'(x)| \cdot 0 = 0 \qquad \blacksquare$$

Now for the algebraic permanence of derivatives.

2.2.5. Proposition. *If f and g are differentiable at a point x in $[a, b]$, then the following hold.*

(a) $f + g$ *is differentiable at x and* $(f + g)'(x) = f'(x) + g'(x)$.
(b) fg *is differentiable at x and* $(fg)'(x) = f'(x)g(x) + f(x)g'(x)$.
(c) *When $g(y) \neq 0$ for all y in $[a, b]$, f/g is differentiable at x and*

$$\left(\frac{f}{g} \right)'(x) = \frac{f'(x)g(x) - f(x)g'(x)}{g(x)^2}$$

Proof. We assume $x \in (a, b)$. The proof of (a) is easy in light of previous results on limits. To establish (b), put $h = fg$ and note that $h(x + t) - h(x) = [f(x + t) - f(x)]g(x + t) + f(x)[g(x + t) - g(x)]$. Dividing by t gives

$$\frac{h(x+t) - h(x)}{t} = \frac{f(x+t) - f(x)}{t}g(x+t) + f(x)\frac{g(x+t) - g(x)}{t}$$

Now let $t \to 0$ and use the fact that g is continuous at x (2.2.4) to obtain part (b).

The proof of (c) proceeds in a similar way, but here we let $h = f/g$. After some algebra we get

$$\frac{h(x+t) - h(x)}{t} = \frac{1}{g(x+t)g(x)}$$
$$\cdot \left[\frac{f(x+t) - f(x)}{t} g(x) - f(x) \frac{g(x+t) - g(x)}{t} \right]$$

Letting $t \to 0$ produces the result. ∎

We might point out that (c) needs the proof given because we have to show that $h = f/g$ is differentiable. If we knew that h were differentiable, however, we could easily find the formula for h' by setting $f = gh$, using (b), and solving for h'.

Recall from calculus the geometric significance of $f'(c)$. If we plot the graph of f near the point $(c, f(c))$, we see that for any x different from a the quotient $(f(x) - f(c))/(x - c)$ is the slope of the straight line L_x going through the points $(c, f(c))$ and $(x, f(x))$. If f has a derivative at c, then as $x \to c$ the line L_x approaches the line tangent to the graph of f. Thus the quotient $(f(x) - f(c))/(x - c)$ approaches the slope of the tangent line. Hence the equation of the line tangent to the graph at the point $(c, f(c))$ is

$$y = f(c) + f'(c)(x - c)$$

2.2.6. Example. Here we sketch the details of showing that polynomials are differentiable. Note that from this and part (c) of the preceding proposition we obtain that the rational functions are differentiable everywhere the denominator does not vanish. We begin with the easy calculation from the definition that for any constant c, $dc/dx = 0$. Also if we are given the identity function, $x \mapsto x$, we quickly get from the definition that $dx/dx = 1$. Now if $n \geq 2$ and $a \in \mathbb{R}$, we have that $x^n - a^n = (x - a)(x^{n-1} + ax^{n-2} + \cdots + a^{n-1})$. Hence if $f(x) = x^n$, forming $[f(x) - f(a)]/(x - a) = (x^n - a^n)/(x - a)$ and letting $x \to a$ we see that

$$f'(a) = na^{n-1}$$

Now let's derive the formulas for differentiating the trig functions. To start we need a lemma.

2.2.7. Lemma. (a) $\lim_{\theta \to 0} \frac{\sin \theta}{\theta} = 1$.
(b) $\lim_{\theta \to 0} \frac{\cos \theta - 1}{\theta} = 0$.

Proof. (a) Since $(\sin(-\theta))/(-\theta) = (\sin \theta)/\theta$, we need only consider the limit as $\theta \to 0+$. To do this return to the proof of Lemma 1.8.1 and the triangle $\triangle AOB$ in Figure 1.8.1. We showed there that Area $\triangle AOB = \frac{1}{2} \sin \theta$. Now extend the line OB until it intersects the line perpendicular to the x-axis at the point A and denote the point of intersection by C. Consider the right triangle $\triangle AOC$ and note that since the length of the segment OA is $|OA| = 1$, we have that $|AC| = \tan \theta$. Hence Area $\triangle AOC = \frac{1}{2} \tan \theta$. Comparing the areas of $\triangle AOB$, the sector of the unit circle subtended by the arc AB, and the area on $\triangle AOC$ we get that $\frac{1}{2}\sin\theta < \frac{1}{2}\theta < \frac{1}{2}\tan\theta$ or

$$\sin\theta < \theta < \frac{\sin\theta}{\cos\theta}$$

Therefore

$$1 < \frac{\theta}{\sin\theta} < \frac{1}{\cos\theta}$$

Now we know that $\cos\theta$ is continuous so if we let $\theta \to 0+$ we get the desired conclusion. (See Exercise 2.1.4.)

(b) Note that

$$\frac{\cos\theta - 1}{\theta} = \frac{[\cos\theta - 1][\cos\theta + 1]}{\theta[\cos\theta + 1]}$$
$$= \frac{-\sin^2\theta}{\theta[\cos\theta + 1]}$$
$$= \theta\left[\frac{-\sin^2\theta}{\theta^2}\right][\cos\theta + 1]$$

By part (a) this must converge to 0 as $\theta \to 0$. ∎

2.2.8. Proposition. $(\sin x)' = \cos x$ and $(\cos x)' = -\sin x$.

Proof. Using the formula for the sine of the sum of two numbers we get

$$\frac{\sin(x+\theta) - \sin x}{\theta} = \frac{\sin x \cos\theta + \sin\theta \cos x - \sin x}{\theta}$$
$$= \sin x \frac{\cos\theta - 1}{\theta} + \cos x \frac{\sin\theta}{\theta}$$

By the preceding lemma this converges to $\cos x$ as $\theta \to 0$. The proof that $(\cos x)' = -\sin x$ is similar and left as Exercise 5. ∎

Deriving the formulas for the derivatives of the remaining trig functions can now proceed by using the preceding proposition with Proposition 2.2.5. For example, since $\tan x = \sin x/\cos x$, we can use (2.2.5(c)) to determine that the tangent function is differentiable whenever $\cos x \neq 0$ and obtain that for such x we have that $(\tan x)' = \sec^2 x$.

The next result is an equivalent formulation of differentiability that is often useful in executing proofs. Its proof is easy.

2.2.9. Proposition. *A function $f : X \to \mathbb{R}$ is differentiable at a point x in X if and only if there there is a function $F : X \to \mathbb{R}$ and a number D such that $\lim_{y\to x} F(y) = 0$ and $f(y) - f(x) = (y - x)[D + F(y)]$ for all y in X. When f is differentiable at x, we have $D = f'(x)$.*

Proof. We only prove half of this result. The proof of the other half is left as Exercise 7. Assume f is differentiable at X, put $D = f'(x)$, and let $F(y) = [f(y) - f(x)]/(y - x) - D$ when $y \neq x$ and $F(x) = 0$. The condition is easily seen to be satisfied. ∎

We'll see the usefulness of the preceding result in this next proof.

2.2.10. Theorem (Chain Rule). *Let $f : (a, b) \to \mathbb{R}$ be differentiable at a point x in (a, b), $f((a, b)) \subseteq (c, d)$, and let $g : (c, d) \to \mathbb{R}$ be a function that is differentiable at $f(x)$. If $h : (a, b) \to \mathbb{R}$ is defined by $h = g \circ f$, then h is differentiable at x and*

$$h'(x) = g'(f(x))f'(x)$$

Proof. Use the preceding proposition to write $f(y) - f(x) = (y - x)[f'(x) + F(y)]$ and $g(\zeta) - g(f(x)) = (\zeta - f(x))[g'(f(x)) + G(\zeta)]$, where $\zeta \in (c, d)$, F and G are functions defined on the appropriate sets, $\lim_{y \to x} F(y) = 0$, and $\lim_{\zeta \to f(x)} G(\zeta) = 0$. Therefore

$$
\begin{aligned}
h(y) - h(x) &= g(f(y)) - g(f(x)) \\
&= [f(y) - f(x)][g'(f(x)) + G(f(y))] \\
&= (y - x)[f'(x) + F(y)][g'(f(x)) + G(f(y))] \\
&= (y - x)[g'(f(x))f'(x) + T(y)]
\end{aligned}
$$

where

$$T(y) = f'(x)G(f(y)) + F(y)g'(f(x)) + F(y)G(f(y))$$

We must show that $T(y) \to 0$ as $y \to x$. Let's consider what happens to each of the summands in the definition of $T(y)$ as $y \to x$. We know that since f is differentiable at x, it is also continuous there. Hence $f(y) \to f(x)$ and so $G(f(y)) \to 0$. Thus the first summand converges to 0. The second summand converges to 0 since $F(y) \to 0$. The third summand converges to 0 by combining the two things we have just established: $G(f(y)) \to 0$ and $F(y) \to 0$. By Proposition 2.2.9 this proves the Chain Rule. ∎

2.2.11. Example. Before studying this example, the reader should look at Example 1.8.3. Define $f : \mathbb{R} \to \mathbb{R}$ by

$$f(x) = \begin{cases} x \sin \frac{1}{x} & \text{if } x \neq 0 \\ 0 & \text{if } x = 0 \end{cases}$$

We will show that f is a continuous function on $\mathbb{R}$ that is differentiable when $x \neq 0$, but it is not differentiable at $x = 0$. The fact that f is continuous when $x \neq 0$ is clear, and continuity when $x = 0$ follows by observing that $|f(x)| \leq |x|$. The differentiability of f when $x \neq 0$ follows from the Chain Rule and a judicious use of Proposition 2.2.5(b). The fact that f is not differentiable at $x = 0$ is seen by observing that for $t \neq 0$,

$$\frac{f(t) - f(0)}{t} = \sin \frac{1}{t}$$

and using Example 1.8.3.

Exercises

(1) State and prove a version of Proposition 2.2.4 when $f : [a, b] \to \mathbb{R}$ is differentiable at a.

(2) Show that if $f : (a, b) \to \mathbb{R}$ and there are constants M and $\alpha > 1$ such that $|f(x) - f(y)| \leq M|x - y|^\alpha$ for all x, y in $[a, b]$, then f is a constant.

(3) Let $a < c < b$ and suppose that $f : (a, c] \to \mathbb{R}$ and $g : [c, b) \to \mathbb{R}$ are continuous functions with $f(c) = g(c)$ and such that f is differentiable on (a, c), g is differentiable on (c, b). If we define $h : (a, b) \to \mathbb{R}$ by

$$h(x) = \begin{cases} f(x) & \text{if } a < x \leq c \\ g(x) & \text{if } c \leq x < b \end{cases}$$

give a necessary and sufficient condition that h is differentiable at $x = c$.

(4) For each positive integer n, find the points x in $\mathbb{R}$ where $f(x) = x^{n-1}|x|$ is differentiable and compute its derivative.

(5) Show that $(\cos x)' = -\sin x$.

(6) For which values of x is $\tan x$ differentiable and derive the formula for its derivative.

(7) Prove the other half of Proposition 2.2.9.

(8) Use Proposition 2.2.5(b) and induction to show that $(x^n)' = nx^{n-1}$ (Example 2.2.6).

(9) Find $f'(x)$ when $f(x) = x^3 \sin(x + 1)^2$.

(10) Are there any values of n in $\mathbb{N}$ such that $f(x) = |x|x^n$ is differentiable at $x = 0$?

(11) For each of the following functions determine where it is differentiable and find a formula for its derivative. (a) $\cos^3 x$. (b) $\tan^2(x + \pi)$. (c) $(\sin^2 x)/(2x + 3)$.

(12) Let $f : \mathbb{R} \to \mathbb{R}$. The function f is an *even function* if $f(-x) = f(x)$ for all x; f is an *odd function* if $f(-x) = -f(x)$ for all x. (a) Give two examples of an even function and two examples of an odd function. (b) If f is a differentiable even function, show that f' is an odd function. (c) If f is a differentiable odd function, what can you say about f'?

2.3. The Sign of the Derivative

2.3.1. Definition. If $f : (a, b) \to \mathbb{R}$ and $x \in (a, b)$, we say that f has a *local maximum* at x if there is a $\delta > 0$ with $(x - \delta, x + \delta) \subseteq (a, b)$ with $f(x) \geq f(y)$ when $|x - y| < \delta$. Similarly we define what it means for f to have a *local minimum* at x.

We note that f has a local maximum at x if and only if $-f$ has a local minimum at x. Also a constant function has both a local maximum and minimum everywhere, while $f(x) = x$ has neither a local maximum nor a local minimum anywhere. Now for a result that should be beloved and familiar to all calculus students.

2.3.2. Theorem. *If f is a differentiable function on (a, b) and f has a local maximum or minimum at a point x, then $f'(x) = 0$.*

Proof. Assume that f has a local maximum at x. (If f has a local minimum at x, we apply the argument that follows to $-f$.) Thus there is a $\delta > 0$ such that $(x - \delta, x + \delta) \subseteq (a, b)$ and $f(y) \leq f(x)$ when $|x - y| < \delta$. In particular when $0 < t < \delta$, $f(x + t) - f(x) \leq 0$. This implies that $[f(x + t) - f(x)]/t \leq 0$. Taking the limit as $t \to 0+$ shows that $f'(x) \leq 0$. Now assume that $-\delta < t < 0$. Again $f(x + t) - f(x) \leq 0$, but now $[f(x + t) - f(x)]/t \geq 0$. Taking the limit as $t \to 0-$ shows that $f'(x) \geq 0$. Therefore we have proved the theorem. ∎

Be aware that an examination of the proof shows that in the statement of the theorem it is required that the interval (a, b) is open. That is, if f is differentiable on $[a, b]$ and has a maximum at b it does not follow that $f'(b) = 0$. A consideration of $f(t) = t$ on $[0, 1]$ makes the point.

2.3.3. Example. The function $f : \mathbb{R} \to \mathbb{R}$ defined by $f(x) = x^3$ satisfies $f'(0) = 0$, but 0 is neither a local maximum nor a local minimum for f. Hence the converse of the preceding theorem is not true.

2.3.4. Theorem (Mean Value Theorem). *If $f : [a, b] \to \mathbb{R}$ is continuous and f is differentiable at each point of (a, b), then there is a point x in (a, b) with*

$$f'(x) = \frac{f(b) - f(a)}{b - a}$$

Proof. Define $h : [a, b] \to \mathbb{R}$ by $h(t) = [f(b) - f(a)]t - (b - a)f(t)$. A little algebra reveals that to prove the theorem it suffices to show there is a point x in (a, b) with $h'(x) = 0$. Some more algebra reveals that $h(a) = h(b)$. If h is constant, $h'(t) = 0$ for every t in (a, b) and we have the result. Suppose there is some point t in $[a, b]$ with $h(t) > h(a)$. Since h is continuous on $[a, b]$, the EVT implies there is a point x where h attains its maximum. Of necessity, $x \in (a, b)$. It must be that this maximum for h is a local maximum; so $h'(x) = 0$ by Proposition 2.3.2. On the other hand if it is the case that there is a point t in (a, b) with $h(t) < h(a)$, then by similar reasoning h attains its minimum value at a point x in (a, b). For this point x we have that $h'(x) = 0$. ∎

We'll refer to the preceding theorem as the MVT.

2.3.5. Corollary. *If f and g are continuous functions on $[a, b]$ and they are differentiable on (a, b), then there is a point x in (a, b) with*

$$[f(b) - f(a)]g'(x) = [g(b) - g(a)]f'(x)$$

Proof. Apply the MVT to the function $F(t) = [f(b) - f(a)]g(t) - [g(b) - g(a)]f(t)$. ∎

This last corollary is sometimes called the *Generalized Mean Value Theorem.* That it is more general than (2.3.6) can be see by letting $g(t) = t$ in the corollary.

2.3.6. Corollary. *If $f : (a, b) \to \mathbb{R}$ is a differentiable function and $f'(x) = 0$ for all x in (a, b), then f is a constant function.*

Proof. If $a < x < y < b$ and $f(x) \neq f(y)$, the MVT implies there is a point c between x and y with $f'(c) = [f(y) - f(x)]/(y - x) \neq 0$, a contradiction. ∎

The student is, undoubtedly, very familiar with the fact that the derivative measures the rate of change of the function. Indeed the very definition of the derivative shows this. The next result formalizes such a discussion.

2.3.7. Proposition. *If f is a differentiable function on (a, b), the following hold.*

(a) *f is an increasing function if and only if $f'(x) \geq 0$ for all x.*
(b) *If $f'(x) > 0$ for all x, then f is strictly increasing.*
(c) *f is a decreasing function if and only if $f'(x) \leq 0$ for all x.*
(d) *If $f'(x) < 0$ for all x, then f is strictly decreasing.*

Proof. The proofs of (c) and (d) follow from their counterparts in (a) and (b) by considering $-f$.

(a) Assume f is increasing. If $x \in (a, b)$, let $\delta > 0$ be such that $[x - \delta, x + \delta] \subseteq (a, b)$. If $0 < t < \delta$, then $f(x + t) - f(t) \geq 0$, and so $[f(x + t) - f(x)]/t \geq 0$. Taking the limit shows that $f'(x) \geq 0$. Conversely assume that $f'(x) \geq 0$ for all x in (a, b). If $a < c < d < b$, then the MVT implies there is an x in (c, d) such that $f(d) - f(c) = f'(x)(d - c) \geq 0$, and we have that f is increasing.

(b) Assume $f'(x) > 0$ for all x, and let $a < c < d < b$. Just as we did in the proof of (a), there is a point x in (c, d) such that $f(d) - f(c) = f'(x)(d - c) > 0$, and we have that f is strictly increasing. ∎

Note that the function $f(x) = x^3$ shows that the converse of part (b) is not true. See Exercise 2.

The next result is a cousin of the preceding one but is unrelated to derivatives. It is presented here for the reader's cultural edification and its use in the further study of differentiable functions.

2.3.8. Proposition. *If I is an interval and f is a continuous injective function defined on I, then $f(I)$ is an interval and f is strictly monotonic.*

Proof. The fact that $f(I)$ is an interval is part of the IVT (1.7.7).

Observe that since f is injective, it suffices to show that it is monotonic. Assume that f is not monotonic; so (Exercise 4) there are points $a, x, b \in I$ with $a < x < b$ such that either $f(a) < f(x)$ and $f(x) > f(b)$ or $f(a) > f(x)$ and $f(x) < f(b)$. As usual we need only consider the first of these since if the second occurs we can consider $-f$.

Let's compare $f(a)$ and $f(b)$: either $f(a) < f(b)$ or $f(a) > f(b)$. If it is the case that $f(a) < f(b)$, then we have that $f(a) < f(b) < f(x)$; that is $f(b)$ is in the interval $(f(a), f(x))$. By the IVT there is a point ζ in (a, x) with $f(\zeta) = f(b)$, contradicting the fact that f is injective. If it is the case that $f(a) > f(b)$, then we have that $f(b) < f(a) < f(x)$; that is, $f(a)$ is in the interval $(f(b), f(x))$. Again an application of the IVT yields a contradiction. These contradictions lead to the conclusion that f must be monotonic. ∎

To set the stage for the next topic, we return to the abstract setting of functions defined between sets that are not necessarily contained in $\mathbb{R}$. If X and Y are sets and $f : X \to Y$ is a bijective function, we can define its *inverse* $f^{-1} : Y \to X$ as follows. If $y \in Y$, then the fact that f is surjective implies there is an x in X such that $f(x) = y$. Because f is also injective, the point x is the only such point in X. Therefore we can define $f^{-1}(y) = x$ and we have a function $f^{-1} : Y \to X$. We note that $f(f^{-1}(y)) = y$ for all y in Y, and $f^{-1}(f(x)) = x$ for all x in X.

Now to return to subsets of $\mathbb{R}$. If I is an interval in $\mathbb{R}$ and $f : I \to \mathbb{R}$ is a continuous function that is also injective, then Proposition 2.3.8 above implies that $J = f(I)$ is also an interval and f is monotonic. Thus $f^{-1} : J \to I$ is also monotonic.

2.3.9. Proposition. *If I is an interval, $f : I \to \mathbb{R}$ is an injective continuous function, and $J = f(I)$, then $f^{-1} : J \to \mathbb{R}$ is a continuous function.*

Proof. f is strictly monotonic by the preceding proposition. We only consider the case where it is increasing. Suppose $\alpha \in J$ and let $a = f^{-1}(\alpha)$. First let's assume that a is not an endpoint of I; the case where it is an endpoint will be treated later. There are points b and c in I such that $b < a < c$; put $\beta = f(b)$, $\gamma = f(c)$. Since f is strictly increasing, $\beta < \alpha < \gamma$. To show that f^{-1} is continuous at α we need to show that if $\{\alpha_n\}$ is a sequence in (β, γ) that converges to α, then $f^{-1}(\alpha_n) \to a$. But if $a_n = f^{-1}(\alpha_n)$, then $\{a_n\} \subseteq [b, c]$. By the Bolzano–Weierstrass Theorem there is a subsequence $\{a_{n_k}\}$ and a point a' in $[b, c]$ such that $a_{n_k} \to a'$. Since f is continuous, $\alpha_{n_k} = f(a_{n_k}) \to f(a')$. But since $\{\alpha_{n_k}\}$ is a subsequence of the convergent sequence $\{\alpha_n\}$, it must be that $f(a') = \alpha$; thus $a' = a$. We have shown that every convergent subsequence of $\{a_n\}$ converges to a. By Exercise 1.3.9, $a_n \to a$.

Now for the case that a is an endpoint of I; assume it is the left-hand endpoint. Since f is increasing, α must be the left-hand endpoint of J. Choose a point c in I with $a < c$. It follows that $\gamma = f(c) \in J$ and $\alpha < \gamma$. The proof of this part follows the lines of the argument used to prove the case where a was not an endpoint and is left to the reader as Exercise 7. ∎

2.3.10. Proposition. *If $f : (a, b) \to (\alpha, \beta)$ is a differentiable function that is bijective, then the function $f^{-1} : (\alpha, \beta) \to (a, b)$ is differentiable and*

$$(f^{-1})'(\zeta) = \frac{1}{f'(f^{-1}(\zeta))}$$

for every ζ in (α, β).

Proof. Fix ζ in (α, β). By Proposition 2.2.9 we want to find a function $G(\omega)$ defined on a small interval about ζ such that $G(\omega) \to 0$ as $\omega \to \zeta$ and such that

$$f^{-1}(\omega) - f^{-1}(\zeta) = (\omega - \zeta)\{[f'(f^{-1}(\zeta))]^{-1} + G(\omega)\}$$

Put $x = f^{-1}(\zeta)$. By the differentiability of f at x we have the existence of a function $F(y)$ defined in a small interval about x such that $F(y) \to 0$ as $y \to x$ and

$$f(y) - f(x) = (y - x)[f'(x) + F(y)]$$

If ω is given and $y = f^{-1}(\omega)$, then

$$
\begin{aligned}
f^{-1}(\omega) - f^{-1}(\zeta) &= (y - x) \\
&= \frac{f(y) - f(x)}{f'(x) + F(y)} \\
&= [f(y) - f(x)]\left[\frac{1}{f'(x)} + \left[\frac{1}{f'(x) + F(y)} - \frac{1}{f'(x)}\right]\right] \\
&= (\omega - \zeta)\left[\frac{1}{f'(f^{-1}(\zeta))} + G(\omega)\right]
\end{aligned}
$$

where

$$G(\omega) = \frac{1}{f'(x) + F(f^{-1}(\omega))} - \frac{1}{f'(x)}$$

Now by the preceding proposition we know that f^{-1} is a continuous function. Thus as $\omega \to \zeta$, $f^{-1}(\omega) \to f^{-1}(\zeta) = x$; hence $F(f^{-1}(\omega)) \to 0$. It follows that $G(\omega) \to 0$ as required. ∎

We know several functions that are differentiable bijections and so we can apply the preceding result. For example, $\sin : (-\frac{\pi}{2}, \frac{\pi}{2}) \to (-1, 1)$ is a differentiable bijection, and we can use (2.3.10) to find the derivative of its inverse, $\arcsin : (-1, 1) \to (-\frac{\pi}{2}, \frac{\pi}{2})$. See Exercise 9.

Exercises

(1) Give the details of the proof of Corollary 2.3.5.

(2) Try to prove the converse of Proposition 2.3.7(b) by assuming f is strictly increasing and starting the argument used to prove the first part of (a) of that proposition. Where does it break down?

(3) Let $f : [0, \infty) \to \mathbb{R}$ be a continuously differentiable function such that $f(0) > 0$ and there is a constant C with $|f'(x)| \le C < 1$ for all x. (a) Use the Mean Value Theorem to show that $f(x) \le f(0) + Cx$ for all $x \ge 0$. (b) Prove that $\lim_{x \to \infty}[f(x) - x] = -\infty$. (c) Show that there is a $x_0 > 0$ such that $f(x_0) = x_0$. (d) Prove that there is only one point $x_0 > 0$ satisfying $f(x_0) = x_0$.

(4) If I is an interval and $f : I \to \mathbb{R}$ is injective, show that f is increasing if and only if when $a, b, x \in I$ with $a < x < b$ we have that $f(a) < f(x) < f(b)$.

(5) Let n be an integer with $n > 2$. If $f(x) = x^n + ax + b$, show that there are at most three numbers x with $f(x) = 0$.

(6) Show that if I and J are intervals and $f : I \to J$ is a continuous bijection, then $f^{-1} : J \to I$ is strictly monotonic. (See Proposition 2.3.8.)

(7) Fill in the details of the proof of Proposition 2.3.9.

(8) Why can't you prove Proposition 2.3.10 by applying the Chain Rule to the fact that $f(f^{-1}(\zeta)) = \zeta$ for all ζ in (α, β)?

(9) Find an appropriate domain on which the trig functions sin, cos, and tan are bijective and use Proposition 2.3.10 to calculate the derivatives of their inverses.

2.4. Critical Points

2.4.1. Definition. If f is a differentiable function on (a, b), a *critical point* is a point x in (a, b) with $f'(x) = 0$.

Why is it called a critical point?

Because interesting things happen there. Of course if the derivative is zero on an entire interval, then it is constant there; from one point of view there is not much critical happening there, from another there's a lot that's going on. Suppose it is zero at an isolated point; then three possible things might happen. The derivative could change from being positive to negative; change from being negative to being positive; or it could have the same sign on both sides of the place where it is zero. An example of this last phenomenon is the function x^3. In the first two cases something truly interesting happens. In fact recall Theorem 2.3.2 where it is shown that when a function has a local maximum or minimum it has a critical point. We saw that x^3 shows the converse of this is false, but below we present the partial converse of (2.3.2), something the reader has encountered in calculus.

2.4.2. Theorem. *Suppose $f : (a, b) \to \mathbb{R}$ is a differentiable function and x_0 is a critical point of f.*

(a) *If there is a $\delta > 0$ such that $a < x_0 - \delta < x_0 + \delta < b$, $f'(x) > 0$ when $x_0 - \delta < x < x_0$, and $f'(x) < 0$ when $x_0 < x < x_0 + \delta$, then f has a local maximum at x_0.*

(g) *If there is a $\delta > 0$ such that $a < x_0 - \delta < x_0 + \delta < b$, $f'(x) < 0$ when $x_0 - \delta < x < x_0$, and $f'(x) > 0$ when $x_0 < x < x_0 + \delta$, then f has a local minimum at x_0.*

Proof. The proof is intuitively clear, but we must translate that intuition into precise mathematics. We only prove (a). If $x_0 - \delta < x < x_0$, then the MVT implies there is a y with $x < y < x_0$ and $f(x_0) - f(x) = f'(y)(x_0 - x)$. But the assumption on $f'(y)$ when $x_0 - \delta < y < x_0$ implies that $f(x_0) - f(x) > 0$. If $x_0 < x < x_0 + \delta$, the MVT implies there is a z with $x_0 < z < x$ and $f(x) - f(x_0) = f'(z)(x - x_0)$. Since $f'(z) < 0$ by hypothesis and $(x - x_0) > 0$ we have that $f(x) - f(x_0) < 0$. Thus f has a local maximum at x_0. ∎

There is a small drawback to the preceding theorem. Namely we assume that the $\delta > 0$ exists. There is, however, a simple extra condition on f that helps in this process. To set the stage, realize that if f is a differentiable function on the open interval (a, b), then the derivative defines another function, $f' : (a, b) \to \mathbb{R}$.

2.4.3. Definition. Say that f is a *continuously differentiable function* on (a, b) if it is a differentiable function and $f' : (a, b) \to \mathbb{R}$ is a continuous function. The collection of all continuously differentiable functions on (a, b) is denoted by $C^{(1)}(a, b) = C'(a, b)$. Functions in $C'(a, b)$ are also called *smooth functions*. Now f' can also be differentiable, in which case we say that f is a *twice differentiable* function. The *second derivative* is denoted by f''. If $f'' : (a, b) \to \mathbb{R}$ is continuous, we say that f is a *twice continuously differentiable function*; denote the collection of all twice continuously differentiable functions on (a, b) by $C^{(2)}(a, b) = C''(a, b)$. This continues with the definition of higher derivatives denoted by $f^{(3)}, \ldots, f^{(n)}, \ldots$ If $f^{(n)}$ defines a continuous function, we say that f is *n-times continuously differentiable* and we denote the space of all such functions on (a, b) by $C^{(n)}(a, b)$. If we can form all the derivatives $f^{(n)}$ for $n \geq 1$, we say that f is *infinitely differentiable* and denote the space of all such functions as $C^{(\infty)}(a, b)$. In some places in the literature the term "smooth function" is reserved for functions in $C^{(\infty)}(a, b)$, where we have used the term for functions in $C'(a, b)$.

See Exercise 1.

2.4.4. Example. (a) Recall Example 2.2.11 and define $f_2 : \mathbb{R} \to \mathbb{R}$ by

$$f_2(x) = \begin{cases} x^2 \sin \frac{1}{x} & \text{if } x \neq 0 \\ 0 & \text{if } x = 0 \end{cases}$$

Clearly f_2 is differentiable at $x \neq 0$. In fact, $f_2'(x) = 2x \sin(x^{-1}) - \cos(x^{-1})$ when $x \neq 0$. Also note that

$$\frac{f_2(t) - f_2(0)}{t} = t \sin \frac{1}{t} \to 0$$

as $t \to 0$, so that f_2 is differentiable at $x = 0$ and $f'(0) = 0$. However, using Example 2.2.11 we have that f_2 is not continuously differentiable on all of $\mathbb{R}$ though it is on any open interval not containing 0. Since f_2' is not continuous at 0, f_2 is not twice differentiable.

(b) Define $f_3 : \mathbb{R} \to \mathbb{R}$ by

$$f_3(x) = \begin{cases} x^3 \sin \frac{1}{x} & \text{if } x \neq 0 \\ 0 & \text{if } x = 0 \end{cases}$$

It follows that $f_3 \in C'(\mathbb{R})$ but it is not twice differentiable at $x = 0$. The reader is asked to show this in Exercise 4 as well as explore additional functions.

We now present the second derivative test for a critical point.

2.4.5. Theorem. *Suppose $f \in C''(a, b)$ and f has a critical point at x_0.*

(a) *If $f''(x_0) < 0$, then f has a local maximum at x_0.*
(b) *If $f''(x_0) > 0$, then f has a local minimum at x_0.*
(c) *If $f''(x_0) = 0$, then the nature of the critical point at x_0 cannot be determined.*

Proof. (a) Since $f \in C''(a, b)$, there is a $\delta > 0$ such that $(x_0 - \delta, x_0 + \delta) \subseteq (a, b)$ and $f''(x) < 0$ for $|x - x_0| < \delta$. Since $f'(x_0) = 0$, this means that we can apply Theorem 2.4.2(a) and conclude that f has a local maximum at x_0. The proof of (b) is similar.

(c) Note that when $f(x) = x^3$, $f''(0) = 0$, and f has nether a local maximum nor a local minimum at 0. If $f(x) = x^4$, $f''(0) = 0$, and f has a local minimum at $x = 0$; if $f(x) = -x^4$, $f''(0) = 0$, and f has a local maximum at $x = 0$. ∎

The next curious result says that if the function is differentiable, then it has the intermediate value property even if we do not assume that the derivative is a continuous function.

2.4.6. Theorem (Darboux's[1] Theorem). *If $f : (a, b) \to \mathbb{R}$ is differentiable, $[c, d] \subseteq (a, b)$, and y is a point between $f'(c)$ and $f'(d)$, then there is an x in $[c, d]$ with $f'(x) = y$.*

Proof. Without loss of generality we can assume that $y \neq f'(c), f'(d)$; for convenience assume that $f'(c) < y < f'(d)$. If we define $g : (a, b) \to \mathbb{R}$ by $g(t) = f(t) - yt$, then $g'(t) = f'(t) - y$ and $g'(d) > 0 > g'(c)$. Since g is continuous there is a point x in $[c, d]$ with $g(x) = \min\{g(t) : t \in [c, d]\}$. Now $0 > g'(c) = \lim_{t \to c+}[g(t) - g(c)]/[t - c]$, and so there is a point $t_1 > c$ with $g(c) > g(t)$ for $c < t < t_1$. Hence $x \geq t_1$. Similarly there is a $t_2 < d$ with $x \leq t_2$. This shows that the point x is in (c, d). By Theorem 2.3.2, $0 = g'(x) = f'(x) - y$. ∎

We conclude this section with a discussion of using what we have done to find the global maximum and minimum of a differentiable function $f : [a, b] \to \mathbb{R}$. By the EVT we know these exist. To find them we can find the critical points of f in the interval, compute the value $f(x_0)$ for each critical point x_0, then compute the values $f(a)$ and $f(b)$. Comparing all these values will then reveal the extreme values as well as their location. See the exercises.

---

[1] Jean Gaston Darboux was born in 1842 in Nimes in the Languedoc region of France. His early education was in Nimes and the nearby city of Montpellier. After this, in 1861, he entered the École Polytechnique and then the École Normale Supérieure. While still a student he published his first paper on orthogonal surfaces, a subject that became his main research focus. These results were part of his doctoral dissertation, which he was awarded in 1866. He had several academic positions until he was appointed as *suppléant* to Liouville at the Sarbonne in 1873. He continued in this position for five years. In 1878 he became *suppléant* to Chasles in the chair of higher geometry at the Sorbonne. Two years later Chasles died and Darboux was appointed as his successor. He held this chair for the rest of his life with a term as dean of the Faculty of Science from 1889 to 1903. He made many contributions to geometry and there is also the Darboux Integral. He was also an excellent teacher and administrator. Among his doctoral students are Émile Borel, Elie Cartan, Édouard Goursat, and Émile Picard.

Exercises

(1) Show that $C^{(n)}(a, b)$ is an *algebra*; that is, it is a vector space over $\mathbb{R}$ that also has the property that the product of any two elements is again in the vector space and all the usual distributive laws hold.

(2) If f and g are functions on (a, b) such that each has derivatives there of order k for $1 \leq k \leq n$, show that fg has a derivative of order n and

$$(fg)^{(n)} = \sum_{k=0}^{n} C_{n,k} f^{(k)} g^{(n-k)}$$

where $C_{n,k} = n!/(n-k)!k!$

(3) For each of the following functions f, decide whether $x = 0$ is a local maximum or local minimum of f or neither. (a) $f(x) = (1 + x^2)/(1 + x^3)$. (b) $f(x) = (1 + x^3)/(1 + x^2)$. (c) $f(x) = x^2 \sin^3 x + x^2 \cos x$.

(4) Define $f_n : \mathbb{R} \to \mathbb{R}$ by

$$f_n(x) = \begin{cases} x^n \sin \frac{1}{x} & \text{if } x \neq 0 \\ 0 & \text{if } x = 0 \end{cases}$$

(a) Show that $f_3 \in C'(\mathbb{R})$, but f_3 is not twice differentiable at $x = 0$. (b) Show that $f_5 \in C''(\mathbb{R})$, but is not three-times differentiable at $x = 0$. (c) Show that $f_{2n+1} \in C^{(n)}(\mathbb{R})$, but it is not $(n + 1)$-times differentiable at $x = 0$. (Warning: This is a complicated computation.)

(5) Let f be a C'' function on (a, b) and suppose there is a point x_0 in (a, b) with $f(x_0) = f'(x_0) = f''(x_0) = 0$. Show that there is a continuous function g on (a, b) with $f(x) = (x - x_0)^2 g(x)$ for all x in (a, b).

(6) Let $f(x) = x^3 - 3x^2 + x + 1$. Find the critical points of f and diagnose their nature.

(7) Let $f(x) = 3x^5 - 5x^3 + 3$. Find the critical points of f and diagnose their nature.

(8) Find the global maximum and minimum values of the function $f(x) = x^3 - 3x$ on the interval $[-2, 2]$ and find the point x where they occur.

(9) Find the global maximum and minimum values of the function $f(x) = 6x^{4/3} - 3x^{1/3}$ on the interval $[-1, 1]$ and find the point x where they occur.

2.5. Some Applications

In this section we will use what we have learned about derivatives to establish two results, L'Hôpital's Rule and Taylor's Theorem.

2.5.1. Theorem (L'Hôpital's[2] Rule). *Assume $a < c < b$ and f and g are differentiable functions on $(a, b)\backslash\{c\}$ such that $g'(x) \neq 0$ for all x. If $\lim_{x\to c} f(x) = 0 = \lim_{x\to c} g(x)$ and there is an L in $\mathbb{R}$ such that*

$$\lim_{x\to c} \frac{f'(x)}{g'(x)} = L$$

then

$$\lim_{x\to c} \frac{f(x)}{g(x)} = L$$

Proof. We begin by showing that $\lim_{x\to c+}[f(x)/g(x)] = L$. Let $\epsilon > 0$ and choose $\delta > 0$ such that $|[f'(z)/g'(z)] - L| < \epsilon$ when $c < z < c + \delta$. Corollary 2.3.5 implies that when $c < y < x < c + \delta$, there is a z with $y < z < x$ and $f'(z)/g'(z) = [f(x) - f(y)]/[g(x) - g(y)]$. Hence when $c < y < x < c + \delta$, we have that

$$\left|\frac{f(x) - f(y)}{g(x) - g(y)} - L\right| = \left|\frac{f'(z)}{g'(z)} - L\right| < \epsilon$$

Now $f(y) \to 0$ and $g(y) \to 0$ as $y \to c+$. So holding x fixed in $(c, c + \delta)$ and letting $y \to c+$ in the above inequality, we get that $|[f(x)/g(x)] - L| \leq \epsilon$ when $c < x < c + \delta$. Since ϵ was arbitrary, this shows that $\lim_{x\to c+}[f(x)/g(x)] = L$. The proof that $\lim_{x\to c-}[f(x)/g(x)] = L$ is similar and left to the reader. ∎

There are variations on L'Hôpital's Rule, which are also called by the same name. Usually these involve having $\pm\infty$ involved. Here is one.

2.5.2. Theorem (L'Hôpital's Rule). *Assume $a < c < b$ and f and g are differentiable functions on $(a, b)\backslash\{c\}$ such that $g'(x) \neq 0$ for all x. If $\lim_{x\to c} g(x) = \infty$ and there is an L in $\mathbb{R}$ such that*

$$\lim_{x\to c} \frac{f'(x)}{g'(x)} = L$$

----

[2] Guillaume François Antoine Marquis de L'Hôpital was born in Paris in 1661. An earlier version of his family name is l'Hospital and you will sometimes see this used today when his result is presented. His family was prominent in France for centuries, and his father was a general in the King's army. L'Hôpital's mathematical story is complicated; it is cloaked in obscurity, and, for present day eyes, strangeness. This is partly due to the fact that he lived so long ago, a fact that means that many documents are forever lost and also that social behavior was so different from what it is today. From an early age L'Hôpital displayed mathematical ability, but he would have probably never been known to calculus students except for his meeting Johann Bernoulli, who at the age of 24 in 1691 had just arrived in Paris. Bernoulli was an expert in the differential calculus of Leibniz. (Realize that Newton was just 18 years older than L'Hôpital and had published his *Principia Mathematica* in 1683; also at the same time Leibniz was developing calculus in Paris.) L'Hôpital and Bernoulli became friends but later had a falling out due to a priority dispute. In 1694 L'Hôpital wrote a letter to Bernoulli promising him money in return for his working on problems of L'Hôpital's choice. Needless to say he added the proviso that Bernoulli not publish this work independently. There is no copy of Bernoulli's reply, but apparently he accepted the proposal and later letters record the exchange of ideas. (The practices of mathematicians as well as the rest of society have changed since then. Others have written on this agreement with more authority than I can muster for this footnote.) In 1696 L'Hôpital published the first textbook on differential calculus; it contains L'Hôpital's Rule. This book was extremely influential with new editions published until 1781. L'Hôpital was a married man and had four children. He died in Paris in 1704.

then

$$\lim_{x \to c} \frac{f(x)}{g(x)} = L$$

Proof. The proof begins as in the proof of the preceding theorem. Let $\epsilon > 0$ and choose $\delta > 0$ such that $|[f'(z)/g'(z)] - L| < \epsilon/2$ when $c < z < c + \delta$. Corollary 2.3.5 implies that when $c < y < x < c + \delta$, there is a z with $y < z < x$ and $f'(z)/g'(z) = [f(y) - f(x)]/[g(y) - g(x)]$. Hence when $c < y < x < c + \delta$, we have that

2.5.3 $$L - \frac{\epsilon}{2} < \frac{f(y) - f(x)}{g(y) - g(x)} < L + \frac{\epsilon}{2}$$

Fix x with $c < x < c + \delta$. Since $g(y) \to \infty$ as $y \to c$, there is a $\delta_1 > 0$ with $c + \delta_1 < x$ such that when $c < y < c + \delta_1$ we have both $g(y) > 0$ and $g(y) > g(x)$; note that for such a y, $[g(y) - g(x)]/g(y) > 0$. So if we multiply the inequalities (2.5.3) by this quotient we obtain the following two inequalities, separated for convenience and valid for all y with $c < y < c + \delta_1$

2.5.4 $$\frac{f(y) - f(x)}{g(y) - g(x)} \left[\frac{g(y) - g(x)}{g(y)} \right] < \left(L + \frac{\epsilon}{2} \right) \left[\frac{g(y) - g(x)}{g(y)} \right]$$

and

2.5.5 $$\left(L - \frac{\epsilon}{2} \right) \left[\frac{g(y) - g(x)}{g(y)} \right] < \frac{f(y) - f(x)}{g(y) - g(x)} \left[\frac{g(y) - g(x)}{g(y)} \right]$$

Doing some algebra (2.5.4) becomes

$$\frac{f(y) - f(x)}{g(y)} < \left(L + \frac{\epsilon}{2} \right) \left[\frac{g(y) - g(x)}{g(y)} \right]$$

or

$$\frac{f(y)}{g(y)} < \left(L + \frac{\epsilon}{2} \right) - \left(L + \frac{\epsilon}{2} \right) \frac{g(x)}{g(y)} + \frac{f(x)}{g(y)}$$

Remember that x is fixed. Since $g(y) \to \infty$ as $y \to c+$, we can find a δ_2 with $0 < \delta_2 < \delta_1$ so that when $c < y < c + \delta_2$ we have that

$$- \left(L + \frac{\epsilon}{2} \right) \frac{g(x)}{g(y)} + \frac{f(x)}{g(y)}$$

can be made as small as we want. Thus we can find such a $\delta_2 > 0$ so that for $c < y < c + \delta_2$

$$\frac{f(y)}{g(y)} < L + \epsilon$$

Now for (2.5.5). Doing similar algebra we arrive at the inequality

$$\left(L - \frac{\epsilon}{2} \right) - \left(L - \frac{\epsilon}{2} \right) \frac{g(x)}{g(y)} + \frac{f(x)}{g(y)} < \frac{f(y)}{g(y)}$$

We again use the hypothesis to find a δ_3 with $0 < \delta_3 < \delta_2$ such that when $c < y < c + \delta_3$ we have that

$$L - \epsilon < \frac{f(y)}{g(y)}$$

Combining our inequalities we get that when $c < y < c + \delta_3$, $|L - f(y)/g(y)| < \epsilon$. This establishes that $\lim_{y \to c+} f(y)/g(y) = L$. In similar fashion we get that $\lim_{y \to c-} f(y)/g(y) = L$. ∎

You should also see [11] where there are several variations on the theme of L'Hôpital's Rule as well as a number of exercises. As we progress we'll use these results, referred to by the name but possibly with small variations.

2.5.6. Theorem (Taylor's[3] Theorem). *If $n \geq 1$, $f \in C^{n-1}(a, b)$, f is n-times differentiable at the point c in (a, b), and $P(x)$ is the polynomial*

$$P(x) = \sum_{k=0}^{n-1} \frac{f^{(k)}(c)}{k!}(x - c)^k$$

then there is a function $h : (a, b) \to \mathbb{R}$ such that $h(x) \to 0$ as $x \to c$ and

$$f(x) = P(x) + h(x)(x - c)^n$$

Moreover for every x in (a, b) there is a point d between x and c such that

$$h(x) = \frac{f^{(n)}(d)}{n!}$$

Proof. Let $P(x)$ be the polynomial defined in the statement of the theorem. If we set

$$h(x) = \begin{cases} \frac{f(x)-P(x)}{(x-c)^n} & \text{when } x \neq c \\ 0 & \text{when } x = c \end{cases}$$

then $f(x) = P(x) + h(x)(x - c)^n$. Let's show that $\lim_{x \to c} h(x) = 0$. This is accomplished by repeated use of L'Hôpital's Rule. Note that applying the rule once won't suffice when $n \geq 2$. In fact this becomes clear once we calculate that

---

[3] Brook Taylor was born in 1685 in Edmonton, England to a financially comfortable family. He was schooled at home until he entered Cambridge. His first paper of note was published in 1714 and was on mechanics using Newton's calculus. In 1721 he married, but his wife died in 1723 during childbirth; the baby also died. He married again in 1725, but his second wife also died in childbirth, though the baby survived. Besides the present theorem Taylor discovered integration by parts and invented the field of finite differences. He made contributions to a variety of sciences as well as the mathematical study of perspective in art. Unfortunately he became distracted as he was virulently involved in the debate over whether the credit for inventing calculus belonged to Newton or Leibniz. This involved nationalism to a degree that is surprising to a modern mathematician. He died in 1731 in London.

$f^{(k)}(a) = P^{(k)}(a)$ for $0 \le k < n$. So L'Hôpital's rule repeated yields

$$\lim_{x \to c} h(x) = \lim_{x \to c} \frac{f(x) - P(x)}{(x - c)^n}$$

$$= \frac{1}{n} \lim_{x \to c} \frac{[f(x) - P(x)]'}{(x - c)^{n-1}}$$

$$= \frac{1}{n(n - 1)} \lim_{x \to c} \frac{[f(x) - P(x)]''}{(x - c)^{n-2}}$$

$$\vdots$$

$$= \frac{1}{n!} \lim_{x \to c} \frac{[f^{(n-1)}(x) - f^{(n-1)}(c)]}{(x - c)}$$

$$= 0$$

Now fix x in (a, b); we want to show there is a point d between x and c such that $h(x) = \frac{1}{n!} f^{(n)}(d)$. To do this introduce the function $g(y) = f(y) - P(y) - M(y - c)^n$, where

$$M = \frac{f(x) - P(x)}{(x - c)^n} = h(x)$$

From the definition of the polynomial $P(y)$ we can compute that $g^{(n)}(y) = f^{(n)}(y) - n!M = f^{(n)}(y) - n!h(x)$; so we want to produce a point d between x and c with $g^{(n)}(d) = 0$. We will do this by applying the MVT to the successive derivatives of $g(y)$.

To begin we use the fact that for $0 \le k \le n - 1$, $P^{(k)}(c) = f^{(k)}(c)$ and we conclude that $g(c) = g'(c) = \cdots = g^{(n-1)}(c) = 0$. Now from the definition of M we get that $g(x) = 0$. Thus the MVT applied to g shows there is a point y_1 between x and c with $g'(y_1) = 0$. But $g'(c) = 0$; so applying the MVT to g' shows there is a point y_2 between y_1 and c with $g''(y_2) = 0$. Now apply the MVT to g'' and so on. Eventually we show the existence of a point y_n between y_{n-1} and c such that $g^{(n)}(y_n) = 0$. Therefore if we let $d = y_n$, then it lies between x and c and $h(x) = \frac{f^{(n)}(d)}{n!}$.   ∎

Again I can recommend [11] for more on Taylor's Theorem.

Exercises

(1) In Theorem 2.5.1 assume $L = \infty$ and prove that the theorem remains valid. Can you do the same for Theorem 2.5.2?

(2) Can you state and prove a version of L'Hôpital's Rule covering limits as $x \to \infty$?

(3) Assume $f, g : (a, b) \to \mathbb{R}$ are differentiable functions such that $g'(x) \ne 0$ for all x in (a, b). If $\lim_{x \to a+} g(x) = \infty = \lim_{x \to a+} f(x)$ and there is an L in $\mathbb{R} \cup \{\infty\}$ such that $\lim_{x \to a+} f'(x)/g'(x) = L$, then $\lim_{x \to a+} f(x)/g(x) = L$.

(4) For each of the following specified functions find the stated limit if it exists, where existing includes the possibility that the limit is $\pm\infty$. (a) $\lim_{x\to 0} \frac{\sin x}{x}$. (b) $\lim_{x\to 0} \frac{\sin x - x + x^3/6}{x^5}$.

(5) Let f be an infinitely differentiable function on an open interval (a, b) and suppose $a < c < b$. If $n \in \mathbb{N}$ and $f^{(k)}(c) = 0$ for $0 \leq k \leq n - 1$ and $f^{(n)}(c) \neq 0$, show that there is an infinitely differentiable function $g : (a, b) \to \mathbb{R}$ such that $f(x) = (x - c)^n g(x)$ and $g(c) \neq 0$.

3 Integration

This chapter gives a mathematical foundation of the theory of integration the reader began in calculus. Unlike when the student took calculus, there will not be as much emphasis on computing integrals. Before the chapter concludes we'll see some topics never covered in calculus as well as establish a foundation for future embellishments and extensions. Integration is a vast, powerful, and useful subject. The development given in calculus, as well as the material on this subject contained in this book, only introduces the subject.

3.1. The Riemann Integral

Throughout this section we work with a closed and bounded interval, $[a, b]$. A *partition* of $[a, b]$ is a finite, ordered subset of the form $P = \{a = x_0 < x_1 < \cdots < x_n = b\}$. (This is the usual notation for a partition where both the points and the ordering are listed.) Say that a partition Q is a *refinement* of a partition P if $P \subseteq Q$. Hence Q adds additional points to P, and each subinterval of $[a, b]$ determined by two consecutive elements of P is the union of one or more of the intervals determined by the elements of Q. For example $P = \{0 < .2 < .4 < .7 < 1\}$ is a partition of $[0, 1]$. Both $\{0 < .1 < .2 < .4 < .6 < .7 < 1\}$ and $\{0 < .2 < .4 < .6 < .7 < 1\}$ are refinements of P as is P itself. An observation that will be used often as we proceed is that if P and Q are two partitions of $[a, b]$, the partition $P \cup Q$ is a refinement of both P and Q.

If $f : [a, b] \to \mathbb{R}$ is any bounded function and $P = \{a = x_0 < x_1 < \cdots < x_n = b\}$, then for $1 \leq j \leq n$ define

$$M_j = M_j^P = \sup\{f(x) : x_{j-1} \leq x \leq x_j\}$$
$$m_j = m_j^P = \inf\{f(x) : x_{j-1} \leq x \leq x_j\}$$

and let

$$U(f, P) = \sum_{j=1}^{n} M_j(x_j - x_{j-1}) \text{ and } L(f, P) = \sum_{j=1}^{n} m_j(x_j - x_{j-1})$$

The term $U(f, P)$ is called the *upper sum* of f for the partition P; $L(f, P)$ is called the *lower sum*. If you remember the way the integral was introduced in (many) calculus courses, for a positive function f, $L(f, P)$ is the sum of the areas of rectangles placed below the graph of the function f while $U(f, P)$ is the total of the areas of the rectangles placed above the graph of f.

3.1.1. Example. (a) Assume $f : [a, b] \to \mathbb{R}$ is a constant function: $f(x) = c$ for all x in $[a, b]$. For any partition P and with the notation as above, $m_j = c = M_j$ for each j. Thus for any partition P, $L(f, P) = c(b - a) = U(f, P)$.
   (b) Now define $f : [a, b] \to \mathbb{R}$ by

$$f(x) = \begin{cases} 1 & \text{if } x \text{ is a rational number} \\ 0 & \text{if } x \text{ is an irrational number} \end{cases}$$

By the Density Property (1.2.1), for any partition $P = \{a = x_0 < x_1 < \cdots < x_n = b\}$ and $1 \leq j \leq n$, $m_j = 0$ and $M_j = 1$. This implies that $L(f, P) = 0$ and $U(f, P) = b - a$.

   Of course we'll see many other examples, but not until we develop a little more of the theory. Before we start we might call attention to Exercise 1, which can often be used to derive a result for the lower sum from a result for the upper sum.

3.1.2. Proposition. *If f is a bounded function on $[a, b]$ with $m \leq f(x) \leq M$ for all x in $[a, b]$ and P and Q are partitions of $[a, b]$, then the following hold.*

(a) $m(b - a) \leq L(f, P) \leq U(f, Q) \leq M(b - a)$.
(b) *If Q is a refinement of P, then*

$$L(f, P) \leq L(f, Q) \leq U(f, Q) \leq U(f, P).$$

(c) *If Q is a refinement of P, then*

$$0 \leq U(f, Q) - L(f, Q) \leq U(f, P) - L(f, P).$$

Proof. (b) Let $P = \{a = x_0 < x_1 < \cdots < x_n = b\}$ and, as usual, define $m_j = \inf\{f(x) : x_{j-1} \leq x \leq x_j\}$ for $1 \leq j \leq n$. Let's consider the case that Q is obtained by adding one point to P; in fact assume that $Q = \{a = x_0 < x^* < x_1 < \cdots < x_n = b\}$. (The case where x^* is between two other points is similar and only involves more complicated notation.) Let $m_1' = \inf\{f(x) : x_0 \leq x \leq x^*\}, m_1'' = \inf\{f(x) : x^* \leq x \leq x_1\}$. Note that $m_1 = \min\{m_1', m_1''\}$. Thus $m_1(x_1 - x_0) \leq m_1'(x^* - x_0) + m_1''(x_1 - x^*)$ and so $L(f, P) \leq L(f, Q)$. The fact that this inequality holds when Q is obtained by adding any number of points is an argument using induction; the details are left to the reader. The proof that $U(f, Q) \leq U(f, P)$ is analogous and left to the reader as Exercise 2.
   (c) This is immediate from (b) and is stated only for convenience.
   (a) For the moment consider just one partition P. Using the usual definitions of m_j and M_j for $1 \leq j \leq n$, we have that $m \leq m_j \leq M_j \leq M$. From here we easily get that $m(b - a) \leq L(f, P) \leq U(f, P) \leq M(b - a)$. Now assume we have two partitions, P and Q. Note that $P \cup Q$ is a refinement of both P and Q. Applying

(b) twice we get that $m(b-a) \leq L(f, P) \leq L(f, P \cup Q) \leq U(f, P \cup Q) \leq U(f, Q) \leq M(b-a)$. $\blacksquare$

3.1.3. Definition. Using the preceding proposition we have that

$$\sup\{L(f, P) : P \text{ is a partition of } [a, b]\} \leq$$
$$\inf\{U(f, Q) : Q \text{ is a partition of } [a, b]\}$$

When these two expressions are equal, we have a desirable event and we set

$$\int_a^b f = \int_a^b f(x)dx$$
$$= \sup\{L(f, P) : P \text{ is a partition of } [a, b]\}$$
$$= \inf\{U(f, Q) : Q \text{ is a partition of } [a, b]\}$$

In this situation we say that f is *Riemann*[1] *integrable* or simply *integrable*. The set of all Riemann integrable functions on $[a, b]$ is denoted by $\mathcal{R}[a, b]$.

A brief word about notation. Ordinarily we will use the notation $\int_a^b f$ rather than $\int_a^b f(x)\, dx$ as seen in calculus. The use of the notation involving dx will be limited to those occasions where there might be some confusion as to the variable of integration. This will be our practice partly because the x is redundant, but mainly to emphasize that we are integrating a function. The notation for a function is f, while $f(x)$ is the value of the function at the point x in its domain.

Note that the function f appearing in Example 3.1.1(b) is not integrable, while the constant functions are. The next result gives a necessary and sufficient condition for integrability that is more convenient than the definition and that will usually be employed when we prove results about integrability.

3.1.4. Proposition. *If f is a bounded function on $[a, b]$, then f is Riemann integrable if and only if for every $\epsilon > 0$ there is a partition P of $[a, b]$ such that $U(f, P) - L(f, P) < \epsilon$. Moreover $\int_a^b f$ is the unique number such that for every refinement Q of P we have*

$$L(f, Q) \leq \int_a^b f \leq U(f, Q)$$

---

[1] Georg Friedrich Bernhard Riemann was born in 1826 in Breselenz, Germany. His early schooling was closely supervised by his father. When he entered the university at Göttingen in 1846, at his father's urging he began to study theology. Later, with his father's blessing, he switched to the Faculty of Philosophy so he could study mathematics. In 1847 he transferred to Berlin where he came under the influence of Dirichlet, which influence was permanent. He returned to Göttingen in 1849 where he completed his doctorate in 1851, working under the direction of Gauss. He took up a lecturer position there and in 1862 he married a friend of his sister. In the autumn of that same year he contracted tuberculosis. This began a period of ill health and he went between Göttingen and Italy, where he sought to recapture his health. He died in 1866 in Selasca, Italy on the shores of beautiful Lake Maggiore. Riemann is one of the giants of analysis and geometry. He made a series of significant discoveries and initiated theories. Besides this integral there are many other things named after him, including the Riemann zeta function, Riemann surfaces, the Riemann Mapping Theorem.

Proof. Define

$$L = \sup\{L(f, P) : P \text{ is a partition of } [a, b]\}$$
$$U = \inf\{U(f, Q) : Q \text{ is a partition of } [a, b]\}$$

We always have that $L \leq U$. Suppose that for every $\epsilon > 0$ we can find a partition P with $U(f, P) - L(f, P) < \epsilon$. But $U - L \leq U(f, P) - L(f, P) < \epsilon$. Since ϵ was arbitrary we have that $f \in \mathcal{R}[a, b]$. Conversely, assume f is Riemann integrable and let $\epsilon > 0$. Choose partitions P_1, P_2 such that $0 \leq L - L(f, P_1) < \epsilon/2$ and $0 \leq U(f, P_2) - U < \epsilon/2$. If we put $P = P_1 \cup P_2$, then Proposition 3.1.2 implies that $U(f, P) - L(f, P) < \epsilon$. If Q is a refinement of P, then Proposition 3.1.2(c) implies that $U(f, Q) - L(f, Q) \leq U(f, P) - L(f, P) < \epsilon$. Since ϵ was arbitrary we have that there can be only one number between $L(f, Q)$ and $U(f, Q)$ for every such refinement. By definition, this unique number must be $\int_a^b f$. ∎

The uniqueness part of the last proposition is there mainly for emphasis, since when a function is integrable there can be no other number between all the lower and upper sums. There is, however, some small benefit in using only partitions Q that are refinements of P as seen in the proof of Proposition 3.1.6 below.

3.1.5. Corollary. *If f is a bounded function on $[a, b]$, then f is Riemann integrable if and only if there is a sequence of partitions $\{P_k\}$ of $[a, b]$ such that each P_{k+1} is a refinement of P_k and $U(f, P_k) - L(f, P_k) \to 0$ as $k \to \infty$. When this happens we have that*

$$\int_a^b f = \lim_{k \to \infty} U(f, P_k) = \lim_{k \to \infty} L(f, P_k)$$

Proof. If such a sequence of partitions exists, then it is immediate from the proposition that $f \in \mathcal{R}[a, b]$. Now assume that f is integrable. By the proposition for each $k \geq 1$ there is a partition Q_k with $U(f, Q_k) - L(f, Q_k) < 1/k$. If $P_k = Q_1 \cup \cdots \cup Q_k$, then P_k is a refinement of Q_k and so Proposition 3.1.2(c) implies that $U(f, P_k) - L(f, P_k) < 1/k$. The evaluation of the integral as the limit of the upper and lower sums follows from the definition of the integral. ∎

3.1.6. Proposition. *If $f : [a, b] \to \mathbb{R}$ is a bounded function and $\{a = x_0 < x_1 < \cdots < x_n = b\}$ is a partition such that f is Riemann integrable on $[x_{j-1}, x_j]$ for $1 \leq j \leq n$, then f is Riemann integrable on $[a, b]$ and*

$$\int_a^b f = \sum_{j=1}^n \int_{x_{j-1}}^{x_j} f$$

Proof. Let $\epsilon > 0$ and let P_j be a partition of $[x_{j-1}, x_j]$ such that $U(f, P_j) - L(f, P_j) < \epsilon/n$. If $P = \bigcup_{j=1}^n P_j$, then P is a partition of $[a, b]$ and

$$U(f, P) - L(f, P) = \sum_{j=1}^n [U(f, P_j) - L(f, P_j)] < \epsilon$$

Thus $f \in \mathcal{R}[a, b]$.

To finish the proof let Q be any refinement of P and set $Q \cap [x_{j-1}, x_j] = Q_j$. It follows that $Q = \bigcup_{j=1}^{n} Q_j$ and each Q_j is a refinement of P_j. We have that $L(f, Q_j) \le \int_{x_{j-1}}^{x_j} f \le U(f, Q_j)$ for $1 \le j \le n$. Thus

$$L(f, Q) = \sum_{j=1}^{n} L(f, Q_j) \le \sum_{j=1}^{n} \int_{x_{j-1}}^{x_j} f$$
$$\le \sum_{j=1}^{n} U(f, Q_j) = U(f, Q)$$

By the uniqueness part of Proposition 3.1.4 it follows that

$$\int_a^b f = \sum_{j=1}^{n} \int_{x_{j-1}}^{x_j} f$$

∎

Now we see a useful sufficient condition for integrability.

3.1.7. Theorem. *If $f : [a, b] \to \mathbb{R}$ is a bounded function that is continuous at all but a finite number of points, then f is Riemann integrable.*

Proof. Suppose $-M \le f(x) \le M$ for all x in $[a, b]$. We begin with a special case.

3.1.8. Claim. *If f is continuous at every point, then f is Riemann integrable.*

Let $\epsilon > 0$. Recall that Theorem 1.7.16 says that $f : [a, b] \to \mathbb{R}$ is uniformly continuous. Hence there is a $\delta > 0$ such that $|x - y| < \delta$ implies that $|f(x) - f(y)| < \epsilon/(b - a)$. Let $P = \{a = x_0 < x_1 < \cdots < x_n = b\}$ such that $x_j - x_{j-1} < \delta$ for $1 \le j \le n$. Since $|x - y| < \delta$ when $x_{j-1} \le x, y \le x_j$, it follows that $M_j - m_j < \epsilon/(b - a)$. (Why?) Therefore

$$U(f, P) - L(f, P) = \sum_{j=1}^{n} (M_j - m_j)(x_j - x_{j-1}) < \frac{\epsilon}{b - a} \sum_{j=1}^{n} (x_j - x_{j-1}) = \epsilon$$

and so $f \in \mathcal{R}[a, b]$ by the Proposition 3.1.4.

3.1.9. Claim. *If f has a single point of discontinuity in $[a, b]$, then $f \in \mathcal{R}[a, b]$.*

We will assume the point of discontinuity is at the left-hand endpoint, a. (The proof of the case where the discontinuity is at b is practically identical. The proof of the case when the discontinuity occurs at an interior point is Exercise 3.) Let $\epsilon > 0$ and choose the point x_1 with $x_1 - a < \epsilon/4M$. Since f is continuous on $[x_1, b]$ there is a partition of this interval, $P_1 = \{x_1 < x_2 < \cdots < x_n = b\}$, with $U(f, P_1) - L(f, P_1) < \epsilon/2$. If $P = \{a = x_0 < x_1 < x_2 < \cdots < x_n = b\}$, then

$$U(f, P) - L(f, P) = (M_1 - m_1)(x_1 - a) + U(f, P_1) - L(f, P_1)$$
$$< 2M \frac{\epsilon}{4M} + \frac{\epsilon}{2}$$
$$= \epsilon$$

By (3.1.4) $f \in \mathcal{R}[a, b]$.

Now assume that f is continuous on $[a, b]$ except for the points $c_1 < c_2 < \cdots < c_m$. Pick points $a = x_0 < x_1 < \cdots < x_m = b$ in $[a, b]$ such that $[a, b] = \bigcup_{k=1}^{m} [x_{k-1}, x_k]$ and each subinterval $[x_{k-1}, x_k]$ contains c_k but $c_j \notin [x_{k-1}, x_k]$ when $j \neq k$. Hence Claim 3.1.9 implies $f \in \mathcal{R}[x_{k-1}, x_k]$ for $1 \leq k \leq m$. By Proposition 3.1.6, f is Riemann integrable on $[a, b]$. ∎

With the last theorem we have a large supply of integrable functions. In §3.5 below a necessary and sufficient condition for Riemann integrability is presented, but the preceding theorem covers the cases we will see in the rest of this book. Now it is time to develop some of the properties of an integrable function.

3.1.10. Proposition. *If $f, g \in \mathcal{R}[a, b]$, then the following hold.*

(a) *If $f(x) \leq g(x)$ for all x in $[a, b]$, then*

$$\int_a^b f \leq \int_a^b g.$$

(b) *If $|f(x)| \leq M$ for all x in $[a, b]$, then*

$$\left| \int_a^b f \right| \leq M(b - a).$$

Proof. (a) From the definition of the upper sum we have that for any partition P, $U(f, P) \leq U(g, P)$. The result follows from Corollary 3.1.5.

(b) Since $-M \leq f(x) \leq M$ for all x in $[a, b]$, part (b) follows from part (a). ∎

3.1.11. Proposition. *$\mathcal{R}[a, b]$ is a vector space over $\mathbb{R}$. Moreover if $f, g \in \mathcal{R}[a, b]$ and $\alpha, \beta \in \mathbb{R}$, then*

$$\int_a^b (\alpha f + \beta g) = \alpha \int_a^b f + \beta \int_a^b g$$

Proof. If $E \subseteq [a, b]$, then note that $\sup\{f(x) + g(x) : x \in E\} \leq \sup\{f(x) : x \in E\} + \sup\{g(x) : x \in E\}$ and $\inf\{f(x) + g(x) : x \in E\} \geq \inf\{f(x) : x \in E\} + \inf\{g(x) : x \in E\}$. Thus for any partition P of $[a, b]$ we have that

$$U(f + g, P) - L(f + g, P) \leq [U(f, P) + U(g, P)] - [L(f, P) + L(g, P)]$$
$$= [U(f, P) - L(f, P) + [U(g, P) - L(g, P)]$$

It follows from Proposition 3.1.4 that if $f, g \in \mathcal{R}[a, b]$, then $f + g \in \mathcal{R}[a, b]$. Now use Corollary 3.1.5 to find a sequence of partitions $\{P_k\}$ such that each P_{k+1} is a refinement of P_k and simultaneously $U(f + g, P_k) - L(f + g, P_k) \to 0$, $U(f, P_k) - L(f, P_k) \to 0$, and $U(g, P_k) - L(g, P_k) \to 0$. (Supply the details needed.)

From Corollary 3.1.5 (see Exercise 8(a)) we have that

$$\int_a^b (f + g) = \lim_k U((f + g), P_k)$$

$$= \lim_k [U(f, P_k) + U(g, P_k)]$$

$$= \int_a^b f + \int_a^b g$$

The rest of the proof is Exercise 8(b).   ∎

Finally we introduce what might be considered as a definition or a convention. If $a < b$ and $f \in \mathcal{R}[a, b]$, then

3.1.12
$$\int_b^a f = -\int_a^b f$$

Let's also add the agreement that

$$\int_a^a f = 0$$

In some ways this is unnecessary and follows from the definition of the integral. On the other hand we have, at least implicitly, defined everything to do with the integral under the assumption that $a < b$.

Exercises

(1) If f is a bounded function on $[a, b]$ and P is any partition of $[a, b]$, show that $L(f, P) = -U(-f, P)$.

(2) Prove that $U(f, Q) \le U(f, P)$ when Q is a refinement of P. (Proposition 3.1.2(b).)

(3) Prove (3.1.9) when the discontinuity occurs at an interior point of $[a, b]$. (Hint: Use Proposition 3.1.6.)

(4) If $f : [a, b] \to \mathbb{R}$ is a monotonic function, show that f is integrable.

(5) If $f \in \mathcal{R}[a, b]$ and $g : [a, b] \to \mathbb{R}$ is such that $\{x \in [a, b] : f(x) \ne g(x)\}$ is finite, show that $g \in \mathcal{R}[a, b]$ and $\int_a^b g = \int_a^b g$.

(6) Let $\{a_0, a_1, \dots\}$ be an increasing sequence in $[a, b]$ such that $a_0 = a$ and $a_n \to b$. If f is a bounded function on $[a, b]$ that is integrable on $[a_n, a_{n+1}]$ for all $n \ge 0$, is f integrable on $[a, b]$?

(7) Show that if $f : [a, b] \to \mathbb{R}$ is a continuous function, $f(x) \ge 0$ for all x, and $\int_a^b f = 0$, then $f(x) = 0$ for all x in $[a, b]$.

(8) (a) Supply the missing details in the proof of Proposition 3.1.11 needed to show that $\int_a^b (f + g) = \int_a^b f + \int_a^b g$. (b) Show that when $f \in \mathcal{R}[a, b]$ and $\alpha \in \mathbb{R}$, $\alpha f \in \mathcal{R}[a, b]$ and $\int_a^b (\alpha f) = \alpha \int_a^b f$.

(9) (a) Show that if $f \in \mathcal{R}[a, b]$, then $f^2 \in \mathcal{R}[a, b]$. (b) Use the identity $(f + g)^2 - (f - g)^2 = 4fg$ and part (a) to show that if $f, g \in \mathcal{R}[a, b]$, then $fg \in \mathcal{R}[a, b]$.

(10) Using (3.1.12) show that if $a, b, c \in \mathbb{R}$ and f is integrable on some interval including these three numbers, then $\int_a^b f = \int_a^c f + \int_c^b f$.

3.2. The Fundamental Theorem of Calculus

There are several versions of what people call the Fundamental Theorem of Calculus. Here is the most general.

3.2.1. Theorem (Fundamental Theorem of Calculus). *If f is a bounded Riemann integrable function on $[a, b]$ and $F : [a, b] \to \mathbb{R}$ is defined by*

$$F(x) = \int_a^x f(t)dt$$

then F is a continuous function. If f is continuous at a point c in $[a, b]$, then F is differentiable at c and $F'(c) = f(c)$.

Proof. If $|f(x)| \le M$ for all x in $[a, b]$, then for $a \le x \le y \le b$,

$$|F(y) - F(x)| = \left| \int_a^y f - \int_a^x f \right|$$

$$= \left| \int_x^y f \right| \text{ (3.1.6)}$$

$$\le M|y - x| \text{ (3.1.10)(b)}$$

So not only is f continuous, it is a Lipschitz function.

Now assume f is continuous at c, which we assume is an interior point of the interval. If $\epsilon > 0$ there is a $\delta > 0$ such that $|f(x) - f(c)| < \epsilon$ when $|x - c| < \delta$. Thus when $0 < |t| < \delta$,

$$\left| \frac{F(c + t) - F(c)}{t} - f(c) \right| = \left| \left(\frac{1}{t} \int_c^{c+t} f \right) - f(c) \right|$$

$$= \left| \frac{1}{t} \int_c^{c+t} [f(x) - f(c)]dx \right|$$

$$\le \frac{1}{|t|} \epsilon |t| \text{ (3.1.10)(b)}$$

$$\le \epsilon$$

By definition $F'(c)$ exists and equals $f(c)$. When c is one of the endpoints of the interval the proof is similar and left to the reader (Exercise 1). ∎

Because of its frequent use in the rest of the book, we refer to this result as well as its corollary below as the FTC. The function $F : [a, b] \to \mathbb{R}$ is called the *indefinite integral* of f. It is also sometimes referred to as the *anti-derivative* of f, though that is a moniker that comes from the following consequence of the preceding theorem.

3.2.2. Corollary. *If $f : [a, b] \to \mathbb{R}$ is a continuous function and F is its indefinite integral, then F is a continuously differentiable function on $[a, b]$ and $F' = f$. Equivalently, if $F : [a, b] \to \mathbb{R}$ is a continuously differentiable function and $F' = f$, then $\int_a^b f = F(b) - F(a)$.*

Proof. The first statement in the corollary is immediate from the theorem. For the second we note that if G is the indefinite integral of $f = F'$, then $G' = F'$, so $G - F$ is the constant function. Thus $\int_a^b f = G(b) - G(a) = F(b) - F(a)$. ∎

We can use the FTC to calculate several integrals.

3.2.3. Example. (a) If $n \in \mathbb{N}$, then $\int_a^b x^n = (n+1)^{-1}(b^{n+1} - a^{n+1})$. In fact, $(x^{n+1})' = (n+1)x^n$ (2.2.6), so this is immediate from the FTC.
(b) $\int_a^b \sin x = -(\cos b - \cos a)$ and $\int_a^b \cos x = \sin b - \sin a$.
(c) To find $F(x) = \int_a^{x^3} \sin y \, dy$, note that $F(x) = G(\phi(x))$, where $G(y) = \int_a^y \sin x$ and $\phi(x) = x^3$. So the Chain Rule and the FTC imply that $F'(x) = G'(\phi(x))\phi'(x) = (\sin x^3)(3x^2) = 3x^2 \sin x^3$.

We now apply the FTC to obtain other important results.

3.2.4. Theorem (Mean Value Theorem). *If $f : [a, b] \to \mathbb{R}$ is a continuous function, then there is a point c in $[a, b]$ such that*

$$\int_a^b f = f(c)(b - a)$$

Proof. Define $F : [a, b] \to \mathbb{R}$ as the indefinite integral of f. By the MVT for derivatives (Theorem 2.3.4) there is a point c in $[a, b]$ with $(b - a)F'(c) = F(b) - F(a) = \int_a^b f$. By the Fundamental Theorem this finishes the proof. ∎

3.2.5. Theorem (Change of Variable Theorem). *Suppose $\phi : [a, b] \to \mathbb{R}$ is a continuously differentiable function and I is an interval such that $\phi([a, b]) \subseteq I$. If $f : I \to \mathbb{R}$ is a continuous function, then*

$$\int_{\phi(a)}^{\phi(b)} f(x) \, dx = \int_a^b f(\phi(t))\phi'(t) \, dt$$

Proof. Define $F : I \to \mathbb{R}$ as the indefinite integral of f:

$$F(u) = \int_{\phi(a)}^u f(x) \, dx$$

By the FTC, $F'(u) = f(u)$. Hence the Chain Rule implies $(F \circ \phi)'(t) = F'(\phi(t))\phi'(t) = f(\phi(t))\phi'(t)$. Again applying the FTC we have that

$$\int_a^b f(\phi(t))\phi'(t)\,dt = \int_a^b (F \circ \phi)'(t)\,dt$$
$$= (F \circ \phi)(b) - (F \circ \phi)(a)$$
$$= F(\phi(b)) - F(\phi(a))$$
$$= \int_{\phi(a)}^{\phi(b)} f(x)\,dx \qquad \blacksquare$$

The preceding result will be referred to as the COV Theorem, which, of course, is used in integration by substitution.

3.2.6. Example. Consider the integral $\int_0^1 x^2 \cos(2 - x^3)\,dx$. If $\phi(t) = 2 - t^3$ for $0 \le t \le 1$, then ϕ maps $[0, 1]$ onto $[1, 2]$ (Why?). So the COV Theorem implies $\int_0^1 x^2 \cos(2 - x^3)\,dx = -\frac{1}{3}\int_0^1 \cos(\phi(t))\phi'(t)\,dt - \frac{1}{3}\int_2^1 \cos x\,dx = -\frac{1}{3}[\sin(1) - \sin(2)] = \frac{1}{3}[\sin(2) - \sin(1)]$. This example emphasizes the importance of correctly placing the limits of integration $\phi(0)$ and $\phi(1)$ and using (3.1.12).

3.2.7. Theorem (Integration by Parts). *If f and g are two continuously differentiable functions on $[a, b]$, then*

$$\int_a^b fg' = f(b)g(b) - f(a)g(a) - \int_a^b f'g$$

Proof. This is one of the easiest proofs of a result labeled as a theorem. Just apply the FTC to obtain that $f(b)g(b) - f(a)g(a) = \int_a^b (fg)' = \int_a^b f'g + \int_a^b fg'$. $\blacksquare$

In calculus this is presented as one of the many techniques of integration. It is much more than that and is a basic tool of analysis.

Exercises

(1) Prove the Fundamental Theorem of Calculus when the point c is an endpoint of $[a, b]$.

(2) Find $F'(x)$ where $F(x) = \int_0^{\cos x} \sin y\,dy$.

(3) Evaluate the following. (a) $\int_0^1 x^2(6x^3 + 5)^{\frac{1}{4}}$. (b) $\int_0^\pi \cos^3 x \sin x$.

(4) If f is a continuously differentiable function on $\mathbb{R}$ such that $f(0) = 0$ and $1 \le f'(x) \le 2$ for all x in $\mathbb{R}$, show that $x \le f(x) \le 2x$ on $\mathbb{R}$.

(5) Compute the following integrals. (a) $\int_0^{\pi/2} x \sin x$. (b) $\int_1^2 \frac{\sqrt{x}-1}{\sqrt{x}}$. (c) $\int_0^\pi x \cos x^2$.

(6) For any continuous function $f : [-1, 1] \to \mathbb{R}$, show that

$$\int_{-1}^1 xf(x^2) = 0.$$

(7) If f is infinitely differentiable on $[a, b]$ and $n \geq 1$, show that

$$f(b) = \sum_{k=0}^{n} \frac{f^{(k)}(a)}{k!}(b-a)^k + \frac{1}{n!}\int_a^b f^{(n+1)}(x)(b-x)^n \, dx.$$

(Hint: Use integration by parts.)

3.3. The Logarithm and Exponential Functions

In this section we introduce the logarithm of a positive number, study the properties of the associated function, and explore its inverse, the exponential function. The logarithm we study is often called the *natural logarithm* so as to distinguish it from the logarithm to base 10, which will be introduced later.

3.3.1. Definition. If $x > 0$, define

$$\log x = \int_1^x \frac{1}{t} dt$$

This is called the *logarithm* of x and the function $\log : (0, \infty) \to \mathbb{R}$ is also called the logarithm function or sometimes the log function.

We note that many people use the notation $\ln x$ instead of $\log x$, which they reserve for the logarithm to base 10. This seems more understandable when these notions are used in an algebra course. The only use I know for the logarithm to base 10 is in performing calculations, a virtue that seems outdated since the advent of computers. The natural logarithm is truly natural, as its definition shows, while the others are somewhat artificial. Hence we use $\log x$ as the natural logarithm.

3.3.2. Theorem. *The logarithm function has the following properties.*

(a) *The logarithm function is differentiable and $(\log x)' = 1/x$.*
(b) $\log 1 = 0$, $\log x > 0$ *when* $x > 1$, *and* $\log x < 0$ *when* $0 < x < 1$.
(c) *If $a, b > 0$, then $\log ab = \log a + \log b$ and $\log(a/b) = \log a - \log b$.*
(d) *The logarithm function is a strictly increasing function that is a bijection between the interval $(0, \infty)$ and the entire real line $\mathbb{R}$.*

Proof. (a) This follows immediately from the FTC.

(b) It's clear that $\log 1 = 0$ and when $x > 1$ the integral defining $\log x$ integrates a strictly positive function. When $0 < x < 1$, $\log x = \int_1^x t^{-1} \, dt = -\int_x^1 t^{-1} \, dt < 0$ (3.1.12).

(c) Fix a in $(0, \infty)$ and consider the function $f(x) = \log(ax) - \log a - \log x$. An easy computation (using the Chain Rule) shows that $f'(x) = 0$ for $0 < x < \infty$. Hence f is a constant function; so for all x, $f(x) = f(1) = 0$. Therefore $0 = f(b)$, yielding the first formula. Using this we have that $0 = \log 1 = \log bb^{-1} = \log b + \log b^{-1}$; so we have that $\log b^{-1} = -\log b$. Now evaluate $f(b^{-1})$ to obtain the second formula in (c).

(d) By (a), $(\log x)' > 0$ so it is strictly increasing and hence injective (2.3.8). Using (c) we have $\log 2^n = n \log 2$; since $\log 2 > \log 1 = 0$, we have that $\log 2^n \to \infty$ as $n \to \infty$. Therefore if $y > 0$, choose n sufficiently large that $\log 2^n > y$. Now consider the log function on the interval $[1, 2^n]$. We have that $\log 1 < y < \log 2^n$. Therefore the IVT implies that there is an x in this interval with $\log x = y$. Similarly if $y < 0$, there is a point x in the interval $(0, 1)$ with $\log x = y$. Thus the logarithm is bijective.   ∎

Since the log function is a bijection from $(0, \infty)$ onto all of $\mathbb{R}$, it has an inverse function from $\mathbb{R}$ onto $(0, \infty)$.

3.3.3. Definition. The inverse of the logarithm function is called the *exponential function* and is denoted by $\exp : \mathbb{R} \to (0, \infty)$.

3.3.4. Theorem. *The exponential function has the following properties.*

(a) *The exponential function is continuously differentiable and* $(\exp x)' = \exp x$.
(b) $\exp 0 = 1$, $\exp x > 1$ *when* $x > 0$, *and* $0 < \exp x < 1$ *when* $x < 0$.
(c) *If* $a, b \in \mathbb{R}$, $\exp(a + b) = (\exp a)(\exp b)$ *and* $\exp(-a) = (\exp a)^{-1}$.
(d) *The exponential function is a strictly increasing function that maps* $\mathbb{R}$ *bijectively onto* $(0, \infty)$.
(e) *We have that*

$$\lim_{x \to \infty} \exp x = \infty \quad \text{and} \quad \lim_{x \to -\infty} \exp x = 0.$$

Proof. As a glance at Theorem 3.3.2 reveals, the parts (a) through (d) in this theorem parallel the corresponding ones of that result and are direct consequences of the fact that the exponential function is the inverse of the log function. The details are left to the reader in Exercise 1. Establishing (e) proceeds as follows. Since the exponential is strictly increasing and maps onto $(0, \infty)$, we must have that $\lim_{x \to \infty} \exp x = \infty$. Similarly we obtain the other half of (e).   ∎

3.3.5. Definition. If $x > 0$ and $a \in \mathbb{R}$ define

$$x^a = \exp(a \log x)$$

First note that this definition makes sense since $x > 0$.

From here we can get the usual laws of exponents by using the previous two theorems. The details are left to the reader in Exercise 4.

3.3.6. Proposition. *If* $x, y > 0$ *and* $a, b \in \mathbb{R}$, *the following hold.*

(a) $x^a x^b = x^{(a+b)}$.
(b) $x^a y^a = (xy)^a$.
(c) $(x^a)^b = x^{(ab)}$.
(d) $x^{(a-b)} = x^a / x^b$.

The number x where $\log x = 1$ is denoted by e. We want to emphasize the following:

$$e^a = \exp a$$

So we frequently will write the exponential function as e^x as does the rest of the mathematical world. The next definition is possible because both the logarithm and exponential functions are bijective.

3.3.7. Definition. If $b, x > 0$, we define $y = \log_b(x)$ to mean that $x = b^y$.

The proof of the next result is left to the reader (Exercise 5).

3.3.8. Proposition. *If $b, x, y > 0$, the following hold.*

(a) $\log_b(xy) = \log_b x + \log_b y$.
(b) $\log_b(x/y) = \log_b x - \log_b y$.
(c) *For any a in $\mathbb{R}$, $\log_b(x^a) = a \log_b x$.*

We return to an examination of the properties of the exponential function.

3.3.9. Proposition. *For any real number a,*

$$e^a = \lim_{n \to \infty} \left(1 + \frac{a}{n}\right)^n$$

Proof. Observe that proving this result is equivalent to showing that

$$\lim_{n \to \infty} \log(1 + a/n)^n = a$$

(Why?) Now if $f(x) = \log(1 + x)$, then $f'(x) = (1 + x)^{-1}$. Thus

$$\lim_{n \to \infty} \log \left(1 + \frac{a}{n}\right)^n = \lim_{n \to \infty} n \log \left(1 + \frac{a}{n}\right)$$

$$= a \lim_{n \to \infty} \left[\frac{\log \left(1 + \frac{a}{n}\right) - \log 1}{\frac{a}{n}}\right]$$

$$= a \lim_{t \to 0} \left[\frac{\log(1 + t) - \log 1}{t}\right]$$

$$= a f'(0) = a \qquad \blacksquare$$

The next result says that exponential decay is more rapid that polynomial decay.

3.3.10. Proposition. *For any $n \geq 1$,*

$$\lim_{x \to \infty} x^n e^{-x} = 0 \text{ and } \lim_{x \to 0+} x^{-n} e^{-\frac{1}{x}} = 0$$

Proof. Note that the second of these equalities follows from the first by substituting x^{-1} for x. The proof of the first equality consists of applying L'Hôpital's Rule n times.

$$\lim_{x \to \infty} x^n e^{-x} = \lim_{x \to \infty} \frac{x^n}{e^x} = \lim_{x \to \infty} \frac{nx^{n-1}}{e^x} = \cdots = \lim_{x \to \infty} \frac{n!}{e^x} = 0$$

by Theorem 3.3.4(e). $\qquad \blacksquare$

In the statement of the preceding proposition we made explicit in the second equality a variation on the theme of the first. There are many other variations that will be used and we will just refer to this proposition. In particular we'll see an additional variation in the following example.

3.3.11. Example. If

$$f(x) = \begin{cases} e^{-\frac{1}{x^2}} & \text{when } x \neq 0 \\ 0 & \text{when } x = 0 \end{cases}$$

then f is infinitely differentiable and $f^{(n)}(0) = 0$ for all $n \geq 1$. To see this put $\xi(x) = -x^{-2}$ so that when $x \neq 0$, $f(x) = \exp(\xi(x))$. By the Chain Rule we have that $f'(x) = \xi'(x)f(x) = 2x^{-3}f(x)$. Also $[f(t) - f(0)]/t = t^{-1}f(t) \to 0$. So (3.3.10) implies that f is differentiable at 0 and $f'(0) = 0$.

Claim. For every $n \geq 1$ there is a polynomial p_n such that when $x \neq 0$, $f^{(n)}(x) = p_n(x^{-1})f(x)$.

This is proved by induction. Just before the statement of the claim we saw that the claim holds when $n = 1$. Assume it holds for n and let's prove it for $n + 1$. In fact when $x \neq 0$

$$\begin{aligned} f^{(n+1)}(x) &= [p_n(x^{-1})f(x)]' \\ &= [p_n(x^{-1})]'f(x) + p_n(x^{-1})f'(x) \\ &= -x^{-2}p_n'(x^{-1})f(x) + p_n(x^{-1})f'(x) \\ &= [-x^{-2}p_n'(x^{-1}) + p_n(x^{-1})(2x^{-3})]f(x) \\ &= p_{n+1}(x^{-1})f(x) \end{aligned}$$

where $p_{n+1}(y) = -y^2 p_n'(y) + 2y^3 p_n(y)$, a polynomial. This establishes the above claim.

Claim. For each $n \geq 1$, $f^{(n)}(0) = 0$.

Again we have already shown this for $n = 1$. Now assume that we have that $f^{(n)}(0) = 0$. The first claim shows that $[f^{(n)}(t) - f^{(n)}(0)]/t = t^{-1}p_n(t^{-1})f(t)$, and by Proposition 3.3.10 this converges to 0 as $t \to 0$.

We have seen that $(e^x)' = e^x$. It turns out that, except for multiplying it by a constant, the exponential function is the only function that has the property that it equals its derivative. The proof is easy.

3.3.12. Proposition. *If $f : (a, b) \to \mathbb{R}$ is a differentiable function such that $f'(x) = f(x)$ for all x in (a, b), then there is a constant c such that $f(x) = ce^x$ on (a, b).*

Proof. In fact for any x in (a, b)

$$\left[\frac{f(x)}{e^x}\right]' = \frac{f'(x)e^x - f(x)e^x}{e^{2x}} = 0$$

Hence f/e^x must be the constant function. ∎

Exercises

(1) Give the details of the proof of parts (a) through (d) in Theorem 3.3.4. (Hint: use Proposition 2.3.10 to establish (a).)

(2) Let $f : (0, 1) \to \mathbb{R}$ be defined by $f(x) = (x - 1)/\log x$. (a) Show that $\lim_{x \to 0+} f(x) = 1$. (b) Show that f is uniformly continuous.

(3) Let $f : (0, \infty) \to \mathbb{R}$ be a continuous function that satisfies $f(x) - f(y) = f(x/y)$ for all x, y in $(0, \infty)$. Show that if f is differentiable at $x = 1$ with $f'(1) = 1$, then f is differentiable everywhere and $f(x) = \log x$ for all x.

(4) Prove Proposition 3.3.6.

(5) Prove Proposition 3.3.8.

(6) What is $(\log_b x)'$?

(7) Which is bigger, e^x or x^e? (Of course you can answer this with a calculator or computer, but try to answer it by an examination of the function $f(x) = \log e^x - \log x^e$.)

(8) For each of the following functions f determine whether $x = 0$ is a local maximum, local minimum, or neither. (a) $f(x) = x^2 e^{x^3}$. (b) $f(x) = x^3 e^{x^2}$. (c) $f(x) = e^{x^2} \sin x$. (d) $f(x) = e^{x^2} \cos x$. (e) $f(x) = e^x \sin x^2$.

(9) Evaluate the following. (a) $\int_1^e x \log x$. (b) $\int_0^\pi x \cos x$.

(10) Find the derivative of $f(x) = x^a$ when $x \in (0, \infty)$ and $a \in \mathbb{R}$.

(11) Let $\mathcal{V}$ be the vector space of all functions from $\mathbb{R}$ into itself and show that any finite collection of functions in $\{x \mapsto e^{ax} : a \in \mathbb{R}\}$ is linearly independent.

(12) If $f : \mathbb{R} \to \mathbb{R}$ is defined by

$$f(x) = \begin{cases} e^{-\frac{1}{x^2}} \sin \frac{1}{x} & \text{when } x \neq 0 \\ 0 & \text{when } x = 0 \end{cases}$$

show that f is infinitely differentiable. Compare this with Example 2.4.4 and ponder this comparison while regarding Proposition 3.3.10.

(13) If $\{x_n\}$ is a sequence of positive real numbers such that $x_n \to 0$, show that $x_n^n \to 0$.

(14) Show that $\lim_{x \to 0+} x \log x = 0$. (Hint: Apply L'Hôpital's Rule as expressed in Exercise 2.5.3 to $x \log x = (\log x)/(x^{-1})$.)

3.4. Improper Integrals

When we defined the Riemann integral at the start of this chapter, we did so only for bounded functions on a closed and bounded interval. We are frequently confronted with situations where we want to integrate unbounded functions over intervals that are not bounded, or not closed, or neither. Here we will deal with such situations, relying on what we know from the previous sections. The material in this section is influenced by the treatment of this subject in [15].

3.4.1. Definition. Suppose $-\infty \leq a < b \leq \infty$ and $f : (a, b) \to \mathbb{R}$. We say f is *locally integrable* on (a, b) if whenever $a < c < d < b$ we have that f is integrable on $[c, d]$. Say that f is *integrable* or *improperly integrable* on (a, b) if it is locally integrable on (a, b) and

$$\lim_{c \to a+} \left[\lim_{d \to b-} \int_c^d f \right]$$

exists and is finite. When f is integrable on $[a, b]$ we denote this iterated limit as $\int_a^b f = \int_a^b f(x)\, dx$.

Let's emphasize that the interval (a, b) could be the whole line and that the function f need not be bounded. Because of the fact that f is assumed to be locally integrable, however, it is required that f is bounded on every closed and bounded interval contained in (a, b). We'll see many examples below, but the first thing we should do is establish that the order of the two limits in the definition above does not matter.

3.4.2. Proposition. *If f is integrable over (a, b), then*

$$\lim_{c \to a+} \left[\lim_{d \to b-} \int_c^d f \right] = \lim_{d \to b-} \left[\lim_{c \to a+} \int_c^d f \right]$$

Proof. Pick some arbitrary point x in (a, b) and observe that

$$\lim_{c \to a+} \left[\lim_{d \to b-} \int_c^d f \right] = \lim_{c \to a+} \left[\int_c^x f + \lim_{d \to b-} \int_x^d f \right]$$

$$= \lim_{c \to a+} \int_c^x f + \lim_{d \to b-} \int_x^d f$$

$$= \lim_{d \to b-} \left[\lim_{c \to a+} \int_c^x f + \int_x^d f \right]$$

$$= \lim_{d \to b-} \left[\lim_{c \to a+} \int_c^d f \right] \qquad \blacksquare$$

Often we will encounter the situation where the difficulty in defining the integral of f on (a, b) only occurs at one of the endpoints. Consider the following.

3.4.3. Proposition. *If $-\infty < a < b \leq \infty$ and $f : [a, b) \to \mathbb{R}$ is a function that is integrable on (a, b), then*

$$\int_a^b f = \lim_{d \to b-} \int_a^d f$$

The proof is Exercise 1.

We record here several facts about integrable functions whose statements and proofs are similar to the corresponding facts about integrable functions on closed and bounded intervals. The proofs are left to the reader's discretion.

3.4.4. Proposition. *If* $-\infty \le a < b \le \infty$ *and* $f : (a, b) \to \mathbb{R}$, *the following hold.*

(a) *If* $a < c < b$ *and* f *is integrable both on* $(a, c]$ *and* $[c, b)$, *then* f *is integrable on* (a, b) *and* $\int_a^b f = \int_a^c f + \int_c^b f$.
(b) *If* f *and* g *are integrable functions on* (a, b) *and* $\alpha, \beta \in \mathbb{R}$, *then* $\alpha f + \beta g$ *is integrable and* $\int_a^b (\alpha f + \beta g) = \alpha \int_a^b f + \beta \int_a^b g$.

The next result could easily be called the Comparison Test for improper integrals.

3.4.5. Proposition. *If* f *and* g *are functions defined on* (a, b) *such that* $0 \le f(x) \le g(x)$ *for all* x *in* (a, b) *and* g *is integrable on* (a, b), *then* f *is integrable on* (a, b) *and*

$$\int_a^b f \le \int_a^b g$$

Proof. Let $a < c < d < b$, temporarily fix c, and define $F(d) = \int_c^d f$, $G(d) = \int_c^d g$; it follows that $F(d) \le G(d)$ whenever $c < d < b$. Moreover because the functions f and g are positive, F and G are increasing. Hence their limits exist as $d \to b-$, and we have that $\int_c^b f = \lim_{d \to b-} F(d) \le \lim_{d \to b-} G(d) = \int_c^b g < \infty$. To complete the proof a similar argument can be used when we let $c \to a+$. ∎

In analogy with infinite series we say that $f : (a, b) \to \mathbb{R}$ is *absolutely integrable* when $|f|$ is integrable.

3.4.6. Proposition. *If* $f : (a, b) \to \mathbb{R}$ *is absolutely integrable, then* f *is integrable and*

$$\left| \int_a^b f \right| \le \int_a^b |f|$$

Proof. Note that $-|f(x)| \le f(x) \le |f(x)|$ for all x in (a, b) and use the preceding proposition. ∎

3.4.7. Example. (a) The integral $\int_1^\infty \frac{1}{x^p}$ converges if and only if $p > 1$. First note that when $p = 1$ we have that $\int_1^d \frac{1}{x} = \log d \to \infty$ as $d \to \infty$, so the integral diverges. We assume that $p \ne 1$. Let $d > 1$ and consider $\int_1^d \frac{1}{x^p} = (-p + 1)^{-1}[\frac{1}{d^{p-1}} - 1]$. Let $d \to \infty$. When $p > 1$, this converges to $\frac{1}{p-1}$; when $p < 1$, this diverges to ∞. Note that since x^{-p} is positive on $[1, \infty)$, it is absolutely integrable if and only if $p > 1$.

(b) $\int_{-\infty}^\infty xe^{-x^2} = \lim_{c \to -\infty}[\lim_{d \to \infty} \int_c^d xe^{-x^2} dx]$. Since we intend to use the substitution $u = x^2$ over an interval involving both positive and negative numbers, it is expeditious to split the integral as $\int_c^d xe^{-x^2} = \int_c^0 xe^{-x^2} + \int_0^d xe^{-x^2}$. Now making the substitution in the second of these integrals gives

$$\lim_{d \to \infty} \int_0^d x e^{-x^2} = \lim_{d \to \infty} \frac{1}{2} \int_0^{d^2} e^{-u} \, du = \lim_{d \to \infty} \left(-\frac{1}{2} [e^{-d^2} - 1] \right) = 1$$

Treating the first integral similarly we get

$$\lim_{c \to -\infty} \int_c^0 x e^{-x^2} = \lim_{c \to -\infty} \frac{1}{2} \int_{c^2}^0 e^{-u} \, du$$

$$= -\lim_{c \to -\infty} \frac{1}{2} \int_0^{c^2} e^{-u} \, du$$

$$= -\lim_{c \to -\infty} \left(-\frac{1}{2} [e^{-c^2} - 1] \right)$$

$$= -1$$

Therefore $x e^{-x^2}$ is integrable on $\mathbb{R}$ and $\int_{-\infty}^{\infty} x e^{-x^2} = 0$.

(c) $\int_0^\infty \sin x = \lim_{d \to \infty} \int_0^d \sin x$. Now $\int_0^d \sin x = (1 - \cos d)$, which has no limit as $d \to \infty$. Hence the sine function is not integrable on $(0, \infty)$.

(d) Note that $\int_{-c}^c \cos x = 0$ for every c since $\cos(-x) = -\cos x$. Hence $\lim_{c \to \infty} \int_{-c}^c \cos x = 0$. However $\cos x$ is not integrable on $\mathbb{R}$ as an argument similar to that used in (c) shows. This provides a cautionary note that as we did in (b) we must use the definition of an improper integral and cannot take shortcuts without justification.

(e) $\int_1^\infty \frac{\cos x}{x^2}$ converges. In fact $|\frac{\cos x}{x^2}| \le x^{-2}$ on $[1, \infty)$, so that the statement follows by Proposition 3.4.6.

(f) $\int_1^\infty \frac{\sin x}{x}$ is integrable. In fact let $1 < d < \infty$. Since the function is continuous here it is integrable over any bounded interval in $[1, \infty)$. Using integration by parts we get

$$\int_1^d \frac{\sin x}{x} = -\left[\frac{\cos d}{d} - \cos(1) \right] - \int_1^d \frac{\cos x}{x^2}$$

By the preceding example, $x^{-1} \sin x$ is integrable. It is not, however, absolutely integrable. See Exercise 6.

3.4.8. Theorem (Integral Test). *If $f : [1, \infty) \to \mathbb{R}$ is a non-negative function that is decreasing, then the series $\sum_{n=1}^\infty f(n)$ converges if and only if the function f is integrable on $[1, \infty)$.*

Proof. Because f is non-negative and decreasing, $f(n+1) \le f(x) \le f(n)$ when $n \le x \le n+1$ and so $f(n+1) \le \int_n^{n+1} f(x) \le f(n)$. Thus if the series converges, then $0 \le \int_1^\infty f \le \sum_{n=1}^\infty f(n) < \infty$, and so f is integrable. If the function is integrable, then $0 \le \sum_{n=1}^\infty f(n+1) \le \sum_{n=1}^\infty \int_n^{n+1} f = \int_1^\infty f$. ∎

3.4.9. Example. The series $\sum_{n=1}^{\infty} \frac{1}{n^p}$ converges if and only if $p > 1$. In fact this follows by using the Integral Test and Example 3.4.7(a). Recall that when $p = 1$ this series is called the harmonic series and diverges (1.4.3). For other values of p it is called the *harmonic p-series*.

Exercises

(1) Prove Proposition 3.4.3.
(2) For which values of p do the following integrals converge?

$$\text{(a)} \int_0^1 \frac{1}{x^p}. \qquad \text{(b)} \int_e^{\infty} \frac{1}{x \log^p x}.$$

(3) Show that $\int_0^1 |\sin x / x^{\frac{1}{3}}|$ converges. (Hint: First show that $|\sin x| \leq |x|$ on $[0, 1]$.)
(4) Justify the statements made in Example 3.4.7(d).
(5) Use integration by parts to show that $\int_0^{\infty} x^n e^{-x} dx = n!$ for every integer $n \geq 0$. (Note that this necessitates combining integration by parts with improper integrals.)
(6) Show that $|\sin x|/x$ is not integrable over $[1, \infty)$. (Hint: Note that

$$\int_1^{n\pi} |\sin x|/x \, dx \geq \frac{1}{k\pi} \sum_{k-2}^{n} \int_{(k-1)\pi}^{k\pi} |\sin x| \, dx$$

and evaluate.)
(7) Show that $\int_0^{\infty} \sqrt{x} e^{-\sqrt{x}} \, dx$ exists and determine its value.
(8) Is $f(x) = x^{-2}(1 - \cos x)$ integrable over $(0, \infty)$?
(9) Show that the product of two improperly integrable functions is not always improperly integrable.
(10) Let $f, g : [a, b) \to \mathbb{R}$ be locally integrable positive functions and assume $L = \lim_{x \to b-} f(x)/g(x)$ exists and satisfies $0 \leq L < \infty$. Show that if the improper integral of g exists on $[a, b)$, then the improper integral of f exists on $[a, b)$.
(11) (a) If f is integrable over $[0, \infty)$ and $\lim_{x \to \infty} f(x)$ exists, show that this limit must be 0. (b) If f is integrable over $[0, \infty)$, show by example that $\lim_{x \to \infty} f(x)$ may not exist.

3.5. Sets of Measure Zero and Integrability*

In this section we will give a necessary and sufficient condition for a bounded function on a closed and bounded interval to be Riemann integrable. This condition is phrased using a concept that has widespread use in a more general setting.

3.5.1. Definition. A subset E of $\mathbb{R}$ has *measure zero* if for every $\epsilon > 0$ there is a sequence of intervals $\{(a_n, b_n)\}$ such that $E \subseteq \bigcup_{n=1}^{\infty}(a_n, b_n)$ and

$$\sum_{n=1}^{\infty}(b_n - a_n) < \epsilon$$

The quantity $b_n - a_n$ is the length of the interval (a_n, b_n), so the definition of a set E of measure zero is that E can be covered by a sequence of intervals the sum of whose lengths is as small as desired. Note that the inequality $< \epsilon$ can be replaced by $\leq \epsilon$. It's clear that $\emptyset$ has measure zero, and here are some more examples.

3.5.2. Example.   (a) Any finite set has measure zero. In fact if $E = \{x_1, \ldots, x_n\}$, then the intervals $(x_1 - \epsilon/3n, x_1 + \epsilon/3n), \ldots, (x_n - \epsilon/3n, x_n + \epsilon/3n)$ work.

(b) Any sequence has measure zero. If $\{x_n\}$ is given, then the sequence of intervals $\{(x_n - \epsilon/2^{n+1}, x_n + \epsilon/2^{n+1})\}$ will work.

(c) Sets of measure zero can be very large in a certain sense. For example $\mathbb{Q}$, the set of rational numbers, has measure zero. In fact by Corollary 1.5.5, $\mathbb{Q}$ is countable so it is possible to arrange $\mathbb{Q}$ as a sequence and it follows from (b) that $\mathbb{Q}$ has measure zero.

From these examples it becomes clear we will use some of the material from §1.5, so the reader has to be familiar with countable sets.

3.5.3. Proposition.   *If $\{E_n\}$ is a sequence of sets of measure zero, then $E = \bigcup_{n=1}^{\infty} E_n$ has measure zero.*

Proof. If $\epsilon > 0$ and $n \geq 1$, let $\{(a_n^k, b_n^k)\}_k$ be a sequence of open intervals covering E_n and satisfying $\sum_{k=1}^{\infty}(b_n^k - a_n^k) < \epsilon/2^n$. It follows that $\{(a_n^k, b_n^k) : k \geq 1, n \geq 1\}$ is countable and can be written as a sequence. Moreover $\sum_{n=1}^{\infty}\sum_{k=1}^{\infty}(b_n^k - a_n^k) < \sum_{n=1}^{\infty}\epsilon/2^n = \epsilon$. ∎

To prove the main result of this section we need to introduce an additional concept. If $f : [a, b] \to \mathbb{R}$ and $I \subseteq [a, b]$, define $\omega_f(I) = \sup\{|f(y) - f(z)| : y, z \in I\}$. It will be useful as we progress to observe that if $J \subseteq I$, then $\omega_f(J) \leq \omega_f(I)$. If $x \in [a, b]$, define the *oscillation* of f at x to be

$$\omega_f(x) = \inf\{\omega_f((x - \delta, x + \delta) \cap [a, b]) : \delta > 0\}$$

(The interested reader can compare this with Exercise 1.7.14.) Note that when x is not an endpoint we can write

$$\omega_f(x) = \inf\{\omega_f(x - \delta, x + \delta) : \delta > 0\}$$

So $\omega_f(x)$ measures how much the values of f vary as points get closer to x (and hence its name). It is not difficult to see that if $f : [0, 1] \to \mathbb{R}$ is defined by letting $f(x) = \sin x^{-1}$ when $x \neq 0$ and $f(0) = 0$, then $\omega_f(0) = 2$.

Keep in mind that we have two types of oscillation for a function: one for intervals and one for points. The pertinence of the oscillation for integrability can be seen in the following.

3.5.4. Proposition. *If $P = \{a = x_0 < x_1 < \cdots < x_n = b\}$ is a partition of $[a, b]$ and $f : [a, b] \to \mathbb{R}$, then*

$$U(f, P) - L(f, P) = \sum_{j=1}^{n} \omega_f([x_{j-1}, x_j])(x_j - x_{j-1})$$

Proof. If $a \le c < d \le b$, then it is easy to verify that $\omega_f([c, d]) = M - m$, where $M = \sup\{f(x) \in [c, d]\}$ and $m = \inf\{f(x) : x \in [c, d]\}$. From here the proposition follows. ∎

The next result gives some basic properties of the oscillation. Particularly notice the last part.

3.5.5. Proposition. *If $f : [a, b] \to \mathbb{R}$, the following hold.*

(a) *For any $t > 0$, $\{x \in [a, b] : \omega_f(x) < t\}$ is open unless it contains one of the endpoints of $[a, b]$. If it contains an endpoint it contains a half-open interval that includes that endpoint.*
(b) *For any $t > 0$, $\{x \in [a, b] : \omega_f(t) \ge t\}$ is closed.*
(c) *The function f is continuous at x if and only if $\omega_f(x) = 0$.*

Proof. The proof of (a) when the set contains an endpoint is similar to that of the statement when it contains no endpoints and is left to the reader (Exercise 1). So assume that $a, b \notin G = \{x : \omega_f(x) < t\}$ and fix x in G. By definition there is a $\delta > 0$ such that $\omega_f(x - \delta, x + \delta) = \sup\{|f(w) - f(z)| : w, z \in (x - \delta, x + \delta)\} < t$. Now for any y in $(x - \delta, x + \delta)$ let $r > 0$ such that $(y - r, y + r) \subseteq (x - \delta, x + \delta)$; we see that $\omega_f(y - r, y + r) < t$. Therefore $\omega_f(y) < t$; that is $(y - r, y + r) \subseteq G$, and so G is open. Part (b) follows since $\{x : \omega_f(t) \ge t\} = [a, b] \backslash \{x : \omega_f(x) < t\}$.
To prove (c) assume f is continuous at x and $\epsilon > 0$. Thus there is a $\delta > 0$ such that $|f(x) - f(y)| < \epsilon/2$ when $|x - y| < \delta$. This says that $\omega_f(x) \le \omega_f(x - \delta, x + \delta) < \epsilon$. Since ϵ was arbitrary, $\omega_f(x) = 0$. Conversely, assume $\omega_f(x) = 0$ and let $\epsilon > 0$. By definition this implies there is a $\delta > 0$ such that $\omega_f(x - \delta, x + \delta) < \epsilon$. That is $|f(x) - f(y)| < \epsilon$ when $|x - y| < \delta$. ∎

Given the basic criterion for integrability (3.1.4) as well as the relationship between oscillation and the continuity of f at a point we just established, the statement of the next theorem might be less surprising.

3.5.6. Theorem (Lebesgue's[2] Theorem). *A bounded function $f : [a, b] \to \mathbb{R}$ is integrable if and only if its set of discontinuities has measure zero.*

....

[2] Henri Léon Lebesgue was born in 1875 in Beauvais, which is in the Picardie region of France just north of Paris. He began his education in Beauvais and entered the École Normale Supérieure in Paris in 1894, where he was awarded the teaching diploma in mathematics in 1897. He remained there for two years, reading mathematics in the library. In 1899 he went to Nancy as a professor at a lycée where he remained for another two years. In 1901 he formulated the theory of measure in a ground breaking paper that defined what we now know as the Lebesgue integral. This formed the basis of his doctoral dissertation and earns him the title of the father of measure theory. The present theorem is important for us, but does not top the list of his most important

Proof. Suppose $f : [a, b] \to \mathbb{R}$ is a function with $|f(x)| \leq M$ for all x in $[a, b]$ and let D denote its set of discontinuities. By Proposition 3.5.5(c), $D = \{x : \omega_f(x) > 0\}$; for each $t > 0$ let $D_t = \{x : \omega_f(x) \geq t\}$. Assume f is Riemann integrable; we must show that D has measure zero. The reader can check that $D = \bigcup_{n=1}^{\infty} D_{1/n}$, so by Proposition 3.5.3 we will be successful if we show that each set D_t has measure zero. Fix $t > 0$ and let $\epsilon > 0$. From Proposition 3.5.4 we have the existence of a partition $P = \{a = x_0 < x_1 < \cdots < x_n = b\}$ such that

$$\sum_{j=1}^{n} \omega_f([x_{j-1}, x_j])(x_j - x_{j-1}) < t\epsilon/2$$

If $J = \{j : (x_{j-1}, x_j) \cap D_t \neq \emptyset\}$, then for each j in J we have that $\omega_f([x_{j-1}, x_j]) \geq t$. (Verify!) Hence

$$\sum_{j \in J}(x_j - x_{j-1}) \leq t^{-1} \sum_{j \in J} \omega_f([x_{j-1}, x_j])(x_j - x_{j-1}) < \frac{\epsilon}{2}$$

But $\bigcup_{j \in J}(x_{j-1}, x_j)$ contains D_t except possibly for some subset of the endpoints of these intervals, which has at most $n + 1$ points. Since each of these points can be included in an interval of length less than $\epsilon/2(n + 1)$ and ϵ was arbitrary, we have that D_t has measure zero.

Now assume that D has measure zero. Let $t > 0$; so D_t has measure zero. (As we see what we need, we will take t to be a multiple of an arbitrary ϵ.) Note that D_t is closed and bounded by Proposition 3.5.5(b). Because of this and the fact that D_t has measure zero we can find a finite number of open intervals $\{(u_1, v_1), \ldots, (u_p, v_p)\}$ such that $D_t \subseteq \bigcup_{j=1}^{p}(u_j, v_j)$ and $\sum_{j=1}^{p}(v_j - u_j) < t$. By replacing any two of the closed intervals $\mathcal{U} = \{[u_1, v_1], \ldots, [u_p, v_p]\}$ that intersect with their union, we can assume that these closed intervals are pairwise disjoint. Put $X = [a, b] \backslash \bigcup_{i=1}^{p}(u_i, v_i)$. Note that X is closed and bounded. Moreover, since X is the complement in $[a, b]$ of the union of a finite number of open intervals, X itself is the union of a finite number of pairwise disjoint closed intervals.

Observe that $\omega_f(x) < t$ for each x in X. Therefore for each x in X there is a closed interval $I = [c, d]$ with x in (c, d) and such that $\omega_f[c, d] < t$. Since X is compact there are a finite number of closed intervals whose union contains X and such that

results. In 1902 he joined the faculty at the University of Rennes in Brittany. He married in 1903 and he and his wife had two children. Unfortunately, the union ended in divorce in 1916. He published two fundamental monographs, *Leçons sur l'intégration et la recherche des fonctions primitives* in 1904 and *Leçons sur les séries trigonométriques* in 1906. These books were unjustly criticized at the time by the classicists. Nevertheless he joined the faculty at the University of Poitiers and finally overcame his critics before joining the faculty at the Sorbonne in 1910. In 1921 he became Professor of Mathematics at the Collège de France and held this position until he died in 1941. Lebesgue received many honors throughout his life and made serious contributions in many areas, but he will forever be associated with measure theory and its impact on analysis.

each satisfies this inequality. By intersecting these intervals with X and using the fact that X is itself the union of a finite number of closed intervals, we obtain closed intervals $\mathcal{I} = \{I_1, \ldots, I_q\}$ having the following properties:

$$\text{(i) } X = \bigcup_{i=1}^{q} I_i,$$

(ii) $\omega_f(I_i) < t$ for $1 \le i \le q$,

(iii) the intervals I_i do not overlap

(When we say the intervals "do not overlap" in (iii), we mean they can only intersect at their endpoints. How do we get (iii)?) Put $I_i = [c_i, d_i]$ for $1 \le i \le q$.

Because of our construction we have that the closed intervals $\mathcal{U} \cup \mathcal{I}$ are a collection of non-overlapping intervals whose union is all of $[a, b]$. Therefore their endpoints form a partition $P = \{a = x_0 < x_1 < \cdots < x_n = b\}$; thus $\mathcal{U} \cup \mathcal{I} = \{[x_0, x_1], \ldots, [x_{n-1}, x_n]\}$. In additon we have that

$$U(f, P) - L(f, P) = \sum_{k=1}^{n} \omega_f([x_{k-1}, x_k])(x_k - x_{k-1})$$

$$= \sum_{j=1}^{p} \omega_f([u_j, v_j])(v_j - u_j) + \sum_{i=1}^{q} \omega_f([c_j, d_j])(d_j - c_j)$$

$$\le \sum_{j=1}^{p} 2M(v_j - u_j) + \sum_{i=1}^{q} t(d_j - c_j)$$

$$< 2Mt + t(b - a)$$

If we are given $\epsilon > 0$, we can choose $t > 0$ such that $2Mt + t(b - a) < \epsilon$ and so we have that f is Riemann integrable by Proposition 3.1.4. ∎

Exercise

(1) Prove Proposition 3.5.5(a) under the assumption that the set contains one of the endpoints.

3.6. The Riemann–Stieltjes Integral*

For a fixed closed, bounded interval $J = [a, b]$ we want to define an extension of the Riemann integral. This extended integral will also assign a number to each continuous function on the interval.

3.6.1. Definition. A function $\alpha : J \to \mathbb{R}$ is of *bounded variation* if there is a constant M such that for every partition $P = \{a = x_0 < \cdots < x_n = b\}$ of J,

$$\sum_{j=1}^{n} |\alpha(x_j) - \alpha(x_{j-1})| \leq M$$

The quantity

$$\mathrm{Var}(\alpha) = \mathrm{Var}(\alpha, J) = \sup \left\{ \sum_{j=1}^{n} |\alpha(x_j) - \alpha(x_{j-1})| : P \text{ is a partition of } J \right\}$$

is called the *total variation* of α over J.

We'll see many examples of functions of bounded variation. The first below is specific and the other two contain collections of such functions.

3.6.2. Example. (a) The "mother of all" functions of bounded variation is $\alpha(x) = x$.

(b) Suppose $\alpha : [a, b] \to \mathbb{R}$ is a smooth function and M is a constant with $|\alpha'(t)| \leq M$ for all t in $[a, b]$. If $a = x_0 < x_1 < \cdots < x_n = b$, then for each j the Mean Value Theorem for derivatives says there is a point t_j in $[x_{j-1}, x_j]$ such that $\alpha(x_j) - \alpha(x_{j-1}) = \alpha'(t_j)(x_j - x_{j-1})$. Hence $\sum_j |\alpha(x_j) - \alpha(x_{j-1})| = \sum_j |\alpha'(t_j)|(x_j - x_{j-1}) \leq M(b - a)$, so that α is of bounded variation.

(c) Any increasing or decreasing function is of bounded variation.

Also see the Exercises.

For any interval $[a, b]$ let $BV[a, b]$ denote the set of all functions of bounded variation defined on $[a, b]$. The proof of the next proposition is left to the reader.

3.6.3. Proposition. *$BV[a, b]$ is a vector space over $\mathbb{R}$, where the algebraic operations on these functions are defined pointwise.*

In light of the preceding proposition any linear combination of increasing functions is a function of bounded variation. The surprising thing is that the converse holds.

3.6.4. Proposition. *If $\alpha : [a, b] \to \mathbb{R}$ is a function of bounded variation, then we can write $\alpha = \alpha_+ - \alpha_-$, where $\alpha_\pm$ are increasing functions.*

Proof. Let $\alpha_+(t) = \frac{1}{2}[\mathrm{Var}(\alpha, [a, t]) + \alpha(t)]$ and $\alpha_-(t) = \frac{1}{2}[\mathrm{Var}(\alpha, [a, t]) - \alpha(t)]$. It is clear that $\alpha = \alpha_+ - \alpha_-$, so what we have to do is show that these functions are increasing. Let $t > s, \epsilon > 0$, and let $a = x_0 < \cdots < x_n = s$ such that $\sum_{j=1}^{n} |\alpha(x_j) - \alpha(x_{j-1})| > \mathrm{Var}(\alpha, [a, s]) - \epsilon$. Now it is easy to verify that

$$|\alpha(t) - \alpha(s)| \pm [\alpha(t) - \alpha(s)] \pm \alpha(s) \geq \pm \alpha(s)$$

(Note that we are not allowed to randomly make a choice of sign each time $\pm$ appears; it must be consistent.) Since $a = x_0 < \cdots < x_n < t$ is a partition of $[a, t]$, we get that

$$\text{Var}(\alpha, [a, t]) \pm \alpha(t) \geq \sum_{j=1}^{n} |\alpha(x_j) - \alpha(x_{j-1})| + |\alpha(t) - \alpha(s)|$$
$$\pm \big([\alpha(t) - \alpha(s)] + \alpha(s) \big)$$
$$= \sum_{j=1}^{n} |\alpha(x_j) - \alpha(x_{j-1})|$$
$$+ |\alpha(t) - \alpha(s)| \pm [\alpha(t) - \alpha(s)] \pm \alpha(s)$$
$$\geq \text{Var}(\alpha, [a, s]) - \epsilon \pm \alpha(s)$$

Since ϵ is arbitrary, we have that $\text{Var}(\alpha, [a, t]) \pm \alpha(t) \geq \text{Var}(\alpha, [a, s]) \pm \alpha(s)$ and so the functions $\alpha_{\pm}$ are increasing. ∎

Now that we have discussed functions of bounded variation, gotten many examples, and discovered a structure of such functions (3.6.4), we might pose a question.

Why the interest?

The important thing for us is that we can define integrals or averaging processes for continuous functions by using a function of bounded variation. These integrals have geometric interpretations as well as applications to the study of various problems in analysis. Let's define the integrals, where the reader will notice a close similarity with the definition of the Riemann integral. Indeed if α is the increasing function $\alpha(t) = t$, then what we do below will result in the Riemann integral over J even though we will bypass the upper and lower sums for a partition. (This is because here we will only integrate continuous functions.)

If α is a function in $BV(J)$, $f : J \to \mathbb{R}$ is some function, and P is a partition, define

$$S_\alpha(f, P) = \sum_{j=1}^{n} f(t_j)[\alpha(x_j) - \alpha(x_{j-1})]$$

where the points t_j are chosen in the subinterval $[x_{j-1}, x_j]$. Yes, the notation does not reflect the dependency of this sum on the choice of the points t_j, but I am afraid we'll just have to live with that; indicating such a dependency is more awkward than any gained benefit. When the function α is the special one, $\alpha(t) = t$, let $S_\alpha(f, P) = S(f, P)$. That is,

$$S(f, P) = \sum_{j=1}^{n} f(t_j)[x_j - x_{j-1}]$$

Define the *mesh* of the partition P to be the number $\|P\| = \max\{|x_j - x_{j-1}| : 1 \leq j \leq n\}$, and for any positive number δ let $\mathcal{P}_\delta$ denote the collection of all partitions P with $\|P\| < \delta$. From the proof of Theorem 3.1.7 in the continuous case, it is seen that when $f : J \to \mathbb{R}$ is a continuous function, then there is a unique number $I = \int_a^b f$ such that for every $\epsilon > 0$ there is a $\delta > 0$ with $|I - S(f, P)| < \epsilon$ whenever $P \in \mathcal{P}_\delta$.

In fact, a look at that proof shows this follows from the fact that f is uniformly continuous. (Verify!) We now start the process of showing that a similar existence result holds if we replace the Riemann sum $S(f, P)$ by the sum $S_\alpha(f, P)$ for an arbitrary function of bounded variation α.

Here is another bit of notation that will simplify matters. For $X \subseteq \mathbb{R}$ and a function $f : X \to \mathbb{R}$, the *modulus of continuity* of f for any $\delta > 0$ is the number $\omega(f, \delta) = \sup\{|f(x) - f(y)| : |x - y| < \delta\}$. (Compare this with the definition of the oscillation of a function defined in the preceding section and Exercise 1.7.14.) This will be infinite for some functions, but the main place we will use it is when X is a closed and bounded interval and f is continuous. In that case f is uniformly continuous so that we have that for any $\epsilon > 0$ there is a δ such that $\omega(f, \delta) < \epsilon$.

Just as in the definition of the Riemann integral, we want to define the integral of a function with respect to a function of bounded variation α. Here is the crucial lemma to get us to that goal.

3.6.5. Lemma. *If α is a function of bounded variation on J and $f : J \to \mathbb{R}$ is a continuous function, then for any $\epsilon > 0$ there is a $\delta > 0$ such that $|S_\alpha(f, P) - S_\alpha(f, Q)| \leq \epsilon$ whenever $P, Q \in \mathcal{P}_\delta$.*

Proof. We start by observing that when P, Q are two partitions and Q is a refinement of P, then $S_\alpha(f, P) \leq S_\alpha(f, Q)$. (Imitate the proof of the appropriate part of Proposition 3.1.2.) In light of the preceding observation and using the same reasoning as in the proof of (3.1.2) it suffices to prove that there is a δ such that when $P, Q \in \mathcal{P}_\delta$ and $P \subseteq Q$, then $|S_\alpha(f, P) - S_\alpha(f, Q)| \leq \epsilon/2$.

Use the uniform continuity of f to find a δ such that $\omega(f, \delta) < \frac{\epsilon}{2} \text{Var}(\alpha)$. Assume that $P \subseteq Q$ and that they belong to $\mathcal{P}_\delta$. To simplify matters we will assume that Q adds only one point to P and that this point lies between x_0 and x_1. That is, we assume $P = \{a = x_0 < x_1 < \cdots < x_n = b\}$ and $Q = \{a = x_0 < x_0^* < x_1 < \cdots < x_n = b\}$. Now for $2 \leq j \leq n$, let $x_{j-1} \leq t_j, s_j \leq x_j, x_0 \leq t_1 \leq x_1, x_0 \leq s_0^* \leq x_0^*, x_0^* \leq s_1^* \leq x_1$. Note that

$$
\begin{aligned}
\Big| f(t_1)[\alpha(x_1) &- \alpha(x_0)] - \big\{ f(s_0^*)[\alpha(x_0^*) - \alpha(x_0)] + f(s_1^*)[\alpha(x_1) - \alpha(x_0^*)] \big\} \Big| \\
&= \Big| f(t_1)[\alpha(x_0^*) - \alpha(x_0) + \alpha(x_1) - \alpha(x_0^*)] \\
&\qquad - \big\{ f(s_0^*)[\alpha(x_0^*) - \alpha(x_0)] + f(s_1^*)[\alpha(x_1) - \alpha(x_0^*)] \big\} \Big| \\
&\leq |f(t_1) - f(s_0^*)| \, |\alpha(x_0^*) - \alpha(x_0)| \\
&\qquad + |f(t_1) - f(s_1^*)| \, |\alpha(x_1) - \alpha(x_0^*)| \\
&\leq \omega(f, \delta) \big[|\alpha(x_0^*) - \alpha(x_0)| + |\alpha(x_1) - \alpha(x_0^*)| \big]
\end{aligned}
$$

We therefore obtain

$$
\begin{aligned}
|S_\alpha(f, P) - S_\alpha(f, Q)| &\leq \omega(f, \delta) \big[|\alpha(x_0^*) - \alpha(x_0)| + |\alpha(x_1) - \alpha(x_0^*)| \big] \\
&\qquad + \sum_{j=2}^n |f(t_j) - f(s_j)| \, |\alpha(x_j) - \alpha(x_{j-1})| \\
&\leq \omega(f, \delta) \text{Var}(\alpha) \\
&< \epsilon/2
\end{aligned}
$$

An inspection of the preceding argument shows that if Q had added more than a single point to P, then the same reasoning would prevail and yielded the same result. ∎

It is important to emphasize that the value of the inequality obtained in the preceding lemma is independent of the choices of transitory points t_j in $[x_{j-1}, x_j]$ that are used to define $S(\alpha, P)$. This amply justifies not incorporating them in the notation used to denote such a sum.

3.6.6. Theorem. *If α is a function of bounded variation on J and $f : J \to \mathbb{R}$ is a continuous function, then there is a unique number I with the property that for every $\epsilon > 0$ there is a $\delta > 0$ such that when $P \in \mathcal{P}_\delta$,*

$$|S_\alpha(f, P) - I| < \epsilon$$

The number I is called the Riemann[3]– Stieltjes[4] integral of f with respect to α, is denoted by

$$I = \int_a^b f \, d\alpha = \int f \, d\alpha$$

and satisfies

$$\left| \int f \, d\alpha \right| \le \mathrm{Var}(\alpha) \max\{|f(t)| : t \in J\}$$

Proof. According to the preceding lemma, for every integer $n \ge 1$ there is a δ_n such that if $P, Q \in \mathcal{P}_{\delta_n}$, $|S_\alpha(f, P) - S_\alpha(f, Q)| < \frac{1}{n}$. We can choose the δ_n so that they are decreasing. Let K_n be the closure of the set of numbers $\{S_\alpha(f, P) : P \in \mathcal{P}_{\delta_n}\}$. If $|f(x)| \le M$ for all x in J, then for any partition P, $|S_\alpha(f, P)| \le M \mathrm{Var}(\alpha)$. So each set K_n is bounded and hence compact. Since the numbers δ_n are decreasing, for all $n \ge 1$ $\mathcal{P}_{\delta_n} \supseteq \mathcal{P}_{\delta_{n+1}}$ and so $K_n \supseteq K_{n+1}$. Finally by the choice of the δ_n, $\mathrm{diam}\, K_n \le n^{-1} \to 0$. Therefore by Cantor's Theorem (Theorem 5.5.3 in the next section) $\bigcap_{n=1}^\infty K_n = \{I\}$ for a single number I. It is now routine to verify that I has the stated properties; its uniqueness is guaranteed by its construction. ∎

A standard example comes, of course, when $\alpha(t) = t$ for all t in J and this is the Riemann integral. The proofs of the next two results are left to the reader as a way of fixing the ideas in his/her head. These results should not come as a surprise.

....

[3] See Definition 3.1.3 for a biographical note.

[4] Thomas Jan Stieltjes was born in 1856 in Zwolle, The Netherlands. He attended the university at Delft, spending most of his time in the library reading mathematics rather than attending lectures. This had the effect of causing him to fail his exams three years in a row and he left the university without a degree. (This phenomenon of talented people having trouble passing exams is not unique and examples exist in the author's personal experience.) The absence of a degree plagued the progress of his career, in spite of the recognition of his mathematical talent by some of the prominent mathematicians of the day. In 1885 he was awarded membership in the Royal Academy of Sciences in Amsterdam. He received his doctorate of science in 1886 for a thesis on asymptotic series. In the same year he secured a position at the University of Toulouse in France. He did fundamental work on continued fractions and is often called the father of that subject. He extended Riemann's integral to the present setting. He died in 1894 in Toulouse, where he is buried.

3.6.7. Proposition. *Let α and β be functions of bounded variation on J, $f, g : J \to \mathbb{R}$ continuous functions, and $s, t \in \mathbb{R}$.*

(a) $\int_a^b (sf + tg)\, d\alpha = s \int_a^b f\, d\alpha + t \int_a^b g\, d\beta.$
(b) *If $f(x) \geq 0$ for all x in J and α is increasing, then $\int_a^b f\, d\alpha \geq 0$.*
(c) $\int_a^b f\, d(s\alpha + t\beta) = s \int_a^b f\, d\alpha + t \int_a^b f\, d\beta.$

3.6.8. Proposition. *If α is a function of bounded variation on J, f is a continuous function on J, and $a < c < b$, then $\int_a^b f\, d\alpha = \int_a^c f\, d\alpha + \int_c^b f\, d\alpha$.*

Here is an important result that enables us to compute some Riemann–Stieltjes integrals from what we know about the Riemann integral.

3.6.9. Theorem. *If α is a function on J that has a continuous derivative at every point of J, then*

$$\int_a^b f\, d\alpha = \int_a^b f(x)\alpha'(x)dx$$

for any continuous function f.

Proof. We already know from Example 3.6.2 that such a function is of bounded variation, so everything makes sense. Fix a continuous function f on J, let $\epsilon > 0$, and choose δ such that simultaneously

$$\left| \int f\, d\alpha - S_\alpha(f, P) \right| < \epsilon/2 \quad \text{and} \quad \left| \int f\alpha'dx - S(f\alpha', P) \right| < \epsilon/2$$

whenever $P \in \mathcal{P}_\delta$. Momentarily fix $P = \{a = x_0 < \cdots < x_n = b\}$ in $\mathcal{P}_\delta$. For each subinterval $[x_{j-1}, x_j]$ defined by P, the MVT for derivatives implies there is a t_j in this subinterval with $\alpha(x_j) - \alpha(x_{j-1}) = \alpha'(t_j)(x_j - x_{j-1})$. Therefore

$$\left| \int_a^b f\, d\alpha - \int_a^b f(x)\alpha'(x)dx \right| \leq \left| \int_a^b f\, d\alpha - \sum_{j=1}^n f(t_j)[\alpha(x_j) - \alpha(x_{j-1})] \right|$$

$$+ \left| \sum_{j=1}^n f(t_j)\alpha'(t_j)(x_j - x_{j-1}) - \int_a^b f(x)\alpha'(x)dx \right| < \epsilon$$

Since ϵ was arbitrary, we have the desired equality. ∎

We now introduce some special increasing functions whose role in the general theory has significance in spite of their simplicity.

3.6.10. Example. Fix s in J. When $a \leq s < b$, define

$$\alpha_s(t) = \begin{cases} 0 & \text{for } t \leq s \\ 1 & \text{for } t > s \end{cases}$$

and

$$\alpha_b(t) = \begin{cases} 0 & \text{for } t < b \\ 1 & \text{for } t = b \end{cases}$$

Each of these functions is increasing. Let $\epsilon > 0$ and choose $\delta > 0$ such that $|\int f \, d\alpha_s - S_{\alpha_s}(f, P)| < \epsilon/2$ whenever $P \in \mathcal{P}_\delta$ and also such that $|f(u) - f(t)| < \epsilon/2$ when $|u - t| < \delta$. Assume $s < b$. (A separate argument is required when $s = b$ and this is left to the reader. See Exercise 6.) Choose a P in $\mathcal{P}_\delta$ that contains s and let s_0 be the point in P that immediately follows s. Using s_0 as the point in $[s, s_0]$ at which to evaluate f, a moment's reflection reveals that $S_{\alpha_s}(f, P) = f(s_0)[\alpha(s_0) - \alpha(s)] = f(s_0)$. Thus $|f(s) - \int f \, d\alpha_s| \le |f(s) - f(s_0)| + |S_{\alpha_s}(f, P) - \int f \, d\alpha_s| < \epsilon$. Since ϵ was arbitrary we have that

$$\int f \, d\alpha_s = f(s)$$

for every continuous function f on J. Similarly, $\int f \, d\alpha_b = f(b)$ for all such f.

Using the preceding example we can calculate the integrals with respect to many functions that only differ from a continuously differentiable one by having jump discontinuities. Consider the following.

3.6.11. Example. Define $\alpha : [0, 1] \to \mathbb{R}$ by $\alpha(t) = t^2$ for $t \le \frac{1}{2}$ and $\alpha(t) = t^2 + 1$ for $t > \frac{1}{2}$. What is $\int_0^1 f(t) \, d\alpha(t)$ for f in $C[0, 1]$? Note that $\alpha(t) = t^2 + \alpha_{\frac{1}{2}}$, where $\alpha_{\frac{1}{2}}$ is defined in the preceding example. So using Proposition 3.6.7(c) and Theorem 3.6.9 we get

$$\int_0^1 f(t) \, d\alpha(t) = \int_0^1 f(t) d(t^2) + \int_0^1 f(t) \, d\alpha_{\frac{1}{2}}(t)$$

$$= 2 \int_0^1 t f(t) dt + f\left(\frac{1}{2}\right)$$

We want to spend a little time studying functions of bounded variation. We begin with the following.

3.6.12. Proposition. *If $\alpha : J \to \mathbb{R}$ is a function of bounded variation, then α has at most a countable number of discontinuities.*

Proof. By Proposition 3.6.4 it suffices to show that the conclusion holds for increasing functions. Hence the result now follows from Proposition 2.1.9. ∎

A function α is called *left-continuous* at c if the left limit $\alpha(c-)$ exists and is equal to $\alpha(c)$. The definition of *right-continuous* is analogous. If $J = [a, b]$ then the left limit of α at a doesn't really make sense, but we will define $\alpha(a-) = \alpha(a)$. Similarly, $\alpha(b+) = \alpha(b)$. So every function α on J is left-continuous at a and right-continuous at b by default. Note that if $a \le s < b$, the function α_s (3.6.10) is left-continuous at s; however, α_b is not left-continuous at b.

3.6.13. Corollary. *If $\alpha : J \to \mathbb{R}$ is increasing and we define $\beta : J \to \mathbb{R}$ by $\beta(t) = \alpha(t-)$, then β is an increasing function that is left-continuous everywhere, has the same discontinuities as α on $[a, b)$, and agrees with α except possibly at its discontinuities.*

Proof. It is clear that β is increasing and left-continuous at each point of J. If c is a point of discontinuity of α, then $\beta(c) = \alpha(c-) < \alpha(c+)$. If $c < b$, then since there are points of continuity for α that approach c from the right, we have that $\beta(c+) = \alpha(c+) > \beta(c)$ and β is discontinuous at c. If $c = b$, then $\beta(b) = \alpha(b-) = \beta(b+)$, so that β is continuous at b irrespective of whether α is continuous at b. (See the function α_b defined in Example 3.6.10.) ∎

Note that the integral of a continuous function with respect to a constant function is 0; hence $\int f d(\alpha + c) = \int f \, d\alpha$ for all continuous functions f. There are other ways that we can produce two functions of bounded variation that yield the same integral. (Compare the function defined in Exercise 3 with the functions in Example 3.6.10.) We want to examine when $\int f \, d\alpha = \int f d\beta$ for two fixed functions of bounded variation and for every continuous function f on J.

For a function of bounded variation α on J and $a \leq t \leq b$, define

3.6.14 $$\widetilde{\alpha}(t) = \alpha(t-) - \alpha(a) + [\alpha(b) - \alpha(b-)]\alpha_b$$

where α_b is defined in Example 3.6.10. Call $\widetilde{\alpha}$ the *normalization* of α. The first thing to observe is that $\alpha(t) - \widetilde{\alpha}(t) = \alpha(a)$ except at the points where α is discontinuous. Also if α is increasing, so is its normalization. Furthermore, if $\alpha = \alpha_+ - \alpha_-$, then $\widetilde{\alpha} = \widetilde{\alpha}_+ - \widetilde{\alpha}_-$. Therefore $\widetilde{\alpha}$ is also a function of bounded variation. Finally note that if α is continuous at b, then $\widetilde{\alpha}(t) = \alpha(t-) - \alpha(a)$.

3.6.15. Proposition. *If α is a function of bounded variation on J and $\widetilde{\alpha}$ is its normalization, then $\int f \, d\alpha = \int f d\widetilde{\alpha}$ for every continuous function f on J.*

Proof. We split this into two cases.

Case 1. α is continuous at b.

Let D be the set of points in J where α is discontinuous – a countable set. Here $\widetilde{\alpha}(t) = \alpha(t-) - \alpha(a)$; fix $\epsilon > 0$ and f in $C([a, b])$. Let $\delta > 0$ such that $|S_\alpha(f, P) - \int f \, d\alpha| < \epsilon/2$ and $|S_{\widetilde{\alpha}}(f, P) - \int f d\widetilde{\alpha}| < \epsilon/2$ whenever $P \in \mathcal{P}_\delta$. Since D is a countable set in $[a, b]$, we can choose $P = \{a = x_0 < \cdots < x_n = b\}$ in $\mathcal{P}_\delta$ such that $x_j \notin D$ for $0 < j \leq n$. We have by definition

$$S_\alpha(f, P) = f(a)[\alpha(x_1) - \alpha(a)] + \sum_{j=2}^{n} f(x_j)[\alpha(x_j) - \alpha(x_{j-1})]$$

and

$$S_{\widetilde{\alpha}}(f, P) = f(a)[\widetilde{\alpha}(x_1) - \widetilde{\alpha}(a)] + \sum_{j=2}^{n} f(x_j)[\widetilde{\alpha}(x_j) - \widetilde{\alpha}(x_{j-1})]$$

For $0 < j \leq n, x_j \notin D$ and so $\alpha(x_j) = \widetilde{\alpha}(x_j) + \alpha(a)$. Since $\widetilde{\alpha}(a) = 0$, we also have $\alpha(a) = \widetilde{\alpha}(a) + \alpha(a)$; thus $\alpha(x_j) - \alpha(x_{j-1}) = \widetilde{\alpha}(x_j) - \widetilde{\alpha}(x_{j-1})$ for $1 \leq j \leq n$. That is, $S_\alpha(f, P) = S_{\widetilde{\alpha}}(f, P)$. Therefore $|\int f \, d\alpha - \int f d\widetilde{\alpha}| < \epsilon$. Since ϵ was arbitrary, this proves Case 1.

Case 2. α is discontinuous at b.

Here $\widetilde{\alpha}(t) = \alpha(t-) + [\alpha(b) - \alpha(b-)]\alpha_b$. Consider the function $\beta = \alpha - [\alpha(b) - \alpha(b-)]\alpha_b$. It follows that β is continuous at b since $\beta(b-) = \alpha(b-) = \beta(b)$. By Case 1, $\int f d\beta = \int f d\widetilde{\beta}$. Moreover since $\alpha_b(t-) = 0$ for all t in J, including $t = b$, it follows that $\widetilde{\beta}(t) = \alpha(t-) = \widetilde{\alpha}(t) - [\alpha(b) - \alpha(b-)]\alpha_b(t)$. Now $\int f \, d\alpha - [\alpha(b) - \alpha(b-)]f(b) = \int f d\beta = \int f d\widetilde{\beta} = \int f d\widetilde{\alpha} - [\alpha(b) - \alpha(b-)]f(b)$. After canceling we get that $\int f \, d\alpha = \int f d\widetilde{\alpha}$. ∎

The converse of the above result holds. To be precise, if α and β are two functions of bounded variation on the interval J, then $\int f \, d\alpha = \int f d\beta$ for every continuous function f on J if and only if $\widetilde{\alpha} = \widetilde{\beta}$. The proof of this requires more advanced techniques and can be found in [3], Proposition 4.5.3.

Exercises

We continue to assume that $J = [a, b]$ unless the interval is otherwise specified.

(1) Show that a function of bounded variation is a bounded function.
(2) Show that the function $x^2 \sin(x^{-1})$ is not of bounded variation.
(3) Define α on the unit interval $[0, 1]$ by $\alpha(\frac{1}{2}) = 1$ and $\alpha(t) = 0$ when $t \neq \frac{1}{2}$. Observe that α is neither left nor right-continuous at $\frac{1}{2}$. Show that α is of bounded variation and find increasing functions $\alpha_\pm$ such that $\alpha = \alpha_+ - \alpha_-$. Compute $\int f \, d\alpha$ for an arbitrary continuous function f on the unit interval. Are you surprised?
(4) If α is an increasing function on J such that $\int f \, d\alpha = 0$ for every continuous function f on J, show that α is constant. Contrast this with Exercise 3. (Hint: First show that you can assume that $\alpha(a) = 0$, then show that α must be continuous at each point of J. Use Example 3.6.10. Now show that α is identically 0.)
(5) Give the details of the proof of Proposition 3.6.7
(6) If α_b is defined as in Example 3.6.10, show that $\int f \, d\alpha_b = f(b)$ for every continuous function f on J.
(7) Suppose a left-continuous increasing function $\alpha : J \to \mathbb{R}$ has a discontinuity at t_0 and $a_0 = \alpha(t_0+) - \alpha(t_0-)$. Let α_{t_0} be the increasing function defined as in Example 3.6.10: $\alpha_{t_0}(t) = 0$ for $t \leq t_0$ and $\alpha_{t_0}(t) = 1$ for $t > t_0$. (Once again if t_0 is the right-hand endpoint of J, a separate argument is required.) (a) Show that $\alpha - a_0\alpha_{t_0}$ is an increasing function that is continuous at t_0. (b) Show that any left-continuous increasing function α on J can be written as $\alpha = \beta + \gamma$, where both β and γ are increasing, β is continuous, and γ has the property that if γ is continuous on the open subinterval (c, d), then γ is constant there.

(8) If γ is an increasing function with the property of γ in Exercise 7, calculate $\int f d\gamma$ for any continuous function f on J.

(9) If A is any countable subset of J, show that there is an increasing function α on J such that A is precisely the set of discontinuities of α. (So, in particular, there is an increasing function on J with discontinuities at all the rational numbers in J.)

(10) Let $\{r_n\}$ denote the set of all rational numbers in the interval J and define $\alpha : J \to \mathbb{R}$ by $\alpha(t) = \sum \frac{1}{2^n}$ where the sum is taken over all n such that $r_n < t$. (a) Show that α is a strictly increasing function that is left-continuous and satisfies $\alpha(a) = 0$ and $\alpha(b) = 1$. (b) Show that α is continuous at each irrational number and discontinuous at all the rational numbers in J.

(11) If α is an increasing function, discuss the possibility of defining $\int f \, d\alpha$ when f has a discontinuity.

(12) Suppose $\alpha : [0, \infty) \to \mathbb{R}$ is an increasing function. Is it possible to define $\int_0^\infty f \, d\alpha$ for a continuous function $f : [0, \infty) \to \mathbb{R}$?

4 Sequences of Functions

In this chapter we present some basic results in analysis that are used throughout the theory and its applications. We also start a process of increased sophistication that eventually migrates to thinking of functions as points in a larger space. As the reader progresses, (s)he will notice a similarity between various aspects of our discussion of sequences of functions and sequences of real numbers. This similarity is no coincidence as the next chapter in this book will reveal.

4.1. Uniform Convergence

In this section we discuss a notion of convergence for sequences and series of functions.

4.1.1. Definition. If $\{f_n\}$ is a sequence of functions from a subset X of $\mathbb{R}$ into $\mathbb{R}$, say that $\{f_n\}$ *converges uniformly* to a function f if for every $\epsilon > 0$ there is an N such that $|f_n(x) - f(x)| < \epsilon$ for all x in X and all $n \geq N$. In symbols this is written as $f_n \to_u f$ on X.

In other words, $f_n \to_u f$ if for every $\epsilon > 0$ it is possible to find an N such that $|f_n(x) - f(x)| < \epsilon$ when $n \geq N$, where the same N works for all x at the same time. We are interested in examining the properties of functions that are preserved under taking uniform limits. That is, if $f_n \to_u f$ on X and each f_n has some property, does f have this same property? We'll see several examples of this shortly. There is, of course, another notion of convergence of a sequence of functions on a subset X of $\mathbb{R}$. Namely, we could investigate what happens when $f_n(x) \to f(x)$ for every x in X. This is called *pointwise convergence*. This has some value but not as much as uniform convergence, as we'll see below in Example 4.1.7 as well as in some of the exercises. Observe that when $f_n \to_u f$, pointwise convergence follows.

4.1.2. Example. (a) Suppose $0 < a < 1$ and $f_n : [0, a] \to \mathbb{R}$ is defined by $f_n(x) = x^n$. Since $|f_n(x)| \leq a^n$ for all x in $[0, a]$ and $a^n \to 0$, we have that $\{f_n\}$ converges uniformly on $[0, a]$ to the constantly zero function. See Exercise 1.

(b) If $X = [0, 1]$ and $f_n : X \to \mathbb{R}$ is defined by $f_n(x) = \frac{1}{n}x^n$, then $f_n(x) \to_u 0$ on X. In fact if $\{g_n\}$ is any sequence of functions on any subset X in $\mathbb{R}$ such that there is a constant M with $|g_n(x)| \leq M$ for all x in X, and $\{a_n\}$ is a sequence of real numbers converging to 0, then $a_n g_n \to_u 0$ on X.

(c) If $X = [0, 1]$ and $f_n(x) = x^n$, then for each x in $[0, 1]$ $f_n(x) \to f(x)$, where f is the function defined by $f(x) = 0$ when $0 \le x < 1$ and $f(1) = 1$. In this case $\{f_n\}$ does not converge uniformly to f. This can be shown directly (Exercise 3), but an easier way is to use what is contained in Example 4.1.7 below.

The next result is often useful in showing that a sequence of functions converges uniformly.

4.1.3. Proposition. *If $\{f_n\}$, $\{g_n\}$, and $\{h_n\}$ are three sequences of functions on X such that $g_n(x) \le f_n(x) \le h_n(x)$ for all x in X and there is a function $f : X \to \mathbb{R}$ such that $g_n \to_u f$ and $h_n \to_u f$ on X, then $f_n \to_u f$ on X.*

Proof. Let $\epsilon > 0$ and choose N such that $|g_n(x) - f(x)| < \epsilon$ and $|h_n(x) - f(x)| < \epsilon$ for all x in X and all $n \ge N$. It follows that for each x in X and all $n \ge N$, $f(x) - \epsilon < g_n(x) \le f_n(x) \le h_n(x) < f(x) + \epsilon$. Hence $|f_n(x) - f(x)| < \epsilon$ for all x in X and $n \ge N$. By definition, $f_n \to_u f$. ∎

The preceding proposition is often called the *Squeeze Principle* or *Sandwich Principle*. It is actually what is happening in Example 4.1.2, parts (a) and (b). Here is another use of the Sandwich Principle.

4.1.4. Example. Define $f_n : [-1, 1] \to \mathbb{R}$ by $f_n(x) = \sqrt{x^2 + n^{-2}}$. Note that $|x|^2 \le f_n(x)^2 \le |x|^2 + \frac{1}{n}^2$ on $[-1, 1]$. So if we let $g_n(x) = |x|$ and $h_n(x) = |x| + \frac{1}{n}$, we have that for all $n \ge 1$, $g_n(x) \le f_n(x) \le h_n(x)$ on $[-1, 1]$. By the preceding proposition, $f_n(x) \to |x|$ uniformly on $[-1, 1]$.

4.1.5. Proposition. *Let $X \subseteq \mathbb{R}$. If $\{f_n\}$ is a sequence of bounded functions on X that converges uniformly to a function $f : X \to \mathbb{R}$, then f is a bounded function and the sequence of functions $\{f_n\}$ is uniformly bounded; that is, there is a constant M such that $|f_n(x)| \le M$ for all x in X and all $n \ge 1$.*

Proof. Let N be such that $|f_n(x) - f(x)| < 1$ for all x in X and all $n \ge N$. If L_N is a number such that $|f_N(x)| \le L_N$ for all x, then for each x in X, $|f(x)| \le |f(x) - f_N(x)| + |f_N(x)| \le 1 + L_N$. Hence f is a bounded function. Note that if $n \ge N$ and $x \in X$, then $|f_n(x)| \le |f_n(x) - f(x)| + |f(x)| \le 2 + L_N$. If $1 \le n \le N$ and $|f_n(x)| \le L_n$ for all x in X, put $M = \max\{L_1, \ldots, L_N, 2 + L_N\}$. It follows that $|f_n(x)| \le M$ for all x in X and all $n \ge 1$. ∎

Also see Exercise 6.

4.1.6. Theorem. *Let $X \subseteq \mathbb{R}$. If $\{f_n\}$ is a sequence of bounded continuous functions on X and $f_n \to_u f$ on X, then $f : X \to \mathbb{R}$ is a continuous function.*

Proof. Fix an arbitrary point a in X; we want to show that f is continuous at a. If $\epsilon > 0$, then by uniform convergence we can choose an integer n such that $|f(x) - f_n(x)| < \epsilon/3$ for all x in X. Since f_n is continuous there is a $\delta > 0$ such that $|f_n(x) - f_n(a)| < \epsilon/3$ when $x \in X$ and $|x - a| < \delta$. Thus for any x in X with $|x - a| < \delta$, $|f(x) - f(a)| \le |f(x) - f_n(x)| + |f_n(x) - f_n(a)| + |f_n(a) - f(a)| < 3(\epsilon/3) = \epsilon$. By definition, f is continuous at a. ∎

4.1.7. Example. If the functions f_n are as in Example 4.1.2(c), the fact that the limit function f is not continuous shows, in light of the preceding result, that $\{f_n\}$ does not converge uniformly on [0, 1] to f.

4.1.8. Theorem. *Let $X \subseteq \mathbb{R}$. If $\{f_n\}$ is a sequence of bounded uniformly continuous functions on X and $f_n \rightarrow_u f$ on X, then $f : X \to \mathbb{R}$ is a uniformly continuous function.*

Proof. This proof is similar to the proof of the preceding result but with a critical difference. If $\epsilon > 0$, then by the uniform convergence we can choose an integer n such that $|f(x) - f_n(x)| < \epsilon/3$ for all x in X. Since f_n is uniformly continuous, there is a $\delta > 0$ such that $|f_n(x) - f_n(y)| < \epsilon/3$ whenever $|x - y| < \delta$. Thus for $|x - y| < \delta$ we have that $|f(x) - f(y)| \le |f(x) - f_n(x)| + |f_n(x) - f_n(y)| + |f(n(y) - f(y)| < 3(\epsilon/3) = \epsilon$. By definition, f is uniformly continuous on X. ∎

4.1.9. Theorem. *If $\{f_n\}$ is a sequence of bounded integrable functions on the interval $[a, b]$ and $f_n \rightarrow_u f$, then f is integrable and*

$$\int_a^b f_n \to \int_a^b f$$

Proof. By Proposition 4.1.5 there is a constant M such that $|f_n(x)| \le M$ for all x in $[a, b]$ and all $n \ge 1$. If $\epsilon > 0$, then by the uniform convergence we can find an integer N such that $|f(x) - f_n(x)| < \epsilon/[9(b - a)]$ for all x in $[a, b]$ and all $n \ge N$. Temporarily fix an $n \ge N$. Since f_n is integrable there is a partition $P = \{a = x_0 < x_1 < \cdots < x_n = b\}$ with $U(f_n, P) - L(f_n, P) < \epsilon/9$. Let $M_j(f) = \sup\{f(x) : x_{j-1} \le x \le x_j\}$ and $M_j(f_n) = \sup\{f_n(x) : x_{j-1} \le x \le x_j\}$. Since $|f(x) - f_n(x)| < \epsilon/[9(b - a)]$ for all x, it follows that $|M_j(f) - M_j(f_n)| \le \epsilon/9(b - a)$ for $1 \le j \le n$. (Verify!) Thus $|U(f, P) - U(f_n, P)| \le \epsilon/9$. Similarly $|L(f, P) - L(f_n, P)| \le \epsilon/9$. Therefore

$$U(f, P) - L(f, P) \le |U(f, P) - U(f_n, P)| + U(f_n, P) - L(f_n, P)$$
$$+ |L(f, P) - L(f_n, P)|$$
$$\le \epsilon/3$$

and so f is integrable. By Proposition 3.1.4, $|\int_a^b f - U(f, P)| \le \epsilon/3$. Similarly $|\int_a^b f_n - U(f_n, P)| \le \epsilon/3$. Hence

$$\left| \int_a^b f - \int_a^b f_n \right| \le \left| \int_a^b f - U(f, P) \right| + |U(f, P) - U(f_n, P)|$$

$$+ \left| U(f_n, P) - \int_a^b f_n \right|$$

$$\le \epsilon$$

Since n was an arbitrarily fixed integer greater than N, this implies $\int_a^b f_n \to \int_a^b f$. ∎

It is a bit unfortunate that uniform convergence has no respect for differentiability as it did for integrability in the preceding theorem. See Example 4.1.11 below. Here is the best we can do.

4.1.10. Proposition. *If $\{f_n\}$ is a sequence of continuously differentiable functions on $[a, b]$, $f_n \to_u f$ on $[a, b]$, and there is a function $g : [a, b] \to \mathbb{R}$ such that $f'_n \to_u g$ on $[a, b]$, then f is continuously differentiable and $f' = g$.*

Proof. This is a consequence of the FTC. Note that $f_n(x) = \int_a^x f'_n + f_n(a)$; also note that Theorem 4.1.6 implies that g is a continuous function on $[a, b]$. By the preceding theorem, $\int_a^x f'_n \to \int_a^x g$ for every x in $[a, b]$. By hypothesis, this implies that $f(x) = \lim_n f_n(x) = \int_a^x g + f(a)$. Again using the FTC, we have that f is differentiable and $f' = g$. ∎

4.1.11. Example. (a) For each $n \geq 1$ define $f_n : [0, 2\pi] \to \mathbb{R}$ by $f_n(x) = \frac{1}{n} \sin(nx)$. It is easy to see that $f_n \to_u 0$ on $[0, 2\pi]$. However $f'_n(x) = \cos(nx)$, so that $\{f'_n\}$ does not converge uniformly even though the limit function of the original sequence is continuously differentiable.

(b) Examine the sequence of functions $\{f_n\}$ on $[-1, 1]$ in Example 4.1.4. This shows that when a sequence $\{f_n\}$ of continuously differentiable functions converges uniformly, it does not follow that the limit function is differentiable.

Exercises

(1) Let $f_n : X \to \mathbb{R}$ and suppose there is a real number a_n with $|f_n(x)| \leq a_n$ for all x in X and each $n \geq 1$. Show that if $a_n \to 0$, then $f_n \to_u 0$.

(2) Find a sequence of bounded functions $\{f_n\}$ on $[0, 1]$ and a function $f : [0, 1] \to \mathbb{R}$ such that $f_n(x) \to f(x)$ for all x in $[0, 1]$, but where f is not a bounded function.

(3) Using only the definition of uniform convergence, show that the sequence $\{f_n\}$ in Example 4.1.2(c) does not converge uniformly to f.

(4) Let $g : [0, 1] \to \mathbb{R}$ be a continuous function with $g(1) = 0$. Show that $x^n g(x) \to 0$ uniformly on $[0, 1]$.

(5) Say that $\{f_n\}$ is a *uniformly Cauchy sequence* on a subset X of $\mathbb{R}$ if each f_n is bounded and for every $\epsilon > 0$ there is an integer N such that $|f_n(x) - f_m(x)| < \epsilon$ for all x in X and all $m, n \geq N$. (a) Show that if $\{f_n\}$ is a uniformly Cauchy sequence, then there is a constant M such that $|f_n(x)| \leq M$ for all x in X and all $n \geq 1$. (b) Show that if $\{f_n\}$ is a uniformly Cauchy sequence on X, then there is a function $f : X \to \mathbb{R}$ such that $f \to_u f$ on X.

(6) If $f_n : X \to \mathbb{R}$, $f_n \to_u f$ on X, and f is a bounded function, show that there is a constant M and an integer N such that $|f_n(x)| \leq M$ for all x in X and all $n \geq N$.

(7) For each $n \geq 1$ let $f_n : [-1, 1] \to \mathbb{R}$ be a continuous function with $f_n(x) \geq 0$ and assume that $\lim_n \int_{-1}^1 f_n = 1$; also assume that $g : [-1, 1] \to \mathbb{R}$ is a continuous function. If for each $c > 0$, $f_n \to_u 0$ on $[-1, -c] \cup [c, 1]$, show that $\lim_n \int_{-1}^1 g f_n = g(0)$.

4.2. Power Series

We start with a definition that is a bit more general than the topic of this section.

4.2.1. Definition. If $X \subseteq \mathbb{R}$ and for each $n \geq 1$ there is a function $f_n : X \to \mathbb{R}$, then say that the series of functions $\sum_{n=1}^{\infty} f_n$ converges *uniformly* on X if the sequence of partial sums $\{\sum_{k=1}^{n} f_k\}$ converges uniformly on X.

The reader can go through the last section and state and prove a collection of results about series of functions. For example if each function f_n is continuous on X and $f(x) = \sum_{n=1}^{\infty} f_n(x)$ converges uniformly on X, then f is a continuous function on X. We will use such results as we proceed without reference. Instead we will present one result that implies uniform convergence and then proceed with the development of power series.

4.2.2. Theorem (Weierstrass[1] M-Test). *Let $X \subseteq \mathbb{R}$ and for each $n \geq 1$ suppose $f_n : X \to \mathbb{R}$ and there is a constant M_n with $|f_n(x)| \leq M_n$ for all x in X. If $\sum_{n=1}^{\infty} M_n$ converges, then $\sum_{n=1}^{\infty} f_n$ converges uniformly on X.*

Proof. Put $M = \sum_{n=1}^{\infty} M_n$ and let $\epsilon > 0$. By Proposition 1.4.5 we can choose N such that $\sum_{k=m-1}^{n} M_k < \epsilon$ whenever $n > m \geq N$. If $F_n(x) = \sum_{k=1}^{n} f_k(x)$ and $n > m > N$, then for every x in X, $|F_n(x) - F_m(x)| \leq \sum_{k=m-1}^{n} |F_k(x)| \leq \sum_{k=m-1}^{n} M_k < \epsilon$ for $n > m > N$. This says that the sequence $\{F_n(x)\}$ is a Cauchy sequence in $\mathbb{R}$. Thus $F(x) = \lim_n F_n(x)$ exists and defines a function $F : X \to \mathbb{R}$. Now $|F(x) - F_m(x)| \leq |F(x) - F_n(x)| + |F_n(x) - F_m(x)|$. So if x is any point in X and $n > m > N$, we have that $|F(x) - F_m(x)| \leq |F(x) - F_n(x)| + \epsilon$. If we hold m fixed but larger than N and let $n \to \infty$, this shows that $|F(x) - F_m(x)| \leq \epsilon$ for all x in X and all $m > N$. That is, $F_m \to_u F$ on X. ∎

Now we come to the focus of this section.

4.2.3. Definition. If $c \in \mathbb{R}$, a *power series* about the point c is an infinite series of the form

$$\sum_{n=0}^{\infty} a_n(x - c)^n$$

where $\{a_n\}$ is some sequence of real numbers.

We note two things about a power series. First, the infinite sum begins with $n = 0$. This is no big deal and, indeed, some power series will begin the sum with $n = 1$. Just be conscious of this. Second is the "center" of the power series, c. If we translate the center to another number, we don't change any of the convergence properties of the power series except for those that involve the center. Thus the basic properties of a power series depend more on the coefficients $\{a_n\}$ than the center. In fact many proofs about an arbitrary power series will be proved under the assumption that

[1] See the footnote to Theorem 1.3.11 for a biographical note.

$c = 0$, and many examples will have $c = 0$. An example of a power series about 0 is the geometric series (1.4.4), in which case each $a_n = 1$. Another is the series $\sum_{n=0}^{\infty} x^n/n!$ (1.4.15), where $a_n = 1/n!$. The first result we prove about power series is the fundamental fact that will form the basis for all that follows on this topic.

4.2.4. Theorem. *For a given power series $\sum_{n=0}^{\infty} a_n(x - c)^n$ define the extended real number R, $0 \le R \le \infty$, by*

$$\frac{1}{R} = \limsup |a_n|^{\frac{1}{n}}$$

(a) *If $|x - c| < R$, the series converges absolutely.*
(b) *If $|x - c| > R$, the series diverges.*
(c) *If $0 < r < R$, the series converges uniformly on $\{x : |x - c| \le r\}$.*
(d) *R is the unique number having properties* (a) *and* (b).

Proof. Without loss of generality we may assume that $c = 0$.

(a) and (b). Note that $\limsup |a_n x^n|^{\frac{1}{n}} = |x| \limsup |a_n|^{\frac{1}{n}} = |x|R^{-1}$. Thus both (a) and (b) follow by the Root Test (1.4.12).

(c) If $|x| \le r$, then $|a_n x^n| \le |a_n r^n| = M_n$. By (b), $\sum_{n=0}^{\infty} M_n$ converges, so the Weierstrass M-Test implies $\sum_{n=0}^{\infty} a_n x^n$ converges uniformly for x in $\{x : |x| \le r\}$.

(d) This is routine and left to the reader. In fact it is more of an observation than a result. ∎

The number R obtained in the last theorem is called the *radius of convergence* of the power series.

4.2.5. Example. (a) The radius of convergence of the geometric series is $R = 1$ since each coefficient $a_n = 1$.

(b) The power series $\sum_{n=0}^{\infty} x^n/n!$ has radius of convergence $R = \infty$. To see this it is better not to use the formula for R. Instead we invoke Example 1.4.15 where it is shown as a consequence of the Ratio Test that the series converges for all x in $\mathbb{R}$. By part (d) of the theorem, $R = \infty$.

(c) Here is an example that says you have to be a little careful when using the Ratio Test to determine the radius of convergence. Consider the series

$$\sum_{n=1}^{\infty} \frac{(-1)^n}{n2^n} x^{2n}$$

Because there are "gaps" in the powers of x we cannot apply the Ratio Test directly. If however we let $y = x^2$, we get the series $\sum_{n=1}^{\infty} \frac{(-1)^n}{n2^n} y^n$. Now we can use the Ratio Test and we get that

$$\left| \frac{\frac{(-1)^{n+1}}{(n+1)2^{n+1}} y^{n+1}}{\frac{(-1)^n}{n2^n} y^n} \right| = \left| \frac{n2^n}{(n+1)2^{n+1}} \right| |y| \to \frac{1}{2}|y|$$

So the original power series converges when $\frac{1}{2}|x|^2 < 1$ and so the radius of convergence is $\sqrt{2}$. See Exercise 1.

4.2.6. Theorem. *Assume the power series $\sum_{n=0}^{\infty} a_n(x-c)^n$ has positive radius of convergence $R > 0$. If $f : (c - R, c + R) \to \mathbb{R}$ is defined by $f(x) = \sum_{n=0}^{\infty} a_n(x - c)^n$, then the following hold.*

(a) *The function f is infinitely differentiable, and for $|x - c| < R$ and $k \geq 0$*

4.2.7
$$f^{(k)}(x) = \sum_{n=k}^{\infty} n(n-1) \cdots (n - k + 1)a_n(x - c)^{n-k}$$

(b) *The radius of convergence of the power series (4.2.7) is also R.*
(c) *For every $n \geq 0$*

$$a_n = \frac{1}{n!} f^{(n)}(c)$$

Proof. Begin by observing that part (c) is an immediate consequence of (a). To prove (a) and (b) we again assume that $c = 0$ and start by establishing the following.

Claim. $\limsup |a_n|^{1/(n-1)} = R^{-1}$.

Put R_1 equal to the the reciprocal of the lim sup in the claim so that R_1 is the radius of convergence of the power series $\sum_{n=1}^{\infty} a_n x^{n-1} = \sum_{n=0}^{\infty} a_{n+1} x^n$. Notice that $x \sum_{n=0}^{\infty} a_{n+1} x^n + a_0 = \sum_{n=0}^{\infty} a_n x^n$. Hence when $|x| < R_1$ we have that $\sum_{n=0}^{\infty} |a_n x^n| = |a_0| + |x| \sum_{n=0}^{\infty} |a_{n+1}||x|^n < \infty$. By the uniqueness of the radius of convergence this shows that $R_1 \leq R$. On the other hand, if $0 < |x| < R$, then $\sum_{n=1}^{\infty} |a_n||x|^{n-1} = |x|^{-1} \sum_{n=1}^{\infty} |a_n||x|^n < \infty$. Hence $R \leq R_1$, establishing the claim.

If (a) and (b) are established for $k = 1$, then, by applying them to the power series $\sum_{n=0}^{\infty} n a_n x^n$, we will obtain (a) and (b) for $k = 2$. By repeating this we arrive at the validity of (a) and (b) for an arbitrary $k \geq 0$. So to prove (a) and (b) we can restrict our attention to the case where $k = 1$. We start this by showing that $R^{-1} = \limsup |n a_n|^{1/(n-1)}$. Now L'Hôpital's rule shows that $\lim_{n \to \infty} (\log n)/(n - 1) = 0$. Hence $n^{1/(n-1)} = \exp[(\log n)/(n - 1)] \to 1$. By Exercise 1.3.16 and the claim, $\limsup |n a_n|^{1/(n-1)} = \limsup |a_n|^{1/(n-1)} = R^{-1}$, giving (b).

To get (a) when $k = 1$ define the functions g, s_n, and R_n on the interval $(-R, R)$ by

$$g(x) = \sum_{n=1}^{\infty} n a_n x^{n-1}, \quad s_n(x) = \sum_{k=0}^{n} a_n x^n, \quad R_n(x) = \sum_{k=n+1}^{\infty} a_n x^n$$

(The definition of g is valid by part (b).) Fix a number y with $|y| < R$; we want to show that $f'(y)$ exists and $f'(y) = g(y)$. Choose r with $|y| < r < R$ and let $\delta > 0$ be arbitrary except that $(y - \delta, y + \delta) \subseteq (-r, r)$. (Later in this proof we will impose an additional restriction on δ.) Let $|x - y| < \delta$. For any integer $n \geq 1$ (this integer will be specified later) we can write

$$\left| \frac{f(x) - f(y)}{x - y} - g(y) \right| \leq \left| \frac{s_n(x) - s_n(y)}{x - y} - s_n'(y) \right| + |s_n'(y) - g(y)|$$
$$+ \left| \frac{R_n(x) - R_n(y)}{x - y} \right|$$

Let $\epsilon > 0$. We want to show that if we take δ sufficiently small and n sufficiently large, we can make each of the summands on the right side of the preceding inequality less than $\frac{\epsilon}{3}$. This will prove that f is differentiable and $f'(y) = g(y)$. The most cumbersome of these three terms to handle is the last, so let's tackle it first. For any choice of n

$$\frac{R_n(x) - R_n(y)}{x - y} = \frac{1}{x - y} \sum_{k=n+1}^{\infty} a_k(x^k - y^k)$$
$$= \sum_{k=n+1}^{\infty} a_k \left(\frac{x^k - y^k}{x - y} \right)$$

But

$$\left| \frac{x^k - y^k}{x - y} \right| = |x^{k-1} + x^{k-2}y + \cdots + y^{k-1}| \leq kr^{k-1}$$

so that

$$\left| \frac{R_n(x) - R_n(y)}{x - y} \right| \leq \sum_{k=n+1}^{\infty} |a_k| kr^{k-1}$$

Now since $r < R$, (b) implies that $\sum_{k=1}^{\infty} |a_k| kr^{k-1}$ converges. Thus there is an integer N_1 such that for all $n \geq N_1$ and for any x in the interval $(-r, r)$,

$$\left| \frac{R_n(x) - R_n(y)}{x - y} \right| \leq \frac{\epsilon}{3}$$

Now $\{s_n'(y)\}$ is just the sequence of partial sums of the series $g(y)$. So there is an integer N_2 such that when $n \geq N_2$, $|s_n'(y) - g(y)| < \frac{\epsilon}{3}$.

Let $n = \max\{N_1, N_2\}$. (So we have now fixed n.) Because s_n is a polynomial and differentiable, there is a $\delta > 0$ such that

$$\left| \frac{s_n(x) - s_n(y)}{x - y} - s_n'(y) \right| < \frac{\epsilon}{3}$$

when $|x - y| < \delta$. (Be aware that the choice of δ depends on the fixed value of n.)

With these choices we have that

$$\left| \frac{f(x) - f(y)}{x - y} - g(y) \right| < \epsilon$$

when $|x - y| < \delta$. This completes the proof of (a) and of the theorem. ∎

4.2.8. Example. (a) Theorem 4.2.6 says that when a function is defined by a power series it is infinitely differentiable. The converse of this is not true. In fact in Example 3.3.11 we saw that the function defined by $f(x) = e^{-\frac{1}{x^2}}$ when $x \neq 0$ and $f(0) = 0$ is infinitely differentiable on $\mathbb{R}$ with $f^{(n)}(0) = 0$ for all $n \geq 0$. By (4.2.6(c)) we have that this function cannot be written as a power series centered at $c = 0$.

(b) The function e^x can be written as a power series. In fact

$$e^x = \sum_{n=0}^{\infty} \frac{x^n}{n!}$$

This can be seen as follows. In Example 4.2.5(b) we saw that this power series has radius of convergence $R = \infty$. If $f(x)$ denotes this power series, then Theorem 4.2.6 implies

$$f'(x) = \sum_{n=0}^{\infty} n \frac{x^{n-1}}{n!} = \sum_{n=1}^{\infty} \frac{x^{n-1}}{(n-1)!} = \sum_{n=0}^{\infty} \frac{x^n}{n!} = f(x)$$

By Proposition 3.3.12 there is a constant c such that $f(x) = ce^x$. Since $f(0) = 1 = e^0$, this constant c must be 1.

When we know that a function can be represented as a power series, this gives us additional power. In fact Theorem 4.2.6 can be phrased as saying that the derivative of a power series is obtained by differentiating the individual terms of the power series. (This is often referred to as differentiating term-by-term.) Also since the convergence of the power series is uniform on any closed interval contained in $(c - R, c + R)$, Theorem 4.1.9 says we can find the integral of this function over this interval by integrating the terms of the series. Since simply being infinitely differentiable does not guarantee the representation as a power series, we seek a criterion to show that such a function can be so represented. The primary tool for this is Taylor's Theorem (2.5.6).

4.2.9. Theorem. *Let $a < b$ and suppose $f \in C^{(\infty)}(a, b)$. If there is a constant M such that $|f^{(n)}(x)| \leq M^n$ for all x in (a, b) and all $n \geq 1$, then for any c in (a, b)*

$$f(x) = \sum_{n=0}^{\infty} \frac{f^{(n)}(c)}{n!}(x - c)^n$$

and this power series has radius of convergence R with $R \geq \min\{c - a, b - c\}$.

Proof. Let $r = \min\{c - a, b - c\}$, and for each $n \geq 1$ let

$$P_n(x) = \sum_{k=0}^{n-1} \frac{f^{(k)}(c)}{k!}(x - c)^k$$

By Taylor's Theorem (2.5.6) for each x in $(c - r, c + r)$ there is a point d between c and x such that

$$f(x) - P_n(x) = \frac{f^{(n)}(d)}{n!}(x - c)^n$$

By the hypothesis this implies

$$|f(x) - P_n(x)| \le r^n \frac{M^n}{n!}$$

Now an application of the Ratio Test reveals that $\sum_{n=1}^{\infty} r^n \frac{M^n}{n!}$ converges (Verify!) and so the n-th term $r^n \frac{M^n}{n!} \to 0$. Therefore $P_n(x) \to f(x)$ whenever $c - r < x < c + r$, and so f has the power series expansion with radius of convergence at least r. ∎

The preceding theorem won't cover all cases, but it will cover many that are of interest. Let's look at a few.

4.2.10. Example. (a) The function $\sin x$ has the power series expansion

$$\sin x = \sum_{n=0}^{\infty} \frac{(-1)^n}{(2n + 1)!} x^{2n+1}$$

The even derivatives of the sine function are $\pm \sin x$, so that $f^{(2n)}(0) = 0$ for all $n \ge 0$. The odd derivatives are $\pm \cos x$, so that $f^{(2n+1)}(0) = \pm 1$ for all $n \ge 0$. Thus we can apply the preceding theorem with $M = 1$. The exact coefficients that appear in the above power series are determined by finding the correct formula for $f^{(2n+1)}(0)$. The reader can fill in the details.

(b) The function $\cos x$ has the power series expansion

$$\cos x = \sum_{n=0}^{\infty} \frac{(-1)^n}{(2n)!} x^{2n}$$

See Exercise 6.

(c) The natural logarithm has the power series expansion

$$\log x = \sum_{n=1}^{\infty} \frac{(-1)^{n+1}}{n}(x - 1)^n$$

and the radius of convergence is 1. Note that here we are finding the power series expansion centered at $c = 1$. A small induction argument shows that when $f(x) = \log x$ and $n \ge 1$

$$f^{(n)}(x) = (-1)^{n+1}(n - 1)! x^{-n}$$

So $|f^{(n)}(x)|$ cannot be bounded as required in Theorem 4.2.9. We have to do something different. Here we have that for $|x-1|<1$

$$\log x = \int_1^x \frac{1}{t}\,dt$$

$$= \int_1^x \frac{1}{1-(1-t)}\,dt$$

$$= \int_1^x \sum_{n=0}^{\infty}(1-t)^n\,dt$$

Now the geometric series under the integral sign converges uniformly on $\{t : |1-t| \le |1-x|\}$, so Theorem 4.1.9 implies

$$\log x = \sum_{n=0}^{\infty}\int_1^x (1-t)^n\,dt = \sum_{n=1}^{\infty}\frac{(-1)^{n+1}}{n}(x-1)^n$$

Since this convergence holds when $|x-1|<1$, we have that $R \ge 1$. Since $\log x$ is not defined at $x=0$, it must be that $R=1$.

Exercises

(1) Show that if we have a power series $\sum_{n=0}^{\infty}a_n x^n$ with radius of convergence R and if $\lim_n |a_{n+1}/a_n|$ exists, then the value of this limit is R.

(2) Find the radius of convergence of the following: (a) $\sum_{n=0}^{\infty}\frac{x^n}{2^n}$; (b) $\sum_{n=1}^{\infty}\frac{x^n}{n^3}$; and (c) $\sum_{n=0}^{\infty}\frac{2^n}{n^2+1}x^n$.

(3) Find the radius of convergence of the following: (a) $\sum_{n=1}^{\infty}(\log n)x^n$; (b) $\sum_{n=1}^{\infty}2^n n^a x^n$, where a is some positive constant; and (c) $\sum_{n=1}^{\infty}\frac{n^n}{n!}x^n$.

(4) Show that the radius of convergence of the power series

$$\sum_{n=1}^{\infty}(-1)^n x^{n(n+1)}$$

is 1 and discuss convergence of the series when $x=\pm 1$.

(5) Find the radius of convergence of the power series

$$\sum_{n=0}^{\infty}\frac{\sin(n\pi/6)}{2^n}x^n.$$

(6) (a) For Example 4.2.10(a), find the precise formula for $f^{(2n+1)}(0)$. (b) Fill in the details needed to verify Example 4.2.10(b).

(7) Fill in the details to obtain Example 4.2.10(c).

(8) If $f \in C^{(\infty)}(a, b)$ and $c \in (a, b)$, show that f can be represented as a power series in an interval about c if and only if there is an $r > 0$ with $(c - r, c + r) \subseteq (a, b)$ such that $\frac{f^{(n)}(c)}{n!}(x - c)^n \to 0$ when $|x - c| < r$.

5 Metric and Euclidean Spaces

In this chapter we begin the study of p-dimensional Euclidean space, $\mathbb{R}^p$, but in this beginning we will carry it a step further. We want to discuss differentiation and integration on $\mathbb{R}^p$, but first we need to extend the notions of sequential convergence, the properties of sets, and the concept of continuity to the higher dimensional spaces. The effort to explore these concepts in $\mathbb{R}^p$, however, is not greater than what is required to explore these notions in what are called metric spaces. In many respects the abstract spaces are easier to cope with than $\mathbb{R}^p$. Moreover some of what we have already done is properly couched in metric spaces. Indeed, the material of §4.1 can be set there with little additional effort. Nevertheless during this venture into abstraction the main set of examples will be Euclidean space.

We start with the concept of distance between points. This must be general enough to encompass a variety of circumstances, but it should conform to the intuitive notion we all have of what is meant by distance. Since this is done at the start of the first section, it would be profitable before proceeding for the reader to reflect on what properties (s)he thinks should be included in an abstract concept of distance; then you can compare your thoughts with the definition that starts the following section.

The treatment of metric spaces here is based on Chapter 1 of [4].

5.1. Definitions and Examples

5.1.1. Definition. A *metric space* is a pair of objects, (X, d), where X is a set and d is a function $d : X \times X \to [0, \infty)$ called a *metric*, that satisfies the following for all x, y, z in X:

(a) $d(x, y) = d(y, x)$;
(b) $d(x, y) = 0$ if and only if $x = y$; and
(c) (Triangle Inequality) $d(x, y) \leq d(x, z) + d(z, y)$.

Condition (a) is sometimes called the symmetric property and says that the distance from x to y is the same as the distance from y to x. The second property says the obvious: the distance from a point to itself is 0 and the only point at a distance zero from x is x itself. The third, the triangle property, says that the shortest distance between two points is the direct one – not a distance involving a third point. In $\mathbb{R}^p$ this is phrased by saying that the shortest distance between two points is a straight line. In the abstract setting we have no concept of straight lines. Though you might have thought of other properties for an idea of distance, these three are usually part of what most people intuitively associate with the concept. In fact, I think the properties above are the minimal ones. There are several particular situations where additional axioms for a distance are assumed; those are more specialized theories, and what we are now going to explore is the basic one. Here are some examples of metric spaces.

5.1.2. Example. (a) Let $X = \mathbb{R}$, the set of real numbers, and define $d(x, y) = |x - y|$. See Exercise 1.

(b) Let $X = \mathbb{R}^2$, the plane, and define $d((x_1, y_1), (x_2, y_2)) = [(x_1 - x_2)^2 + (y_1 - y_2)^2]^{\frac{1}{2}}$. The reader knows from the Pythagorean Theorem that this is the straight-line distance and (s)he can use geometry to verify that this standard notion of the distance between two points satisfies the axioms in the preceding definition.

(c) We define p-dimensional *Euclidean space*, $\mathbb{R}^p$, to be

$$\mathbb{R}^p = \{(x_1, \ldots, x_p) : x_n \in \mathbb{R} \text{ for } 1 \leq n \leq p\}$$

For $x = (x_1, \ldots, x_p)$ and $y = (y_1, \ldots, y_p)$ in $\mathbb{R}^q$, define

$$d(x, y) = \left[\sum_{n=1}^{p} (x_n - y_n)^2 \right]^{\frac{1}{2}}$$

This is a metric on $\mathbb{R}^p$; however to prove this satisfies the triangle inequality requires some effort that we'll do below (5.1.6).

(d) Let $X = \mathbb{R}^p$ and for x, y in $\mathbb{R}^p$ define

$$d(x, y) = \sum_{n=1}^{p} |x_n - y_n|$$

This is also a metric on $\mathbb{R}^p$ but it is easier to verify this than the previous example (Exercise 2).

(e) Again let $X = \mathbb{R}^p$ and now define

$$d(x, y) = \max\{|x_n - y_n| : 1 \leq n \leq p\}$$

Once again $(\mathbb{R}^p, d)$ is a metric space (Exercise 3). It is worth observing that in each of the last three examples, when $p = 1$, all these metrics are the standard absolute value on $\mathbb{R}$.

(f) Let X be any set and define

$$d(x, y) = \begin{cases} 0 & \text{if } x = y \\ 1 & \text{if } x \neq y \end{cases}$$

It is a simple exercise to verify that (X, d) is a metric space. This is called the *discrete metric* on X. You won't encounter this much except in the study of metric spaces, as it is a nice example to test the concepts.

(g) An important class of examples arise as follows. Suppose (X, d) is a given metric space. If Y is a non-empty subset of X, then (Y, d) is a metric space and is referred to as a *subspace*. As a specific instance of this we can take $X = \mathbb{R}$ and $Y = [a, b]$. We saw this often in $\mathbb{R}$ and will see this often in $\mathbb{R}^p$ when we consider subsets of Euclidean space and examine them as metric spaces.

The next result will be quite useful in our discussion of metric spaces. It is a direct consequence of the triangle inequality and is often called the *Reverse Triangle Inequality*.

5.1.3. Proposition. *If (X, d) is a metric space and $x, y, z \in X$, then*

$$|d(x, y) - d(y, z)| \leq d(x, z)$$

Proof. The triangle inequality implies that $d(x, y) - d(y, z) \leq d(x, z)$. Now reverse the roles of x and z in this inequality and we get that $d(z, y) - d(y, x) \leq d(z, x)$, from which we also have that $d(y, z) - d(x, y) \leq d(x, z)$. That is, $\pm[d(x, y) - d(y, z)] \leq d(x, z)$ from which the proposition follows. ∎

Now let's show that the function d given in Example 5.1.2(c) is a metric. To do this we need a famous inequality. To facilitate the proof, introduce the helpful notation that for vectors $x = (x_1, \ldots, x_p)$ and $y = (y_1, \ldots, y_p)$ in $\mathbb{R}^p$, $\langle x, y \rangle = \sum_{n=1}^{p} x_n y_n$. Actually this is more than just "helpful" notation as it denotes the inner or "dot" product in the vector space $\mathbb{R}^p$. This connection will not be explored here, where we will only regard this as notation. Of course some of you may have explored the inner product in linear algebra. It is useful to observe the following properties for all vectors x, y, z in $\mathbb{R}^p$ and all real numbers t.

5.1.4
$$\begin{cases} \langle x, x \rangle \geq 0 \\ \langle x, y \rangle = \langle y, x \rangle \\ \langle tx + z, y \rangle = t\langle x, y \rangle + \langle z, y \rangle \\ \langle x, y + tz \rangle = \langle x, y \rangle + t\langle x, z \rangle \end{cases}$$

5.1.5. Theorem (Cauchy[1]–Schwarz[2] Inequality). *For $x = (x_1, \ldots, x_p)$ and $y = (y_1, \ldots, y_p)$ in $\mathbb{R}^p$ we have*

$$\left[\sum_{n=1}^{p} x_n y_n \right]^2 \leq \left[\sum_{n=1}^{p} x_n^2 \right] \left[\sum_{n=1}^{p} y_n^2 \right]$$

---

[1] See the footnote in Definition 1.3.12 for a biographical note.

[2] Hermann Amandus Schwarz was a German mathematician born in 1843 in Hermsdorf, Silesia, which at present is part of Poland. He began his studies at Berlin in Chemistry, but switched to mathematics and received his doctorate in 1864 under the direction of Weierstrass. He held positions at Halle, Zurich, Göttingen, and Berlin. His work centered on various geometry problems that were deeply connected to analysis. This included work on surfaces and conformal mappings in analytic function theory, any student of which will see his name in prominence. He died in Berlin in 1921.

Proof. First note that using the inner product notation introduced above, the sought after inequality becomes

$$\langle x, y \rangle^2 \le \langle x, x \rangle \langle y, y \rangle$$

Using (5.1.4) we have that

$$
\begin{aligned}
0 &\le \langle x - ty, x - ty \rangle \\
&= \langle x, x \rangle - t \langle y, x \rangle - t \langle x, y \rangle + t^2 \langle y, y \rangle \\
&= \langle x, x \rangle - 2t \langle x, y \rangle + t^2 \langle y, y \rangle \\
&= \gamma - 2\beta t + \alpha t^2 \equiv q(t)
\end{aligned}
$$

where $\gamma = \langle x, x \rangle$, $\beta = \langle x, y \rangle$, $\alpha = \langle y, y \rangle$. Thus $q(t)$ is a quadratic polynomial in the variable t. Since $q(t) \ge 0$ for all t, the graph of $q(t)$ stays above the x-axis except that it might be tangent at a single point; that is, $q(t) = 0$ has at most one real root. From the quadratic formula we get that $0 \ge 4\beta^2 - 4\alpha\gamma = 4(\beta^2 - \alpha\gamma)$. Therefore

$$0 \ge \beta^2 - \alpha\gamma = \langle x, y \rangle^2 - \langle x, x \rangle \langle y, y \rangle$$

proving the inequality. ∎

5.1.6. Corollary. *If $d : \mathbb{R}^p \times \mathbb{R}^p \to [0, \infty)$ is defined as in Example 5.1.2(b), then d is a metric.*

Proof. We begin by noting that $d(x, y) = \sqrt{\langle x - y, x - y \rangle}$. Using the Cauchy–Schwarz Inequality, (5.1.4), and the vector space properties of $\mathbb{R}^p$ we get that

$$
\begin{aligned}
d(x, y)^2 &= \langle x - y, x - y \rangle \\
&= \langle (x - z) + (z - y), (x - z) + (z - y) \rangle \\
&= \langle x - z, x - z \rangle + 2 \langle x - z, z - y \rangle + \langle z - y, z - y \rangle \\
&\le d(x, z)^2 + 2 \sqrt{\langle x - z, x - z \rangle} \sqrt{\langle z - y, z - y \rangle} + d(z, y)^2 \\
&= d(x, z)^2 + 2d(x, z)d(z, y) + d(z, y)^2 \\
&= [d(x, z) + d(z, y)]^2
\end{aligned}
$$

Taking square roots shows that the triangle inequality holds. The remainder of the proof that d defines a metric is straightforward. (Verify!) ∎

Notice. Whenever we are discussing $\mathbb{R}^p$ we assume that the metric under consideration is that defined in Example 5.1.2(c).

We close with the following, whose proof is Exercise 5.

5.1.7. Proposition. *If (X, d) and (Z, ρ) are two metric spaces and we define the function $\delta : (X \times Z) \times (X \times Z) \to \mathbb{R}$ by*

$$\delta((x_1, z_1), (x_2, z_2)) = d(x_1, x_2) + \rho(z_1, z_2)$$

then δ is a metric on $X \times Z$.

(1) Verify the statement in Example 5.1.2(a).
(2) Verify the statement in Example 5.1.2(d).
(3) Verify the statement in Example 5.1.2(e).
(4) In the Cauchy–Schwarz Inequality, show that equality holds if and only if the vectors x and y are linearly dependent.
(5) Prove Proposition 5.1.7.

5.2. Sequences and Completeness

For the remainder of this chapter (X, d) is a given metric space.

Now that we have a generalization of the concept of distance, we can introduce convergent sequences. As in Chapter 1, a sequence is just a way of enumerating some points in the space X: $x_1, x_2, \ldots$; this is denoted by $\{x_n\}$. Precisely, this is a function from the natural numbers $\mathbb{N}$ into X: $n \mapsto x_n$. As when we discussed sequences of real numbers, we'll sometimes change the domain of this function to $\mathbb{N} \cup \{0\}$ so that we get a sequence $\{x_0, x_1, \ldots\}$; or maybe it might be changed to get $\{x_2, x_3, \ldots\}$. There is no real difference; we have a specific beginning and a countably infinite following. This means that the set of all integers, $\mathbb{Z}$, is not permitted as an indexing set for sequences.

Unlike when we considered sequences in $\mathbb{R}$ we have no concept of an interval so we cannot use the definition of convergence we used in $\mathbb{R}$. Instead we use the generalization of the equivalent formulation from Proposition 1.3.3.

5.2.1. Definition. A sequence $\{x_n\}$ in (X, d) *converges* to x if for every $\epsilon > 0$ there is an integer N such that $d(x, x_n) < \epsilon$ when $n \geq N$. The notation for this is $x_n \to x$ or $x = \lim_n x_n$.

5.2.2. Example. (a) A sequence in $\mathbb{R}$ converges in the sense of §1.3 if and only if it converges in $\mathbb{R}$ considered as a metric space. (This is the content of (1.3.3).)

(b) If $x_n = (x_n^1, \ldots, x_n^p)$ and $(x^1, \ldots, x^p)$ are in $\mathbb{R}^p$, then $x_n \to x$ if and only if $x_n^j \to x^j$ in $\mathbb{R}$ for $1 \leq j \leq p$. In fact if $x_n \to x$, then we need only note that for $1 \leq j \leq p$, $|x_n^j - x^j| \leq d(x_n, x)$ to conclude that $x_n^j \to x^j$. Now assume that $x_n^j \to x^j$ for $1 \leq j \leq p$. Let $\epsilon > 0$ and for $1 \leq j \leq p$ choose N_j such that $|x_n^j - x^j| < \epsilon/\sqrt{p}$ when $n \geq N_j$. If we put $N = \max\{N_1, \ldots, N_p\}$ and $n \geq N$, then

$$\sum_{j=1}^{p} \left(x_n^j - x^j\right)^2 < \sum_{j=1}^{p} \frac{\epsilon^2}{p} = \epsilon^2$$

Hence $d(x_n, x) < \epsilon$ for $n \geq N$ and we have that $x_n \to x$ in $\mathbb{R}^p$.

(c) If (X, d) is the discrete metric space, then a sequence $\{x_n\}$ in X converges to x if and only if there is an integer N such that $x_n = x$ whenever $n \geq N$. In words, a sequence in X converges if and only if it is eventually constant.

As in $\mathbb{R}$, if $\{x_n\}$ is a sequence in (X, d) and $\{n_k\}$ is a sequence of positive integers with $n_1 < n_2 < \cdots$, then $\{x_{n_k}\}$ is called a *subsequence* of $\{x_n\}$. As we saw in Proposition 1.3.10, if a sequence in (X, d) converges to x, every subsequence converges to x. Unlike $\mathbb{R}$ it makes no sense to talk of monotonic sequences in an abstract metric space. So the results proved in §1.3 about monotonic sequences have no analogy here. A concept from our study of $\mathbb{R}$ that had little significance there, however, will have great significance now, though its true prominence will not arise until §5.3.

5.2.3. Definition. A sequence $\{x_n\}$ in (X, d) is called a *Cauchy sequence* if for every $\epsilon > 0$ there is an integer N such that $d(x_m, x_n) < \epsilon$ when $m, n \geq N$. The discrete metric space (X, d) is said to be *complete* if every Cauchy sequence converges.

5.2.4. Example. (a) As in $\mathbb{R}$, every convergent sequence in (X, d) is a Cauchy sequence in (X, d) (Exercise 2).

(b) $\mathbb{R}$ is a complete metric space (1.3.13) as is $\mathbb{R}^p$ (Exercise 3).

(c) If we furnish $\mathbb{Q}$ with the metric it has as a subspace of $\mathbb{R}$, then it is not a complete metric space. In fact just take a sequence $\{x_n\}$ in $\mathbb{Q}$ that converges to $\sqrt{2}$ in $\mathbb{R}$. It is a Cauchy sequence in $(\mathbb{Q}, d)$ but does not converge to a point in $\mathbb{Q}$.

(d) The discrete metric space is complete. Indeed a sequence in the discrete metric space is a Cauchy sequence if and only if it is eventually constant.

5.2.5. Proposition. *If $\{x_n\}$ is a Cauchy sequence and some subsequence of $\{x_n\}$ converges to x, then $x_n \to x$.*

Proof. Suppose $x_{n_k} \to x$ and let $\epsilon > 0$. Choose an integer N_1 such that $d(x_{n_k}, x) < \epsilon/2$ for $n_k \geq N_1$, and choose an integer N_2 such that $d(x_n, x_m) < \epsilon/2$ when $m, n \geq N_2$. Put $N = \max\{N_1, N_2\}$ and let $n \geq N$. Fix any $n_k \geq N$. Since we have that $n_k \geq N_1$ and both n and n_k are larger than N_2, we get that $d(x, x_n) \leq d(x, x_{n_k}) + d(x_{n_k}, x_n) < \epsilon/2 + \epsilon/2 = \epsilon$. ∎

Exercises

(1) Suppose $\{x_n\}$ is a sequence in X that converges to x and $z_1, \ldots, z_m$ is a finite collection of points in X. Define a new sequence $\{y_n\}$ in X by letting $y_k = z_k$ for $1 \leq k \leq m$ and $y_k = x_{k-m}$ when $k \geq m + 1$. Show that $y_n \to x$.

(2) Verify the statement in Example 5.2.4(a).

(3) Show that $\mathbb{R}^p$ is complete. (Hint: Use the method in Example 5.2.2(b) to show that a sequence in $\mathbb{R}^p$ is a Cauchy sequence if and only if each of the sequences of its coordinates is a Cauchy sequence in $\mathbb{R}$.)

(4) Show that a sequence $\{(x_1^n, x_2^n)\}$ in $\mathbb{R}^p \times \mathbb{R}^q$ (see Proposition 5.1.7) converges to (x_1, x_2) if and only if the same thing happens when we consider the sequence as belonging to $\mathbb{R}^{p+q}$.

(5) Let (X, d) be the cartesian product of the two metric spaces (X_1, d_1) and (X_2, d_2) as in Proposition 5.1.7. (a) Show that a sequence $\{(x_n^1, x_n^2)\}$ in X is a Cauchy sequence in X if and only if $\{x_n^1\}$ is a Cauchy sequence in X_1 and $\{x_n^2\}$ is a Cauchy sequence in X_2. (b) Show that X is complete if and only if both X_1 and X_2 are complete.

5.3. Open and Closed Sets

As we did for $\mathbb{R}$ in §1.6, we define open and closed sets in a metric space. In an abstract metric space these sets will play an even bigger role than they did in $\mathbb{R}$.

5.3.1. Definition. A subset F of (X, d) is said to be *closed* if whenever $\{x_n\}$ is a sequence of points in F and $x_n \to x$ we have that $x \in F$. A subset G is said to be *open* if $X \backslash G$ is a closed set.

Again we note that as in the case that $X \subseteq \mathbb{R}$, the entire space X and the empty set are simultaneously open and closed; also any finite subset of (X, d) is closed. Also see Exercise 1. The first two parts of the next proposition are restatements in the case of (X, d) of the first two parts of Proposition 1.6.3. The exact same proof used to demonstrate (1.6.3) can be used word for word to prove these. The other two parts of the next proposition are analogous to the last two parts of Proposition 1.6.3 with one big difference. Unlike when we discuss open and closed subsets of $\mathbb{R}$ we need to talk about arbitrary unions and intersections of sets, not just their countable versions. Here are the definitions.

Suppose I is an arbitrary non-empty set, sometimes called in this circumstance an *index set*, and assume that for each i in I there is a subset A_i of X. We define the *union* and *intersection* of the collection $\{A_i : i \in I\}$ by

$$\bigcup_{i \in I} A_i = \{x \in X : x \in A_i \text{ for some } i \text{ in } I\}$$

$$\bigcap_{i \in I} A_i = \{x \in X : x \in A_i \text{ for every } i \text{ in } I\}$$

These operations satisfy the customary rules enjoyed by their finite versions. In particular, there is a version of De Morgan's Laws that is valid (Exercise 2).

5.3.2. Proposition. *Let (X, d) be a metric space.*

(a) *If $F_1, \ldots, F_n$ are closed sets in (X, d), then $\bigcup_{k=1}^n F_k$ is closed.*
(b) *If $G_1, \ldots, G_n$ are open sets in (X, d), then $\bigcap_{k=1}^n G_k$ is open.*
(c) *If $\{F_i : i \in I\}$ is a collection of closed sets, then $\bigcap_{i \in I} F_i$ is closed.*
(d) *If $\{G_i : i \in I\}$ is a collection of open sets, then $\bigcup_{i \in I} G_i$ is open.*

Proof. As we said the proofs of (a) and (b) are verbatim copies of the proofs of the similar statements for $\mathbb{R}$. The proofs of (c) and (d) are similar to the proofs for the analogous statements for sequences of sets in $\mathbb{R}$ given in Proposition 1.6.3. The details are left to the reader. ∎

When $x \in X$ and $r > 0$, introduce the notation

$$B(x; r) = \{y \in X : d(x, y) < r\}$$
$$\overline{B}(x; r) = \{y \in X : d(x, y) \leq r\}$$

The set $B(x; r)$ is called the *open ball* of radius r about x, or centered at x; $\overline{B}(x; r)$ is called the *closed ball* of radius r about x. If $X = \mathbb{R}$, then $B(x; r)$ is the open interval $(x - r, x + r)$ and $\overline{B}(x; r)$ is the closed interval $[x - r, x + r]$. If $X = \mathbb{R}^2$, then $B(x; r)$ is the so-called "open" ball or disk centered at x of radius r that does not include the bounding circle; and $\overline{B}(x; r)$ is the corresponding "closed" disk that does include the bounding circle. The use of the words open and closed here will be justified momentarily. Meanwhile notice the trivial but useful observation that when $s < r$, $\overline{B}(x; s) \subseteq B(x; r)$.

The next result is a restatement of Proposition 1.6.4 for an arbitrary metric space. Small modifications of the proof of (1.6.4) will furnish a proof of the present result.

5.3.3. Proposition. *A subset G of (X, d) is open if and only if for each x in G there is an $r > 0$ such that $B(x; r) \subseteq G$.*

The preceding proposition quantifies what it means for a set to be open and as such is most useful as we will see.

5.3.4. Example. (a) For any $r > 0$, $\overline{B}(x; r)$ is closed. In fact let $\{a_n\}$ be a sequence in $\overline{B}(x; r)$ that converges to a. Thus $d(a, x) \leq d(a, a_n) + d(a_n, x) \leq d(a, a_n) + r$. Since $d(a, a_n) \to 0$, we have that $a \in \overline{B}(x; r)$.

(b) For any $r > 0$, $B(x; r)$ is open. In fact let $F = X \backslash B(x; r) = \{y \in X : d(y, x) \geq r\}$. If $\{a_n\}$ is a sequence in F that converges to a, then the reverse triangle inequality implies $|d(x, a) - d(a_n, a)| \geq d(a_n, x) \geq r$. Since $d(a_n, a) \to 0$ we have that $a \in F$.

It is important when discussing open and closed sets to be conscious of the universe.

When (X, d) is a metric space and $Y \subseteq X$, we have that (Y, d) is also a metric space (Example 5.1.2(f)). To say that we have an open set A in (Y, d) does not mean that A is open in X. Note that in such a circumstance when $y \in Y$ and $r > 0$, then the open ball about y of radius r is $B_Y(y; r) = \{z \in Y : d(z, y) < r\} = B(y; r) \cap Y$. This may not be an open set in X. For example if $X = \mathbb{R}$ and $Y = [0, 1]$, then $[0, \frac{1}{2})$ is open as a subset of Y but not as a subset of X; another example: $B_Y(\frac{1}{4}; \frac{1}{3}) = [0, 7/12)$. When we want to emphasize the open and closed sets in the subspace metric (Y, d), we'll use the terms *open relative to Y* or *relatively open* in Y and *closed relative to Y* or *relatively closed* in Y. The proof of the next proposition is Exercise 3.

5.3.5. Proposition. *Let (X, d) be a metric space and let Y be a subset of X.*

(a) *A subset G of Y is relatively open in Y if and only if there is an open subset U in X with $G = U \cap Y$.*

(b) *A subset F of Y is relatively closed in Y if and only if there is a closed subset D in X such that $F = D \cap Y$.*

Now to introduce certain concepts that had little significance when we studied $\mathbb{R}$ but will have importance for $\mathbb{R}^p$ as well as any other metric space.

5.3.6. Definition. Let A be a subset of X. The *interior* of A, denoted by $\operatorname{int} A$, is the set defined by

$$\operatorname{int} A = \bigcup \{G : G \text{ is open and } G \subseteq A\}$$

The *closure* of A, denoted by $\operatorname{cl} A$, is the set defined by

$$\operatorname{cl} A = \bigcap \{F : F \text{ is a closed and } A \subseteq F\}$$

The *boundary* of A, denoted by ∂A, is the set defined by

$$\partial A = \operatorname{cl} A \cap \operatorname{cl}(X \backslash A)$$

Let's note that there is always an open set contained in any set A – namely the empty set, $\emptyset$. It may be, however, that $\emptyset$ is the only open set contained in A, in which case $\operatorname{int} A = \emptyset$. Similarly X is a closed set containing any set A; but it may be the only such set, in which case $\operatorname{cl} A = X$. (We'll have more to say about this latter case below.) It follows from what we have established that $\operatorname{int} A$ is open (though possibly empty) and $\operatorname{cl} A$ is closed (though possible equal to X). We also have that $\operatorname{int} \emptyset = \emptyset = \operatorname{cl} \emptyset$ and $\operatorname{int} X = X = \operatorname{cl} X$. Before looking at more interesting examples, it is profitable to first prove some properties of the closure and interior of a set. We start with a characterization of these sets using open balls.

5.3.7. Proposition. *Let $A \subseteq X$.*

(a) $x \in \operatorname{int} A$ *if and only if there is an $r > 0$ such that $B(x; r) \subseteq A$.*
(b) $x \in \operatorname{cl} A$ *if and only if for every $r > 0$, $B(x; r) \cap A \neq \emptyset$.*
(c) $x \in \partial A$ *if and only if for every $r > 0$ we have that $B(x; r) \cap A \neq 0$ and $B(x; r) \cap (X \backslash A) \neq \emptyset$.*

Proof. (a) If $B(x; r) \subseteq A$, then since $B(x; r)$ is open we have that $B(x; r) \subseteq \operatorname{int} A$; hence $x \in \operatorname{int} A$. Now assume that $x \in \operatorname{int} A$. So there is an open set G such that $x \in G \subseteq A$. But since G is open, there is a radius $r > 0$ with $B(x; r) \subseteq G$ and we have established the converse.

(b) Suppose $x \in \operatorname{cl} A$. If $r > 0$, then $B(x; r)$ is open and $X \backslash B(x; r)$ is closed. It cannot be that $A \subseteq X \backslash B(x; r)$ since, by definition, this implies $\operatorname{cl} A \subseteq X \backslash B(x; r)$, contradicting the fact that $x \in \operatorname{cl} A$. Thus $B(x; r) \cap A \neq \emptyset$. Now assume that $x \notin \operatorname{cl} A$; that is, $x \in X \backslash \operatorname{cl} A$, an open set. By Proposition 5.3.3 there is a radius $r > 0$ such that $B(x; r) \subseteq X \backslash \operatorname{cl} A$. So for this radius, $B(x; r) \cap A = \emptyset$.

(c) This is immediate from (b). ∎

The preceding proposition is very useful as it provides a concrete, one-point-at-a-time method to determine the closure and the interior of a set. We'll see this in the following example.

5.3.8. Example. (a) Sometimes things can become weird with interiors and closures. Consider the metric space $\mathbb{R}$ and the subset $\mathbb{Q}$ of all rational numbers. If $x \in \mathbb{R}$, then $B(x; r) = (x - r, x + r)$ and this interval must contain a rational number. By the preceding proposition, $x \in \operatorname{cl} \mathbb{Q}$. We also have that $\operatorname{int} \mathbb{Q} = \emptyset$. To see this again use the preceding proposition and the fact that between any two real numbers there is an irrational number. This means that when $x \in \mathbb{Q}$, no open ball $B(x; r)$ can be contained in $\mathbb{Q}$ so that $\operatorname{int} \mathbb{Q} = \emptyset$. Using the same reasoning we see that $\operatorname{cl}[\mathbb{R}\backslash\mathbb{Q}] = \mathbb{R}$ and $\operatorname{int}[\mathbb{R}\backslash\mathbb{Q}] = \emptyset$. (Verify!) It follows that $\partial \mathbb{Q} = \mathbb{R}$.

(b) Here is a cautionary tale. Since $\overline{B}(x; r)$ is closed, we have that $\operatorname{cl} B(x; r) \subseteq \overline{B}(x; r)$. It may be, however, that $\operatorname{cl} B(x; r) \neq \overline{B}(x; r)$. In fact suppose that $X = \{(0, 0)\} \cup \{(a, b) : a^2 + b^2 = 1\} \subseteq \mathbb{R}^2$. So X consists of the origin in the plane together with the unit circle centered at the origin. Give X the metric it inherits as a subset of $\mathbb{R}^2$. In this case $\{(0, 0)\} = \operatorname{cl} B((0, 0); 1) \neq \overline{B}((0, 0); 1) = X$. It is also true that when X is the discrete metric space, then $\{x\} = B(x; 1) = \operatorname{cl} B(x; 1)$ whereas $\overline{B}(x; 1) = X$.

(c) We haven't said much about the boundary of a set, but note that for any x in $\mathbb{R}^p$ and $r > 0$ we have that $\partial B(x; r) = \partial \overline{B}(x; r) = \{y \in X : d(x, y) = r\}$. Is this true in every metric space? (Exercise 5).

The next proposition contains some useful information about closures and interiors of sets. Its proof is left as Exercise 6.

5.3.9. Proposition. *Let A be a subset of X.*

(a) *A is closed if and only if $A = \operatorname{cl} A$.*
(b) *A is open if and only if $A = \operatorname{int} A$.*
(c) *$\operatorname{cl} A = X\backslash[\operatorname{int}(X\backslash A)]$, $\operatorname{int} A = X\backslash\operatorname{cl}(X\backslash A)$, and $\partial A = \operatorname{cl} A\backslash\operatorname{int} A$.*
(d) *If $A_1, \ldots, A_n$ are subsets of X, then $\operatorname{cl}[\bigcup_{k=1}^n A_k] = \bigcup_{k=1}^n \operatorname{cl} A_k$.*

Part (d) of the preceding proposition does not hold for the interior. For example if $X = \mathbb{R}$, $a < b < c$, $A = (a, b]$, and $B = [b, c)$, then $\operatorname{int}(A \cup B) = (a, c)$ while $\operatorname{int} A \cup \operatorname{int} B = (a, b) \cup (b, c)$. Also see Exercises 7 and 8.

5.3.10. Definition. A subset E of a metric space (X, d) is *dense* if $\operatorname{cl} E = X$. A metric space (X, d) is *separable* if it has a countable dense subset.

We made a reference to this concept just after defining the closure of a set. So a set E is dense if and only if X is the only closed subset of X that contains E.

5.3.11. Example. (a) Every metric space is dense in itself.

(b) The rational numbers form a dense subset of $\mathbb{R}$ as do the irrational numbers. This is a rephrasing of Example 5.3.8(a). We note that this implies that $\mathbb{R}$ is separable since $\mathbb{Q}$ is countable (1.5.5).

(c) The set of all points in $\mathbb{R}^p$ with rational coordinates is dense in $\mathbb{R}^p$. This follows from the preceding example and it also says that $\mathbb{R}^p$ is separable by Corollary 1.5.6.

(d) If X is any set and d is the discrete metric on X (5.1.2(e)), then the only dense subset of X is X itself. In fact if E is a dense subset of (X, d) and $x \in X$, then it must be that $B(x; 1/2) \cap E \neq \emptyset$; but from the definition of the discrete metric it follows that $B(x; 1/2) = \{x\}$. Hence the discrete metric space is separable if and only if it is countable.

In part (d) of the preceding example we used the next result, and we record it here for future reference. Also it is a consequence of Proposition 5.3.7(b).

5.3.12. Proposition. *A set E is dense in (X, d) if and only if for every x in X and every $r > 0$, $B(x; r) \cap E \neq \emptyset$.*

5.3.13. Definition. If $A \subseteq X$, a point x in X is called a *limit point* of A if for every $\epsilon > 0$ there is a point a in $B(x; \epsilon) \cap A$ with $a \neq x$.

See Exercise 1.6.11, where this concept was defined in $\mathbb{R}$.

The emphasis here is that no matter how small we take ϵ we can find such a point a different from x that belongs to $B(x; \epsilon) \cap A$. It is not required that x belongs to A for it to be a limit point (more on this later). If x is not a limit point of A and, in addition, belongs to A, then it is called an *isolated point* of A.

5.3.14. Example. (a) Let $X = \mathbb{R}$ and $A = (0, 1) \cup \{2\}$. Every point in $[0, 1]$ is a limit point of A but 2 is not. In fact, 2 is an example of an isolated point of A, the only isolated point of A.

(b) If $X = \mathbb{R}$ and $A = \mathbb{Q}$, then every point of X is a limit point of A and A has no isolated points.

(c) If $X = \mathbb{R}$ and $A = \{n^{-1} : n \in \mathbb{N}\}$, then 0 is a limit point of A while the points n^{-1} are all isolated points.

5.3.15. Proposition. *Let A be a subset of the metric space X.*

(a) *A point x is a limit point of A if and only if there is a sequence of distinct points in A that converges to x.*
(b) *A is a closed set if and only if it contains all its limit points.*
(c) *$\operatorname{cl} A = A \cup \{x : x \text{ is a limit point of } A\}$.*

Proof. In Exercise 1.6.11 the reader was asked to prove parts (a) and (b) when $X = \mathbb{R}$.

(a) Suppose $\{a_n\}$ is a sequence of distinct points in A such that $a_n \to x$. If $\epsilon > 0$, then there is an N such that $a_n \in B(x; \epsilon)$ for $n \geq N$. Since the points in $\{a_n\}$ are distinct, there is at least one different from x. Thus x is a limit point. Now assume that x is a limit point. Let $a_1 \in A \cap B(x; 1)$ such that $a_1 \neq x$. Let $\epsilon_2 = \min\{2^{-1}, d(x, a_1)\}$; so there is a point $a_2 \in A \cap B(x; \epsilon_2)$ with $a_2 \neq x$. Note that $a_2 \neq a_1$.

Claim. There is a sequence of positive numbers $\{\epsilon_n\}$ and a sequence of distinct points $\{a_n\}$ in A such that: (i) $\epsilon_n \leq n^{-1}$; (ii) $a_n \neq x$ for all $n \geq 1$; and (iii) $d(x, a_n) < \epsilon_n$.

We already have a_1. Assume that $a_1, \ldots, a_{n-1}$ have been chosen. Since the numbers $d(x, a_1), \ldots, d(x, a_{n-1})$ are all positive we can choose $0 < \epsilon_n < n^{-1}$ and also smaller than any of these numbers. Let $a_n \in A \cap B(x, \epsilon_n)$ with $a_n \neq x$. This proves the claim and finishes the proof of (a).

(b) *and* (c). Clearly (b) will follow once we prove (c). Let B denote the set on the right-hand side of the equation in (c). By the definition of a closed set and part (a) we have that $B \subseteq \operatorname{cl} A$. On the other hand, if $x \in \operatorname{cl} A$, (5.3.7) implies that for every $n \geq 1$ there is a point a_n in A with $d(a_n, x) < n^{-1}$. Either $\{a_n\}$ has an infinite number of distinct terms or a finite number. In the first case there is a subsequence $\{a_{n_k}\}$ of distinct terms; by (a), x is a limit point. Thus $x \in B$. In the case that $\{a_n\}$ has only a finite number of points, there is a subsequence $\{a_{n_k}\}$ that is constant; thus $a_{n_k} = x$ for all $k \geq 1$ and so $x \in B$. ∎

5.3.16. Definition. If $A \subseteq X$ and $x \in X$, the *distance* from x to A is

$$\operatorname{dist}(x, A) = \inf\{d(x, a) : a \in A\}$$

Clearly when $x \in A$, $\operatorname{dist}(x, A) = 0$. But it is possible for the distance from a point to a set to be 0 when the point is not in the set as we now see.

5.3.17. Proposition. *If $A \subseteq X$, then* $\operatorname{cl} A = \{x \in X : \operatorname{dist}(x, A) = 0\}$.

Proof. If $x \in \operatorname{cl} A$, then there is a sequence $\{a_n\}$ in A such that $a_n \to x$; so $\operatorname{dist}(x, a_n) \to 0$ and it follows that $\operatorname{dist}(x, A) = 0$. Conversely if $\operatorname{dist}(x, A) = 0$, then there is a sequence $\{a_n\}$ in A such that $d(x, a_n) \to 0$. Thus $a_n \to x$ and so $x \in \operatorname{cl} A$. ∎

Also see Exercise 14. Now we extend Cantor's Theorem (1.6.8) to the setting of a complete metric space. First we need to extend the definition of the diameter of set from that given in §1.6 to the setting of a metric space; a little thought will show this is the natural extension. If $E \subseteq X$, define the *diameter* of E to be

$$\operatorname{diam} E = \sup\{d(x, y) : x, y \in E\}$$

5.3.18. Theorem (Cantor's Theorem). *A metric space (X, d) is complete if and only if whenever $\{F_n\}$ is a sequence of non-empty subsets satisfying: (i) each F_n is closed; (ii) $F_1 \supseteq F_2 \supseteq \cdots$; and (iii) $\operatorname{diam} F_n \to 0$, then $\bigcap_{n=1}^{\infty} F_n$ is a single point.*

Proof. Assume (X, d) is complete and $\{F_n\}$ is as in the statement of the theorem. For each n let $x_n \in F_n$. If $\epsilon > 0$, let N be such that $\operatorname{diam} F_n < \epsilon$ for $n \geq N$. So if $m, n \geq N$, (ii) implies $x_n, x_m \in F_N$ and so $d(x_n, x_m) \leq \operatorname{diam} F_N < \epsilon$. Thus $\{x_n\}$ is a Cauchy sequence. Since (X, d) is complete, there is an x in X such that $x_n \to x$. But each F_n is closed, so $x \in \bigcap_{n=1}^{\infty} F_n$. If there is another point y in $\bigcap_{n=1}^{\infty} F_n$, then $d(x, y) \leq \operatorname{diam} F_n$ for each $n \geq 1$. By (iii), $y = x$.

Now assume that (X, d) satisfies the stated conditions and $\{x_n\}$ is a Cauchy sequence. Put $F_n = \operatorname{cl}\{x_n, x_{n+1}, \ldots\}$. Clearly (i) and (ii) are satisfied. If $\epsilon > 0$, let N be such that $d(x_n, x_m) < \epsilon$ for $m, n \geq N$. But for $k \geq N$, $\operatorname{diam} F_k = \sup\{d(x_n, x_m) : m, n \geq k\} \leq \epsilon$. Thus $\{F_n\}$ satisfies the three conditions so that $\bigcap_{n=1}^{\infty} F_n = \{x\}$ for some point x. But for any $n \geq 1$, $d(x, x_n) \leq \operatorname{diam} F_n \to 0$. Therefore $x_n \to x$ and (X, d) is complete. ∎

Notice that we did not state this extension as we did Theorem 1.6.8. What happens in the preceding version of Cantor's Theorem if we do not assume diam $F_n \to 0$? See Exercise 16.

The proof of the next proposition is Exercise 17.

5.3.19. Proposition. *If (X, d) is a complete metric space and $Y \subseteq X$, then (Y, d) is complete if and only if Y is closed in X.*

We close this section by dwelling a bit on the concept of the diameter of a set.

5.3.20. Definition. Say that a subset A of (X, d) is *bounded* if diam $A < \infty$.

5.3.21. Proposition.

(a) *A subset A of (X, d) is bounded if and only if for any x in X there is an $r > 0$ such that $A \subseteq B(x; r)$.*
(b) *The union of a finite number of bounded sets is bounded.*
(c) *A Cauchy sequence in (X, d) is a bounded set.*

Proof. (a) If $A \subseteq B(x; r)$, then diam $A \leq 2r$, so that A is bounded. Conversely assume that A is bounded with δ as its finite diameter. Fix a point x in X and some point a_0 in A. For any point a in A, $d(x, a) \leq d(x, a_0) + d(a_0, a) \leq d(x, a_0) + \delta$. If we let $r = 2[d(a_0, x) + \delta]$, then $A \subseteq B(x; r)$.

(b) If A_k is bounded for $1 \leq k \leq n$ and $x \in X$, let $r_k > 0$ such that $A_k \subseteq B(x; r_k)$. If we set $r = \max\{r_1, \ldots, r_n\}$, then $A_1 \cup \cdots \cup A_n \subseteq B(x; r)$.

(c) If $\{x_n\}$ is a Cauchy sequence, there is an $N \geq 1$ such that $d(x_n, x_m) < 1$ for $m, n \geq N$. If $B = \{x_n : n \geq N\}$, this says that diam $B \leq 1$ so that B is bounded. On the other hand, $A = \{x_1, \ldots, x_N\}$ is bounded since finite sets are bounded. By part (b), $\{x_n\} = A \cup B$ is bounded. ∎

The concept of boundedness in an arbitrary metric space is not as useful as it was in $\mathbb{R}$ and does not have the same consequences. We will see, however, that the benefits of boundedness we had in $\mathbb{R}$ carry over to $\mathbb{R}^p$.

Exercises

(1) Show that the following sets F are closed. (a) $F = \{(x, y) \in \mathbb{R}^2 : -1 \leq x \leq 1$ and $-1 \leq y \leq 1\}$. (b) $F = \{(x, y) \in \mathbb{R}^2 : y \geq x^{-1}\}$. (c) $F = \{(x, y, z) \in \mathbb{R}^3 : y \leq x \leq z\}$.

(2) Prove the following version of De Morgan's Laws for any collection of subsets $\{A_i : i \in I\}$, where each $A_i \subseteq X$:

$$X \backslash \bigcup_{i \in I} A_i = \bigcap_{i \in I} (X \backslash A_i)$$

$$X \backslash \bigcap_{i \in I} A_i = \bigcup_{i \in I} (X \backslash A_i).$$

(3) Prove Proposition 5.3.5. (Hint: If G is a relatively open subset of Y, for each y in G let $r_y > 0$ such that $B_Y(y; r_y) \subseteq G$. Now consider $\bigcup\{B(y; r_y) : y \in G\}$.)

(4) If Y is a subset of X, consider the metric space (Y, d) and suppose $Z \subseteq Y$. (a) Show that H is a relatively open subset of Z if and only if there is a relatively open subset H_1 of Y such that $H = H_1 \cap Z$. (b) Show that D is a relatively closed subset of Z if and only if there is a relatively closed subset D_1 of Y such that $D = D_1 \cap Z$. (Hint: Use Proposition 5.3.5.)

(5) Is it true that $\partial B(x; r) = \partial \bar{B}(x; r) = \{y \in X : d(x, y) = r\}$ in an arbitrary metric space?

(6) Prove Proposition 5.3.9.

(7) Show that if $A_1, \ldots, A_n$ are subsets of X, then int $[\bigcap_{k=1}^n A_k] = \bigcap_{k=1}^n \text{int } A_k$.

(8) Show that (5.3.9(d)) does not hold for infinite unions and the preceding exercise does not hold for infinite intersections.

(9) If $E \subseteq X$ and $x \in \partial E$, show that x is either an isolated point of E or a limit point.

(10) If A is a subset of X that is simultaneously open and closed, show that $\partial A = \emptyset$.

(11) See Corollary 1.5.6 for the definition of $X_1 \times X_2$. Define $\rho : X_1 \times X_2 \to [0, \infty)$ by

$$\rho((x_1, x_2), (y_1, y_2)) = \max\{d_1(x_1, y_1), d_2(x_2, y_2)\}.$$

(a) Show that ρ is a metric on $X_1 \times X_2$. (b) Show that a set G is open in $(X_1 \times X_2, \rho)$ if and only if for every (x_1, x_2) in G there are open sets G_1 in X_1 and G_2 in X_2 such that $x_1 \in G_1, x_2 \in G_2$, and $G_1 \times G_2 \subseteq G$.

(12) Let ℓ^∞ denote the set of all bounded sequences of real numbers; that is, ℓ^∞ consists of all sequences $\{x_n\}$ such that $x_n \in \mathbb{R}$ for all $n \geq 1$ and $\sup_n |x_n| < \infty$. If $x = \{x_n\}, y = \{y_n\} \in \ell^\infty$, define $d(x, y) = \sup_n |x_n - y_n|$. (a) Show that d defines a metric on ℓ^∞. (b) If e_n denotes the sequence with a 1 in the n-th place and zeros elsewhere, show that $B(e_n; \frac{1}{2}) \cap B(e_m; \frac{1}{2}) = \emptyset$ when $n \neq m$. (c) Is the set $\{e_n : n \geq 1\}$ closed?

(13) Show that the metric space ℓ^∞ defined in Exercise 12 is complete.

(14) If $A \subseteq X$, show that int $A = \{x : \text{dist}(x, X \backslash A) > 0\}$. Can you give an analogous characterization of ∂A?

(15) (a) If $A \subseteq X$, show that $x \in \text{cl } A$ if and only if x is either a limit point of A or an isolated point of A. (b) Show that if a set has no limit points, it is closed. (c) Give an example of an infinite subset of $\mathbb{R}$ that has no limit points.

(16) If (X, d) is a complete metric space, show that if $\{F_n\}$ is a sequence of bounded closed subsets of X satisfying (i) and (ii) in Cantor's Theorem, then $\bigcap_{n=1}^\infty F_n \neq \emptyset$.

(17) Prove Proposition 5.3.19.

5.4. Continuity

Here we will extend the concept of a continuous function seen in Chapter 1 to a mapping between two metric spaces and investigate the properties of such functions.

5.4.1. Definition. If (X, d) and (Z, ρ) are two metric spaces, a function $f : X \to Z$ is *continuous at a point a* in X if whenever $\{x_n\}$ is a sequence in X that converges to a, $f(x_n) \to f(x)$ in Z. f is said to be a *continuous function* if it is continuous at each point of X

Note that if $X = Z = \mathbb{R}$, then this is the same as the definition of continuity found in §1.7. The most common situation we will encounter in the remainder of this book is where X is a subset of $\mathbb{R}^p$ and $Z = \mathbb{R}^q$. To establish results in such situations, however, requires no more effort than to prove them in the setting described in the definition. The next few results should have a familiar ring from §1.7.

5.4.2. Proposition. *If (X, d) and (Z, ρ) are metric spaces and $f : X \to Z$, then f is continuous at a if and only if for every $\epsilon > 0$ there is a $\delta > 0$ such that when $d(a, x) < \delta$ it follows that $\rho(f(a), f(x)) < \epsilon$.*

Proof. Suppose f is continuous at a and, to the contrary, there is an $\epsilon > 0$ such that for every $\delta > 0$ there is at least one x with $d(x, a) < \delta$ and $\rho(f(x), f(a)) \geq \epsilon$. In particular taking $\delta = n^{-1}$ we have that for every $n \geq 1$ there is an x_n with $d(x_n, a) < n^{-1}$ and $\rho(f(x_n), f(a)) \geq \epsilon$. But this says that $x_n \to a$ and $\{f(x_n)\}$ does not converge to $f(a)$, contradicting the assumption of continuity.

Now assume that for every $\epsilon > 0$ there is a $\delta > 0$ as stated in the proposition. If $x_n \to a$ in X, and $\epsilon > 0$, let $\delta > 0$ such that when $d(a, x) < \delta$ it follows that $\rho(f(a), f(x)) < \epsilon$. Let N be an integer such that $d(x_n, a) < \delta$ when $n \geq N$. So $\rho(f(x_n), f(a)) < \epsilon$ when $n \geq N$. Since ϵ was arbitrary, this says that $f(x_n) \to f(a)$ and f is continuous at a. ∎

Let's mention that the previous result is often taken as the definition of a continuous function. It is this equivalent formulation of continuity that generalizes to more abstract structures than a metric space.

We won't spend any time investigating functions continuous at a single point, but we will have much to say about functions continuous on the entire metric space, especially when that metric space is some subset of $\mathbb{R}^p$.

5.4.3. Theorem. *If (X, d) and (Z, ρ) are metric spaces and $f : X \to Z$, then the following statements are equivalent.*

(a) *f is a continuous function on X.*
(b) *If U is an open subset of Z, then $f^{-1}(U)$ is an open subset of X.*
(c) *If D is a closed subset of Z, then $f^{-1}(D)$ is a closed subset of X.*

Proof. (b) *is equivalent to* (c). Note that

$$f^{-1}(Z\backslash U) = X\backslash f^{-1}(U) \text{ and } f^{-1}(Z\backslash D) = X\backslash f^{-1}(D)$$

From these equalities the equivalence of the two statements is straightforward.

(a) *implies* (b). Let $a \in f^{-1}(U)$ so that $\alpha = f(a) \in U$. Since U is open there is an $\epsilon > 0$ such that $B(\alpha; \epsilon) \subseteq U$. Since f is continuous there is a $\delta > 0$ such that

$d(a, x) < \delta$ implies $\rho(f(a), f(x)) < \epsilon$. In other words, $B(a; \delta) \subseteq f^{-1}(B(\alpha; \epsilon)) \subseteq f^{-1}(U)$. Since a was an arbitrary point in $f^{-1}(U)$, this says that $f^{-1}(U)$ is open.

(b) *implies* (a). If $a \in X$ and $\epsilon > 0$, then $B(f(a); \epsilon)$ is open; so by (b) we have that $f^{-1}(B(f(a); \epsilon))$ is an open set in X that contains a. Thus there is a $\delta > 0$ such that $B(a; \delta) \subseteq f^{-1}(B(f(a); \epsilon))$. That is, $d(a, x) < \delta$ implies $\rho(f(a), f(x)) < \epsilon$ and so f is continuous at a. ∎

See Exercise 5. The next result extends Proposition 1.7.4 to the present setting. The same proof given there applies here as the reader will see when (s)he does a step-by-step process of translation.

5.4.4. Proposition. *The composition of two continuous functions is also continuous.*

The proof of the next proposition is elementary.

5.4.5. Proposition. *If (X, d) is a metric space, $f, g : X \to \mathbb{R}^p$ are continuous, and $\alpha \in \mathbb{R}$, then:*

(a) $(f + g) : X \to \mathbb{R}^p$ *defined by* $(f + g)(x) = f(x) + g(x)$ *is continuous;*
(b) $(\alpha f) : X \to \mathbb{R}^p$ *defined as* $(\alpha f)(x) = \alpha f(x)$ *is continuous; and*
(c) *the function from X into $\mathbb{R}$ defined by* $x \mapsto \langle f(x), g(x) \rangle$ *is continuous.*

Now to generate some examples of continuous functions. Of course §1.7 furnishes several examples of continuous functions from the real line into itself. We also have the trigonometric functions, the exponential, and logarithm.

5.4.6. Example. (a) The function from $\mathbb{R}^p \times \mathbb{R}^p \to \mathbb{R}^p$ defined by $(x, y) \mapsto x + y$ (vector addition) is continuous.

(b) The function from $\mathbb{R} \times \mathbb{R}^p \to \mathbb{R}^p$ defined by $(t, x) \mapsto tx$ (scalar multiplication) is continuous.

(c) If $f_k : X \to \mathbb{R}$ is a continuous function for $1 \le k \le p$, then $f : X \to \mathbb{R}^p$ defined by $f(x) = (f_1(x), \ldots, f_p(x))$ is a continuous function. (Exercise 7(b)).

(d) If $f_k : X \to \mathbb{R}^p$ is a continuous function for $1 \le k \le m$, then $f : X \to \mathbb{R}^p$ defined by $f(x) = \sum_{k=1}^m f_k(x)$ is a continuous function. (Exercise 7(c)).

(e) If (X, d) is any metric space, $x_0 \in X$, and we define $f : X \to \mathbb{R}$ by $f(x) = d(x, x_0)$, then f is continuous. In fact the reverse triangle inequality (5.1.3) says that $|f(x) - f(y)| \le d(x, y)$, from which continuity follows by either the definition or Proposition 5.4.2.

(f) If (X, d) is the discrete metric space, then the only continuous functions from $[0, 1]$ into (X, d) are the constant functions. On the other hand, for any metric space (Z, ρ) every function $f : (X, d) \to (Z, \rho)$ is continuous.

(g) Consider $\mathbb{R}^p$, $1 \le q < p$, and integers $1 \le n_1 < n_2 < \cdots n_q \le p$. The function $\pi : \mathbb{R}^p \to \mathbb{R}^q$ defined by $\pi(x_1, \ldots, x_p) = (x_{n_1}, \ldots, x_{n_q})$ is continuous. (Exercise 7(e)).

Recall the definition of the distance from a point to a set A, dist (x, A) (5.3.16).

5.4.7. Proposition. *If (X, d) is a metric space and $A \subseteq X$, then*

$$|\text{dist}\,(x, A) - \text{dist}\,(y, A)| \leq d(x, y)$$

for all x, y in X.

Proof. If $a \in A$, $d(x, a) \leq d(x, y) + d(y, a)$; so taking the infimum over all a in A we get $\text{dist}\,(x, A) \leq \inf\{d(x, y) + d(y, a) : a \in A\} = d(x, y) + \text{dist}\,(y, A)$. Reversing the roles of x and y we have $\text{dist}\,(y, A) \leq d(x, y) + \text{dist}\,(x, A)$, whence we get the inequality. ∎

5.4.8. Corollary. *If A is a non-empty subset of X, then $f : X :\to \mathbb{R}$ defined by $f(x) = \text{dist}\,(x, A)$ is a continuous function.*

In the next proof we'll use the preceding results to construct continuous functions with specified behavior.

5.4.9. Theorem (Urysohn's[3] Lemma). *If A and B are two disjoint closed subsets of X, then there is a continuous function $f : X \to \mathbb{R}$ having the following properties:*

(a) $0 \leq f(x) \leq 1$ *for all x in X;*
(b) $f(x) = 0$ *for all x in A; and*
(c) $f(x) = 1$ *for all x in B.*

Proof. Define $f : X \to \mathbb{R}$ by

$$f(x) = \frac{\text{dist}\,(x, A)}{\text{dist}\,(x, A) + \text{dist}\,(x, B)}$$

which is well-defined since the denominator never vanishes (Why?). It is easy to check that f has the desired properties. ∎

Don't be fooled by the simple proof; the result is powerful. We did a lot of work before this theorem to be able to concoct such a simple proof.

5.4.10. Corollary. *If F is a closed subset of X and G is an open set containing F, then there is a continuous function $f : X \to \mathbb{R}$ such that $0 \leq f(x) \leq 1$ for all x in X, $f(x) = 1$ when $x \in F$, and $f(x) = 0$ when $x \notin G$.*

Proof. In Urysohn's Lemma, take A to be the complement of G and $B = F$. ∎

Now we extend the concept of a uniformly continuous function to the metric space setting. The definition is just a transliteration of the definition in the case of $\mathbb{R}$.

[3] Pavel Samuilovich Urysohn was born in 1898 in Odessa, Ukraine. He was awarded his habilitation in June 1921 from the University of Moscow, where he remained as an instructor. He began his work in analysis but switched to topology where he made several important contributions, especially in developing a theory of dimension. His work attracted the attention of the mathematicians of the day, and in 1924 he set out on a tour of the major universities in Germany, Holland, and France, meeting with Hausdorff, Hilbert, and others. That same year, while swimming off the coast of Brittany, France, he drowned. He is buried in Batz-sur-Mer in Brittany. In just three years he left his mark on mathematics.

5.4.11. Definition. A function $f : (X, d) \to (Z, \rho)$ between two metric spaces is *uniformly continuous* if for every $\epsilon > 0$ there is a δ such that $\rho(f(x), f(y)) < \epsilon$ when $d(x, y) < \delta$.

5.4.12. Example. (a) We also extend the concept of a Lipschitz function. A function $f : (X, d) \to (Z, \rho)$ is a *Lipschitz function* if there is a constant $M > 0$ such that $\rho(f(x), f(y)) \leq Md(x, y)$ for all x, y in X. A ready collection of examples occurs by letting I be an interval in $\mathbb{R}$ and letting $f : I \to \mathbb{R}$ be a continuously differentiable function with $|f'(x))| \leq M$ for all x in I. Thus $|f(x) - f(y)| = |\int_y^x f'(t)dt| \leq \int_y^x |f'(t)|dt \leq M|x - y|$. As we saw in Proposition 1.7.14, every Lipschitz function is uniformly continuous and the same is true for functions on a metric space.

(b) We note that when $A \subseteq X$, the function $x \mapsto \operatorname{dist}(x, A)$ is a Lipschitz function by Proposition 5.4.7. Thus the distance function gives rise to a plentiful source of uniformly continuous functions on any metric space.

Exercises

(1) Let (X, d) and (Z, ρ) be metric spaces, $A \subseteq X$, $f : A \to (X, d)$, and let a be a limit point of A. Say that $\lim_{x \to a} f(x) = z$ if for every $\epsilon > 0$ there is a $\delta > 0$ such that $\rho(f(x), z) < \epsilon$ when $0 < d(x, a) < \delta$. Fix the set A and its limit point a. (a) If $a \in A$, show that f is continuous at a if $\lim_{x \to a} f(x) = f(a)$. (b) Show that $\lim_{x \to a} f(x) = z$ if and only if for every sequence $\{x_n\}$ in A that converges to a we have that $f(x_n) \to z$.

(2) If (X, d) is a metric space, $f : B(a; r) \to \mathbb{R}$ is continuous at a with $f(a) = 0$, and $g : B(a; r) \to \mathbb{R}$ is a function satisfying $|g(x)| \leq M$ for some constant M and all x in X (but not necessarily continuous), then fg is continuous at a.

(3) If $f : (X, d) \to (Z, \rho)$ is continuous, A is a dense subset of X, and $z \in Z$ such that $f(a) = z$ for every a in A, show that $f(x) = z$ for every x in X.

(4) If $f : (X, d) \to (Z, \rho)$ is both continuous and surjective and A is a dense subset of X, show that $f(A)$ is a dense subset of Z.

(5) Prove the equivalence of (b) and (c) in Theorem 5.4.3 by using only the sequential definition of continuity.

(6) In Theorem 5.4.3, give an independent proof that shows that conditions (a) and (c) are equivalent. (Here, "independent" means that the proof should not use the equivalence of (a) and (b) or of (b) and (c).)

(7) (a) Prove the statements made in parts (a) and (b) of Example 5.4.6. (b) Prove (5.4.6(c)). (Hint: Write the function as the composition of continuous functions.) (c) Prove (5.4.6(d)). (d) Prove the statement made in Example 5.4.6(f). (e) Prove (5.4.6(g)).

(8) If $f : \mathbb{R}^2 \to \mathbb{R}^2$ is defined by

$$f(x, y) = \begin{cases} \frac{x^2 - y^2}{x - y} & \text{if } x \neq y \\ x - y & \text{if } x = y \end{cases}$$

where is f continuous?

(9) If f and g are continuous functions from (X, d) into $\mathbb{R}$ prove that $f \vee g(x) = \max\{f(x), g(x)\}$ and $f \wedge g(x) = \min\{f(x), g(x)\}$ are continuous. (See Proposition 1.7.11.)

(10) Is the composition of two uniformly continuous functions a uniformly continuous function?

(11) Note that Proposition 5.4.5 says that if (X, d) is a metric space and we define the collection of all continuous functions $f : X \to \mathbb{R}$ by $C(X)$, then $C(X)$ is a vector space over $\mathbb{R}$. Show that $C(X)$ is a finite dimensional vector space if and only if X is a finite set. (Hint: use Urysohn's Lemma.)

(12) Let (X, d) be a metric space and let $\mathcal{U}$ denote the set of all uniformly continuous functions from X into $\mathbb{R}$. (a) If $f, g \in \mathcal{U}$, show that $f + g \in \mathcal{U}$. In words, $\mathcal{U}$ is a vector space over $\mathbb{R}$. (b) If $f, g \in \mathcal{U}$, show by an example that it does not necessarily follow that $fg \in \mathcal{U}$. If, however, the functions are also bounded, then $fg \in \mathcal{U}$. (A function $f : X \to (Z, \rho)$ is *bounded* if $f(X)$ is a bounded subset of Z.) (c) Can you give some conditions under which the quotient of two functions in $\mathcal{U}$ is uniformly continuous?

(13) Is the function π defined in Example 5.4.6(g) uniformly continuous?

(14) If X, Y, Z are metric spaces and $f : X \times Y \to Z$, say that f is *separately continuous* at (a, b) if the following hold: (i) the function from Y into Z defined by $y \mapsto f(a, y)$ is continuous at b; (ii) the function from X into Z defined by $x \mapsto f(x, b)$ is continuous at a. (a) Show that if f is continuous at (a, b), then it is separately continuous at (a, b). (b) Show that the function $f : \mathbb{R}^2 \to \mathbb{R}$ defined by

$$f(x, y) = \begin{cases} \frac{xy}{x^2+y^2} & \text{if } (x, y) \neq (0, 0) \\ 0 & \text{if } (x, y) = (0, 0) \end{cases}$$

is separately continuous at $(0, 0)$ but not continuous there.

5.5. Compactness

To a large extent this section is concerned with extending the Bolzano–Weierstrass Theorem to the setting of metric spaces. But the concept of compactness involves much more. In fact as we progress we will develop other ideas that are useful in the study of $\mathbb{R}^p$ and beyond.

If $\mathcal{G}$ is a collection of subsets of X and $E \subseteq X$, then $\mathcal{G}$ is a *cover* of E if $E \subseteq \bigcup\{G : G \in \mathcal{G}\}$. A *subcover* of E is a subset $\mathcal{G}_1$ of $\mathcal{G}$ that is also a cover of E. Finally we say that $\mathcal{G}$ is an *open cover* of E if $\mathcal{G}$ is a cover and every set in the collection $\mathcal{G}$ is open.

5.5.1. Definition. A subset K of the metric space (X, d) is said to be *compact* if every open cover of K has a finite subcover.

We mention that the term "open cover" in this definition can be replaced by "cover by subsets of K that are relatively open." See Exercise 2.

It is easy to find examples of sets that are not compact. Specifically, the open interval $(0, 1)$ is not compact. In fact if we put $G_n = (n^{-1}, 1)$, then $\mathcal{G} = \{G_n : n \in \mathbb{N}\}$ is an open cover of the interval that has no finite subcover. Similarly $\mathbb{R}$ is not compact since $\{(-n, n) : n \in \mathbb{N}\}$ is an open cover of $\mathbb{R}$ that has no finite subcover. We can easily see that every finite subset of X is compact, but finding non-trivial examples of compact sets requires us to first prove some results.

5.5.2. Proposition. *Let (X, d) be a metric space.*

(a) *If K is a compact subset of X, then K is closed and bounded.*
(b) *If K is compact and F is a closed set contained in K, then F is compact.*
(c) *The continuous image of a compact subset is a compact subset. That is, if $f : (X, d) \to (Z, \rho)$ is continuous and K is a compact subset of X, then $f(K)$ is a compact subset of Z.*

Proof. (a) If $x \notin K$, then for each z in K let $r_z, s_z > 0$ such that $B(z; r_z) \cap B(x; s_z) = \emptyset$. Now $\{B(z; r_z) : z \in K\}$ is an open cover of K. Since K is compact, there are points $z_1, \ldots, z_n$ in K such that $K \subseteq \bigcup_{k=1}^n B(z_k; r_{z_k})$. Let $s = \min\{s_{z_k} : 1 \le k \le n\}$. Note that $B(x; s) \cap K = \emptyset$; in fact, if there is a y in $B(x; s) \cap K$, then there is a k such that $y \in B(x; s) \cap B(z_k; r_{z_k}) \subseteq B(x; s_{z_k}) \cap B(z_k; r_{z_k})$, which contradicts the choice of the numbers s_{z_k} and r_{z_k}. Therefore $B(x; s) \subseteq X \backslash K$. Since x was arbitrary, this says that $X \backslash K$ is open and so K is closed. Also for any point x_0 in X, $\{B(x_0; n) : n \in \mathbb{N}\}$ is an open cover of K; hence there is a finite subcover. But the sets in this cover are increasing, so there is an integer n such that $K \subseteq B(x_0; n)$ and so K is bounded.

(b) Let $\mathcal{G}$ be an open cover of F and observe that since F is closed, $\{X \backslash F\} \cup \mathcal{G}$ is an open cover of K. The existence of a finite subcover of K implies there is a finite sub-collection of $\mathcal{G}$ that covers F.

(c) Let $f : X \to (Z, \rho)$ be a continuous function and assume K is a compact subset of X; we want to show that $f(K)$ is a compact subset of Z. Let $\mathcal{U}$ be an open cover of $f(K)$ in (Z, ρ). Since f is continuous it follows that $\mathcal{G} = \{f^{-1}(U) : U \in \mathcal{U}\}$ is an open cover of K (5.4.3(b)). Therefore there are sets $U_1, \ldots, U_n$ in $\mathcal{U}$ such that $K \subseteq \bigcup_{k=1}^n f^{-1}(U_k)$. It follows that $f(K) \subseteq \bigcup_{k=1}^n U_k$. ∎

We can use the preceding proposition to prove the EVT for continuous functions on a compact metric space.

5.5.3. Corollary. *If (X, d) is a compact metric space and $f : X \to \mathbb{R}$ is a continuous function, then there are points a and b in X such that $f(a) \le f(x) \le f(b)$ for all x in X.*

Proof. We have from Proposition 5.5.2 that $f(X)$ is a closed and bounded subset of $\mathbb{R}$. Put $\alpha = \inf\{f(x) : x \in X\}$, $\beta = \sup\{f(x) : x \in X\}$. Since $f(X)$ is closed, $\alpha, \beta \in f(X)$; this proves the corollary. ∎

Before extending the Bolzano–Weierstrass Theorem, we need two more definitions.

5.5.4. Definition. Say that a subset K of the metric space (X, d) is *totally bounded* if for any radius $r > 0$ there are points $x_1, \ldots, x_n$ in K such that $K \subseteq \bigcup_{k=1}^{n} B(x_k; r)$. A collection $\mathcal{F}$ of subsets of K has the *finite intersection property* or FIP if whenever $F_1, \ldots, F_n \in \mathcal{F}, \bigcap_{k=1}^{n} F_k \neq \emptyset$.

The following is the main result on compactness in metric spaces.

5.5.5. Theorem. *The following statements are equivalent for a closed subset K of a metric space (X, d).*

(a) *K is compact.*
(b) *If $\mathcal{F}$ is a collection of closed subsets of K having the FIP, then*

$$\bigcap_{F \in \mathcal{F}} F \neq \emptyset.$$

(c) *Every sequence in K has a convergent subsequence.*
(d) *Every infinite subset of K has a limit point.*
(e) *(K, d) is a complete metric space that is totally bounded.*

Proof. (a) *implies* (b). Let $\mathcal{F}$ be a collection of closed subsets of K having the FIP. Suppose $\bigcap \{F : F \in \mathcal{F}\} = \emptyset$. If $\mathcal{G} = \{X \backslash F : F \in \mathcal{F}\}$, then it follows that $\mathcal{G}$ is an open cover of X and therefore of K. By (a), there are $F_1, \ldots, F_n$ in $\mathcal{F}$ such that $K \subseteq \bigcup_{j=1}^{n} (X \backslash F_j) = X \backslash [\bigcap_{j=1}^{n} F_j]$. But since each F_j is a subset of K this implies $\bigcap_{j=1}^{n} F_j = \emptyset$, contradicting the fact that $\mathcal{F}$ has the FIP.

(b) *implies* (a). Let $\mathcal{G}$ be an open cover of K and put $\mathcal{F} = \{K \backslash G : G \in \mathcal{G}\}$. Since $\mathcal{G}$ covers K, $\bigcap \{K \backslash G : G \in \mathcal{G}\} = \emptyset$. Thus $\mathcal{F}$ cannot have the FIP and there must be a finite number of sets $G_1, \ldots, G_n$ in $\mathcal{G}$ with $\emptyset = \bigcap_{j=1}^{n} (K \backslash G_j)$. But this implies that $\{G_1, \ldots, G_n\}$ is a finite cover of K. Hence K is compact.

(d) *implies* (c). Assume $\{x_n\}$ is a sequence of distinct points in K. By (d), $\{x_n\}$ has a limit point; since K is closed, that limit point must be in K. We are tempted here to invoke (5.3.15(a)), but we have to manufacture an actual subsequence of the original sequence. That is, we must find $\{x_{n_k}\}$ such that $x_{n_k} \to x$ and $n_1 < n_2 < \cdots$ This takes a little bit of care and effort, which we leave to the interested reader.

(c) *implies* (d). If S is an infinite subset, then S has a sequence of distinct points $\{x_n\}$; by (c) there is a subsequence $\{x_{n_k}\}$ that converges to some point x. It follows that x is a limit point of S. (Details?)

(a) *implies* (d). Assume that (d) is false. So there is an infinite subset S of K with no limit point; it follows that there is an infinite sequence $\{x_n\}$ in S with no limit point. Thus for each $n \geq 1$, $F_n = \{x_k : k \geq n\}$ contains all its limit points and is therefore closed. Also $\bigcap_{n=1}^{\infty} F_n = \emptyset$. But each finite subcollection of $\{F_1, F_2, \ldots\}$ has non-empty intersection, contradicting (b), which is equivalent to (a).

(a) *implies* (e). First let $\{x_n\}$ be a Cauchy sequence in K. Since (a) implies (d), which is equivalent to (c), there is an x in K and a subsequence $\{x_{n_k}\}$ such that $x_{n_k} \to x$. But this implies $x_n \to x$ by Proposition 5.2.5. Hence (K, d) is complete. To show that K is totally bounded, just note that $\{B(x; r) : x \in K\}$ is an open cover of K for any $r > 0$.

(e) *implies* (c). Fix an infinite sequence $\{x_n\}$ in K and let $\{\epsilon_n\}$ be a decreasing sequence of positive numbers such that $\epsilon_n \to 0$. By (e) there is a covering of K by a finite number of balls of radius ϵ_1. Thus there is a ball $B(y_1; \epsilon_1)$ that contains an infinite number of points from $\{x_n\}$; let $\mathbb{N}_1 = \{n \in \mathbb{N} : d(x_n, y_1) < \epsilon_1\}$. Now consider the sequence $\{x_n : n \in \mathbb{N}_1\}$ and balls of radius ϵ_2. As we just did, there is a point y_2 in K such that $\mathbb{N}_2 = \{n \in \mathbb{N}_1 : d(y_2, x_n) < \epsilon_2\}$ is an infinite set. Using induction we can show that for each $k \geq 1$ we get a point y_k in K and an infinite set of positive integers $\mathbb{N}_k$ such that $\mathbb{N}_{k+1} \subseteq \mathbb{N}_k$ and $\{x_n : n \in \mathbb{N}_k\} \subseteq B(y_k; \epsilon_k)$. If $F_k = \text{cl}\,\{x_n : n \in \mathbb{N}_k\}$, then $F_{k+1} \subseteq F_k$ and $\text{diam}\, F_k \leq 2\epsilon_k$. Since K is complete, Cantor's Theorem implies that $\bigcap_{k=1}^{\infty} F_k = \{x\}$ for some point x in X. Now, using a small induction argument, pick integers n_k in $\mathbb{N}_k$ such that $n_k < n_{k+1}$. It follows that $\{x_{n_k}\}$ is a subsequence of the original sequence and $x_{n_k} \to x$.

(e) *implies* (a). We first prove the following.

5.5.6. Claim. If K satisfies (c) and $\mathcal{G}$ be an open cover of K, then there is an $r > 0$ such that for each x in K there is a G in $\mathcal{G}$ such that $B(x; r) \subseteq G$.

Let $\mathcal{G}$ be an open cover of K and suppose the claim is false; so for every $n \geq 1$ there is an x_n in K such that $B(x_n; n^{-1})$ is not contained in any set G in $\mathcal{G}$. By (c) there is an x in K and a subsequence $\{x_{n_k}\}$ such that $x_{n_k} \to x$. Since $\mathcal{G}$ is a cover, there is a G in $\mathcal{G}$ such that $x \in G$; choose a positive ϵ such that $B(x; \epsilon) \subseteq G$. Let $n_k > 2\epsilon^{-1}$ such that $x_{n_k} \in B(x; \epsilon/2)$. If $y \in B(x_{n_k}; n_k^{-1})$, then $d(x, y) \leq d(x, x_{n_k}) + d(x_{n_k}, y) < \epsilon/2 + n_k^{-1} < \epsilon$, so that $y \in B(x; \epsilon) \subseteq G$. Thus $B(x_{n_k}; n_k^{-1}) \subseteq G$, contradicting the restriction imposed on x_{n_k}. This establishes the claim.

From here it is easy to complete the proof. We know that (e) implies (c), so for an open cover $\mathcal{G}$ of K let $r > 0$ be the number guaranteed by Claim 5.5.6. Now let $x_1, \ldots, x_n \in K$ such that $K \subseteq \bigcup_{k=1}^{n} B(x_k; r)$ and for $1 \leq k \leq n$ let $G_k \in \mathcal{G}$ such that $B(x_k; r) \subseteq G_k$. $\{G_1, \ldots, G_n\}$ is the sought after finite subcover.

(c) *implies* (e). If $\{x_n\}$ is a Cauchy sequence in K, then (c) implies it has a convergent subsequence; by Proposition 5.2.5 the original sequence converges. Thus (K, d) is complete. Now fix an $r > 0$. Let $x_1 \in K$; if $K \subseteq B(x_1; r)$, we are done. If not, then there is a point x_2 in $K \backslash B(x_1; r)$. Once again, if $K \subseteq B(x_1; r) \cup B(x_2; r)$, we are done; otherwise pick an x_3 in $K \backslash [B(x_1; r) \cup B(x_2; r)]$. Continue. If this process does not stop after a finite number of steps, we produce an infinite sequence $\{x_n\}$ in K with $d(x_n, x_m) \geq r$ whenever $n \neq m$. But this implies that this sequence can have no convergent subsequence, contradicting (c). ∎

As an application we tweak Cantor's Theorem (5.3.18) and address a question that arose in connection with it.

5.5.7. Corollary. *If (X, d) is compact and $\{F_n\}$ is a sequence of non-empty closed subsets of X such that $F_1 \supseteq F_2 \supseteq \cdots$, then $\bigcap_{n=1}^{\infty} F_n \neq \emptyset$.*

The proof of this corollary is clear since a decreasing sequence of closed non-empty sets has the FIP.

Compactness is one of the most important properties in mathematics. Many good things follow from it as we shall see in this book and the reader will continue to see as (s)he continues his/her career. The point is that compact sets are "almost" finite in a very precise sense, and this approximation of finiteness often suffices to allow us to carry out an argument for a compact set that we can easily make for a finite set. We already saw this in Corollary 5.5.3.

We might observe that the Bolzano–Weierstrass Theorem together with Theorem 5.5.5 show that every closed and bounded interval in $\mathbb{R}$ is a compact set. By Proposition 5.5.2 the converse is also true. The next result extends this to $\mathbb{R}^p$.

5.5.8. Theorem (Heine[4]–Borel[5] Theorem). *A subset of $\mathbb{R}^p$ is compact if and only if it is closed and bounded.*

Proof. If K is compact, then K is closed and bounded by Proposition 5.5.2. Now assume that K is closed and bounded. It follows that there are bounded intervals $[a_1, b_1], \ldots, [a_p, b_p]$ in $\mathbb{R}$ such that $K \subseteq [a_1, b_1] \times \cdots \times [a_p, b_p]$. Suppose $\{x_n\}$ is a sequence in K with $x_n = (x_n^1, \ldots, x_n^p)$. Thus $\{x_n^1\}$ is a sequence in $[a_1, b_1]$, so the Bolzano–Weierstrass Theorem implies it has a convergent subsequence. The notation in this proof could become grotesque if we do the standard things, so we depart from the standard. Denote the convergent subsequence by $\{x_n^1 : n \in \mathbb{N}_1\}$, where $\mathbb{N}_1$ is an infinite subset of $\mathbb{N}$ with its natural ordering. We have that the limit exists, so put $x^1 = \lim_{n \in \mathbb{N}_1} x_n^1$. Now consider the sequence $\{x_n^2 : n \in \mathbb{N}_1\}$ in $[a_2, b_2]$. It has a convergent subsequence $\{x_n^2 : n \in \mathbb{N}_2\}$ with $x^2 = \lim_{n \in \mathbb{N}_2} x_n^2$. Continue and we get $\mathbb{N}_p \subseteq \cdots \subseteq \mathbb{N}_1 \subseteq \mathbb{N}$ and $x^k = \lim_{n \in \mathbb{N}_k} x_n^k$ for $1 \le k \le p$. It follows that $\{x_n : n \in \mathbb{N}_p\}$ is a subsequence of the original sequence and it converges to $x = (x^1, \ldots, x^p)$; it must be that $x \in K$ since K is closed. By Theorem 5.5.5, K is compact. ∎

5.5.9. Example. For the metric space $\mathbb{Q}$, if $a, b \in \mathbb{Q}, a < b$, the set $F = \{x \in \mathbb{Q} : a \le x \le b\} = \mathbb{Q} \cap [a, b]$ is closed and bounded but not compact.

....

[4] Heinrich Eduard Heine was born in 1821 in Berlin, the eighth of nine children. He received his doctorate in 1842 from Berlin; in 1844 he received the habilitation at Bonn where he was appointed a privatdozent. In 1856 he was made professor at the University of Halle, where he remained for the rest of his career. In 1850 he married Sophie Wolff from Berlin, and, over the years, they had five children. He worked on partial differential equations and then on special functions – Legendre polynomials, Lam functions, and Bessel functions. He made significant contributions to spherical harmonics and introduced the concept of uniform continuity. It was in 1872 that he gave a proof of the present theorem. It requires some scholarship to discover the differences in the contribution to this result between him and Borel, who published it in 1895. He died in 1881 in Halle.

[5] Emile Borel was born in Saint Affrique in the south of France in 1871. He published his first two papers in 1890, two years before receiving his doctorate in Paris and joining the faculty at Lille. He returned to Paris in 1897. In 1909 a special chair in the Theory of Functions was created for him at the Sorbonne. During World War I he was very supportive of his country and was put in charge of a central department of research. He also spent time at the front and in 1918 he was awarded the Croix de Guerre. In 1928 he set up the Institute Henri Poincaré. He was one of the founders of the modern theory of functions along with Baire and Lebesgue and he also worked on divergent series, complex variables, probability, and game theory. He continued to be very active in the French government, serving in the French Chamber of Deputies (1924–36) and as Minister of the Navy (1925–40). He died in 1956 in Paris.

The next theorem extends Theorem 1.7.16. In fact the proof of that result will also extend to the present setting. The details are left to the reader in Exercise 9.

5.5.10. Theorem. *If (X, d) is a compact metric space and $f : X \to (Z, \rho)$ is a continuous function, then f is uniformly continuous.*

5.5.11. Proposition. *A compact metric space is separable and complete.*

Proof. The completeness of (X, d) is explicit in Theorem 5.5.5(e). To prove that (X, d) is separable again Theorem 5.5.5(e) implies that for each natural number n we can find a finite set F_n such that $X = \bigcup \{B(x; n^{-1}) : x \in F_n\}$. Put $F = \bigcup_{n=1}^{\infty} F_n$; we will show that this countable set F is dense in X. In fact if x_0 is an arbitrary point in X and $\epsilon > 0$, choose n such that $n^{-1} < \epsilon$. Thus there is a point x in $F_n \subseteq F$ with $d(x_0, x) < n^{-1} < \epsilon$, proving that $x_0 \in \text{cl } F$.   ∎

In light of the preceding proposition any closed and bounded subset of $\mathbb{R}^p$ is separable.

5.5.12. Theorem. *Let (X, d) and (Z, ρ) be compact metric spaces and assume D is a dense subset of X. If $f : D \to Z$ is uniformly continuous, then there is a uniformly continuous function $F : X \to Z$ that extends f; that is, the function F satisfies $F(x) = f(x)$ for all x in D.*

Proof. If $x \in X$, let $\{x_n\}$ be a sequence in D that converges to x. We claim that $\{f(x_n)\}$ is a Cauchy sequence in Z. In fact, if $\epsilon > 0$, let $\delta > 0$ such that $\rho(f(y), f(w)) < \epsilon$ when $y, w \in D$ and $d(y, w) < \delta$. Now there is an integer N with $d(x_n, x_m) < \delta$ for $m, n \geq N$. Hence $\rho(f(x_n), f(x_m)) < \epsilon$ when $m, n \geq N$. Thus $\{f(x_n)\}$ is a Cauchy sequence in Z. By Proposition 5.5.11, (Z, ρ) is complete, so there is a ζ in Z such that $f(x_n) \to \zeta$. If $\{y_n\}$ is another sequence in D that converges to x, then the same argument shows there is a ζ' in Z such that $f(y_n) \to \zeta'$. But we have that $d(x_n, y_n) \to 0$, so the uniform continuity of f implies $\rho(f(x_n), f(y_n)) \to 0$ (Why?). Thus $\zeta = \zeta'$. This means we can define a function $F : X \to Z$ by letting $F(x) = \lim_n f(x_n)$ for any sequence $\{x_n\}$ in D that converges to x. Clearly F is an extension of f.

Now to show that this function F is uniformly continuous. Let $\epsilon > 0$. Choose $\delta > 0$ such that $\rho(f(x), f(y)) < \epsilon$ when $x, y \in D$ and $d(x, y) < \delta$. Fix points x and y in X with $d(x, y) < \delta/3$ and let $\{x_n\}$ and $\{y_n\}$ be sequences in D that converge to x and y. Choose N such that $d(x_n, x) < \delta/3$ and $d(y_n, y) < \delta/3$ when $n \geq N$. So for all $n \geq N$ we have that $d(x_n, y_n) \leq d(x_n, x) + d(x, y) + d(y, y_n) < \delta$. Hence $\rho(f(x_n), f(y_n)) < \epsilon$ when $n \geq N$. it follows that $\rho(F(x), F(y)) \leq \epsilon$. Therefore F is uniformly continuous.   ∎

The above proposition fails if f is not uniformly continuous. See Example 5.6.13 in the next section.

5.5.13. Theorem (Dini's[6] Theorem).   *If (X, d) is compact, $\{f_n\}$ is an increasing sequence of continuous functions on X, and f is a continuous function on X such that $f_n(x) \to f(x)$ for all x in X, then $f_n \to_u f$ on X.*

Proof. Let $\epsilon > 0$ and for each $n \geq 1$ let $F_n = \{x \in X : f(x) \geq f_n(x) + \epsilon\}$. We note that because both f and f_n are continuous, F_n is a closed subset of X. Also because the sequence of functions is increasing, $F_1 \supseteq F_2 \supseteq \cdots$. We want to show that there is an integer N such that $F_N = \emptyset$. In fact if this is done, then for all $n \geq N$ we have that $f_n(x) \leq f(x) < f_n(x) + \epsilon$ for all x in $[a, b]$. That is, $|f(x) - f_n(x)| < \epsilon$ for all x in X and all $n \geq N$, establishing that $f_n \to_u f$ on X. But if there is an x in $\bigcap_{n=1}^{\infty} F_n$, then $x \in F_n$ for every $n \geq 1$. This says that $f(x) \geq f_n(x) + \epsilon$ for all n, contradicting the fact that $f_n(x) \to f(x)$. ∎

Exercises

(1) Show that the union of a finite number of compact sets is compact.

(2) If K is a subset of (X, d), show that K is compact if and only if every cover of K by relatively open subsets of K has a finite subcover.

(3) Show that the closure of a totally bounded set is totally bounded.

(4) Show that a totally bounded set is bounded. Is the converse true?

(5) If $\{E_n\}$ is a sequence of totally bounded sets such that $\operatorname{diam} E_n \to 0$, show that $\bigcup_{n=1}^{\infty} E_n$ is totally bounded.

(6) If (X, d) is a complete metric space and $E \subseteq X$, show that E is totally bounded if and only if $\operatorname{cl} E$ is compact.

(7) (a) If G is an open set and K is a compact set with $K \subseteq G$, show that there is a $\delta > 0$ such that $\{x : \operatorname{dist}(x, K) < \delta\} \subseteq G$. (b) Find an example of an open set G in a metric space X and a closed, non-compact subset F of G such that there is no $\delta > 0$ with $\{x : \operatorname{dist}(x, F) < \delta\} \subseteq G$.

---

[6] Ulisse Dini was born in 1845 in Pisa. He attended Scuola Normale Superiore in Pisa, a teaching preparatory college. In 1865 he won a scholarship for study abroad, which he used to go to Paris for a year. During this time he was very active in research, eventually publishing seven papers based on the work he had done. He returned to Pisa and an academic position at the university. Dini's life span was a period of myriad political developments in Italy as the country worked its way toward unification. This is not a period for the casual historian. In 1859 there was a war with Austria and in 1861 the Kingdom of Italy was formed, though it did not include Venice and Rome. (Can you imagine Italy without Venice and Rome?) It was not until 1866 that Venice became part of the kingdom and Rome had to wait until 1870. The turmoil affected Dini as he progressed in both his academic as well as a political career. In 1871 he took over Betti's chair of analysis, and that same year he was elected to the Pisa City Council. In 1877 he was appointed to a second chair in mathematics, and in 1880 he was elected as a representative of Pisa to the national assembly. In 1883 he was appointed rector of the university, holding the position for two years. In 1892 he was elected a senator in the Italian Parliament; and in 1908 he became director of the Scuola Normale Superiore, a position he held for the rest of his life. This was a period of development in mathematical analysis when the turmoil seemed to be trying to parody the events in Italy; mathematicians sought rigorous proofs of results that had only casually been established, and they sought the boundaries of validity for these results. Dini seemed to flourish in this undertaking. In addition to the present result, there is one in Fourier series that bears his name. He also wrote several influential texts. He died in 1918 in Pisa.

(8) If K is a compact set in (X, d) and G is an open set such that $K \subseteq G$, show that there is an open set G_1 such that cl G_1 is compact and $K \subseteq G_1 \subseteq$ cl $G_1 \subseteq G$.

(9) Give the details of the proof of Theorem 5.5.10.

(10) For two subsets A and B of X, define the distance from A to B by dist $(A, B) = \inf\{d(a, b) : a \in A, b \in B\}$. (a) Show that dist $(A, B) =$ dist $(B, A) =$ dist (cl A, cl B). (b) If A and B are two disjoint closed subsets of X such that B is compact, then dist $(A, B) > 0$. (c) Give an example of two disjoint closed subsets A and B of the plane $\mathbb{R}^2$ such that dist $(A, B) = 0$. (d) Is this exercise related to Exercise 7?

(11) Consider the metric space ℓ^∞ (see Exercise 5.3.12) and show that

$$\left\{ x = \{x_n\} \in \ell^\infty : \sup_n |x_n| \leq 1 \right\}$$

is not totally bounded and, therefore, not compact.

(12) Say that a metric space is *σ-compact* if it can be written as the union of a countable number of compact sets. (a) Give three examples of σ-compact metric spaces that are not compact. (b) Show that a σ-compact metric space is separable.

5.6. Connectedness

Consider the following two examples of subsets of $\mathbb{R}$. The first is the set $X = [0, 1] \cup (2, 3)$ and the second is $Y = [0, 1] \cup (1, 2)$. In X there are two distinct "parts," $[0, 1]$ and $(2, 3)$. In the second we have written Y as the union of two disjoint sets, but these two sets are not really separate "parts." (The term "parts" will be made technically precise soon, though we won't use that term.) In a sense writing Y as the union of those two sets is just accidental. We could just as well have written $Y = [0, 1) \cup [1, 2)$ or $Y = [0, \frac{1}{2}] \cup (\frac{1}{2}, 2)$ or even $Y = [0, 2)$. What's the true difference between the two sets X and Y?

Note that in the metric space (X, d) of the last paragraph, the set $[0, 1]$ is simultaneously both an open and a closed subset of X. For example, $B_X(1; \frac{1}{2}) = \{x \in X : |x - 1| < \frac{1}{2}\} = (\frac{1}{2}, 1] \subseteq [0, 1]$. The set X is an example of what we now define as a set that is not connected, or, more succinctly, a disconnected set.

5.6.1. Definition. A metric space (X, d) is *connected* if there are no subsets of X that are simultaneously open and closed other than X and $\emptyset$. If $E \subseteq X$, we say that E is connected if (E, d) is connected. If E is not connected we will say that it is *disconnected* or a *non-connected* set.

An equivalent formulation of connectedness is to say that (X, d) is connected provided that when $X = A \cup B$ where $A \cap B = \emptyset$ and both A and B are open (or closed), then either $A = \emptyset$ or $B = \emptyset$. This is the sense of our use of the term "parts" in the introduction of this section; for the set X there, $A = [0, 1]$ and $B = (2, 3)$ are two disjoint, non-trivial sets that are both open and closed in X. Let's collect some of these equivalent formulations of connectedness. The proof is left to the reader (Exercise 1); we will often use this result without referencing it.

5.6.2. Proposition. *If (X, d) is a metric space, the following statements are equivalent.*

(a) *X is connected.*
(b) *If $X = A \cup B$ where $A \cap B = \emptyset$ and both A and B or open (or closed), then either $A = \emptyset$ or $B = \emptyset$.*
(c) *If $A \subseteq X$, $A \neq \emptyset$, and A is both open and closed, then $A = X$.*

The next result can be considered an example, but it is much more than that.

5.6.3. Proposition. *A subset of $\mathbb{R}$ is connected if and only if it is an interval.*

Proof. Assume that $X = [a, b]$ and let's show that X is connected. (The proof that other types of intervals are connected is Exercise 2.) Assume that $[a, b] = A \cup B$, where both A and B are open and $A \cap B = \emptyset$. One of these sets contains the point a; suppose $a \in A$. Note that A is also closed. We want to show that $A = X$. Since A is open there is an $\epsilon > 0$ such that $[a, a + \epsilon) \subseteq A$. Put $r = \sup\{\epsilon : [a, a + \epsilon) \subseteq A\}$. We claim that $[a, a + r) \subseteq A$. In fact if $a \leq x < a + r$ then the definition of the supremum implies there is an $\epsilon > 0$ such that $\epsilon < r$, $[a, a + \epsilon) \subseteq A$, and $a \leq x < a + \epsilon$; thus $x \in A$. Now A is also closed and $[a, a + r) \subseteq A$, so it must also be that $a + r \in A$. If $a + r \neq b$, then the fact that A is open implies that there is a $\delta > 0$ such that $(a + r - \delta, a + r + \delta) \subseteq A$. But this means that $[a, a + r + \delta) = [a, a + r) \cup (a + r - \delta, a + r + \delta) \subseteq A$, contradicting the definition of r. Therefore $a + r = b$ and we have that $A = [a, b] = X$. (So $B = \emptyset$.)

Recall that by Lemma 1.6.6 a subset I of $\mathbb{R}$ is an interval if and only if whenever $a, b \in I$ with $a < b$ it follows that $[a, b] \subseteq I$. Assume that X is a non-empty connected subset of $\mathbb{R}$ and $a, b \in X$ with $a < b$; suppose $a < c < b$. If $c \notin X$, let $A = X \cap (-\infty, c)$, $B = X \cap (c, \infty)$. Clearly A and B are open subsets relative to X; since $c \notin X$ we also have that $A = X \cap (-\infty, c]$, $B = X \cap [c, \infty)$, and they are also closed relative to X. Since $a \in A$ and $b \in B$, neither is empty, contradicting the assumption that X is connected. Hence $c \in X$. That is, $[a, b] \subseteq X$ and by Lemma 1.6.6 X is an interval. ∎

5.6.4. Theorem. *The continuous image of a connected set is connected.*

Proof. Let $f : (X, d) \to (Z, \rho)$ be a continuous function and E a connected subset of X; we want to show that $f(E)$ is a connected subset of Z. By replacing X with E, we may assume $X = E$ is connected; by replacing Z with $f(E)$, we may assume that f is surjective. We must now show that Z is connected. If D is a subset of Z that is both open and closed, then the continuity of f implies $f^{-1}(D)$ is both open and closed in X. Since X is connected, $f^{-1}(D)$ is either $\emptyset$ or X. But since f is surjective, this implies D is either $\emptyset$ or Z. Thus Z is connected. ∎

The preceding theorem together with Proposition 5.6.3 allows us to deduce the IVT (1.7.7) in this setting.

5.6.5. Corollary (Intermediate Value Theorem). *If $f : (X, d) \to \mathbb{R}$ is continuous, X is connected, $a, b \in f(X)$ with $a < b$, then for any number c in the interval $[a, b]$ there is a point x in X with $f(x) = c$.*

Proof. We know that $f(X)$ is a connected subset of $\mathbb{R}$ so that it must be an interval (5.6.3). Since $a, b \in f(X)$, it must be that $[a, b] \subseteq f(X)$. ∎

5.6.6. Example. (a) If $x, y \in \mathbb{R}^p$, then the straight line segment $[x, y] \equiv \{ty + (1 - t)x : 0 \le t \le 1\}$ is connected. In fact $t \mapsto ty + (1 - t)x$ is a continuous function from the unit interval into $\mathbb{R}^p$. Since the unit interval is connected, so is its image under this continuous mapping.

(b) In $\mathbb{R}^p$ the balls $B(x; r)$ are connected. In fact let A be a non-empty subset of $B(x; r)$ that is both relatively open and closed and fix a point y in A. If $z \in B(x; r)$, then the line segment $[y, z] \subseteq B(x; r)$ and $[y, z]$ is connected by part (a). But it follows that $A \cap [y, z]$ is a non-empty subset of this line segment that is both relatively open and relatively closed. Thus $[y, z] \subseteq A$; in particular the arbitrary point z from $B(x; r)$ belongs to A so that $A = B(x; r)$.

(c) Any circle in $\mathbb{R}^2$ is connected. In fact if $X = \{(x, y) \in \mathbb{R}^2 : (x - a)^2 + (y - b)^2 = r\}$, then $\gamma : [0, 2\pi] \to \mathbb{R}^2$ defined by $\gamma(\theta) = (a + r \cos \theta, b + r \sin \theta)$ is a continuous function and $\gamma([0, 2\pi]) = X$. By Theorem 5.6.4, X is connected.

(d) If (X, d) is the discrete metric space, then X is not connected if X has more than one point. In fact each singleton set $\{x\}$ is a non-empty set that is both open and closed.

5.6.7. Proposition. *Let (X, d) be a metric space.*

(a) *If $\{E_i : i \in I\}$ is a collection of connected subsets of X such that $E_i \cap E_j \ne \emptyset$ for all i, j in I, then $E = \bigcup_{i \in I} E_i$ is connected.*
(b) *If $\{E_n : n \ge 1\}$ is a sequence of connected subsets of X such that $E_n \cap E_{n+1} \ne \emptyset$ for each n, then $E = \bigcup_{n=1}^{\infty} E_n$ is connected.*

Proof. (a) Let A be a non-empty subset of E that is relatively open and closed. If $i \in I$, then $A \cap E_i$ is a relatively closed and open subset of E_i; if $A \cap E_i \ne \emptyset$, then the fact that E_i is connected implies $E_i \subseteq A$. Now since A is non-empty, there is at least one i such that $E_i \subseteq A$. But then for every j in I, the hypothesis implies there is a point in E_j that belongs to A; thus $E_j \subseteq A$. Therefore $A = E$ and E must be connected.

(b) Let A be a non-empty relatively open and closed subset of E. Since $A \ne \emptyset$, there is some integer N with $A \cap E_N \ne \emptyset$. But $A \cap E_N$ is both relatively open and closed in E_N, so $E_N \subseteq A$ by the connectedness of E_N. By hypothesis, $E_{N-1} \cap E_N \ne \emptyset$, so $E_{N-1} \cap A \ne \emptyset$ and it follows that $E_{N-1} \subseteq A$. Continuing we get that $E_n \subseteq A$ for $1 \le n \le N$. Since $E_N \cap E_{N+1} \ne \emptyset$, similar arguments show that $E_{N+1} \subseteq A$. Continuing we get that $E_n \subseteq A$ for all $n \ge 1$. That is, $E = A$ and so E is connected. ∎

5.6.8. Corollary. *The union of two intersecting connected subsets of a metric space is connected.*

See Exercise 8.

5.6.9. Definition. If (X, d) is a metric space, a *component* of X is a maximal connected subset of X.

The word "maximal" in the definition means that there is no connected set that properly contains it. So if C is a component of X and D is a connected subset of X with $C \subseteq D$, then $D = C$.

A component is the correct interpretation of the word "part" used in the introduction of this section. The set X in that introduction has two components. Notice that a connected metric space has only one component. In the discrete metric space each singleton set is a component.

5.6.10. Proposition. *For any metric space, every connected set is contained in a component, distinct components are disjoint, and the union of all the components is the entire space.*

Proof. Fix a connected subset D of X and let $\mathcal{C}_D$ denote the collection of all connected subsets of X that contain D. According to Proposition 5.6.7(a), $C = \bigcup\{A : A \in \mathcal{C}_D\}$ is connected. Clearly C is a component and contains D. By taking $D = \{x\}$ in what was just established, we have that every point of X is contained in a component so that the union of all the components is X. Finally note that if C and D are two components and $C \cap D \neq \emptyset$, then $C \cup D$ is connected by Corollary 5.6.8; so it has to be that $C = C \cup D = D$ by the maximality of C and D. That is, distinct components are disjoint sets. ∎

A consequence of the preceding proposition is that the components form a partition of X – they divide the space X into a collection of pairwise disjoint connected sets. The next result says that the components are all closed, the proof of which emphasizes once again that, when discussing relatively open and closed sets, you must be aware of what the universe is.

5.6.11. Proposition. *If C is a connected subset of the metric space X and $C \subseteq Y \subseteq \operatorname{cl} C$, then Y is connected.*

Proof. Let A be a non-empty subset of Y that is both relatively open and closed, and fix a point x_0 in A. By Proposition 5.3.5 there is an open subset G of X such that $A = Y \cap G$. Since $x_0 \in \operatorname{cl} C$ and $x_0 \in G$, there must be a point x in $G \cap C = A \cap C$; that is, $A \cap C$ is a non-empty relatively open subset of C. Since A is relatively closed in Y, Proposition 5.3.5 and an analogous argument implies that $A \cap C$ is also relatively closed in C. Since C is connected, $C = C \cap A \subseteq A$. That is, $C \subseteq A \subseteq Y \subseteq \operatorname{cl} C$ so that A is both closed in Y and dense in Y; hence $A = Y$, and it must be that Y is connected. ∎

5.6.12. Corollary. *The closure of a connected set is connected and each component is closed.*

In light of the preceding proposition and Example 5.6.6(b), if $x \in \mathbb{R}^p$ and $B(x; r) \subseteq E \subseteq \overline{B}(x; r)$, then E is connected. Here is an example that will illustrate additional properties as we proceed. In fact this example is used so often it has a name, the *topologist's sine curve*.

5.6.13. Example. $X = \{(x, \sin x^{-1}) \in \mathbb{R}^2 : 0 < x \leq 1\} \cup \{(0, 0)\}$ is connected. In fact, $f : (0, 1] \to X$ defined by $f(x) = (x, \sin x^{-1})$ is a continuous function, so $C = f((0, 1])$ is connected. Since $C \subseteq X \subseteq \operatorname{cl} C$, X is connected by Proposition 5.6.11. The space X consists of the graph of the function $\sin x^{-1}$ for $0 < x \leq 1$ together with the origin. Note that instead of the origin we could have added to the graph any subset of $\{(0, y) : -1 \leq y \leq 1\}$ and the resulting set would still be connected.

We now focus on $\mathbb{R}^p$. Recall Proposition 1.6.7 where it is proved that every open subset of $\mathbb{R}$ is the union of a countable number of pairwise disjoint open intervals. We can interpret this as saying that the components of an open subset G of $\mathbb{R}$ are open intervals, which are not closed subsets of $\mathbb{R}$. They are, however, relatively closed in G. We want to extend Proposition 1.6.7 to $\mathbb{R}^p$; of course the extension is not literal. We start with an idea in $\mathbb{R}^p$ that is useful.

5.6.14. Definition. If $E \subseteq \mathbb{R}^p$, $x, y \in E$, and $\epsilon > 0$, say that there is an ϵ-*chain* from x to y in E when there are a finite number of points $x_1, \ldots, x_n$ in E such that: (i) for $1 \leq k \leq n$, $B(x_k; \epsilon) \subseteq E$; (ii) for $2 \leq k \leq n$, $x_{k-1} \in B(x_k; \epsilon)$; (iii) $x_1 = x$ and $x_n = y$.

The proof of the next lemma is Exercise 11.

5.6.15. Lemma. *If $r > 0$ and $z \in \mathbb{R}^p$, then for any pair of points x and y in $B(z; r)$ and all sufficiently small ϵ there is an ϵ-chain in $B(z; r)$ from x to y.*

5.6.16. Proposition. *Consider the metric space $\mathbb{R}^p$.*

(a) *If G is an open subset of $\mathbb{R}^p$, then every component of G is open and there are countably many components.*
(b) *An open subset G of $\mathbb{R}^p$ is connected if and only if for any x, y in G there is an $\epsilon > 0$ such that there is an ϵ-chain in G from x to y.*

Proof. (a) If H is a component of G and $x \in H$, choose $r > 0$ such that $B(x; r) \subseteq G$. Since $B(x; r)$ is also connected (5.6.6), Corollary 5.6.8 implies $H \cup B(x; r)$ is connected. Since this is also a subset of G it follows that $H = H \cup B(x; r)$, so $B(x; r) \subseteq H$ and H is open. Because $\mathbb{R}^p$ is separable, there is a countable dense subset D. Now since each component is open, each component contains an element of D and different components contain different points. If there are uncountably many components this would show there is an uncountable subset of D (Why?), which is nonsense.

(b) Assume that the open set G satisfies the stated condition and let's prove that G is connected. Fix x and let H be the component of G that contains x; we want to show that $H = G$. We know that H is open by part (a). If $y \in G$ there is an $\epsilon > 0$ such that there is an ϵ-chain $x_1, \ldots, x_n$ in G from x to y. Since $x_{k-1} \in B(x_{k-1}; \epsilon) \cap B(x_k; \epsilon)$ for $2 \le k \le n$, Proposition 5.6.7(b) says $B = \bigcup_{k=1}^{n} B(x_k; \epsilon)$ is connected. Condition (i) of the definition of an ϵ-chain implies $B \subseteq G$, and so $B \subseteq H$. In particular, $y \in H$. Since y was arbitrary, $H = G$ and G is connected.

Now assume that G is connected. Fix a point x in G and let

$$D = \{y \in G : \text{ there is an } \epsilon > 0 \text{ and an } \epsilon\text{-chain in } G \text{ from } x \text{ to } y\}$$

The strategy of the proof will be to show that D is both relatively open and closed in G; since it is not empty ($x \in D$), it will then follow that $D = G$ and so G will have been shown to satisfy the condition. If $y \in D$, let $\epsilon > 0$ and let $x_1, \ldots, x_n$ be an ϵ-chain from x to y. It follows from the definition of an ϵ-chain that $B(y; \epsilon) \subseteq D$. Thus D is open. Now suppose $z \in G \cap \operatorname{cl} D$ – this is the relative closure of D in G. (Why?) Choose $r > 0$ such that $B(z; r) \subseteq G$; so $B(z; r) \cup D \subseteq G$. Since $z \in \operatorname{cl} D$, there is a point y in $B(z; r) \cap D$. Let $x_0 = x, x_1, \ldots, x_n$ be an ϵ-chain from x to y. By Exercise 10 there is an ϵ'-chain from x to y in G whenever $0 < \epsilon' < \epsilon$. Applying this with $0 < \epsilon' < \min\{\epsilon, r\}$, we may assume $\epsilon < r$. Using Lemma 5.6.15 we see that this implies there is an ϵ-chain in G from x to z. Thus $z \in D$ and so D is relatively closed in G. ∎

We note that (i) of the definition of an ϵ-chain was used to establish that the condition in part (b) was sufficient for connectedness. Without this, the result is false as we see in the following example.

5.6.17. Example. Let $X = \{(x, y) \in \mathbb{R}^2 : y > x^{-1}\} \cup \{(x, y) \in \mathbb{R}^2 : y < 0\}$. Clearly X is not connected; in fact it has two components. It is also easy to see that if $\bar{x} = (x, y)$ with $y < 0$ and $\bar{y} = (w, z)$ with $z > w^{-1}$, then for all sufficiently small ϵ there are points $\bar{x}_1, \ldots, \bar{x}_n$ in X such that for $2 \le k \le n, \bar{x}_{k-1} \in B(\bar{x}_k; \epsilon)$, and $\bar{x}_1 = \bar{x}$ and $\bar{x}_n = \bar{y}$.

Unlike open subsets of $\mathbb{R}^p$, components in an arbitrary metric space are not necessarily open. (Exercise 13.)

Exercises

(1) Prove Proposition 5.6.2.
(2) Prove that open and half-open intervals are connected, completing the proof of half of Proposition 5.6.3.
(3) If (X, d) is connected and $f : X \to \mathbb{R}$ is a continuous function such that $|f(x)| = 1$ for all x in X, show that f must be constant.
(4) Show that X is connected if and only if whenever a, b are two points in X, there is a connected subset E of X such that $a, b \in E$.

(5) If A is a subset of X, define the *characteristic function* of A as the function $\chi_A :$ $X \to \mathbb{R}$ such that $\chi_A(x) = 1$ when $x \in A$ and $\chi_A(x) = 0$ when $x \notin A$. Show that A is simultaneously open and closed if and only if χ_A is continuous.

(6) Look at the preceding exercise for the definition of the characteristic function on a set X. (a) If A and B are subsets of X, which function is $\chi_A \chi_B$? (b) Which function is $\chi_A + \chi_B$? (c) What is the characteristic function of the empty set? (d) Is $1 - \chi_A$ the characteristic function of some set?

(7) Can you think of any way to generalize Proposition 5.6.7(a) and obtain a theorem whose conclusion is that $E = \bigcup \{E_i : i \in I\}$ is connected?

(8) Give an example of two connected sets whose intersection is not connected.

(9) Let $E \subseteq \mathbb{R}^2$ and let

$$E_1 = \{x \in \mathbb{R} : \text{ there is a } y \in \mathbb{R} \text{ such that } (x, y) \in E\}$$

(a) Show that if E is compact, then E_1 is compact. (b) Give an example of a set E such that E_1 is compact, but E is not. (c) Show that if E is connected, then E_1 is connected. (d) Give an example of a set E such that E_1 is connected, but E is not.

(10) If E is a subset of $\mathbb{R}^p$, $x, y \in E$, and $\epsilon > 0$ such that there is an ϵ-chain from x to y, show that for any ϵ' with $0 < \epsilon' < \epsilon$ there is an ϵ'-chain from x to y.

(11) Prove Lemma 5.6.15.

(12) A *polygon* $[x, x_1, \ldots, x_{n-1}, y]$ is the union of straight line segments in $\mathbb{R}^p$ of the form

$$[x, x_1], [x_1, x_2], \ldots, [x_{n-1}, y]$$

(a) Show that a polygon is a connected subset of $\mathbb{R}^p$. (b) Show that an open subset G of $\mathbb{R}^p$ is connected if and only if for any two points x and y in G there is a polygon $[x, x_1, \ldots, x_{n-1}, y]$ contained in G.

(13) Give an example of a metric space (X, d) such that the components are not all open.

5.7. The Space of Continuous Functions

In this section we extend the definition of uniform convergence encountered in §4.1 to sequences of continuous functions defined on a metric space. After extending the definition and proving some basic propositions, we will shift the point of view by considering the set of all bounded continuous functions as a metric space. We continue to assume that (X, d) is a fixed metric space.

5.7.1. Definition. If $\{f_n\}$ is a sequence of functions from (X, d) into $\mathbb{R}$, say that $\{f_n\}$ *converges uniformly* to a function f if for every $\epsilon > 0$ there is an N such that $|f_n(x) - f(x)| < \epsilon$ for all x in X and all $n \geq N$. In symbols this is written as $f_n \to_u f$ on X.

I suspect that many readers could have guessed this was the definition of uniform convergence. We could have extended the definition further by having the sequence

$\{f_n\}$ consist of functions from (X, d) into a second metric space (Z, ρ). We will not explore this here, however. The next result extends Proposition 4.1.3 to the present context. The same proof will work here.

5.7.2. Proposition. *If $\{f_n\}$, $\{g_n\}$, and $\{h_n\}$ are three sequences of functions from X into $\mathbb{R}$ such that $g_n(x) \leq f_n(x) \leq h_n(x)$ for all x in X and there is a function $f : X \to \mathbb{R}$ such that $g_n \to_u f$ and $h_n \to_u f$ on X, then $f_n \to_u f$ on X.*

The next proposition extends Proposition 4.1.5 and the same proof will work.

5.7.3. Proposition. *If $\{f_n\}$ is a sequence of bounded functions from X into $\mathbb{R}$ that converges uniformly to a function $f : X \to \mathbb{R}$, then f is a bounded function and the sequence of functions $\{f_n\}$ is uniformly bounded; that is, there is a constant M such that $|f_n(x)| \leq M$ for all x in X and all $n \geq 1$.*

The next result extends Theorem 4.1.6. The same proof works provided we make a small modification, which is carried out below.

5.7.4. Theorem. *If $\{f_n\}$ is a sequence of bounded continuous functions from X into $\mathbb{R}$ and $f_n \to_u f$ on X, then $f : X \to \mathbb{R}$ is a continuous function.*

Proof. Fix a point a in X; we want to show that f is continuous at a. If $\epsilon > 0$, then by the uniform convergence we can choose an integer n such that $|f(x) - f_n(x)| < \epsilon/3$ for all x in X. Since f_n is continuous there is a $\delta > 0$ such that $|f_n(x) - f_n(a)| < \epsilon/3$ when $d(x, a) < \delta$. Thus for any x in X with $d(x, a) < \delta$, $|f(x) - f(a)| \leq |f(x) - f_n(x)| + |f_n(x) - f_n(a)| + |f_n(a) - f(a)| < 3(\epsilon/3) = \epsilon$. By definition, f is continuous at a. ∎

The next result extends Theorem 4.1.8 to uniformly continuous functions defined on a metric space. A slight modification of that proof will work here. The details are left to the reader.

5.7.5. Theorem. *Let $X \subseteq \mathbb{R}$. If $\{f_n\}$ is a sequence of bounded uniformly continuous functions on X and $f_n \to_u f$ on X, then $f : X \to \mathbb{R}$ is a uniformly continuous function.*

Now we change the context of the discussion.

5.7.6. Definition. We denote the set of all continuous functions from X into $\mathbb{R}$ by $C(X)$. $C_b(X)$ denotes the subset of all those functions f in $C(X)$ that are bounded. That is, $C_b(X)$ consists of all those continuous functions f from X into $\mathbb{R}$ such that

$$\|f\| \equiv \sup\{|f(x)| : x \in X\} < \infty$$

Here we call upon the student to have a somewhat different point of view and to begin to think of continuous functions as points in the space $C(X)$. We'll see this repeatedly as we progress.

An *algebra* is a vector space $\mathcal{A}$ over $\mathbb{R}$ in which there is a multiplication and all the usual distributive and associative laws hold: for all a, b, c in $\mathcal{A}$ and t in $\mathbb{R}$

$$a(b+c) = ab + ac$$
$$(b+c)a = ba + ca$$
$$a(bc) = (ab)c$$
$$t(ab) = (ta)b = a(tb)$$

In other words an algebra is both a vector space and a ring such that the distributive laws hold. If there is a multiplicative identity, it is denoted by 1. Note that when the algebra $\mathcal{A}$ has an identity, it contains a replica of the constants: $t = t1$ for every t in $\mathbb{R}$.

5.7.7. Proposition. *For any metric space (X, d) the following hold.*

(a) *$C(X)$ and $C_b(X)$ are algebras.*
(b) *If we set $d(f, g) = \|f - g\|$ for f and g in $C_b(X)$, then d defines a metric. Whenever we discuss $C_b(X)$ as a metric space we refer to this metric.*
(c) *A sequence $\{f_n\}$ in $C_b(X)$ converges to f in the metric space $C_b(X)$ if and only if $f_n \to_u f$.*
(d) *$C_b(X)$ is a complete metric space.*

Proof. Part (a) was already proved in Proposition 5.4.5. (Also see Exercise 5.4.11.) Part (b) is left as Exercise 1.

(c) If $f_n \to_u f$ and $\epsilon > 0$, let n be an integer such that $|f_n(x) - f(x)| < \epsilon$ for all $n \geq N$. It follows that $d(f_n, f) = \|f_n - f\| \leq \epsilon$ for $n \geq N$; so $f_n \to f$ in $C_b(X)$. For the converse assume that when $n \geq N$, $\|f_n - f\| = \sup\{|f_n(x) - f(x)| : x \in X\} < \epsilon$. It follows that for $n \geq N$ we have that $|f_n(x) - f(x)| < \epsilon$ for every x in X; so $f_n \to_u f$.

(d) If $\{f_n\}$ is a Cauchy sequence in $C_b(X)$, let $\epsilon > 0$ and choose N such that $\|f_n - f_m\| < \epsilon$ when $m, n \geq N$. Thus for each x in X, $\{f_n(x)\}$ is a Cauchy sequence in $\mathbb{R}$; hence it has a limit. Put $f(x) = \lim_n f_n(x)$ so that $f : X \to \mathbb{R}$ defines a function. Let $\epsilon > 0$, and choose N such that $\|f_n - f_m\| < \epsilon/6$ when $m, n \geq N$. For $m, n \geq N$ and x an arbitrary point in X we have that

5.7.8 $$|f(x) - f_n(x)| \leq |f(x) - f_m(x)| + |f_m(x) - f_n(x)| < \frac{2\epsilon}{6}$$

Now hold $n \geq N$ fixed. Since the above inequality holds for any x in X it also holds for any y in X and so

$$|f(x) - f(y)| \leq |f(x) - f_n(x)| + |f_n(x) - f_n(y)| + |f_n(y) - f(y)|$$
$$< \frac{2\epsilon}{3} + |f_n(x) - f_n(y)|$$

But f_n is continuous so there is a $\delta > 0$ such that $|f_n(x) - f_n(y)| < \epsilon/3$ when $d(x, y) < \delta$. Putting all this together gives that $|f(x) - f(y)| \leq \epsilon$ when $d(x, y) < \delta$;

that is, f is a continuous function. By (5.7.8) $f_n \to_u f$, so $f \in C_b(X)$ (Proposition 5.7.3). ∎

What we have done so far in our study of $C(X)$ and $C_b(X)$ can be thought of as fundamental orientation and housekeeping. Now we begin the process of establishing one of the important results about $C(X)$ when X is compact. We begin with a lemma about $[0, 1]$.

5.7.9. Lemma. *There is a sequence of polynomials $\{p_n\}$ such that $p_n \to \sqrt{x}$ in $C[0, 1]$.*

Proof. We define the sequence of polynomials inductively by letting $p_1(x) = 0$ and $p_{n+1}(x) = p_n(x) + \frac{1}{2}[x - p_n(x)^2]$. It is an easy job to show that all these polynomials are positive on $[0, 1]$ (Exercise 4).

Claim. For all $n \geq 1$ and all x in $[0, 1]$, $p_n(x) \leq \sqrt{x}$.

In fact this is clear for $n = 1$, so assume it is true for n. Since $p_n(x) \leq \sqrt{x} \leq 1$ when $x \in [0, 1]$, we have that $\frac{1}{2}[\sqrt{x} + p_n(x)] \leq 1$. Hence

$$p_{n+1}(x) = p_n(x) + \frac{1}{2}[x - p_n(x)^2]$$
$$= p_n(x) + \frac{1}{2}[\sqrt{x} - p_n(x)][\sqrt{x} + p_n(x)]$$
$$\leq p_n(x) + [\sqrt{x} - p_n(x)]$$
$$= \sqrt{x}$$

establishing the claim.

In light of this claim $x - p_n(x)^2 \geq 0$ so that $p_n(x) \leq p_{n+1}(x)$. Thus for each x in $[0, 1]$ we have that $\{p_n(x)\}$ is an increasing sequence in $\mathbb{R}$ that is bounded above by $\sqrt{x}$; hence there is a t in $\mathbb{R}$ such that $p_n(x) \to t \leq \sqrt{x}$. Therefore

$$t = \lim_{n\to\infty} p_{n+1}(x) = \lim_{n\to\infty} p_n(x) + \lim_{n\to\infty} \frac{1}{2}[x - p_n(x)^2] = t + \frac{1}{2}[x - t^2]$$

implying that $t = \sqrt{x}$. Thus the sequence $\{p_n\}$ is increasing and converges pointwise to the continuous function $f(x) = \sqrt{x}$ on $[0, 1]$. By Dini's Theorem (5.5.13), the convergence is uniform. ∎

There is a part of me that doesn't like the preceding proof.

It works, so what's not to like and why am I raising the issue? The rub is that it's not clear how the sequence $\{p_n\}$ is thought up and maybe this disturbs some readers as well. The source of the polynomials is connected to Taylor's Theorem (2.5.6) and Newton's method from calculus. An explication of this would be a distraction. So let's just say we obtain one approximation $p_n(x)$, add on half the amount by which $p_n(x)^2$ fails to equal x, then we keep our fingers crossed. Lo and behold, it works.

Recall the definition of the maximum and minimum of two functions (1.7.9), $f \vee g$ and $f \wedge g$. (That definition was given for functions defined on subsets of $\mathbb{R}$ but the same definition extends.)

5.7.10. Lemma. *If $\mathcal{A}$ is a closed subalgebra of $C_b(X)$ that contains the identity and $f, g \in \mathcal{A}$, then $f \vee g, f \wedge g \in \mathcal{A}$.*

Proof. In light of Exercise 3, it suffices to show that $|f| \in \mathcal{A}$ whenever $f \in \mathcal{A}$. To do this fix f in $\mathcal{A}$; we may assume that f is not the function 0. Put $a = \|f\|$ and observe that $a^{-2}f^2 \in \mathcal{A}$ and takes its values in $[0, 1]$. Let $\{p_n\}$ be the sequence of polynomials from the preceding lemma such that $p_n \to_u \sqrt{x}$ on $[0, 1]$. Since $\mathcal{A}$ is an algebra that contains the constant functions, $p_n \circ (a^{-2}f^2) = p_n((a^{-2}f^2)) \in \mathcal{A}$ for each $n \geq 1$. It follows that in $C_b(X)$, $p_n((a^{-2}f^2)) \to \sqrt{a^{-2}f^2} = a^{-1}|f|$. (Why?) Because $\mathcal{A}$ is a closed subalgebra of $C_b(X)$ we have that $|f| \in \mathcal{A}$. ∎

A collection of functions $\mathcal{S}$ in $C(X)$ is said to *separate the points* of X if whenever $x, y \in X$ and $x \neq y$, there is a function f in $\mathcal{S}$ with $f(x) \neq f(y)$. Both $C_b(X)$ and $C(X)$ separate the points of X. (You can show this by using Urysohn's Lemma, but it can be done directly.) If X is any subset of $\mathbb{R}$, the collection of all polynomials separates the points of X.

5.7.11. Theorem (Stone[7]–Weierstrass[8] Theorem). *If X is compact and $\mathcal{A}$ is a closed subalgebra of $C(X)$ that separates the points of X and contains the constant functions, then $\mathcal{A} = C(X)$.*

Proof. Fix an arbitrary f in $C(X)$ and let's show that $f \in \mathcal{A}$. We start with a simple claim.

Claim. If x and y are distinct points in X, there is a function $h_{x,y}$ in $\mathcal{A}$ such that $h_{x,y}(x) = f(x)$ and $h_{x,y}(y) = f(y)$.

....

[7] Marshall H Stone was born in 1902 in New York. His father was Harlan Stone who, after time as the dean of the Columbia Law School, became a member of the US Supreme Court, including a term as its chief justice. Marshall Stone entered Harvard in 1919 intending to study law. He soon diverted to mathematics and received his doctorate in 1926 under the direction of David Birkhoff. Though he had brief appointments at Columbia and Yale, most of his early career was spent at Harvard. His initial work continued the direction it took under Birkhoff, but in 1929 he started working on hermitian operators. His American Mathematical Society book, *Linear Transformations in Hilbert space and Their Applications to Analysis*, became a classic. Indeed, a reading of that book today shows how the arguments and clarity might easily lead to the conclusion that it is a contemporary monograph. During World War II he worked for the Navy and the War Department and in 1946 he left Harvard to become the chairman of the mathematics department at the University of Chicago. He himself said that this decision was arrived at because of "my conviction that the time was also ripe for a fundamental revision of graduate and undergraduate mathematical education." Indeed he transformed the department at Chicago. The number of theorems that bear his name is truly impressive. Besides the present theorem there is the Stone–Čech compactification, the Stone–von Neumann Theorem, the Stone Representation Theorem in Boolean algebra, and Stone's Theorem on one-parameter semigroups. He stepped down as chair at Chicago in 1952 and retired in 1968, but then went to the University of Massachusetts where he taught in various capacities until 1980. He loved to travel and on a trip to India in 1989 he died in Madras. He had 14 doctoral students.

[8] See Theorem 1.3.11 for a biographical note.

The proof is easy. Since $\mathcal{A}$ separates points there is a function g in $\mathcal{A}$ with $g(x) \neq g(y)$. Set

$$h_{x,y} = f(x) + [f(y) - f(x)]\frac{g - g(x)}{g(y) - g(x)}$$

The reader can verify that this function has the desired properties.

There are many functions in $\mathcal{A}$ that take on the values $f(x)$ and $f(y)$ at x and y. (Name one more.) For every such pair x, y, however, we fix one such function $h_{x,y}$. Now fix x in X and let $\epsilon > 0$. For each y in X put $G(y) = \{z \in X : h_{x,y}(z) < f(z) + \epsilon\}$. Note that the continuity of f and $h_{x,y}$ implies that $G(y)$ is open. Using the properties of $h_{x,y}$ we have that $x, y \in G(y)$. Hence $\{G(y) : y \in X\}$ is an open cover of X. By compactness there are points $y_1, \ldots, y_n$ such that $X = \bigcup_{j=1}^n G(y_j)$. Put

$$h_x = h_{x,y_1} \wedge \cdots \wedge h_{x,y_n}$$

Lemma 5.7.10 implies that $h_x \in \mathcal{A}$. Since $h_{x,y}(x) = f(x)$ for every y in X, we have that $h_x(x) = f(x)$; because the sets $\{G(y_j) : 1 \leq j \leq n\}$ cover X, we have that $h_x(z) < f(z) + \epsilon$ for every z in X.

For each x in X let $H(x) = \{z \in X : h_x(z) > f(z) - \epsilon\}$. Once again we note that $x \in H(x)$ and so $\{H(x) : x \in X\}$ is an open cover of X; let $\{H(x_i) : 1 \leq i \leq m\}$ be a finite subcover, and put

$$h = h_{x_1} \vee \cdots \vee h_{x_m}$$

Once again Lemma 5.7.10 implies $h \in \mathcal{A}$. It follows (Verify!) that for every z in X, $f(z) - \epsilon < h(z) < f(z) + \epsilon$. That is, $\|f - h\| < \epsilon$. since ϵ was arbitrary and $\mathcal{A}$ is closed, $f \in \mathcal{A}$. ∎

The proof of the next two corollaries is required in Exercise 5.

5.7.12. Corollary (Weierstrass Theorem). *For any closed and bounded interval $[a, b] \subseteq \mathbb{R}$, the polynomials are dense in $C[a, b]$.*

The preceding corollary is the reason that Weierstrass's name is attached to (5.7.11).

5.7.13. Corollary. *If X is a closed and bounded subset of $\mathbb{R}^p$, the set of polynomials in the p-variables $x_1, \ldots, x_p$ is dense in $C(X)$.*

The reader should examine Exercises 6 and 7.

Now we obtain another important result about $C(X)$ when X is compact: the characterization of the compact subsets of $C(X)$. We start with a definition.

5.7.14. Definition. If $\mathcal{F}$ is a subset of $C_b(X)$, then $\mathcal{F}$ is said to be *equicontinuous* if for every $\epsilon > 0$ and every point x_0, there is a neighborhood U of x_0 such that $|f(x) - f(x_0)| < \epsilon$ for all x in U and every f in $\mathcal{F}$.

Note that if we were considering a single f, the fact that for each ϵ and x_0 there is such a neighborhood U is just the fact that f is continuous at x_0. So every finite set in $C_b(X)$ is equicontinuous. The salient point in the definition is that one open neighborhood U of x_0 works for every function in the family $\mathcal{F}$. That is, the family of functions $\mathcal{F}$ is sufficiently constrained that there is a uniformity in the way that they are continuous — hence the prefix "equi".

5.7.15. Theorem (Arzelà[9]–Ascoli[10] Theorem). *If X is compact, then a subset $\mathcal{F}$ of $C(X)$ is totally bounded if and only if $\mathcal{F}$ is bounded and equicontinuous.*

Proof. Assume that $\mathcal{F}$ is totally bounded. Automatically $\mathcal{F}$ is bounded (Exercise 5.5.4). To establish equicontinuity let $\epsilon > 0$. So there are $f_1, \ldots, f_n$ in $\mathcal{F}$ such that $\mathcal{F} \subseteq \bigcup_{k=1}^{n}\{f \in C(X) : \|f - f_k\| < \epsilon/3\}$. If $x_0 \in X$, let U be a neighborhood of x_0 such that for $1 \le k \le n$, $|f_k(x) - f_k(x_0)| < \epsilon/3$ when $x \in U$. Thus if $f \in \mathcal{F}$ and we choose f_k with $\|f - f_k\| < \epsilon/3$, then $|f(x) - f(x_0)| \le |f(x) - f_k(x)| + |f_k(x) - f_k(x_0)| + |f_k(x_0) - f(x_0)| < \epsilon$. Hence $\mathcal{F}$ is equicontinuous.

Now assume that $\mathcal{F}$ is equicontinuous and bounded. Without loss of generality we can assume that $\|f\| \le 1$ for every f in $\mathcal{F}$. Fix $\epsilon > 0$; by equicontinuity we have that for each x in X there is an open neighborhood U_x of x such that $|f(y) - f(x)| < \epsilon/3$ whenever $y \in U_x$ and $f \in \mathcal{F}$. It follows that $\{U_x : x \in X\}$ is an open cover of X; by the compactness of X we can extract a finite subcover $\{U_{x_1}, \ldots, U_{x_n}\}$. Now choose $\alpha_1, \ldots, \alpha_m$ in $[-1, 1]$ such that $[-1, 1] \subseteq \bigcup_{j=1}^{m}\{\alpha : |\alpha - \alpha_j| < \epsilon/6\}$. Note that the collection of all ordered n-tuples of elements from the set $\{\alpha_1, \ldots, \alpha_m\}$ is finite. We don't want to consider all such n-tuples, however, but only the set B of those ordered n-tuples $b = (\beta_1, \ldots, \beta_n)$ with $\beta_1, \ldots, \beta_n \in \{\alpha_1, \ldots, \alpha_m\}$ such that there is a f_b in $\mathcal{F}$ with $|f_b(x_j) - \beta_j| < \epsilon/6$. Since for each f in $\mathcal{F}$ we have that $f(X) \subseteq [-1, 1]$ we see that $B \ne \emptyset$. For each b in B fix one such function f_b in $\mathcal{F}$; so $\{f_b : b \in B\}$ is a finite subset of $\mathcal{F}$. The fact that $\mathcal{F}$ is totally bounded holds once we establish the following.

Claim. $\mathcal{F} \subseteq \bigcup_{b \in B}\{f \in C(X) : \|f - f_b\| < \epsilon\}$.

If $f \in \mathcal{F}$, $\{f(x_1), \ldots, f(x_n)\} \subseteq [-1, 1]$ and so there is a $b = (\beta_1, \ldots, \beta_n)$ in B with $|f(x_k) - \beta_k| < \epsilon/6$ for $1 \le k \le n$. Thus $|f(x_k) - f_b(x_k)| < \epsilon/3$ for $1 \le k \le n$. For each x in X choose x_k with x in U_{x_k}. Hence $|f(x) - f_b(x)| \le |f(x) - f(x_k)| + |f(x_k) - f_b(x_k)| + |f_b(x_k) - f_b(x)| < \epsilon$. Since x was arbitrary, the claim is established. ∎

5.7.16. Corollary. *If X is compact and $\mathcal{F} \subseteq C(X)$, then $\mathcal{F}$ is compact if and only if $\mathcal{F}$ is closed, bounded, and equicontinuous.*

[9] Cesare Arzelà was born in 1847 in La Spezia, Italy. He received his doctorate from the university in Pisa under the direction of Enrico Betti. He held positions at Florence and Palermo before he became a professor in Bologna in 1880. His most famous work is this result, where he established the condition as sufficient for compactness. He died in 1912 in La Spezia.

[10] Giulio Ascoli was born in Trieste in 1843. He received his degree from Pisa and became a professor at Milan in 1872. He had a distinguished career and this theorem is his most notable result. He died in 1896 in Milan.

Exercises

(1) Prove Proposition 5.7.7(b).

(2) (a) Extend the definition of a uniformly Cauchy sequence given in Exercise 4.1.5 to the context of continuous functions from X into $\mathbb{R}$. (b) Show that a sequence $\{f_n\}$ in $(C_b(X), d)$ is a Cauchy sequence if and only if it is a uniformly Cauchy sequence.

(3) Show that if $f, g : X \to \mathbb{R}$, then $f \vee g = \frac{1}{2}(f + g + |f - g|)$ and $f \wedge g = \frac{1}{2}(f + g - |f - g|)$. (Hint: first prove it for numbers.)

(4) Show that each of the polynomials p_n in the proof of Lemma 5.7.9 satisfies $p_n(x) \geq 0$ for all x in $[0, 1]$. (Hint: Use induction.)

(5) Prove Corollaries 5.7.12 and 5.7.13.

(6) Let X be compact and assume $\mathcal{A}$ is a closed subalgebra of $C(X)$. Assume $\mathcal{A}$ contains the constants and there are two points x_1 and x_2 such that if $x, y \in X \backslash \{x_1, x_2\}$ then there is a function f in $\mathcal{A}$ with $f(x) \neq f(y)$.

    (a) Show that $\{f \in C(X) : f(x_1) = f(x_2)\} \subseteq \mathcal{A}$, with equality if and only if $f(x_1) = f(x_2)$ for every f in $\mathcal{A}$.

    (b) Show that $\mathcal{A} = \{f \in C[0, 2\pi] : f(0) = f(2\pi)\}$ is the closure of the set of all polynomials in $\sin x$ and $\cos x$. (Such polynomials are called *trigonometric polynomials*.)

(7) In the Stone–Weierstrass Theorem consider each possible pair of the conditions: (a) $\mathcal{A}$ is an algebra; (b) $\mathcal{A}$ separates points; and (c) $\mathcal{A}$ contains the constant functions. For each possible pair of these three conditions, find an example of a compact space X and a closed algebra $\mathcal{A} \subseteq C(X)$ such that $\mathcal{A}$ satisfies those two conditions but $\mathcal{A} \neq C(X)$.

(8) If $f \in C[0, 1]$ such that $\int_0^1 f(x)x^n \, dx = 0$ for all $n \geq 0$, show that $f = 0$.

(9) If (X, d) is a compact metric space and $\mathcal{F} \subseteq C(X)$, show that $\mathcal{F}$ is equicontinuous if and only if for every $\epsilon > 0$ there is a $\delta > 0$ such that when $d(x, y) < \delta$, $|f(x) - f(y)| < \epsilon$ for every f in $\mathcal{F}$

(10) If $\mathcal{F}$ is a family of functions in $C_b(X)$ that is equicontinuous at each point of X, does it follow that $\mathcal{F}$ is a bounded set? What happens if you also assume that X is compact?

6 Differentiation in Higher Dimensions

Here we want to extend the theory of differentiation to situations involving several variables. First we'll assume only the range of the function has more than one variable, then only the domain, and finally where both domain and range have more than a single dimension. After doing the necessary work to define the concept of a differentiable function in each of these three cases, we'll see some applications.

6.1. Vector-valued Functions

This section starts the chapter by discussing functions γ defined on a subset of $\mathbb{R}$ and taking values in $\mathbb{R}^p$ with $p \geq 1$; in other words, vector-valued functions of a single variable. Defining differentiability in this situation presents few difficulties and the resulting theory does not differ very much from the case where $p = 1$ that was presented in Chapter 2. The first task is to define $\lim_{t \to x} \gamma(t)$ when $\gamma : X \to \mathbb{R}^p$ for some subset X of $\mathbb{R}$. This is not a major problem since $\mathbb{R}^p$ is a metric space. To keep the discussion similar to that in Chapter 2, however, we start by finding a substitute for the absolute value of a real number. For any x in $\mathbb{R}^p$ let

$$\|x\| = \left[\sum_{j=1}^{p} x_j^2 \right]^{\frac{1}{2}}$$

The quantity $\|x\|$ is called the *norm* of x. Recall that $\|x\|^2 = \langle x, x \rangle$, where this latter symbol is the inner product in $\mathbb{R}^p$. See (5.1.4). So when $x, y \in \mathbb{R}^p$, $\|x - y\|$ is precisely the distance from x to y as defined in (5.1.2(c)) and makes $\mathbb{R}^p$ into a metric space. We note from Corollary 5.1.6 that the norm satisfies the triangle inequality: $\|x + y\| \leq \|x\| + \|y\|$.

Now when $X \subseteq \mathbb{R}$, a function $\gamma : X \to \mathbb{R}^p$ is a function from one metric space into another, so the definition of limits and continuity is inherited from that more general setting. (Also see Exercise 2.) Specifically $\lim_{t \to x} \gamma(t) = z$ means that for every $\epsilon > 0$ there is a $\delta > 0$ such that

$$\|\gamma(t) - z\| < \epsilon \text{ when } 0 < |t - x| < \delta$$

and $t \in X$. See Exercise 5.4.1.

(The author asks the reader's indulgence for a certain ambiguity in the notation here. We will use letters like x, y, etc. both for numbers in $\mathbb{R}$ and vectors in $\mathbb{R}^p$. This is to maintain some consistency with the past and future and hopefully will not cause confusion. Some people employ the convention of using boldface letters for the vectors, but that strikes me as cumbersome and not justified. It is also the case that in more advanced courses no such use of boldface is made. My policy, however, will call for a bit of extra attention and awareness of the context when reading this material. Maybe that's an extra benefit of such ambiguity.)

With the concept of limit there is almost no problem defining the derivative of such a function.

6.1.1. Definition. A function $\gamma : (a, b) \to \mathbb{R}^p$ is *differentiable* at x if

$$\lim_{t \to 0} \frac{\gamma(x + t) - \gamma(t)}{t}$$

exists. The value of this limit is denoted by $\gamma'(x)$ and is called the *derivative* of γ at x. Note that when the derivative exists, $\gamma'(x) \in \mathbb{R}^p$. γ is said to be *differentiable* on (a, b) if it is differentiable at each point. Observe that if $\gamma : [a, b] \to \mathbb{R}^p$ we can define $\gamma'(a)$ and $\gamma'(b)$ as we did in (2.2.2).

Also see Exercise 3.

Notice that when $\gamma : [a, b] \to \mathbb{R}^p$ is differentiable, then $\gamma' : [a, b] \to \mathbb{R}^p$, so we can speak of γ being continuously differentiable or twice differentiable or having any number of derivatives. We refer to a function that is continuously differentiable as a *smooth function*.

6.1.2. Definition. A *curve* in $\mathbb{R}^p$ is a continuous function $\gamma : [a, b] \to \mathbb{R}^p$. Say that the curve γ is *smooth* if γ is continuously differentiable. The *trace* of the curve is its image in $\mathbb{R}^p$ and is denoted by $\{\gamma\} = \{\gamma(t) : a \leq t \leq b\}$.

It would not surprise me if some readers are a bit uncomfortable when we define a curve as a function. Most think of a curve as a set of points, what we call the trace of the curve. For example, we might say the top arc of the circle in $\mathbb{R}^2$ centered at the origin and having radius r is a curve. This language is not, however, sufficiently precise to do mathematics. Defining it as a function allows us to better introduce and use the analysis we are developing to study curves. For example, we eventually want to examine the direction of the curve. This is easy to do when we define the curve as the function $\gamma : [a, b] \to \mathbb{R}^p$ as γ has the natural direction it inherits from the interval $[a, b]$; we simply look at the direction of $\gamma(t)$ as t goes from a to b. As another example, in (7.1.2) we define the length of a smooth curve, something that is awkward to formulate if a curve is defined as a set of points.

When the curve $\gamma : [a, b] \to \mathbb{R}^p$ is smooth, its derivative is a tangent vector to the curve. We'll examine this in a more general context in §6.7 below.

We state a version of Proposition 2.2.9 for the present situation.

6.1.3. Proposition. *A function $\gamma : (a, b) \to \mathbb{R}^p$ is differentiable at a point x in (a, b) if and only if there there is a function $\Gamma : (a, b) \to \mathbb{R}^p$ and a vector G in $\mathbb{R}^p$ such that $\lim_{y \to x} \Gamma(y) = 0$ and*

$$\gamma(y) - \gamma(x) = (y - x)[G + \Gamma(y)]$$

for all y in (a, b). When γ is differentiable at x, we have $G = \gamma'(x)$.

As in the case of a scalar-valued function, the proof is straightforward; see the proof of Proposition 2.2.9.

Most of the results for the derivatives of scalar-valued functions obtained in §2.2 carry over to this situation. In particular the derivative of the sum of two vector-valued functions is the sum of the derivatives. The straight product of two functions with values in $\mathbb{R}^p$ makes no sense, though we can employ the inner product of two such functions. Recall the inner product notation on $\mathbb{R}^p$ introduced when we proved the Cauchy–Schwarz Inequality:

$$\langle x, y \rangle = \sum_{j=1}^{p} x_j y_j$$

The essential properties of the inner product were listed in (5.1.4).

6.1.4. Proposition. *If γ and τ are two functions from (a, b) into $\mathbb{R}^p$ that are differentiable at x, then the function $f : (a, b) \to \mathbb{R}$ defined by $f(t) = \langle \gamma(t), \tau(t) \rangle$ is differentiable at x and*

$$f'(x) = \langle \gamma'(x), \tau(x) \rangle + \langle \gamma(x), \tau'(x) \rangle$$

Proof. The proof proceeds much like the proof of Proposition 2.2.5(b) except that we use the properties of the inner product to get

$$\frac{f(x+t) - f(x)}{t} = \frac{\langle \gamma(x+t), \tau(x+t) \rangle - \langle \gamma(x), \tau(x) \rangle}{t}$$

$$= \frac{\langle \gamma(x+t) - \gamma(x), \tau(x+t) \rangle}{t} + \frac{\langle \gamma(x), \tau(x+t) - \tau(x) \rangle}{t}$$

Now we let $t \to 0$ and appeal to Exercise 1 to get the result. ∎

In Exercises 2 and 3 we show that discussing the differentiability of a function $\gamma : (a, b) \to \mathbb{R}^p$ can be reduced to discussing the functions $t \mapsto \langle \gamma(t), e_j \rangle$, where $e_1, \ldots, e_p$ is the standard basis. In fact what we have done is demonstrate that there is little distinction between the theory of differentiation of functions with values in $\mathbb{R}$ and those with values in $\mathbb{R}^p$.

> *There is, however, a result for derivatives of functions from $[a, b]$ into $\mathbb{R}$ that is dramatically lacking when we consider vector-valued functions:* the Mean Value Theorem (2.3.4).

See Exercise 4.

Exercises

(1) If $\phi, \psi : (a, b) \to \mathbb{R}^p$, $\lim_{y \to x} \phi(y) = \Phi$, and $\lim_{y \to x} \psi(y) = \Psi$, show $\lim_{y \to x} \langle \phi(y), \psi(y) \rangle = \langle \Phi, \Psi \rangle$.

(2) For $\gamma : (a, b) \to \mathbb{R}^p$ and $1 \le j \le p$, let $\gamma_j : (a, b) \to \mathbb{R}$ be defined by $\gamma_j(t) = \langle \gamma(t), e_j \rangle$, where $e_1, \ldots, e_p$ are the standard basis vectors. (a) Show that if $x \in (a, b)$, then $\lim_{t \to x} \gamma(t) = G$ if and only if $\lim_{t \to x} \gamma_j(t) = \langle G, e_j \rangle$ for $1 \le j \le p$. (b) Show that γ is continuous on (a, b) if and only if each γ_j is continuous on (a, b).

(3) Use the notation established in the preceding exercise. (a) Show that γ is differentiable on (a, b) if and only if each γ_j is differentiable on (a, b). (b) If γ is differentiable on (a, b), show that $\gamma'(t) = (\gamma_1'(t), \ldots, \gamma_p'(t))$.

(4) Show that the function $\gamma : [0, 1] \to \mathbb{R}^2$ defined by $\gamma(t) = (t^2, t^3)$ does not satisfy $\gamma(1) - \gamma(0) = \gamma'(c)$ for any point c in $[0, 1]$.

6.2. Differentiable Functions, Part 1

Now we confront the differentiability of functions from an open subset of $\mathbb{R}^p$ into $\mathbb{R}$. In this situation the difficulties in defining differentiability are significant. We start by examining the partial derivatives of such functions, something everyone has brushed up against in calculus. Here we renew that acquaintance and perhaps broaden and deepen the encounter.

For a variety of reasons, including ease of notation, we will continue to denote the elements of $\mathbb{R}^p$ as vectors x rather than a p-tuple of real numbers $x = (x_1, \ldots, x_p)$ unless there is a reason for such specificity. As usual the *standard basis* of the vector space $\mathbb{R}^p$ will be denoted by $e_1, \ldots, e_p$. In other words $e_1 = (1, 0, \ldots, 0)$, $e_2 = (0, 1, 0, \ldots, 0), \ldots, e_p = (0, \ldots, 0, 1)$.

In this section, G will always denote an open subset of $\mathbb{R}^p$.

6.2.1. Definition. If $f : G \to \mathbb{R}$, $x \in G$, and $1 \le j \le p$, say that the *j-th partial derivative* of f exists at x provided

$$\lim_{t \to 0} \frac{f(x + te_j) - f(x)}{t}$$

exists. When this limit exists, it is denoted by

$$\frac{\partial f}{\partial x_j}(x) = \partial_j f(x)$$

There is a possible confusion in the term "j-th partial derivative." For example, the term "second partial derivative" could mean $\partial_2 f$ as defined above or the same phrase might mean $\partial^2 f/\partial x^2$ or $\partial^2 f/\partial x \partial y$. The context should indicate which we mean and the notation is certainly different.

What is happening here is that for x fixed in G we are considering the scalar-valued function $t \mapsto f(x + te_j)$ defined on a small interval about 0 and we are differentiating this with respect to t. This is why we insist that f be defined on an open set G so that for $1 \le j \le p$ there is an $\epsilon > 0$ such that $t \mapsto f(x + te_j)$ is defined for t in $(-\epsilon, \epsilon)$ and we can discuss its derivative. We are assuming the reader remembers from calculus how to carry out the partial differentiation: just treat all the variables $\{x_k : k \ne j\}$ as constants and differentiate the remainder as though it were a function of the variable x_j alone.

If the j-th partial derivative of $f : G \to \mathbb{R}$ exists throughout G, then this gives us another function $\partial_j f : G \to \mathbb{R}$. We can discuss the continuity of $\partial_j f$ as well as the existence of its partial derivatives. When $\partial_j f$ has a partial derivative with respect to x_i for some i we write that derivative as

$$\partial_{ji} f = \partial_i(\partial_j f)$$

Note the order of the indices i and j in $\partial_{ji} f$ as this is important. The question immediately arises whether $\partial_{ji} f = \partial_{ij} f$. This is not always the case.

6.2.2. Example. Define $f : \mathbb{R}^2 \to \mathbb{R}$ by

$$f(x, y) = \begin{cases} \frac{xy(x^2 - y^2)}{x^2 + y^2} & \text{when } (x, y) \ne (0, 0) \\ 0 & \text{when } (x, y) = (0, 0) \end{cases}$$

(In the calculations that follow, the reader is asked in Exercise 2 to supply the details.) It follows that $\partial_2 f(x, 0) = x$. Hence

$$\partial_{21} f(0, 0) = \lim_{x \to 0} \frac{\partial_2 f(x, 0) - \partial_2 f(0, 0)}{x} = 1$$

On the other hand $\partial_1 f(0, y) = -y$ and so $\partial_{12} f(0, 0) = -1$.

The key is whether those second partial derivatives are continuous. The next proof uses the MVT for derivatives of functions of one variable (2.3.4).

6.2.3. Proposition. *If $f : G \to \mathbb{R}$, $1 \le j, i \le p$, and both $\partial_i \partial_j f$ and $\partial_j \partial_i f$ exist and are continuous at x in G, then $\partial_i \partial_j f(x) = \partial_j \partial_i f(x)$.*

Proof. Fix i and j and fix x_k when $k \ne i, j$. By considering the function $(x_i, x_j) \mapsto f(x_1, \ldots, x_p)$, we see that if we prove the proposition when $p = 2$, we prove the proposition for any value of p. So without loss of generality assume $p = 2$.

Fix a point (a, b) in G; we want to show that $\partial_{12} f(a, b) = \partial_{21} f(a, b)$. Fix $r > 0$ such that $B = B((a, b); r) \subseteq G$, and let δ be any positive number with $0 < \delta < r$. (δ will be further specified later.) Put $B_\delta = B((a, b); \delta)$. Consider a point (x, y) in the disk B_δ with $x \ne a$ and $y \ne b$. Apply the MVT for derivatives to the function of one variable $t \mapsto [f(t, y) - f(t, b)]$ to obtain a point ξ_1 between x and a such that

$$[f(x, y) - f(x, b)] - [f(a, y) - f(a, b)] = [\partial_1 f(\xi_1, y) - \partial_1 f(\xi_1, b)](x - a)$$

Now apply the MVT to the function $t \mapsto \partial_1 f(\xi_1, t)$ (Is the hypothesis of the MVT satisfied?) to obtain a point η_1 between y and b such that

$$\partial_1 f(\xi_1, y) - \partial_1 f(\xi_1, b) = \partial_{12} f(\xi_1, \eta_1)(y - b)$$

Combining these equations we see that for any (x, y) in the disk B_δ there is a point (ξ_1, η_1) in B_δ with

6.2.4 $\quad [f(x, y) - f(x, b)] - [f(a, y) - f(a, b)] = \partial_{12} f(\xi_1, \eta_1)(x - a)(y - b)$

Interchanging the roles of the first and second variable in the preceding argument we see that when $(x, y) \in B_\delta$ with $x \neq a$ and $y \neq b$, there is a point (ξ_2, η_2) in B_δ with ξ_2 between x and a and η_2 between y and b such that

6.2.5 $\quad [f(x, y) - f(a, y)] - [f(x, b) - f(a, b)] = \partial_{21} f(\xi_2, \eta_2)(x - a)(y - b)$

But a quick inspection of the left-hand sides of (6.2.4) and (6.2.5) shows they are equal. Hence $\partial_{12} f(\xi_1, \eta_1)(x - a)(y - b) = \partial_{21} f(\xi_2, \eta_2)(x - a)(y - b)$. But since $(x - a)(y - b) \neq 0$ we have that

$$\partial_{12} f(\xi_1, \eta_1) = \partial_{21} f(\xi_2, \eta_2)$$

Now we use the hypothesis that both $\partial_{12} f$ and $\partial_{21} f$ are continuous. If $\epsilon > 0$ there is a $\delta > 0$ such that $|\partial_{12} f(x, y) - \partial_{12} f(a, b)| < \epsilon/2$ and $|\partial_{21} f(x, y) - \partial_{21} f(a, b)| < \epsilon/2$ when $(x, y) \in B_\delta$. Combining this with the last equation we see that $|\partial_{12} f(a, b) - \partial_{21} f(a, b)| < \epsilon$ for every positive ϵ, so the conclusion of the proposition follows. ∎

There is something disagreeable about all this.

If we want to use partial derivatives to study the behavior of a function, we cannot do it one variable at a time. Exercise 5.4.14 shows that there are functions defined on $\mathbb{R}^p$ that are continuous in each variable separately, but not continuous. In fact that example illustrates an additional anomaly. In Exercise 3 below the reader is asked to show that that function has partial derivatives at the origin with $\partial_1 f(0, 0)) = \partial_2 f(0, 0) = 0$ even though it is not continuous at the origin. We must have a concept of the differentiability of a function $f : G \to \mathbb{R}$ that simultaneously incorporates the influence of all the variables on f. We will now see such a concept of differentiability for a real-valued function defined on an open subset of $\mathbb{R}^p$ that implies the existence of all the partial derivatives and also has the psychologically satisfying property that it implies continuity.

To do this recall Proposition 2.2.9 where it is shown that $f : (a, b) \to \mathbb{R}$ is differentiable at x if and only if there there is a a number D and a function $F : (a, b) \to \mathbb{R}$ such that $f(y) - f(x) = D(y - x) + F(y)(y - x)$ for all y in X and $\lim_{y \to x} F(y) = 0$. (When this is the case, $D = f'(x)$.) We also saw a similar result for functions $\gamma : (a, b) \to \mathbb{R}^p$ in Proposition 6.1.3. We modify these results for functions f defined on an open subset of $\mathbb{R}^p$. The first thing to observe is that should we write such an equation when f is defined on a subset of $\mathbb{R}^p$, the left-hand side of this equation, $f(y) - f(x)$, is a number while on the right-hand side the vector $y - x$ in $\mathbb{R}^p$ appears. Therefore we have to a find suitable way to interpret D and the function F.

We start with $F(y)(y - x)$; finding a substitute for D will come shortly. If we return to the case of a scalar-valued function we see that we could change the F-term to $F(y)|y - x|$ without changing the definition of differentiability. So in our present case we can use $F(y)\|x - y\|$ instead of $F(y)(y - x)$, where F remains a scalar-valued function $F : G \to \mathbb{R}$. But now we need an appropriate definition of the limiting process as $y \to x$ in $\mathbb{R}^p$. We need look no further than what we did in the preceding chapter since such a function F maps one metric space to another. Namely, if $x \in G$ and $F : G \to \mathbb{R}$, say that $\lim_{y \to x} F(y) = A$ if for every $\epsilon > 0$ there is a $\delta > 0$ such that $|F(y) - A| < \epsilon$ whenever $0 < \|y - x\| < \delta$.

It is more involved to find a replacement for the number D that appears in the equation $f(y) - f(x) = D(y - x) + F(y)(y - x)$ in Proposition 2.2.9. Unlike with the term $F(y)(y - x)$, we cannot replace $y - x$ with $|y - x|$ in the one-variable situation without changing the definition of differentiability. (Why?) So the replacement for D cannot be a number. What to do? Don't forget that in the case of a function f defined on an open interval it was the case that $D = f'(x)$. This argues, perhaps, that the correct replacement for D should be some form of a vector though it must be able to interact with $y - x$ and produce a number. It is possible to once again resort to the inner product. But for a variety of reasons the solution is that D will be replaced by a linear functional $L : \mathbb{R}^p \to \mathbb{R}$ from linear algebra. We review here some of the pertinent facts about linear functionals and present some additional things needed for our discussion of differentiability. We start by recalling the definition.

6.2.6. Definition. A *linear functional* on $\mathbb{R}^p$ is a function $L : \mathbb{R}^p \to \mathbb{R}$ that satisfies $L(\alpha x + \beta y) = \alpha L(x) + \beta L(y)$ for all α, β in $\mathbb{R}$ and all x, y in $\mathbb{R}^p$. Denote the collection of all linear functionals on $\mathbb{R}^p$ by $\mathbb{R}^{p*}$.

We recall that $\mathbb{R}^{p*}$ is also a vector space where for two linear functionals L_1 and L_2, $(L_1 + L_2)(x) = L_1(x) + L_2(x)$. Since the unit vectors $e_1, \ldots, e_p$ form a basis for $\mathbb{R}^p$, the numbers $L(e_1), \ldots, L(e_p)$ determine any linear functional L. That is, if $\epsilon_j = L(e_j)$ and $x = (x_1, \ldots, x_p)$, then $L(x) = \sum_{j=1}^{p} \epsilon_j x_j$. (Also see Exercise 4.) We rephrase this a bit differently.

6.2.7. Proposition. *If* $1 \le j \le p$ *and we define* $E_j : \mathbb{R}^p \to \mathbb{R}$ *by*

$$E_j(x_1, \ldots, x_p) = x_j$$

then $E_j \in \mathbb{R}^{p*}$. *Moreover,* $\{E_1, \ldots, E_p\}$ *is a basis for* $\mathbb{R}^{p*}$.

Proof. We leave it to the reader to show that each E_j is a linear functional. To show that $\{E_1, \ldots, E_p\}$ is linearly independent, first observe that for $1 \le j, k \le p$,

$E_j(e_k) = 0$ when $j \neq k$ and $E_j(e_j) = 1$. Thus if $0 = \sum_{j=1}^{p} \alpha_j E_j$ in $\mathbb{R}^{p*}$, we have that for each k, $0 = \sum_{j=1}^{p} \alpha_j E_j(e_k) = \alpha_k$. So the linear functionals $\{E_1, \ldots, E_p\}$ are linearly independent. If $L \in \mathbb{R}^{p*}$ and $\alpha_j = L(e_j)$, then for $1 \leq j \leq q$

$$\left(L - \sum_{j=1}^{p} \alpha_j E_j \right)(e_k) = L(e_k) - \alpha_k E_k(e_k) = 0$$

Since $\{e_1, \ldots, e_p\}$ is a basis for $\mathbb{R}^p$, this shows that $L = \sum_{j=1}^{p} \alpha_j E_j$. ∎

Now we relate linear functionals to the inner product.

6.2.8. Proposition. *If $L \in \mathbb{R}^{p*}$, then there is a unique vector a in $\mathbb{R}^p$ such that $L(x) = \langle x, a \rangle$. In fact $a = (L(e_1), \ldots, L(e_p))$.*

Proof. If L is given, set $a_j = L(e_j)$ and put $a = (a_1, \ldots, a_p)$ in $\mathbb{R}^p$. It follows that

$$L(x) = L\left(\sum_{j=1}^{p} x_j e_j \right) = \sum_{j=1}^{p} x_j L(e_j) = \sum_{j=1}^{p} x_j a_j = \langle x, a \rangle$$

The proof of uniqueness is Exercise 5. ∎

We need the next result for making estimates. The key here is to set things up so we can apply the Cauchy–Schwarz Inequality. Recall that using the inner product the Cauchy–Schwartz Inequality becomes

$$\langle x, y \rangle^2 \leq \langle x, x \rangle \langle y, y \rangle$$

for x, y in $\mathbb{R}^p$.

6.2.9. Proposition. *If we define*

$$\|L\| = \left[\sum_{j=1}^{p} L(e_j)^2 \right]^{\frac{1}{2}}$$

then for every x in $\mathbb{R}^p$ we have that $|L(x)| \leq \|L\| \|x\|$. Moreover there is a vector x in $\mathbb{R}^p$ with $\|x\| = 1$ and $L(x) = \|L\|$. Therefore

$$\|L\| = \sup\{|L(x)| : \|x\| \leq 1\}$$

Proof. From the preceding proposition $L(x) = \langle x, a \rangle$, where a is the vector whose j-th coordinate is $a_j = L(e_j)$. Thus the quantity $\|L\| = \|a\|$, and so $|L(x)|^2 = \langle x, a \rangle^2 \leq \langle x, x \rangle \langle a, a \rangle = \|L\|^2 \|x\|^2$. Also setting $x = \|a\|^{-1} a$ gives that $\|x\| = 1$ and $L(x) = \|L\|$. The fact that $\|L\| = \sup\{|L(x)| : \|x\| \leq 1\}$ is now straightforward. ∎

The quantity $\|L\|$ is called the *norm* of the linear functional L. We are now in a position to define a differentiable function following the discussion that precedes the definition of a linear functional.

6.2.10. Definition. If G is an open subset of $\mathbb{R}^p$, $f : G \to \mathbb{R}$, and $x \in G$ say that f is *differentiable* at x if there is a linear functional $L : \mathbb{R}^p \to \mathbb{R}$ and a function $F : G \to \mathbb{R}$ such that

$$f(y) - f(x) = L(y - x) + F(y)\|y - x\|$$

for all y in G and $\lim_{y \to x} F(y) = 0$. We define the *derivative* of f at x to be the linear functional

$$Df(x) = f'(x) = L$$

If f is differentiable at every point of G then f is said to be *differentiable* on G.

Again you might want to show that this coincides with the definition of a function $f : (a, b) \to \mathbb{R}$ being differentiable at a point x by examining Proposition 2.2.9. For such a function what is the linear functional $L : \mathbb{R} \to \mathbb{R}$?

The introduction of the notation $Df = f'$ may seem capricious, but I'll ask the reader to be patient. There are times that Df is more convenient than f'. This is especially true when we later consider functions from $\mathbb{R}^p$ into $\mathbb{R}^q$. This notation Df is in wide use and rather standard. In fact it seems to me more convenient. The reader might want to look at Exercise 6.

In the atmosphere of the preceding definition, since $f'(x) = Df(x)$ is a linear functional it makes sense to write $f'(x)(y) = Df(x)(y)$ for any y in $\mathbb{R}^p$. Indeed $f' : G \to \mathbb{R}^{p*}$.

6.2.11. Proposition. *If $f : G \to \mathbb{R}^p$ and $g : G \to \mathbb{R}^p$ are differentiable at x then for any scalars α and β, the function $\alpha f + \beta g$ is differentiable at x and $D(\alpha f + \beta g)(x) = \alpha Df(x) + \beta Dg(x)$.*

Proof. Exercise 7 ■

6.2.12. Theorem. *If G is an open subset of $\mathbb{R}^p$, $x \subset G$, and $f : G \to \mathbb{R}$ is differentiable at x, then the following hold.*

(a) *f is continuous at x.*
(b) *For $1 \leq j \leq p$, the j-th partial derivative of f at x exists.*
(c) *If we define the vector*

$$\nabla f(x) = \left(\frac{\partial f}{\partial x_1}(x), \ldots, \frac{\partial f}{\partial x_p}(x) \right)$$

then for every y in $\mathbb{R}^p$ we have

$$Df(x)(y) = \langle y, \nabla f(x) \rangle$$

(d) *The linear functional L in the definition of differentiability is unique.*

As in calculus we call $\nabla f(x)$ the *gradient* of f at x.

Proof. The theorem contains a lot of information, but part-by-part the proof is not difficult. We use the notation of Definition 6.2.10.

(a) We have that $|f(y) - f(x)| \le \|L\|\|y - x\| + |F(y)|\|y - x\|$, so the continuity is immediate.

(b) From Definition 6.2.10 we have that for $t \ne 0$ and $1 \le j \le p$

$$\lim_{t\to 0} \frac{f(x + te_j) - f(x)}{t} = L(e_j) + \lim_{t\to 0} \frac{F(x + te_j)\|te_j\|}{t}$$

$$= L(e_j) + \lim_{t\to 0} F(x + te_j)\frac{|t|}{t}$$

$$= L(e_j)$$

since $|t|/t$ remains bounded by 1. Thus $\frac{\partial f}{\partial x_j}(x)$ exists and equals $L(e_j)$.

(c) This is immediate from Proposition 6.2.8.

(d) This follows from part (c) since the partial derivatives are unique. ∎

Here is a convenient sufficient condition for differentiability. If you are presented with a function, you can often use it to establish this.

6.2.13. Theorem. *If $f : G \to \mathbb{R}$ is such that each partial derivative of f exists and is continuous, then f is differentiable.*

Proof. Fix a point x in G. We have to find a linear functional L such that

$$\lim_{y\to x} \|y - x\|^{-1}[f(y) - f(x) - L(y - x)] = 0$$

That is we must find L with the property that for every $\epsilon > 0$ there is a $\delta > 0$ such that when $\|h\| < \delta$,

$$|f(x + h) - f(x) - L(y - x)| < \epsilon\|h\|$$

Finding the linear functional L is easy since we know from the preceding theorem that if the derivative exists, it must be given by the gradient: $L(y) = \langle y, \nabla f(x)\rangle$, which the hypothesis implies exists. Since each $\partial_j f$ is continuous and there are only a finite number of these derivatives, for every $\epsilon > 0$ there is a $\delta > 0$ such that

$$|\partial_j f(y) - \partial_j f(x)| < \frac{\epsilon}{\sqrt{p}}$$

when $\|y - x\| < \delta$ and $1 \le j \le p$.

Assume $h \in \mathbb{R}^p$ with $\|h\| < \delta$ and write $h = (h_1, \ldots, h_p) = \sum_{j=1}^p h_j e_j$. Define vectors $y_0, y_1, \ldots, y_p$ in $\mathbb{R}^p$ by $y_0 = 0$ and $y_k = h_1 e_1 + \cdots + h_k e_k$ for $1 \le k \le p$. Note that $f(x + h) - f(x) = \sum_{j=1}^p [f(x + y_j) - f(x + y_{j-1})]$. (Verify!) Now the hypothesis says we can apply the MVT to the function defined on $[0, 1]$ by $t \mapsto f(x + y_{j-1} + th_j e_j)$. This yields a t_j in $[0, 1]$ such that

$$f(x + y_j) - f(x + y_{j-1}) = \frac{d}{dt}[f(x + y_{j-1} + th_j e_j)](t_j)$$

$$= h_j[\partial_j f(x + y_{j-1} + t_j h_j e_j)]$$

Because $\|(x + y_{j-1} + t_j h_j e_j) - x\| = \|y_{j-1} + t_j h_j e_j\| = \|\sum_{i=1}^{j-1} h_i e_i + t_j h_j e_j\| \leq$
$\|h\| < \delta$, we have that $|(\partial_j f)(x + y_{j-1} + t_j h_j e_j) - \partial_j f(x)| < \epsilon/\sqrt{p}$. Therefore using the Cauchy–Schwarz Inequality we have

$$|f(x+h) - f(x) - \langle \nabla f(x), h \rangle|$$

$$= \left| \sum_{j=1}^{p} [f(x + y_j) - f(x + y_{j-1})] - \sum_{j=1}^{p} \partial_j f(x) h_j \right|$$

$$= \left| \sum_{j=1}^{p} h_j [(\partial_j f)(x + y_{j-1} + t_j h_j e_j) - \partial_j f(x)] \right|$$

$$\leq \sum_{j=1}^{p} |h_j| |(\partial_j f)(x + y_{j-1} + t_j h_j e_j) - \partial_j f(x)|$$

$$\leq \|h\| \left[\sum_{j=1}^{p} |(\partial_j f)(x + y_{j-1} + t_j h_j e_j) - \partial_j f(x)|^2 \right]^{\frac{1}{2}}$$

$$\leq \epsilon \|h\| \qquad \blacksquare$$

Because we have so many different situations where functions are defined with values in $\mathbb{R}^p$ or domain in $\mathbb{R}^p$ we will have many Chain Rules. Here is the first.

6.2.14. Theorem (Chain Rule). *Let $\gamma : (a, b) \to \mathbb{R}^p$ with range contained in G and suppose $f : G \to \mathbb{R}$. If $t_0 \in (a, b)$, γ is differentiable at t_0, $x_0 = \gamma(t_0)$, and f is differentiable at x_0, then $f \circ \gamma : (a, b) \to \mathbb{R}$ is differentiable at t_0 and*

$$(f \circ \gamma)'(t_0) = \langle \nabla f(\gamma(t_0)), \gamma'(t_0) \rangle$$

Proof. We use the notation of Definition 6.2.10. Also let $\gamma(t) - \gamma(t_0) = (t - t_0)[\gamma'(t_0) + \chi(t)]$, where $\chi : (a, b) \to \mathbb{R}^p$ such that $\chi(t) \to 0$ as $t \to t_0$. If $t \in (a, b)$ with $t - t_0 \neq 0$ and we put $x_0 = \gamma(t_0)$, then

$$\frac{f \circ \gamma(t) - f \circ \gamma(t_0)}{t - t_0} = \left\langle \nabla f(x_0), \frac{\gamma(t) - x_0}{t - t_0} \right\rangle + F(\gamma(t)) \frac{\|\gamma(t) - x_0\|}{t - t_0}$$

Now

$$\frac{\|\gamma(t) - x_0\|}{t - t_0} = \left\| \frac{\gamma(t) - x_0}{t - t_0} \right\| \frac{|t - t_0|}{t - t_0}$$

Thus as $t \to t_0$, this remains bounded while $F(\gamma(t)) \to 0$. Hence

$$\lim_{t \to t_0} \frac{f \circ \gamma(t) - f \circ \gamma(t_0)}{t - t_0} = \lim_{t \to t_0} \left\langle \nabla f(x_0), \frac{\gamma(t) - x_0}{t - t_0} \right\rangle$$

$$= \langle \nabla f(\gamma(t_0)), \gamma'(t_0) \rangle \qquad \blacksquare$$

6.2.15. Theorem (Mean Value Theorem). *If $B(a; r) \subseteq \mathbb{R}^p$, $f : B(a; r) \to \mathbb{R}$ is differentiable, and $b \in B(a; r)$, then there is a point x on the line segment $[a, b]$ such that*

$$Df(x)(a - b) = \langle \nabla f(x), b - a \rangle = f(b) - f(a)$$

Proof. Define the function $g : [0, 1] \to \mathbb{R}$ by $g(t) = f(tb + (1 - t)a)$. Using the just established Chain Rule with $\gamma(t) = tb + (1 - t)a$ we have that $g'(t) = \langle \nabla f(tb + (1 - t)a), \gamma'(t) \rangle = \langle \nabla f(tb + (1 - t)a), b - a \rangle$. Now apply the MVT to g. ∎

6.2.16. Definition. If G is an open subset of $\mathbb{R}^p$, $f : G \to \mathbb{R}$, $x \in G$, and $d \in \mathbb{R}^p$, then f is said to have a *directional derivative* at x in the direction d provided the function $t \mapsto f(x + td)$ is differentiable at $t = 0$. When the directional derivative exists we denote it by

$$\nabla_d f(x) = \lim_{t \to 0} \frac{f(x + td) - f(x)}{t}$$

If you consult the literature you may sometimes find a small variation on the definition of the directional derivative where the vector d is constrained to be a unit vector. This difference is conceptually minor though certain formulas will be different. Note that if the vector $d = e_j$, then the function f has a directional derivative at x in the direction e_j precisely when the corresponding partial derivative of f at x exists, and in this case $\nabla_{e_j} f(x) = \partial_j f(x)$. Also observe that if in the definition of the directional derivative we choose $r > 0$ such that $x + td \in G$ for $|t| < r$ and define $\gamma : (-r, r) \to G$ by $\gamma(t) = x + td$, then we can apply the version of the Chain Rule proved above. With this observation the proof of the next proposition is an immediate consequence of Theorem 6.2.14

6.2.17. Proposition. *If $x \in G$ and $f : G \to \mathbb{R}$ is differentiable at x, then $\nabla_d f(x)$ exists for any direction d and*

$$\nabla_d f(x) = \langle \nabla f(x), d \rangle$$

6.2.18. Proposition. *If G is a connected open subset of $\mathbb{R}^p$, $f : G \to \mathbb{R}$ is differentiable at every point of G, and $f'(x) = 0$ for all x in G, then f is a constant function.*

Proof. If $a \in G$ and we consider $f - f(a)$ instead of f, we may assume there is a point a in G with $f(a) = 0$. Set $X = \{x \in G : f(x) = 0\}$; so we want to show that $X = G$. We do this by showing that X is non-empty and both open and relatively closed in G, so that the equality will follow from the connectedness of G.

We know that $X \neq \emptyset$ since it contains a; since f is continuous on G, it follows that X is relatively closed in G. We want to show that X is also open. Fix c in X and pick $r > 0$ such that $B(c; r) \subseteq G$. If $b \in B(c; r)$, let d be the vector $d = b - c$ and define $\gamma(t) = c + td$. By Theorem 6.2.14 for all t where $\gamma(t) \in G$, $(f \circ \gamma)'(t) = \langle \nabla f(\gamma(t)), \gamma'(t) \rangle = 0$. By Theorem 2.3.2, $f \circ \gamma$ is constant. Hence $f(b) = f \circ \gamma(1) = f \circ \gamma(0) = f(c) = 0$. Since b was an arbitrary point in $B(c; r)$ we have that $B(c; r) \subseteq X$ and so X is open. ∎

We extend some definitions given in §2.4.

6.2.19. Definition. If $a \in G$ and $f : G \to \mathbb{R}$ is differentiable at a, then f has a *critical point* at a provided $\nabla f(a) = 0$; equivalently, if $f'(a) = 0$. We say that f has a *local maximum* at a if there is a $\delta > 0$ such that $f(a) \geq f(x)$ when $\|a - x\| < \delta$. Similarly we say that f has a *local minimum* at a if there is a $\delta > 0$ such that $f(a) \leq f(x)$ when $\|a - x\| < \delta$. If f has either a local maximum or local minimum at a, then we say f has a *local extremum* there. If there is a vector x in $\mathbb{R}^p$ such that the function $t \mapsto f(a + tx)$ defined on $[-1, 1]$ has a local maximum at $t = 0$ and there is a vector y in $\mathbb{R}^p$ such that the function $t \mapsto f(a + ty)$ defined on $[-1, 1]$ has a local minimum at $t = 0$, then f is said to have a *saddle point* at a.

6.2.20. Theorem. *If $a \in G$ and $f : G \to \mathbb{R}$ is differentiable on G and has a local extremum at a, then $\nabla f(a) = 0$.*

Proof. Let $d \in \mathbb{R}^p$ and consider the function $t \mapsto f(a + td)$. Since f has a local extremum at a, this function has a local extremum at $t = 0$. By Theorem 2.3.2, the derivative of this function at $t = 0$ must vanish. According to Proposition 6.2.17 this means that $\langle \nabla f(a), d \rangle = 0$. Since d was an arbitrary vector in $\mathbb{R}^p$, this proves the theorem. ∎

We can extend Theorem 2.4.2 to the present case.

6.2.21. Theorem. *Let G be an open subset of $\mathbb{R}^p$, $a \in G$, and let $f : G \to \mathbb{R}$ be a differentiable function on all of G.*

(a) *If for each vector d in $\mathbb{R}^p$ there is $\delta > 0$ such that $\nabla_d f(a - td) > 0$ and $\nabla_d f(a + td) < 0$ for $0 < t < \delta$, then f has a local maximum at a.*
(b) *If for each vector d in $\mathbb{R}^p$ there is $\delta > 0$ such that $\nabla_d f(a - td) < 0$ and $\nabla_d f(a + td) > 0$ for $0 < t < \delta$, then f has a local minimum at a.*

Proof. The idea is to look at the behavior of f near a in each direction d and apply Theorem 2.4.2. That is, for each vector d in $\mathbb{R}^p$ consider the function $t \mapsto f(a + td)$ and apply Theorem 2.4.2 to this function. Exercise 9 asks the reader to supply the details. ∎

We'll say more about critical points in §6.6.

6.2.22. Proposition. *Let G be an open subset of $\mathbb{R}^p$, let $a \in G$, and let $f : G \to \mathbb{R}$ be a differentiable function on all of G. If d is any unit vector in $\mathbb{R}^p$, then $|\nabla_d f(a)| \leq \|\nabla f(a)\|$. Equality is achieved when $d = \|\nabla f(a)\|^{-1} \nabla f(a)$.*

Proof. According to Proposition 6.2.17 $\nabla_d f(a) = \langle \nabla f(a), d \rangle$. Therefore by the Cauchy–Schwarz Inequality, $|\nabla_d f(a)| \leq \|\nabla f(a)\|$. The statement on equality is immediate. ∎

In light of the preceding proposition we conclude that for any a the vector $\nabla f(a)$ points in the direction in which the scalar-valued function f is changing the fastest.

Exercises

(1) Define $f : \mathbb{R}^2 \to \mathbb{R}$ by

$$f(x, y) = \begin{cases} \frac{(x^2 y + xy^2)\sin(x-y)}{x^2+y^2} & \text{when } (x, y) \neq (0, 0) \\ 0 & \text{when } (x, y) = (0, 0) \end{cases}$$

Find $\partial_x f(x, y)$ and $\partial_y f(x, y)$ for every point (x, y) in $\mathbb{R}^2$. (Be careful when $(x, y) = (0, 0)$.)

(2) Supply the missing details in Example 6.2.2.

(3) Show that the function f defined in Exercise 5.4.14 has partial derivatives at the origin with $\partial_1 f(0, 0) = \partial_2 f(0, 0) = 0$ even though it fails to be continuous at $(0, 0)$.

(4) Show that if $L, K \in \mathbb{R}^{p*}$ and $L(e_j) = K(e_j)$ for $1 \leq j \leq p$, then $L(x) = K(x)$ for all x in $\mathbb{R}^p$.

(5) (a) Show that if $x \in \mathbb{R}^p$ and $\langle x, y \rangle = 0$ for every y in $\mathbb{R}^p$, then $x = 0$. (b) Prove the uniqueness statement in Proposition 6.2.8.

(6) Let $f : \mathbb{R}^p \to \mathbb{R}$ and let $x \in \mathbb{R}^p$. Suppose we say that the function f has Property D at x if

$$\lim_{y \to x} \frac{f(y) - f(x)}{\|y - x\|}$$

exists. (a) Show that if f is differentiable at x, it has property D at x. (b) Does Property D at x imply that f is differentiable at x? (Prove or give a conterexample.)

(7) Prove Proposition 6.2.11.

(8) (a) Let $f(x, y, z) = \exp(x^2 + z + \cos y)$ and find $\nabla_d f(x, y, z)$ when $d = (1, \pi, 0)$. (b) Let $f : \mathbb{R}^3 \to \mathbb{R}$ be defined by $f(x, y, z) = xyz - x^2 + 3x^3$ and $d = (-1, 2, 1)$ and calculate $\nabla_d f$.

(9) Supply the required details for the proof of Theorem 6.2.21.

6.3. Orthogonality

This section and the next are a linear algebra interlude, one that is required to go further in our study of differentiability in $\mathbb{R}^p$. Specifically, when we study the differentiability of functions from open subsets of $\mathbb{R}^p$ into $\mathbb{R}^q$, we need to explore linear transformations and the inner product more extensively than we have.

6.3.1. Definition. If $x, y \in \mathbb{R}^p$ we say that x and y are *orthogonal* if $\langle x, y \rangle = 0$; in symbols we write $x \perp y$. Say that the vectors in a non-empty subset $\mathcal{S}$ of $\mathbb{R}^p$ are *pairwise orthogonal* if $x \perp y$ whenever x and y are distinct points in $\mathcal{S}$. Say that the set $\mathcal{S}$ is *orthonormal* if $\mathcal{S}$ is pairwise orthogonal and also $\|x\| = 1$ for all x in $\mathcal{S}$. An *orthonormal basis* for $\mathbb{R}^p$ is a basis for $\mathbb{R}^p$ that is also an orthonormal set.

If $\mathcal{S} \subseteq \mathbb{R}^p$, write $x \perp \mathcal{S}$ when $x \perp y$ for all y in $\mathcal{S}$. Let $\mathcal{S}^\perp$ denote the set of all vectors that are orthogonal to the set $\mathcal{S}$. Two non-empty subsets $\mathcal{S}$ and $\mathcal{T}$ of $\mathbb{R}^p$ are said to be orthogonal if every vector in $\mathcal{S}$ is orthogonal to every vector in $\mathcal{T}$; in symbols, $\mathcal{S} \perp \mathcal{T}$.

First note that $x \perp x$ if and only if $x = 0$. Now observe that if $\mathcal{S}$ is a pairwise orthogonal set of non-zero vectors, then the vectors in $\mathcal{S}$ are linearly independent. In fact, if $x_1, \ldots, x_m$ are distinct vectors in $\mathcal{S}$ and if $\alpha_1, \ldots, \alpha_m$ are scalars such that $\sum_{j=1}^{m} \alpha_j x_j = 0$, then for a fixed k, $1 \le k \le m$,

$$0 = \left\langle \sum_{j=1}^{m} \alpha_j x_j, x_k \right\rangle = \sum_{j=1}^{m} \alpha_j \langle x_j, x_k \rangle = \alpha_k \|x_k\|^2$$

So each α_k is zero. Since $\mathbb{R}^p$ has dimension p, $\mathcal{S}$ cannot have more than p vectors. Examples of pairwise orthogonal sets abound. To start, any non-empty subset of the standard basis for $\mathbb{R}^p$ is an orthonormal set, and the standard basis is an example of an orthonormal basis. There are other orthonormal bases as we will see in Corollary 6.3.4 below.

The reader may recall some of this from linear algebra. Assume the vectors x and y are linearly independent and consider the two-dimensional subspace of $\mathbb{R}^p$ spanned by them. These vectors are orthogonal if the straight lines they define form a right angle. Note that for any x, y we have

$$\|x + y\|^2 = \langle x + y, x + y \rangle = \|x\|^2 + \langle x, y \rangle + \|y\|^2$$

This is called the *polar identity*. From here we get the following.

6.3.2. Proposition (Pythagorean[1] Theorem). *If $x_1, \ldots, x_m$ are pairwise orthogonal vectors in $\mathbb{R}^p$, then*

$$\|x_1 + \cdots + x_m\|^2 = \|x_1\|^2 + \cdots + \|x_m\|^2$$

Proof. If $m = 2$, then this easily follows from the polar identity. The proof can now be completed using induction (Exercise 1). ∎

. . . .

[1] Pythagoras of Samos was born around 569 BC in Samos, a city in Ionia. Little is known about him, especially since none of the work he wrote still exists. He was the leader of a society dedicated to science and religion, and some of the ancient books claim that he had divine powers. When he was a young man Thales introduced him to mathematics, and it is likely that Anaximander, a student of Thales, gave a series of mathematical lectures that Pythagoras attended. All dates here are dubious, but around 535 BC he traveled to Egypt. From here he adopted many practices of Egyptian priests, which he later imposed on the members of his society. These included the refusal to wear anything made from an animal skin and an abstinence from eating beans. In 525 BC the Persians invaded Egypt and brought Pythagoras to Babylon. In 520 BC he left there and returned to Samos; there is no explanation how he obtained his freedom. After founding his society, in 518 BC he went to southern Italy. Apparently while he was in Samos he used the "symbolic method of teaching" and the Samians did not approve. In what is present day Crotone, half-way up the heel of Italy, he founded a mathematical and philosophical society. One of their beliefs was that at its deepest level, reality is mathematical in nature. This present theorem (in two-dimensions) was known to the Babylonians 1000 years before Pythagoras, though it is likely he was the first to prove it. He introduced abstraction into mathematics and made many other discoveries in geometry, including that the sum of the interior angles of a triangle equals two right angles. He was also intensely interested in numbers and discovered the irrationals. He died about 475 BC. The Pythagorean Society continued after his death but they made powerful enemies. Eventually they were attacked and some 50 or 60 members were killed. Those who survived took refuge at Thebes and other places.

We observed earlier that a set of pairwise orthogonal vectors is linearly indepen-dent. A similar argument shows that if $x \in \mathbb{R}^p$ and $x \perp S$, then $x \perp \bigvee S$, the linear span of S (Exercise 3).

6.3.3. Theorem (Gram[2]–Schmidt[3] Process). *If $x_1, \ldots, x_m$ are linearly indepen-dent, then there are orthonormal vectors $y_1, \ldots, y_m$ such that for $1 \le j \le m$, y_j is in the linear span of $\{x_1, \ldots, x_j\}$. Consequently,*

$$\bigvee \{x_1, \ldots, x_j\} = \bigvee \{y_1, \ldots, y_j\}$$

for $1 \le j \le m$.

Proof. Observe that the last statement of the theorem follows from the first. In fact since the first part of the proposition implies $\bigvee \{y_1, \ldots, y_j\} \subseteq \bigvee \{x_1, \ldots, x_j\}$, to show equality we need only show that these two subspaces have the same dimension. But since $\{y_1, \ldots, y_j\}$ are orthonormal and $\{x_1, \ldots, x_j\}$ are linearly independent, both subspaces have dimension j.

The proof of the first part is by induction on m. When $m = 1$ just take $y_1 = \|x_1\|^{-1} x_1$. Now assume the proposition is true for some $k < m$ and that $x_1, \ldots, x_{k+1}$ are linearly independent. By the induction hypothesis there are orthonormal vectors $y_1, \ldots, y_k$ such that for $1 \le j \le k$, y_j is in the linear span of $\{x_1, \ldots, x_j\}$. Consider $y = x_{k+1} - \sum_{i=1}^{k} \langle x_{k+1}, y_i \rangle y_i$. If $1 \le j \le k$, then $\langle y, y_j \rangle = \langle x_{k+1}, y_j \rangle - \sum_{i=1}^{k} \langle x_{k+1}, y_i \rangle \langle y_i, y_j \rangle = \langle x_{k+1}, y_j \rangle - \langle x_{k+1}, y_j \rangle = 0$. Also note that $y \ne 0$ since $x_{k+1} \notin \bigvee \{x_1, \ldots, x_k\} = \bigvee \{y_1, \ldots, y_k\}$. If we set $y_{k+1} = \|y\|^{-1} y$, then $\{y_1, \ldots, y_{k+1}\}$ is an orthonormal set. It is left to the reader to check that $y_{k+1} \in \bigvee \{x_1, \ldots, x_{k+1}\}$. ∎

....

[2] Jorgen Pedersen Gram was the son of a farmer, born in 1850 in Nustrup, Denmark. In 1868 he began his uni-versity education, receiving a masters degree in 1873. When he received this degree he had already published his first paper in algebra. In 1875 he began work with an insurance company, but this work led him into a study of probability and numerical analysis. He soon published a paper on these topics. As a result of this paper he was awarded a doctorate in 1879. His work here, which he also applied to forestry, led him to study abstract problems in number theory. He continued working at the insurance company but was invited to lecture in the Danish Mathematical Society. For his work the Videnskabernes Society awarded him their Gold Medal in 1884. He married twice. The first time in 1879 and the second in 1896, just over a year after his first wife's death. He died in 1916 in Copenhagen; he was on his way to a meeting of the Videnskabernes Society and was struck by a bicycle.

[3] Erhard Schmidt was born in 1876 in what is today Tartu, Estonia. His early education was typical of someone who was born to a professional family. (His father was a medical biologist.) He began his university studies in Berlin and went to Göttingen where he obtained his doctorate under the supervision of Hilbert; his thesis was in integral equations. He went to Bonn and this was followed by several academic posts before he was awarded a professorship at Berlin in 1917. He was quickly drawn into administrative matters involved with filling recently vacated faculty positions. He is credited with establishing applied mathematics at Berlin. Needless to say his role at the university was difficult during the Nazi era. Many of his Jewish colleagues were forced out of their positions. Some have criticized his role in this and others have defended it. In the final analysis his reputation survived. The present result was obtained by Schmidt independently in 1907. There are also the Hilbert–Schmidt operators named after him. He deserves a place as one of several mathematicians who founded the abstract theory of functional analysis. He died in 1959 in Berlin.

We will eventually want to manufacture an orthonormal basis for $\mathbb{R}^p$ different from the standard basis. The next corollary allows us to do this.

6.3.4. Corollary. *If $\mathcal{E}_0$ is a set of orthonormal vectors in $\mathbb{R}^p$, then there is an orthonormal basis $\mathcal{E}$ for $\mathbb{R}^p$ that contains $\mathcal{E}_0$.*

Proof. From linear algebra we know there is a basis $\mathcal{B}$ for $\mathbb{R}^p$ that contains $\mathcal{E}_0$ since $\mathcal{E}_0$ is a linearly independent set. By The Gram–Schmidt Process we can replace $\mathcal{B}$ by orthonormal vectors $\{y_1, \ldots, y_p\}$ with the same span. Hence $\{y_1, \ldots, y_p\}$ is an orthonormal basis. But if the Gram–Schmidt process is examined, we see that the orthonormal vectors $\mathcal{E}_0$ will not be altered by the process. Hence $\mathcal{E}_0 \subseteq \{y_1, \ldots, y_p\}$. ∎

The proof of the next result is Exercise 4.

6.3.5. Proposition. *If $y_1, \ldots, y_p$ is an orthonormal basis for $\mathbb{R}^p$ and $x \in \mathbb{R}^p$, then*

$$x = \sum_{j=1}^{p} \langle x, y_j \rangle y_j$$

The next result is crucial in our study of orthogonality.

6.3.6. Theorem. *If $\mathcal{M}$ is a vector subspace of $\mathbb{R}^p$ and $x \in \mathbb{R}^p$, then there is a unique vector y_0 in $\mathcal{M}$ such that*

$$\|x - y_0\| = \text{dist}\,(x, \mathcal{M}) = \inf\{\|x - y\| : y \in \mathcal{M}\}$$

In addition $x - y_0 \perp \mathcal{M}$.

Proof. Let $\{y_1, \ldots, y_m\}$ be an orthonormal basis for $\mathcal{M}$ and set

$$y_0 = \sum_{j=1}^{m} \langle x, y_j \rangle y_j$$

We want to show that y_0 has the desired properties. It is easily checked that $\langle x - y_0, y_j \rangle = 0$ for $1 \le j \le m$. Since $\{y_1, \ldots, y_m\}$ is a basis for $\mathcal{M}$, this shows that $x - y_0 \perp \mathcal{M}$. Now let y be an arbitrary vector in $\mathcal{M}$. Since $x - y_0 \perp \mathcal{M}$ and $y - y_0 \in \mathcal{M}$, the Pythagorean Theorem implies that

$$\begin{aligned}
\|x - y\|^2 &= \|(x - y_0) - (y - y_0)\|^2 \\
&= \|x - y_0\|^2 + \|y - y_0\|^2 \\
&\ge \|x - y_0\|^2
\end{aligned}$$

This shows that $[\text{dist}\,(x, \mathcal{M})]^2 \ge \|x - y_0\|^2$. Since $y_0 \in \mathcal{M}$, we must have equality.

Before showing the uniqueness of y_0 we first prove the following, which is another way in which y_0 is unique.

Claim. y_0 is the only vector in $\mathcal{M}$ such that $x - y_0 \perp \mathcal{M}$.

Suppose z_0 is another vector in $\mathcal{M}$ such that $x - z_0 \perp \mathcal{M}$. Since $\langle x - y_0, y_i \rangle = 0 = \langle x - z_0, y_i \rangle$, it follows that $\langle z_0, y_i \rangle = \langle y_0, y_i \rangle$ for $1 \leq i \leq m$. Thus $z_0 - y_0$ is both in $\mathcal{M}$ and orthogonal to $\mathcal{M}$. In particular it is orthogonal to itself and thus $z_0 - y_0 = 0$.

Now assume $w_0 \in \mathcal{M}$ such that $\|x - w_0\| = \text{dist}\,(x, \mathcal{M})$. If $y \in \mathcal{M}$, then $w_0 + y \in \mathcal{M}$ so using the polar identity we get that

$$\|x - w_0\|^2 \leq \|x - (w_0 + y)\|^2$$
$$= \|(x - w_0) - y\|^2$$
$$= \|x - w_0\|^2 - 2\langle x - w_0, y \rangle + \|y\|^2$$

Thus $2\langle x - w_0, y \rangle \leq \|y\|^2$ for all y in $\mathcal{M}$. If $y \in \mathcal{M}$, then by substituting $-y$ for y if necessary we may assume that $\langle x - w_0, y \rangle \geq 0$. Letting $t > 0$ and replacing y by ty, the inequality becomes $2t\langle x - w_0, y \rangle \leq t^2 \|y\|^2$, valid for all $t > 0$. Divide both sides by t and letting $t \to 0$ shows that $\langle x - w_0, y \rangle = 0$. That is, $x - w_0 \perp \mathcal{M}$. By the claim, this implies that $w_0 = y_0$. ∎

If x, $\mathcal{M}$, and y_0 are as in the preceding theorem, then y_0 is called the *orthogonal projection* of x onto $\mathcal{M}$. We will return to this concept after we begin the study of linear transformations in the next section. See Definition 6.4.13.

6.3.7. Definition. When $\mathcal{M}$ and $\mathcal{N}$ are two linear subspaces of $\mathbb{R}^p$ and $\mathcal{M} \perp \mathcal{N}$, write

$$\mathcal{M} \oplus \mathcal{N} = \{x + y : x \in \mathcal{M}, y \in \mathcal{N}\}$$

This is actually the linear space $\mathcal{M} + \mathcal{N}$ and the symbol $\oplus$ is used for emphasis of the fact that every vector in $\mathcal{M}$ is orthogonal to every vector in $\mathcal{N}$. In fact we will sometimes want to write $x \oplus y$ when $x \perp y$.

Exercises

(1) Perform the induction argument needed to complete the proof of the Pythagorean Theorem.

(2) Prove the *Parallelogram Law*: $\|x + y\|^2 + \|x - y\|^2 = 2(\|x\|^2 + \|y\|^2)$ for all x and y in $\mathbb{R}^p$. Why is this called the parallelogram law?

(3) Prove that if $\mathcal{S} \subseteq \mathbb{R}^p$ and x is a vector that is orthogonal to $\mathcal{S}$, then $x \perp \bigvee \mathcal{S}$.

(4) Prove Proposition 6.3.5. (Hint: Show that $[x - \sum_{j=1}^p \langle x, y_j \rangle y_j] \perp y_k$ for $1 \leq k \leq p$.)

(5) In $\mathbb{R}^3$ let $x_1 = (1, 0, 2)$, $x_2 = (0, 1, 2)$, and $x_3 = (1, 2, 0)$. Show that $\{x_1, x_2, x_3\}$ is a set of linearly independent vectors and carry out the Gram–Schmidt process to manufacture the corresponding orthonormal basis.

(6) In $\mathbb{R}^4$ let $x_1 = (1, 0, 2, 0)$, $x_2 = (0, 1, 2, 0)$, $x_3 = (1, 2, 0, 0)$, $x_4 = (0, 0, 0, 1)$ and describe the vector space $\bigvee \{x_1, x_3\}^\perp$.

6.4. Linear Transformations

We want to define the derivative of a function $f : G \to \mathbb{R}^q$ when G is an open subset of $\mathbb{R}^p$. When $q = 1$ we saw that the natural object to define as $f'(x)$ was a linear functional. When $q > 1$, we'll see that the derivative is a linear transformation $A : \mathbb{R}^p \to \mathbb{R}^q$. In this section we will discuss some of the essential features of linear transformations. We continue to assume that every reader knows the basics of vector spaces and has been exposed to linear transformations. For some what is presented here may be familiar. If that's the case just be patient or you can skip to the next section, though I suspect most readers will encounter some material here for the first time. In particular I think Theorem 6.4.29 below will be a first-time experience for most readers of this book.

Only linear transformations between $\mathbb{R}^p$ and $\mathbb{R}^q$ are discussed here, rather than the usual linear transformations between arbitrary vector spaces. Recall that a *linear transformation* is a function $A : \mathbb{R}^p \to \mathbb{R}^q$ such that whenever $\alpha, \beta \in \mathbb{R}$ and $x, y \in \mathbb{R}^p$ we have that $A(\alpha x + \beta y) = \alpha A(x) + \beta A(y)$. (A notational point should be made here: we will often use the notation Ax rather than $A(x)$. This practice is widespread and not limited to the author.) The set of all linear transformations from $\mathbb{R}^p$ into $\mathbb{R}^q$ is denoted by $\mathcal{L}(\mathbb{R}^p, \mathbb{R}^q)$. This itself is a vector space where for A, B in $\mathcal{L}(\mathbb{R}^p, \mathbb{R}^q)$ and α, β in $\mathbb{R}$ we define $(\alpha A + \beta B)(x) = \alpha A(x) + \beta B(x)$. We note that $\mathcal{L}(\mathbb{R}^p, \mathbb{R}) = \mathbb{R}^{p*}$. An important special case occurs when $q = p$. In this case we let

$$\mathcal{L}(\mathbb{R}^p) = \mathcal{L}(\mathbb{R}^p, \mathbb{R}^p)$$

Also recall that the *kernel* and *range* of A are defined as

$$\ker A = \{x \in \mathbb{R}^p : Ax = 0\}$$
$$\operatorname{ran} A = \{y \in \mathbb{R}^q : \text{there is an } x \text{ in } \mathbb{R}^p \text{ with } Ax = y\}$$
$$= A(\mathbb{R}^p)$$

So A is injective if and only if $\ker A = (0)$; and A is surjective if and only if $\operatorname{ran} A = \mathbb{R}^q$.

The first step is to represent the linear transformation as a $q \times p$ matrix. To facilitate this and keep things straight, we'll continue to denote the standard basis in $\mathbb{R}^p$ by $e_1, \ldots, e_p$; however, we will denote the standard basis in $\mathbb{R}^q$ by $d_1, \ldots, d_q$. So when $A \in \mathcal{L}(\mathbb{R}^p, \mathbb{R}^q)$ and $1 \le j \le p$,

$$A(e_j) = \sum_{i=1}^{q} a_{ij} d_i = \sum_{i=1}^{q} \langle A(e_j), d_i \rangle d_i$$

This leads to the matrix representation of A

6.4.1
$$\begin{bmatrix} a_{11} & a_{12} & \cdots & a_{1p} \\ a_{21} & a_{22} & \cdots & a_{2p} \\ \vdots & \vdots & \vdots & \vdots \\ a_{q1} & a_{q2} & \cdots & a_{qp} \end{bmatrix}$$

We assume the reader remembers matrix multiplication and that if $x = (x_1, \ldots, x_p) \in \mathbb{R}^p$, then

$$A(x) = \begin{bmatrix} a_{11} & a_{12} & \cdots & a_{1p} \\ a_{21} & a_{22} & \cdots & a_{2p} \\ \vdots & \vdots & \vdots & \vdots \\ a_{q1} & a_{q2} & \cdots & a_{qp} \end{bmatrix} \begin{bmatrix} x_1 \\ x_2 \\ \vdots \\ x_p \end{bmatrix}$$

which is a $q \times 1$ column vector; equivalently, a vector in $\mathbb{R}^q$.

When $A \in \mathcal{L}(\mathbb{R}^p, \mathbb{R}^q)$, define (see Exercise 2) the *norm* of A to be the quantity

$$\|A\| = \sup\{\|Ax\| : x \in \mathbb{R}^p \text{ and } \|x\| \leq 1\}$$

We might point out that when $q = 1$ and the linear transformation A becomes the linear functional L, we defined the norm of L in Proposition 6.2.9. It is left to the reader to prove the following in Exercise 3.

6.4.2. Proposition. *If $A, B \in \mathcal{L}(\mathbb{R}^p, \mathbb{R}^q)$ and $\alpha \in \mathbb{R}$, then the following hold.*

(a) $\|\alpha A + B\| \leq |\alpha| \|A\| + \|B\|$.
(b) $d(A, B) = \|A - B\|$ *defines a metric on* $\mathcal{L}(\mathbb{R}^p, \mathbb{R}^q)$.
(c) *For every x in $\mathbb{R}^p$ we have that* $\|Ax\| \leq \|A\|\|x\|$.
(d) *If $q = p$, then $AB \in \mathcal{L}(\mathbb{R}^p, \mathbb{R}^q)$ and* $\|AB\| \leq \|A\|\|B\|$.

6.4.3. Corollary. *With the metric d on $\mathcal{L}(\mathbb{R}^p, \mathbb{R}^q)$ defined in the proposition, addition and scalar multiplication are continuous functions.*

Proof. See Exercise 4. ∎

6.4.4. Proposition. (a) *With the definition of distance given in the preceding proposition, $\mathcal{L}(\mathbb{R}^p, \mathbb{R}^q)$ is a complete metric space.*
(b) *If $\{A_n\}$ is a sequence in $\mathcal{L}(\mathbb{R}^p, \mathbb{R}^q)$ and $\sum_{n=1}^{\infty} \|A_n\| < \infty$, then the infinite series $\sum_{n=1}^{\infty} A_n$ converges in $\mathcal{L}(\mathbb{R}^p, \mathbb{R}^q)$.*

Proof. (a) If $\{A_n\}$ is a Cauchy sequence in $\mathcal{L}(\mathbb{R}^p, \mathbb{R}^q)$ and $x \in \mathbb{R}^p$, then $\|A_n x - A_m x\| = \|(A_n - A_m)x\| \leq \|x\|\|A_n - A_m\|$; it follows that $\{A_n x\}$ is a Cauchy sequence in $\mathbb{R}^q$. Hence there is a vector y in $\mathbb{R}^q$ such that $A_n x \to y$; define $A : \mathbb{R}^p \to \mathbb{R}^q$ by $Ax = \lim_n A_n x$. It is left to the reader to show that A is a linear transformation. It remains to show that $\|A_n - A\| \to 0$. If $x \in \mathbb{R}^p$ with $\|x\| \leq 1$, then $\|A_n x - Ax\| \leq \|A_n x - A_m x\| + \|A_m x - Ax\| \leq \|A_n - A_m\| + \|A_m x - Ax\|$. If $\epsilon > 0$ then we can choose N such that $\|A_n - A_m\| < \epsilon/2$ when $m, n \geq N$. Thus $\|A_n x - Ax\| < \epsilon/2 + \|A_m x - Ax\|$ when $n, m \geq N$. Now choose $M > N$ such that $\|A_m x - Ax\| < \epsilon/2$ when $m \geq M$. (Note that M depends on x.) This implies that $\|A_n x - Ax\| < \epsilon$ when $n \geq N$ and the value of N is independent of x. Since x was arbitrary with $\|x\| \leq 1$, we have that $\|A_n - A\| < \epsilon$ when $n \geq N$. That is, $A_n \to A$ in $\mathcal{L}(\mathbb{R}^p, \mathbb{R}^q)$.

(b) The proof of this part is like that of the Weierstrass M-Test (4.2.2). Put $M_n = \|A_n\|$ and note that $\|A_n x\| \leq M_n \|x\|$ for each x in $\mathbb{R}^p$. Put $B_n = \sum_{k=1}^{n} A_k$ in

$\mathcal{L}(\mathbb{R}^p, \mathbb{R}^q)$. We have that for $n > m$,

$$\|B_n - B_m\| = \left\| \sum_{k=m+1}^{n} A_k \right\| \le \sum_{k=m+1}^{n} \|A_k\|$$

Thus $\{B_n\}$ is a Cauchy sequence in $\mathcal{L}(\mathbb{R}^p, \mathbb{R}^q)$. By (a) there is a linear transformation A in $\mathcal{L}(\mathbb{R}^p, \mathbb{R}^q)$ such that $B_n \to A$. ∎

When $p = q$ we have an additional property.

6.4.5. Proposition. *If $\{A_n\}$ is a sequence of invertible linear transformations in $\mathcal{L}(\mathbb{R}^p)$, A is an invertible linear transformation in $\mathcal{L}(\mathbb{R}^p)$, and $A_n \to A$, then $A_n^{-1} \to A^{-1}$.*

Proof. We begin with the following.

Claim. If $B \in \mathcal{L}(\mathbb{R}^p)$ and $\|1 - B\| < 1$, then B is invertible and

$$B^{-1} = \sum_{n=0}^{\infty} (1 - B)^n$$

To prove the claim first observe that since there is an r with $\|1 - B\| < r < 1$, we have that $\|(1 - B)^n\| < r^n$ and so the series $\sum_{n=0}^{\infty} \|(1 - B)^n\|$ converges. By part (b) of the preceding proposition this implies that $C = \sum_{n=0}^{\infty} (1 - B)^n$ converges in $\mathcal{L}(\mathbb{R}^p)$. Put $C_n = \sum_{k=0}^{n} (1 - B)^k$. Now

$$C_n B = C_n - C_n(1 - B)$$

$$= \sum_{k=0}^{n} (1 - B)^k - \sum_{k=1}^{n+1} (1 - B)^k$$

$$= 1 - (1 - B)^{n+1}$$

But $\|(1 - B)^{n+1}\| \to 0$, so we have that $CB = 1$. Similarly $BC = 1$ and so the claim is established. (Using linear algebra we could have said that $1 = CB$ means B is left invertible, but a linear transformation on $\mathbb{R}^p$ that is left invertible is invertible.)

Claim. If $\{B_n\}$ is a sequence in $\mathcal{L}(\mathbb{R}^p)$ and $B_n \to 1$, then there is an integer N such that B_n is invertible for all $n \ge N$ and $B_n^{-1} \to 1$.

We choose N_1 such that $\|B_n - 1\| < 1$ when $n \ge N_1$. By the preceding claim B_n is invertible for all such n. If $\delta > 0$ (we'll further specify δ in a moment), choose $N > N_1$ such that $\|1 - B_n\| < \delta$ when $n \ge N$. Again the first claim implies that $B_n^{-1} = [1 - (1 - B_n)]^{-1} = \sum_{k=0}^{\infty} (1 - B_n)^k = 1 + \sum_{k=1}^{\infty} (1 - B_n)^k$. Hence

$$\|B_n^{-1} - 1\| = \left\| \sum_{k=1}^{\infty} (1 - B_n)^k \right\| \le \sum_{k=1}^{\infty} \|1 - B_n\|^k < \frac{\delta}{1 - \delta}$$

Now if $\epsilon > 0$ we can choose $\delta > 0$ with $\delta/(1 - \delta) < \epsilon$ and we have that $\|B_n^{-1} - 1\| < \epsilon$ when $n \ge N$, establishing the claim.

Now to finish the proof. Using the notation in the statement of the proposition, we have that $A^{-1}A_n \to 1$. By the last claim

$$A_n^{-1}A = \left(A^{-1}A_n\right)^{-1} \to 1$$

Therefore $A_n^{-1} = (A_n^{-1}A)A^{-1} \to A^{-1}$ by Exercise 4. ∎

The preceding proposition can be rephrased as follows: If $\mathcal{G}_p$ is the set of all invertible linear transformations from $\mathbb{R}^p$ into itself, then the map $A \mapsto A^{-1}$ is a continuous function from $\mathcal{G}_p$ into itself.

Writing $x = \sum_{j=1}^p \langle x, e_j \rangle e_j$ and using the notation established in (6.4.1) we have

$$A(x) = \sum_{j=1}^p A(\langle x, e_j \rangle e_j) = \sum_{j=1}^p \langle x, e_j \rangle A(e_j)$$

$$= \sum_{j=1}^p \langle x, e_j \rangle \sum_{i=1}^q a_{ij}d_i = \sum_{i,j} \langle x, e_j \rangle a_{ij}d_i$$

Hence by the Cauchy–Schwarz Inequality

$$\|A(x)\|^2 \leq \left[\sum_{i,j} |\langle x, e_j \rangle| |a_{ij}| \|d_i\| \right]^2$$

$$\leq \left[\sum_j |\langle x, e_j \rangle|^2 \right] \left[\sum_{i,j} |a_{ij}|^2 \right]$$

$$= \|x\|^2 \left[\sum_{i,j} |a_{ij}|^2 \right]$$

This implies the following.

6.4.6. Proposition. *If $A \in \mathcal{L}(\mathbb{R}^p, \mathbb{R}^q)$ and A has the matrix (a_{ij}), then*

$$\|A\| \leq \left[\sum_{i,j} |a_{ij}|^2 \right]^{\frac{1}{2}}$$

See Exercise 6.

Let $A \in \mathcal{L}(\mathbb{R}^p, \mathbb{R}^q)$ and fix a vector z in $\mathbb{R}^q$. Observe that the map $x \mapsto \langle A(x), z \rangle$ is a linear functional on $\mathbb{R}^p$. By Proposition 6.2.8 there is a unique vector y_z in $\mathbb{R}^p$ such that $\langle A(x), z \rangle = \langle x, y_z \rangle$ for every x in $\mathbb{R}^p$. Since this vector y_z is unique, we have defined a function from $\mathbb{R}^q$ into $\mathbb{R}^p$: $z \mapsto y_z$. Note that the definition of this map depends on the linear transformation A and so we denote it by A^*; that is $y_z = A^*(z) \in \mathbb{R}^p$ for every z in $\mathbb{R}^p$. From what we have just done we have that for all x in $\mathbb{R}^p$ and all z in $\mathbb{R}^q$,

6.4.7 $$\langle x, A^*(z) \rangle = \langle A(x), z \rangle$$

The function A^* is called the *adjoint* of A. The reader is asked to recall from his/her encounter with linear algebra the definition of the transpose of a matrix. The transpose of a $q \times p$ matrix (a_{ij}) is denoted by $(a_{ij})^t$ and is the $p \times q$ matrix that has as its ji-entry the number a_{ji}. This is needed in the next proposition.

6.4.8. Proposition. *If $A \in \mathcal{L}(\mathbb{R}^p, \mathbb{R}^q)$, then the function $A^* : \mathbb{R}^q \to \mathbb{R}^p$ is a linear transformation. The matrix of the adjoint A^* is the transpose of the matrix for A.*

Proof. The proof that $A^* \in \mathcal{L}(\mathbb{R}^q, \mathbb{R}^p)$ is Exercise 7. Observe that the matrix of A^* is size $p \times q$. If its matrix is given by (b_{ji}) with $1 \le j \le p$, $1 \le i \le q$, then $b_{ji} = \langle A^*(d_i), e_j \rangle$. But using (6.4.7) we have that $b_{ji} = \langle e_j, A^*(d_i) \rangle = \langle A(e_j), d_i \rangle = a_{ij}$. Therefore the matrix (b_{ji}) is precisely $(a_{ij})^t$. ∎

We also have the following. The proof is left to the reader in Exercise 8(a); doing this exercise will help further cement the properties of the adjoint in your mind.

6.4.9. Proposition. *If $A, B \in \mathcal{L}(\mathbb{R}^p, \mathbb{R}^q)$ and $\lambda \in \mathbb{R}$, then $(A + \lambda B)^* = A^* + \lambda B^*$.*

We now focus on linear transformations $A : \mathbb{R}^p \to \mathbb{R}^p$. So when $A \in \mathcal{L}(\mathbb{R}^p)$, its matrix is a square one of size $p \times p$. Recall from linear algebra that in this case the three statements A is invertible, A is injective, and A is surjective are equivalent.

When $A \in \mathcal{L}(\mathbb{R}^p)$ we have that $A^* \in \mathcal{L}(\mathbb{R}^p)$. When $A, B \in \mathcal{L}(\mathbb{R}^p)$ we can form the product AB. In this case we have the following (Exercise 8(b)).

6.4.10. Proposition. *If $A, B \in \mathcal{L}(\mathbb{R}^p)$, then $(AB)^* = B^*A^*$.*

6.4.11. Definition. A linear transformation in $\mathcal{L}(\mathbb{R}^p)$ is *self-adjoint* or *hermitian*[4] if $A = A^*$.

A linear transformation A is hermitian if and only if its matrix is equal to its transpose. That is, $A = A^*$ if and only if $(a_{ij}) = (a_{ij})^t$. (Why?) Thus $A = (a_{ij})$ is hermitian if and only if $a_{ij} = a_{ji}$ for all i and j. In other words, A is hermitian if and only if its matrix is *symmetric*. In the literature many will use the term "symmetric"

....

[4] Charles Hermite was born in 1822 in Dieuze, France, which is east of Paris near the German border. In 1840 he went to school at Collège Louis-le-Grand in Paris, 15 years after Galois studied there. His tendency was to read original papers of mathematicians rather than work to pass the exams. Nevertheless, with a somewhat average performance on the entrance exam, he was admitted to the École Polytechnique. Unfortunately he had a birth defect that resulted in a malformed foot and because of this he was told he had to leave. (From today's perspective, this is truly amazing; but such things happened.) An appeal led to a reversal of the decision, but strict conditions were imposed on him and he decided to leave. On his own he pursued his studies, all the while doing research. In 1847 he passed the exams to receive the baccalauréat. A year later he was appointed to the faculty at École Polytechnique, the same school that had made life difficult for him. He worked on number theory and algebra, orthogonal polynomials, and elliptic functions, with several important contributions. The Hermite polynomials are named after him, and he was the first to prove that the number e is transcendental – that is, it is not the zero of any polynomial with rational coefficients. He had nine doctoral students, including Poincaré and Stieltjes. He died in 1901 in Paris.

for these linear transformations rather than "hermitian." This discrepancy with our usage of hermitian arises because you can also carry out this study of linear transformations and adjoints on vector spaces over the complex numbers and some prefer to reserve the term hermitian for this case. The point is when you use the complex numbers the matrix of a self-adjoint linear transformation is not quite symmetric. If you are curious and/or interested, please investigate the literature.

6.4.12. Example. Denote by $\mathrm{diag}\,(\lambda_1, \ldots, \lambda_p)$ the matrix that has the entries $\lambda_1, \ldots, \lambda_p$ along its main diagonal and zeros everywhere else. Call such a matrix a *diagonal matrix*. A linear transformation whose matrix is diagonal is hermitian.

Warning. *Soon we will prove an important result about hermitian linear transformations (6.4.29). To do this we need to extend the definition of the adjoint, and therefore of hermitian linear transformations, to linear transformations defined on a subspace of $\mathbb{R}^p$ rather than the entirety of Euclidean space. In reality this is a technical matter and not a substantial problem since, as some of you may recall from your course on linear algebra, all finite dimensional vector spaces over $\mathbb{R}$ are isomorphic to some Euclidean space. The details of all this are not presented here as this extension is only used in one place in the proof of Theorem 6.4.29. The interested reader can carry out the details as a project; this would involve taking each result and definition involving the adjoint and making the appropriate modifications so that it holds for linear transformations defined on a linear subspce $\mathcal{M}$ of $\mathbb{R}^p$.*

We now use the main result obtained in the last section (Theorem 6.3.6) to define an important hermitian linear transformation.

6.4.13. Definition. If $\mathcal{M}$ is a linear subspace of $\mathbb{R}^p$ and $x \in \mathbb{R}^p$, then the unique vector $Px = P(x)$ in $\mathcal{M}$ such that $x - Px \perp \mathcal{M}$ is called the *orthogonal projection* of x onto $\mathcal{M}$.

Note that if $x \in \mathcal{M}$, then $Px = x$; if $x \perp \mathcal{M}$, then $Px = 0$. The converses of these two statements are also true.

6.4.14. Proposition. *If $\mathcal{M}$ is a linear subspace of $\mathbb{R}^p$ and, for each x in $\mathbb{R}^p$, Px is the orthogonal projection of x onto $\mathcal{M}$, then the following hold.*

(a) $P : \mathbb{R}^p \to \mathbb{R}^p$ *is a hermitian linear transformation.*
(b) $\|Px\| \le \|x\|$ *for all x in $\mathcal{X}$.*
(c) P *is an idempotent; that is, $P^2 \equiv PP = P$.*
(d) $\ker P = \mathcal{M}^\perp$ *and* $\mathrm{ran}\, P = \mathcal{M} = \{x \in \mathbb{R}^p : Px = x\}$.
(e) *If $y_1, \ldots, y_m$ is any orthonormal basis for $\mathcal{M}$, then*

$$Px = \sum_{j=1}^{m} \langle x, y_j \rangle y_j.$$

Proof. Note that (e) has already been proven when we proved Theorem 6.3.6 and is only included for emphasis. But (e) easily implies that P is linear. Also for any

x, w in $\mathbb{R}^p$

$$\langle Px, w \rangle = \left\langle \sum_{j=1}^{m} \langle x, y_j \rangle y_j, w \right\rangle$$

$$= \sum_{j=1}^{m} \langle x, y_j \rangle \langle y_j, w \rangle$$

$$= \sum_{j=1}^{m} \langle x, \langle w, y_j \rangle y_j \rangle$$

$$= \left\langle x, \sum_{j=1}^{m} \langle w, y_j \rangle y_j \right\rangle$$

$$= \langle x, Pw \rangle$$

Hence $P = P^*$, completing the proof of (a).

Since $Px \perp x - Px$, we have by the Pythagorean Theorem that $\|x\|^2 = \|x - Px + Px\|^2 = \|x - Px\|^2 + \|Px\|^2 \geq \|Px\|^2$, proving (b).

The meaning of (c) is that $P(P(x)) = P(x)$ for all x in $\mathbb{R}^p$. If x is any vector in $\mathbb{R}^p$, then $Px \in \mathcal{M}$. But as we observed, $Py = y$ for all y in $\mathcal{M}$, so $P(Px) = Px$.

Using (e) it is clear that $\operatorname{ran} P \subseteq \mathcal{M}$ and $\mathcal{M}^\perp \subseteq \ker P$. Since $Px = x$ whenever $x \in \mathcal{M}$, it must be that $\operatorname{ran} P = \mathcal{M}$. Also since $\|Px\|^2 = \sum_{j=1}^{m} |\langle x, y_j \rangle|^2$, the only way that x can be in $\ker P$ is for x to be orthogonal to each y_j, $1 \leq j \leq m$. Since the y_j form a basis for $\mathcal{M}$, it follows that $x \perp \mathcal{M}$ whenever $x \in \ker P$. ∎

6.4.15. Definition. An *orthogonal projection* is an idempotent $P : \mathbb{R}^p \to \mathbb{R}^p$ such that $x - Px \perp \operatorname{ran} P$ for every x in $\mathbb{R}^p$.

In a certain sense the preceding definition is redundant. There is a sense, however, in which it is required. Definition 6.4.13 depends on first being given a subspace $\mathcal{M}$ of $\mathbb{R}^p$; Definition 6.4.15 defines what it means for a linear transformation to be an orthogonal projection without first being given such a subspace. On the other hand, if P is as in Definition 6.4.15 and $\mathcal{M} = \operatorname{ran} P$, then P is the orthogonal projection of $\mathbb{R}^p$ onto $\mathcal{M}$ as in Definition 6.4.13.

We need to recall and establish the definition and properties of the determinant for a square matrix. The reader is assumed to be somewhat familiar with this, but my experience is that this material is not fully known to most students who take this course; consequently there follows a presentation of determinants, including the definition. This must be preceded, however, by a discussion of permutations. Most of the results on permutations will be stated without proof. The reader can consult any source on permutations for the missing details, but I have used [2].

For $p \geq 2$ we want to consider an ordered set having p elements, where the word "ordered" is key. An example, of course, is the set of the first p integers, $\{1, \ldots, p\}$. A *permutation* is a reordering of the set. An example when $p = 5$ is given by

$(2, 3, 1, 5, 4)$; the meaning of this notation is that this permutation maps $2 \mapsto 3 \mapsto 1 \mapsto 5 \mapsto 4 \mapsto 2$. Another permutation when $p = 5$ is $(2, 3, 1)(5, 4)$; this permutation maps $2 \mapsto 3 \mapsto 1 \mapsto 2$ and $5 \mapsto 4 \mapsto 5$. Another is $(2, 3, 1)$ where the absence of 4 and 5 means they are left fixed; if you prefer, $(2, 3, 1) = (2, 3, 1)(4)(5)$. In particular (1) denotes the identity permutation that leaves every number fixed. If you prefer, a permutation is a bijection of the set, but where the resulting order is important. We let S_p denote the set of all permutations of the integers $1, \ldots, p$. If you know the concept of a group, S_p is a group under composition: if $\sigma, \tau \in S_p$, then $\sigma\tau$ is the element of S_p defined by $(\sigma\tau)(k) = \sigma(\tau(k))$ for $1 \le k \le p$. In fact it is one of the first examples of a group when you study this subject; it is called the *symmetric group* on p elements. See Exercise 9.

A permutation that switches two of the elements and leaves all the others fixed is called a *transposition*. For example the permutation $(4, 5)$ is a transposition that interchanges 4 and 5 and leaves the other integers fixed. A fact that seems intuitively clear after thinking about it but requires proof (see [2], p. 94) is that every σ in S_p is the product of a finite number of transpositions. As one example $(1, 2, \ldots, p) = (1, 2)(1, 3) \cdots (1, p)$. The way to write σ as a product of transpositions, however, is not unique.

6.4.16. Proposition. *Any permutation σ in S_p can be written as a product of transpositions. If there are two products of transpositions that are both equal to σ, one containing m factors and the other containing n factors, then m and n are either simultaneously even or simultaneously odd.*

A proof of the existence of the factorization could be concocted by using the example factorization of $(1, 2, \ldots, p)$ above. For a complete proof of the result see [2], p. 94.

This enables us to define the sign of a permutation or its *parity*. If $\sigma \in S_p$, define $\text{sign}(\sigma)$ to be $+1$ if σ is the product of an even number of transpositions and $\text{sign}(\sigma) = -1$ otherwise. In light of the preceding proposition, this is well-defined.

6.4.17. Proposition. *If $\sigma, \tau \in S_p$, then $\text{sign}(\sigma\tau) = \text{sign}(\sigma)\text{sign}(\tau)$.*

Proof. Let τ be arbitrary in S_p and let σ be a transposition. Thus the factorization of $\sigma\tau$ as a product of transpositions has one more transposition than a factorization of τ and so $\text{sign}(\sigma\tau) = -\text{sign}(\tau) = \text{sign}(\sigma)\text{sign}(\tau)$. If we repeatedly apply this to an arbitrary σ, we get the result. ∎

We now use this material on permutations to define the determinant of a square $p \times p$ matrix. Let $A \in \mathcal{L}(\mathbb{R}^p)$ with $p \times p$ matrix (a_{ij}); for convenience let $a(i, j) = a_{ij}$.

6.4.18. Definition. If $A = (a_{ij})$ is a $p \times p$ matrix, define the *determinant* of A as

$$\det A = \sum_{\sigma \in S_p} \text{sign}(\sigma)\, a(1, \sigma(1)) \cdots a(p, \sigma(p))$$

It is helpful to have another expression for the sign of a permutation. To state it we need to introduce the *sign of a real number a* as: $si(a) = 1$ if $a > 0$; $si(a) = -1$ if $a < 0$; $si(0) = 0$. (A word about notation. The usual notation for the sign of a real number a is $sign(a)$. I've chosen the different notation $si(a)$ to avoid any confusion with the sign of a permutation.)

6.4.19. Lemma. *For any σ in S_p,*

$$\text{sign}(\sigma) = \prod_{k<m} si(\sigma(m) - \sigma(k))$$

Hence for any $p \times p$ matrix $A = (a_{i,j})$,

$$\det A = \sum s(j_1, \ldots, j_p) a(1, j_i) \cdots a(p, j_p)$$

where $s(j_1, \ldots, j_p) = \prod_{k<m} si(j_m - j_k)$ and the sum is taken over all the distinct p-tuples $(j_1, \ldots, j_p)$ with $1 \leq j_k \leq p$.

Proof. The proof of the formula for $\text{sign}(\sigma)$ can be fashioned from the material in [2], page 98 in the section labeled "Second Proof". The proof of the additional formula for $\det A$ is immediate from the formula once we realize that $s(j_1, \ldots, j_p) = s(\sigma(1), \ldots, \sigma(p))$, where σ is the permutation defined as $\sigma(m) = j_m$. ∎

There are other ways to define the determinant, and if you have defined it differently you should show that the two definitions are the same. You might also try Exercise 10. It will be helpful to regard the determinant as a function of the columns of A. So if $x_1, \ldots, x_p \in \mathbb{R}^p$, we define $\det(x_1, \ldots, x_p) = \det A$, where A is the matrix with column vectors $x_1, \ldots, x_p$.

6.4.20. Theorem. *If $A \in \mathcal{L}(\mathbb{R}^p)$ with columns $x_1, \ldots, x_p$, the following hold.*

(a) $\det I = 1$.
(b) *If $\sigma \in S_p$ and B is the matrix with columns $x_{\sigma(1)}, \ldots, x_{\sigma(p)}$, then $\det B = \text{sign}(\sigma) \det A$.*
(c) *If two columns in A are equal, then $\det A = 0$.*
(d) *For $1 \leq j \leq p$, if the columns $\{x_k : k \neq j\}$ are held fixed, then the function $x_j \mapsto \det(x_1, \ldots, x_p)$ is a linear functional of $\mathbb{R}^p$ into $\mathbb{R}$.*
(e) *For any B in $\mathcal{L}(\mathbb{R}^p)$, $\det(BA) = (\det B)(\det A)$.*
(f) *A is invertible if and only if $\det A \neq 0$, in which case $\det(A^{-1}) = (\det A)^{-1}$.*

Proof. (a) If $A = I$, we have that $a(j, j) = 1$ and $a(i, j) = 0$ when $i \neq j$. It follows that the only permutation σ such that $a(1, \sigma(1)) \cdots a(p, \sigma(p)) \neq 0$ is $\sigma = (1)$, in which case this product is 1. Since the sign of (1) is 1, this proves the result.

(b) Here we will use Lemma 6.4.19 that gives the second formula for $\det A$. We also start by assuming that σ is a transposition. Notice from the definition of $s(j_1, \ldots, j_p)$ that if two of these integers are interchanged, then $s(j_1, \ldots, j_p)$ changes sign. For example $s(j_1, \ldots, j_p) = -s(j_2, j_1, j_3, \ldots, j_p)$. Since interchanging two columns has precisely the effect of interchanging two of these integers, this proves (b) when σ is a transposition. The general form of (b) follows by induction since every permutation is the product of transpositions

(c) If two columns in A are equal, interchanging those columns does not effect A. So (c) is a corollary of (b).

(d) Again use the formula for $\det A$ in (6.4.19). Consider each of the summands $s(j_1, \ldots, j_p)a(1, j_i) \cdots a(p, j_p)$ in Definition 6.4.2. Since the sum in this formula is over all the $p!$ distinct p-tuples and all but the r-th column are held fixed, what is left is a linear function of that column.

(e) The proof of this part is more involved than the preceding parts. Fix B and define $\Delta : \mathcal{L}(\mathbb{R}^p) \to \mathbb{R}$ by $\Delta(A) = \det(BA)$. If $x_1, \ldots, x_p$ are the columns of A, then the columns of BA are $Bx_1, \ldots, Bx_p$. Thus

$$\Delta(A) = \Delta(x_1, \ldots, x_p) = \det(Bx_1, \ldots, Bx_p)$$

Note that Δ also enjoys properties (b), (c), and (d). (Verify!) Hence considering only the first column of A, property (d) implies that

$$\Delta(A) = \Delta\left(\sum_{i=1}^{p} a(i, 1)e_1, x_2, \ldots, x_p\right) = \sum_{i=1}^{p} a(i, 1)\Delta(e_1, x_2, \ldots, x_p)$$

Repeating this argument for all the succeeding columns we get that

6.4.21 $$\Delta(A) = \sum a(i_1, 1) \cdots a(i_p, p)\Delta(e_{i_1}, \ldots, e_{i_p})$$

where the sum is over all p-tuples of integers $(i_1, \ldots, i_p)$ between 1 and p. Using properties (b) and (c) for Δ, we get that for any p-tuple $(i_1, \ldots, i_p)$, $\Delta(e_{i_1}, \ldots, e_{i_p}) = t(i_1, \ldots, i_p)\Delta(e_1, \ldots, e_p)$, where the number $t(i_1, \ldots, i_p)$ equals 0 or ± 1. (For the moment do not worry about the relation between the numbers $t(i_1, \ldots, i_p)$ and $s(i_1, \ldots, i_p)$, where this last number was used in the definition of the determinant. As we will see, it all comes out in the end.) Using the fact that $\Delta(I) = \det B$, using the preceding equalities, and substituting them into (6.4.21) shows that

$$\det(BA) = \Delta(A) = \left[\sum a(i_1, 1) \cdots a(i_p, p)t(i_1, \ldots, i_p)\right]\det B$$

Observe that when $B = I$, the preceding equation becomes $\det A = \Delta(A) = \sum a(i_1, 1) \cdots a(i_p, p)t(i_1, \ldots, i_p)$, and we see that the preceding displayed equation becomes what we want to prove.

(f) If A is invertible, then by (e) we have that $(\det A)(\det(A^{-1})) = \det(AA^{-1}) = \det I = 1$, so $\det A \neq 0$ and $\det(A^{-1}) = (\det A)^{-1}$. Conversely assume that A is not invertible. This implies there is at least one column of A that is a linear combination of the others. (Why?) For notational convenience assume the dependent column is the first. Hence there are real numbers $c_2, \ldots, c_p$ such that $x_1 = \sum_{j=2}^{p} c_j x_j$. Observe that

$$\det(x_1 - c_2x_2, x_2, \ldots, x_p) = \det A - c_2 \det(x_2, x_2, \ldots, x_p) \quad \text{(by (d))}$$
$$= \det A \quad \text{(by (c))}$$

Repeating this argument for successive columns we get that

$$\det A = \det \left(x_1 - \sum_{j=2}^{p} c_j x_j, x_2, \ldots, x_p \right) = \det(0, x_2, \ldots, x_p) = 0 \qquad \blacksquare$$

The following corollary follows by invoking Exercise 12.

6.4.22. Corollary. *If $A \in \mathcal{L}(\mathbb{R}^p)$, the following hold.*

(a) *If B is the matrix obtained by applying the permutation σ in S_p to the rows of A, then $\det B = \text{sign}(\sigma) \det A$.*
(b) *If two rows in A are equal, then $\det A = 0$.*
(c) *For $1 \leq j \leq p$, if the rows $\{y_k : k \neq j\}$ are held fixed and A_{y_j} is the matrix with rows $y_1, \ldots, y_p$, then the function $y_j \mapsto \det(A_{y_j})$ is a linear functional from $\mathbb{R}^p$ into $\mathbb{R}$.*

Also see Exercises 10, 12, and 13.

We will use the above material on permutations and determinants now as well as in §9.2 and elsewhere in Chapter 9.

(Note that here, as well as in many other places in the literature, for a scalar λ, $A - \lambda$ is used for the linear transformation $A - \lambda I$. In other words, when we write λ we are talking about both the scalar λ and the linear transformation λI.)

6.4.23. Proposition. *If $A \in \mathcal{L}(\mathbb{R}^p)$, then the mapping of $\mathbb{R}$ into itself defined by $\lambda \mapsto \det(A - \lambda)$ is a polynomial in λ of degree p.*

Proof. Let $x_1, \ldots, x_p$ be the columns of A. Repeatedly using part (d) of the preceding theorem as well as part (a), we have

$$
\begin{aligned}
\det(A - \lambda) &= \det(x_1 - \lambda e_1, \ldots, x_p - \lambda e_p) \\
&= \det(x_1, x_2 - \lambda e_2, \ldots, x_p - \lambda e_p) - \lambda \det(e_1, x_2 - \lambda e_2, \ldots, x_p) \\
&= \ldots \\
&= \det A + \lambda c_1 \lambda + \cdots + c_{p-1} \lambda^{p-1} + (-1)^p \lambda^p
\end{aligned}
$$

for some choice of constants $c_1, \ldots, c_p$. This completes the proof since the coefficient of λ^p is not zero. $\qquad \blacksquare$

The polynomial $\det(A - \lambda) = \det(A - \lambda I)$ is called the *characteristic polynomial* of A. Recall that if $A \in \mathcal{L}(\mathbb{R}^p)$, an *eigenvalue* of A is a scalar λ such that there is a non-zero vector x called an *eigenvector* with $Ax = \lambda x$. The eigenvector x is said to correspond to λ. So the eigenvectors are precisely the vectors that belong to $\ker(A - \lambda)$. The subspace $\ker(A - \lambda)$ is called the *eigenspace* of A corresponding to λ. Thus λ is an eigenvalue for A if and only if $\ker(A - \lambda) \neq (0)$; equivalently, if and only if $A - \lambda$ is not invertible; equivalently, if and only if $\det(A - \lambda) = 0$. So the eigenvalues of A are precisely the zeros of the characteristic polynomial. The *multiplicity* of the eigenvalue λ is the dimension of its eigenspace. (Be aware that some books give a different definition of multiplicity.) The reader will often encounter in this section and beyond a phrase like the following: let $\lambda_1, \ldots, \lambda_n$ be the eigenvalues of A repeated as often as their multiplicity. This means that if λ is an eigenvalue of A of multiplicity m, then λ appears in the sequence $\lambda_1, \ldots, \lambda_n$ precisely m times. Also realize that since $\det(A - \lambda)$ is a polynomial of degree p and eigenvalues happen only when this polynomial has a zero, A can have at most p eigenvalues counting multiplicity. It is possible that a linear transformation has significantly fewer eigenvalues as the next example shows.

6.4.24. Example. (a) Let $p = 2$ and let

$$A = \begin{bmatrix} 0 & -1 \\ 1 & 0 \end{bmatrix}$$

So $\det(A - \lambda) = \lambda^2 + 1$, which has no zeros. Thus A has no eigenvalues.

(b) If $A = \operatorname{diag}(\lambda_1, \ldots, \lambda_p)$ then each λ_j is an eigenvalue. The multiplicity of λ_j is the number of times it occurs in the finite sequence $\lambda_1, \ldots, \lambda_p$.

6.4.25. Definition. The set of all eigenvalues of A, not counting multiplicity, is called the *spectrum* of A and is denoted by $\sigma(A)$.

From Example 6.4.24(a) we see that it is possible for $\sigma(A)$ to be empty. If A is λ times the identity, $\sigma(A)$ is the singleton $\{\lambda\}$. When A is hermitian, $\sigma(A) \neq \emptyset$ as we will see shortly.

6.4.26. Proposition. *If A is a hermitian linear transformation, then*

$$\|A\| = \sup\{|\langle Ax, x \rangle| : \|x\| = 1\}$$

Proof. Let M denote the supremum in the statement. Since $|\langle Ax, x \rangle| \leq \|Ax\|\|x\| \leq \|A\|\|x\|^2$, we have $M \leq \|A\|$.

If $\|x\| = \|y\| = 1$, using the fact that $A^* = A$ we have the following two equations. (Here when $\pm$ appears more than once in an equation it is always a $+$ in that equation or always a $-$.)

$$\langle A(x \pm y), x \pm y \rangle = \langle Ax, x \rangle \pm \langle Ax, y \rangle \pm \langle Ay, x \rangle + \langle Ay, y \rangle$$
$$= \langle Ax, x \rangle \pm \langle Ax, y \rangle \pm \langle y, A^*x \rangle + \langle Ay, y \rangle$$
$$= \langle Ax, x \rangle \pm \langle Ax, y \rangle \pm \langle A^*x, y \rangle + \langle Ay, y \rangle$$

Since $A = A^*$, when we subtract one of these equations from the other and do some simplifying, we get $\langle A(x + y), x + y \rangle - \langle A(x - y), x - y \rangle = 4\langle Ax, y \rangle$. Now $|\langle Az, z \rangle| \leq M\|z\|^2$ for all z in $\mathbb{R}^p$. When $\|x\| = \|y\| = 1$, the parallelogram law (Exercise 6.3.2) shows that this last equation yields

$$4\langle Ax, y \rangle \leq M(\|x + y\|^2 + \|x - y\|^2)$$
$$= 2M(\|x\|^2 + \|y\|^2)$$
$$= 4M$$

Since $\langle Ax, y \rangle = \pm|\langle Ax, y \rangle|$, substituting $-x$ for x in the above inequality if necessary gives $|\langle Ax, y \rangle| \leq M$ whenever $\|x\| = \|y\| = 1$. If we take the supremum over all y with $\|y\| = 1$, we get $\|Ax\| \leq M$; now take the supremum over all x with $\|x\| = 1$ and the proof is complete. ∎

6.4.27. Corollary. *If A is hermitian and $\langle Ax, x \rangle = 0$ for all x, then $A = 0$.*

The preceding proposition and corollary are decidedly false if the operator A is not hermitian. For example consider the linear transformation defined in Example 6.4.24(a). It is easy to check that in that example $\langle Ax, x \rangle = 0$ for every x.

6.4.28. Proposition. *If A is hermitian, then either $\|A\|$ or $-\|A\|$ is an eigenvalue for A.*

Proof. According to Proposition 6.4.26, $\|A\| = \sup\{|\langle Ax, x \rangle| : \|x\| = 1\}$. But $x \mapsto |\langle Ax, x \rangle|$ is a continuous function from $\mathbb{R}^p \to \mathbb{R}$ (Exercise 15) and $\{x \in \mathbb{R}^p : \|x\| = 1\}$ is a compact set by the Heine–Borel Theorem. Therefore the supremum is attained and there is a vector x_0 with $\|x_0\| = 1$ such that $|\langle Ax_0, x_0 \rangle| = \|A\|$. Let $\lambda = \langle Ax_0, x_0 \rangle$. We will show that λ is an eigenvalue with eigenvector x_0, completing the proof. Indeed, $\|(A - \lambda)x_0\|^2 = \langle Ax_0 - \lambda x_0, Ax_0 - \lambda x_0 \rangle = \|Ax_0\|^2 - 2\lambda\langle Ax_0, x_0 \rangle + \lambda^2\|x_0\|^2 = \lambda^2 - 2\lambda\lambda + \lambda^2 = 0$. Therefore $(A - \lambda)x_0 = 0$. ∎

Now for a very important result in mathematics. Before tackling the proof, the reader should solve Exercise 18.

6.4.29. Theorem (The Spectral Theorem). *Assume A is a hermitian linear transformation on $\mathbb{R}^p$. If $\lambda_1, \dots, \lambda_m$ are the distinct eigenvalues of A and, for $1 \leq j \leq m$, P_j is the orthogonal projection of $\mathbb{R}^p$ onto $\ker(A - \lambda_j)$, then $\ker(A - \lambda_j) \perp \ker(A - \lambda_i)$ for $j \neq i$ and*

6.4.30
$$A = \sum_{j=1}^{m} \lambda_j P_j.$$

Proof. We can assume that $A \neq 0$. According to Proposition 6.4.28, A has an eigenvalue λ_1. Put $\mathcal{M}_1 = \ker(A - \lambda_1)$ and let P_1 be the orthogonal projection of $\mathbb{R}^p$ onto $\mathcal{M}_1$. By Exercise 18 we can consider the restriction $A_1 = A|\mathcal{M}_1^\perp$, a linear transformation of $\mathcal{M}_1^\perp$ into itself. It is left to the reader to verify that A_1 is hermitian. (Here is where we encounter the small wrinkle mentioned in the **Warning** given earlier in this section. We have only discussed hermitian linear transformations on $\mathbb{R}^p$, and $\mathcal{M}_1^\perp$ is a subspace of $\mathbb{R}^p$.) Therefore A_1 has an eigenvalue λ_2 by (6.4.28); clearly λ_2 is also an eigenvalue for A. It is transparent that every eigenvector for A_1 is also an eigenvector for A. Thus we note that λ_2 must be different from λ_1 since all the eigenvectors for A corresponding to λ_1 were disposed of in $\mathcal{M}_1$. Put $\mathcal{M}_2 = \ker(A - \lambda_2)$ and let P_2 be the orthogonal projection of $\mathbb{R}^p$ onto $\mathcal{M}_2$. Note that $\lambda_1 \neq 0$. (Why?) So if $x_1 \in \mathcal{M}_1$ and $x_2 \in \mathcal{M}_2$,

$$\langle x_1, x_2 \rangle = \lambda_1^{-1}\langle Ax_1, x_2 \rangle = \lambda_1^{-1}\langle x_1, Ax_2 \rangle = \lambda_2\lambda_1^{-1}\langle x_1, x_2 \rangle$$

That is, $(1 - \lambda_2\lambda_1^{-1})\langle x_1, x_2 \rangle = 0$. Since $(1 - \lambda_2\lambda_1^{-1}) \neq 0$ and x_1 and x_2 were arbitrarily chosen, it must be that $\mathcal{M}_1 \perp \mathcal{M}_2$.

Let $A_3 = A|(\mathcal{M}_1 \oplus \mathcal{M}_2)^\perp$ and continue the above process. Since $\mathbb{R}^p$ is finite dimensional, this process must stop after a finite number of steps and we obtain distinct eigenvalues $\lambda_1, \ldots, \lambda_m$ of A with P_j the orthogonal projection of $\mathbb{R}^p$ onto $\ker(A - \lambda_j)$, $1 \leq j \leq m$. Just as we showed that $\mathcal{M}_1 \perp \mathcal{M}_2$, we can show any pair of distinct eigenspaces are orthogonal. (Do it!) Now $\ker(A - \lambda_1) \oplus \cdots \oplus \ker(A - \lambda_m) = \mathbb{R}^p$ or we could continue the process still further. Therefore if $x \in \mathbb{R}^p$, $x = \sum_{j=1}^m P_j x$ and so

$$ Ax = A\left(\sum_{j=1}^m P_j x\right) = \sum_{j=1}^m AP_j x = \sum_{j=1}^m \lambda_j P_j x = \left(\sum_{j=1}^m \lambda_j P_j\right) x \qquad \blacksquare $$

When A is hermitian the expression in (6.4.30) is called the *spectral decomposition* of A.

6.4.31. Example. Let $\alpha_1, \ldots, \alpha_p \in \mathbb{R}$ and let A be the linear transformation on $\mathbb{R}^p$ defined by the diagonal matrix with these entries $\alpha_1, \ldots, \alpha_p$ on the main diagonal. So if $e_1, \ldots, e_p$ is the standard basis for $\mathbb{R}^p$, $Ae_j = \alpha_j e_j$ for $1 \leq j \leq p$. To find the spectral decomposition of A, let $\lambda_1, \ldots, \lambda_m$ be the distinct eigenvalues of A. So each λ_k is at least one of the numbers α_j, but it may appear several times in the list $\alpha_1, \ldots, \alpha_p$. If for $1 \leq j \leq m$, $K_j = \{k : \alpha_k = \lambda_j\}$, let $\mathcal{M}_j$ be the linear span of $\{e_k : k \in K_j\}$. If P_j is the orthogonal projection of $\mathbb{R}^p$ onto $\mathcal{M}_j$, then the spectral decomposition of A is $A = \sum_{j=1}^m \lambda_j P_j$.

There are other ways in which The Spectral Theorem is sometimes stated. Here is one, which partially furnishes a converse of the preceding example.

6.4.32. Corollary. *If A is a hermitian linear transformation on $\mathbb{R}^p$, then there is an orthonormal basis for $\mathbb{R}^p$ consisting of eigenvectors for A.*

Proof. Using the notation from The Spectral Theorem, let $\mathcal{M}_j = P_j \mathbb{R}^p = \ker(A - \lambda_j)$. For $1 \leq j \leq m$ pick an orthonormal basis $\mathcal{B}_j$ for $\mathcal{M}_j$. The union of these m bases, $\bigcup_{j=1}^m \mathcal{B}_j$, is an orthonormal basis for $\mathbb{R}^p$ and each vector in this basis is an eigenvector for A. $\qquad \blacksquare$

Exercises

(1) Let $A, B \in \mathcal{L}(\mathbb{R}^p, \mathbb{R}^q)$ and let $y_1, \ldots, y_p$ be some basis for $\mathbb{R}^p$. Show that if $Ay_j = By_j$ for $1 \leq j \leq p$, then $A = B$. In other words, a linear transformation in $\mathcal{L}(\mathbb{R}^p, \mathbb{R}^q)$ is determined by its values on a basis.

(2) In defining the norm of a linear transformation why does the supremum exist?

(3) Prove Proposition 6.4.2.

(4) (a) Prove Corollary 6.4.3. That is, show that the following functions are continuous: (i) the map from $\mathcal{L}(\mathbb{R}^p, \mathbb{R}^q) \times \mathcal{L}(\mathbb{R}^p, \mathbb{R}^q) \to \mathcal{L}(\mathbb{R}^p, \mathbb{R}^q)$ defined by $(A, B) \mapsto A + B$; and (ii) the map from $\mathbb{R} \times \mathcal{L}(\mathbb{R}^p, \mathbb{R}^q) \to \mathcal{L}(\mathbb{R}^p, \mathbb{R}^q)$ defined by $(\alpha, A) \mapsto \alpha A$. (b) Show that when $p = q$, the map from $\mathcal{L}(\mathbb{R}^p) \times \mathcal{L}(\mathbb{R}^p) \to \mathcal{L}(\mathbb{R}^p)$ defined by $(A, B) \mapsto AB$ is continuous.

(5) If $A \in \mathcal{L}(\mathbb{R}^p, \mathbb{R}^q)$, show that A is a continuous function.

(6) Find an example of a linear transformation in $\mathcal{L}(\mathbb{R}^p, \mathbb{R}^q)$ such that the inequality in Proposition 6.4.6 is strict.

(7) Prove that the function A^* in Proposition 6.4.8 is a linear transformation.

(8) (a) Prove Proposition 6.4.9. (b) Prove Proposition 6.4.10.

(9) If S_p is the set of all permutations of $\{1, \ldots, p\}$ and $\sigma, \tau \in S_p$, show that with the definition of multiplication $\sigma \tau$ as composition the following hold. (a) $\sigma \tau \in S_p$. (b) This multiplication is associative. (c) There is a σ^{-1} in S_p such that $\sigma \sigma^{-1} = \sigma^{-1} \sigma$ is the identity permutation. (This shows that S_p is a group under composition.)

(10) (a) Use Definition 6.4.18 to show that when A is either a 2×2 or a 3×3 matrix, the definition gives the expected answer for $\det A$. (b) Use Definition 6.4.18 to show that $\det A$ equals its expansion by minors using the first column of A. (c) If your definition of a determinant is not that given in (6.4.18), prove that formula from your definition.

(11) In Theorem 6.4.20(d) find the unique vector u in $\mathbb{R}^p$ such that the linear functional $x_j \mapsto \det(x_1, \ldots, x_p)$ equals $\langle x_j, u \rangle$,

(12) Let A be a $p \times p$ matrix. (a) Show that $\det A^* = \det A$. (b) If two columns of A are linearly dependent, show that $\det A = 0$. (c) If two rows of A are linearly dependent, show that $\det A = 0$. (d) Show that any statement about the columns of A relative to $\det A$ can also be made about the rows of A relative to $\det A$ since the rows of A are the columns of A^* as stated in Corollary 6.4.22.

(13) When we compute the matrix (a_{ij}) for a linear transformation A in $\mathcal{L}(\mathbb{R}^p)$, recall that we are using the usual basis $e_1, \ldots, e_p$. Recall from linear algebra that for any basis $\mathcal{B}$ of $\mathbb{R}^p$ we can form another matrix of A, which we denote by $[A]_{\mathcal{B}}$. Show that the determinant of this new matrix is the same as $\det A$ as defined in (6.4.18).

(14) If $f(t)$ is a polynomial and A is a linear transformation on $\mathbb{R}^p$ such that $f(A) = 0$, show that $f(\lambda) = 0$ for every λ in $\sigma(A)$. What does this say about the relationship between $f(t)$ and the characteristic polynomial of A?

(15) Prove that for any A in $\mathcal{L}(\mathbb{R}^p)$, $x \mapsto |\langle Ax, x \rangle|$ is a continuous function from $\mathbb{R}^p \to \mathbb{R}$.

(16) Show that if E is an idempotent on $\mathbb{R}^p$, then $\sigma(E) = \{0, 1\}$.

(17) Let $\mathcal{M}$ be a linear subspace of $\mathbb{R}^p$, give $\mathcal{M}$ its subspace topology, and show that the orthogonal projection $P : \mathbb{R}^p \to \mathcal{M}$ is an open mapping; that is, $P(G)$ is an open subset of $\mathcal{M}$ whenever G is open in $\mathbb{R}^p$.

(18) Let A be a hermitian linear transformation on $\mathbb{R}^p$ and let λ be an eigenvalue of A. (a) Show that $A(\ker(A - \lambda)) \subseteq \ker(A - \lambda)$ and $A([\ker(A - \lambda)]^\perp) \subseteq [\ker(A - \lambda)]^\perp$. (b) If P is the orthogonal projection of $\mathbb{R}^p$ onto $\ker(A - \lambda)$, show that $AP = PA$.

6.5. Differentiable Functions, Part 2

6.5.1. Definition. If G is an open subset of $\mathbb{R}^p$, $x \in G$, and $f : G \to \mathbb{R}^q$, then f is *differentiable* at x when there is a linear transformation $A : \mathbb{R}^p \to \mathbb{R}^q$ and a function $F : G \to \mathbb{R}^q$ such that $\lim_{y \to x} F(y) = 0$ and for all y in G

$$f(y) - f(x) = A(y - x) + \|y - x\|F(y)$$

We define the *derivative* of f at x to be the linear transformation $Df(x) = f'(x) = A$. If f is differentiable at every point of G then f is said to be *differentiable* on G.

Once again in this section G is always an open subset of $\mathbb{R}^p$.

6.5.2. Proposition. *Suppose $f : G \to \mathbb{R}^q$ and $x \in G$. If there are A_1, A_2 in $\mathcal{L}(\mathbb{R}^p, \mathbb{R}^q)$ and functions F_1, F_2 from G into $\mathbb{R}^q$ such that $\lim_{y \to x} F_1(y) = 0 = \lim_{y \to x} F_2(y)$ and $f(y) - f(x) = A_1(y - x) + \|y - x\|F_1(y) = A_2(y - x) + \|y - x\|F_2(y)$ whenever $y \in G$, then $A_1 = A_2$. In other words the derivative of f at x, if it exists, is unique.*

Proof. Just following Theorem 6.2.12 we pointed out that that theorem implies the uniqueness of the derivative of a function from G into $\mathbb{R}$. The idea here is to reduce the proof of this proposition to the scalar case.

For any vector d in $\mathbb{R}^q$ let $\langle f, d \rangle : G \to \mathbb{R}$ be defined as

$$\langle f, d \rangle(y) = \langle f(y), d \rangle$$

for all y in G. (The notation is a bit awkward, but we'll just have to live with it. The alternatives all seem awkward.) We note that for $k = 1, 2$,

$$\langle f, d \rangle(y) - \langle f, d \rangle(x) = \langle A_k(y - x), d \rangle + \|y - x\|\langle F_k(y), d \rangle$$

By Definition 6.2.10 $\langle f, d \rangle$ has a derivative at x and that derivative is the linear functional on $\mathbb{R}^p$ defined by $w \mapsto \langle A_k(w), d \rangle$. Since the derivative of $\langle f, d \rangle$ is unique, $\langle A_1(w), d \rangle = \langle A_2(w), d \rangle$ for all w in $\mathbb{R}^p$. Since d was arbitrary in $\mathbb{R}^q$, we have that $A_1 = A_2$. ∎

6.5.3. Example. (a) If $A \in \mathcal{L}(\mathbb{R}^p, \mathbb{R}^q)$ and $f : \mathbb{R}^p \to \mathbb{R}^q$ is defined by $f(x) = A(x)$ for all x in $\mathbb{R}^p$, then f is differentiable everywhere and $Df(x) = A$ for all x in $\mathbb{R}^p$. In fact this f satisfies the definition with $F(y) = 0$ for all y in $\mathbb{R}^p$.

(b) If $f : G \to \mathbb{R}$ is differentiable, then $Df : G \to \mathbb{R}^{p*} = \mathbb{R}^p$. It therefore makes sense to ask whether Df is differentiable. As we progress we'll revisit this idea of the second derivative of a real-valued differentiable function of several variables.

The proof of the next result is left to the reader as Exercise 1.

6.5.4. Proposition. *A function $f : G \to \mathbb{R}^q$ is differentiable at x if and only if there is a linear transformation A in $\mathcal{L}(\mathbb{R}^p, \mathbb{R}^q)$ such that*

$$\lim_{z \to 0} \frac{\|f(x + z) - f(x) - A(z)\|}{\|z\|} = 0$$

We set some notation that will be frequently used. Suppose $f : G \to \mathbb{R}^q$ and $d_1, \ldots, d_q$ is the standard basis for $\mathbb{R}^q$. Recalling the notation introduced in the proof of Proposition 6.5.2, for $1 \leq i \leq q$ define $f_i : G \to \mathbb{R}$ by

6.5.5 $$f_i(x) = \langle f, d_i \rangle(x) = \langle f(x), d_i \rangle$$

We note that if $f_i(x) = 0$ for $1 \leq i \leq q$, then $f(x) = 0$ in $\mathbb{R}^q$. Hence the functions $f_1, \ldots, f_q$ completely determine f. If f is differentiable at x, then each $f_i : G \to \mathbb{R}$ is differentiable at x, with its derivative a linear functional from $\mathbb{R}^p$ into $\mathbb{R}$ such that for each z in $\mathbb{R}^p$,

6.5.6 $$\langle f_i'(x), z \rangle = \langle f'(x)z, d_i \rangle$$

(Verify!)

6.5.7. Theorem. *If G is an open subset of $\mathbb{R}^p$, $x \in \mathbb{R}^p$, and $f : G \to \mathbb{R}^q$ is differentiable at x, then the following hold.*

(a) *f is continuous at x.*
(b) *For $1 \leq i \leq q$ and $1 \leq j \leq p$ the j-th partial derivative of f_i exists at x.*
(c) *The matrix representing the linear transformation $Df(x)$ is*

6.5.8 $$\left[\frac{\partial f_i}{\partial x_j}(x) \right] = \begin{bmatrix} \frac{\partial f_1}{\partial x_1}(x) & \frac{\partial f_1}{\partial x_2}(x) & \cdots & \frac{\partial f_1}{\partial x_p}(x) \\ \frac{\partial f_2}{\partial x_1}(x) & \frac{\partial f_2}{\partial x_2}(x) & \cdots & \frac{\partial f_2}{\partial x_p}(x) \\ \vdots & \vdots & \cdots & \vdots \\ \frac{\partial f_q}{\partial x_1}(x) & \frac{\partial f_q}{\partial x_2}(x) & \cdots & \frac{\partial f_q}{\partial x_p}(x) \end{bmatrix}$$

Proof. The proof of part (a) follows as the analogous part of Theorem 6.2.12 did. The reader is required to show the details in Exercise 2. For (b) we leave it to the reader to show that for $1 \leq i \leq q$, $f_i : G \to \mathbb{R}$ is differentiable at x by using Definition 6.2.10 and the fact that f is differentiable. Thus the partial derivative of f_i with respect to x_j exists by (6.2.12(b)). For (c) we observe that the ij-entry of the matrix representation of $Df(x)$ is, by Theorem 6.2.12,

$$\langle Df(x)e_j, d_i \rangle = (f_i)'(x)(e_j) = \frac{\partial f_i}{\partial x_j}(x) = \partial_j f_i(x) \qquad \blacksquare$$

The next result is the extension of Theorem 6.2.13 to the present situation. It can be proved by using that theorem. The details are left to the reader in Exercise 6.

6.5.9. Theorem. *If $f : G \to \mathbb{R}^q$ such that for $1 \leq i \leq q$ and $1 \leq j \leq p$ the j-th partial derivative of f_i exists and is continuous, then f is differentiable.*

6.5.10. Proposition. *If G is a connected set and $f : G \to \mathbb{R}^q$ is differentiable with $Df(x) = 0$ for all x in G, then f is constant.*

Proof. For $1 \leq i \leq q$, examine the function $f_i : G \to \mathbb{R}$ defined in (6.5.5). As in (6.5.6) we have that the linear functional $f_i'(x)$ on $\mathbb{R}^p$ is given by $z \mapsto \langle f_i'(x), z \rangle = \langle f'(x)z, d_i \rangle = 0$ by hypothesis. By Proposition 6.2.18, each f_i is constant. Thus $f(x) = (f_1(x), \ldots, f_q(x))$ is constant in $\mathbb{R}^q$. $\blacksquare$

6.5.11. Definition. If $f : G \to \mathbb{R}^q$ is differentiable, then f is *continuously differentiable* provided $Df : G \to \mathcal{L}(\mathbb{R}^p, \mathbb{R}^q)$ is continuous.

The metric on $\mathcal{L}(\mathbb{R}^p, \mathbb{R}^q)$ is the one defined in Proposition 6.4.2, so in the above definition the function Df is a mapping between two metric spaces. It is in this sense that the continuity of Df is defined. Also from Theorem 6.5.7(c), if f is continuously differentiable, then all the partial derivatives $\partial f_i / \partial x_j$ are also continuous. (Why?)

The next result is undoubtedly expected. The proof is left to the reader in Exercise 7.

6.5.12. Proposition. *If $f : G \to \mathbb{R}^q$ and $g : G \to \mathbb{R}^q$ are differentiable, then $f + g : G \to \mathbb{R}^q$ is differentiable and $D(f + g) = Df + Dg$. If $\alpha \in \mathbb{R}$, then αf is differentiable and $D(\alpha f) = \alpha Df$.*

6.5.13. Theorem (Chain Rule). *If $f : G \to \mathbb{R}^q$ is differentiable, H is an open subset of $\mathbb{R}^q$ that contains $f(G)$, and $g : H \to \mathbb{R}^d$ is differentiable, then $g \circ f : G \to \mathbb{R}^d$ is differentiable and*

$$D(f \circ g)(x) = Dg(f(x)) \circ Df(x) = Dg(f(x))[Df(x)]$$

or, equivalently,

$$(g \circ f)'(x) = g'(f(x)) \circ f'(x) = g'(f(x))[f'(x)]$$

for all x in G.

Proof. The proof parallels the proof of Theorem 2.2.10. Fix x in G, let $\xi = f(x)$ in H, and set $A = f'(x)$ and $B = g'(\xi) = g'(f(x))$. From the definition of differentiability we know that

$$f(y) - f(x) = A(y - x) + \|y - x\| F(y)$$
$$g(\eta) - g(\xi) = B(\eta - \xi) + \|\eta - \xi\| K(\eta)$$

where $F : G \to \mathbb{R}^q$, $K : H \to \mathbb{R}^d$, and $\lim_{y \to x} F(y) = 0 = \lim_{\eta \to \xi} K(\eta)$. We want to show that

$$(g \circ f)(y) - (g \circ f)(x) = (B \circ A)(y - x) + \|y - x\| \Gamma(y)$$
$$= B[A(y - x)] + \|y - x\| \Gamma(y)$$

where $\lim_{y \to x} \Gamma(y) = 0$.

Now

$$(g \circ f)(y) - (g \circ f)(x) = B(f(y) - f(x)) + \|f(y) - f(x)\|K(f(y))$$
$$= B[A(y - x)$$
$$+ \|y - x\|F(y)] + \|f(y) - f(x)\|K(f(y))$$
$$= B[A(y - x)] + \Gamma_1(y)$$

where $\Gamma_1(y)$ is the vector in $\mathbb{R}^d$ defined by

$$\Gamma_1(y) = \|y - x\|B[F(y)] + \|f(y) - f(x)\|K(f(y))$$
$$= \|y - x\|B[F(y)] + \big\|A(y - x) + \|y - x\|F(y)\big\|K(f(y))$$

Note that

$$\|\Gamma_1(y)\| \le \|y - x\|\|B[F(y)]\| + [\|A\|\|y - x\| + \|y - x\|\|F(y)\|]\|K(y)\|$$
$$= \|y - x\|[\|B[F(y)]\| + \|A\|\|K(y)\| + \|F(y)\|\|K(y)\|]$$

So if we set $\Gamma(y) = \|y - x\|^{-1}\Gamma_1(y)$ we have that $\lim_{y \to x} \Gamma(y) = 0$ and

$$(g \circ f)(y) - (g \circ f)(x) = B[A(y - x)] + \|y - x\|\Gamma(y)$$

proving the theorem. ∎

As we said, when $f : G \to \mathbb{R}^q$ is differentiable, $Df : G \to \mathcal{L}(\mathbb{R}^p, \mathbb{R}^q)$. We have discussed the continuity of the derivative, but can we take a second derivative?

Therein lies the road to complication.

We can make sense of this second derivative, but the effort doesn't seem worthwhile. For example we could identify $\mathcal{L}(\mathbb{R}^p, \mathbb{R}^q)$ with $\mathbb{R}^{pq}$ by identifying each linear transformation in $\mathcal{L}(\mathbb{R}^p, \mathbb{R}^q)$ with its $q \times p$ matrix and then considering $Df : G \to \mathbb{R}^{pq}$. In this setup $D^2F(x) = f''(x) \in \mathcal{L}(\mathbb{R}^p, \mathbb{R}^{pq})$. This is beyond the scope of this course.

Exercises

(1) Prove Proposition 6.5.4.
(2) (a) Prove Theorem 6.5.7(a). (b) Fill in the missing details of the proof of Theorem 6.5.7(b).
(3) If $f : \mathbb{R}^3 \to \mathbb{R}^3$ is given by $f(x, y, z) = (x^2 + \sin z, e^y, xy + \cos z)$, compute Df.
(4) If $f : \mathbb{R}^3 \to \mathbb{R}^2$ is given by $f(x, y, z) = (xye^{xz}, x \sin y \cos z)$, compute Df.
(5) If $f : \mathbb{R}^2 \to \mathbb{R}^3$ is given by $f(x, y) = (xe^y, y \cos xy, x^2 \sin y)$, compute Df.
(6) Prove Theorem 6.5.9.
(7) Prove Proposition 6.5.12.

(8) Let $f : \mathbb{R}^2 \to \mathbb{R}^2$ and $g : \mathbb{R}^2 \to \mathbb{R}^3$ be defined by $f(x, y) = (x \sin y, xy^2)$ and $g(x, y) = (x^2, y, xy^2)$ and compute $D(g \circ f)$ in two different ways: first by using the Chain Rule and then by computing $g \circ f(x, y)$ and performing the calculation of the derivative.

6.6. Critical Points

Here is a theorem found in many Calculus books.

6.6.1. Theorem. *Suppose $G \subseteq \mathbb{R}^2$ and $f : G \to \mathbb{R}$ is twice continuously differentiable. If $a \in G$, $\nabla f(a) = 0$, $\partial_{11} f(a) > 0$, and $\partial_{11} f(a) \partial_{22} f(a) - \partial_{12} f(a)^2 > 0$, then f has a relative minimum at a.*

> *When I first saw this as a student, I was befuddled*

First there was the sense of disappointment. The test for a relative minimum of a function of one variable was so plain and simple, while this test was even difficult to remember. Second, the generalization of this to functions defined on G when G is an open subset of $\mathbb{R}^p$ and $p > 2$ seems formidable if not impossible. If we are to succeed we need a different approach. In this section we'll see this and, I think, see what is really going on in the above theorem.

We continue to assume that G is always an open subset of $\mathbb{R}^p$. In §6.2 we defined a differentiable function $f : G \to \mathbb{R}$ to have a *critical point* at x_0 if $f'(x_0) = 0$. (Recall that the derivative of such a function, $f'(x)$, is a linear functional on $\mathbb{R}^p$ while $\nabla f(x)$ is the vector in $\mathbb{R}^p$ that implements this linear functional via the inner product: $f'(x)(w) = \langle \nabla f(x), w \rangle$ for all w in $\mathbb{R}^p$.) In this section we want to use what we have developed about linear transformations to analyze the behavior of f at such a point. What we do here is analogous to what we did for a differentiable function of one variable in Theorem 2.4.5.

Let's underline that we are only discussing functions that are real-valued. We could just as easily define a differentiable function $f : G \to \mathbb{R}^q$ to have a critical point at x_0 when $Df(x_0) = 0$. However the diagnosis of the behavior of f near a critical point for such a function is not as complete and satisfying as what we can do for a real-valued function.

Making no distinction between $f'(x)$ and the vector $\nabla f(x)$ in $\mathbb{R}^p$ for the moment, we have that $Df : G \to \mathbb{R}^p$. Thus $D^2 f(x) \in \mathcal{L}(\mathbb{R}^p)$. What we will show is that when $f'(x_0) = \nabla f(x_0) = 0$, then f has a local minimum at x_0 when $D^2 f(x_0) > 0$. What does this condition $D^2 f(x_0) > 0$ mean? This is discussed below, but we start by examining Theorem 6.5.7(c) and applying this to $Df(x)$.

6.6.2. Proposition. *If G is an open subset of $\mathbb{R}^p$ and $f : G \to \mathbb{R}$ is a twice differentiable function, then the matrix representing the linear transformation*

$D^2 f(x)$ is

$$
\left[\frac{\partial^2 f}{\partial x_i \partial x_j}(x)\right] =
\begin{bmatrix}
\frac{\partial^2 f}{\partial^2 x_1}(x) & \frac{\partial^2 f}{\partial x_2 \partial x_1}(x) & \cdots & \frac{\partial^2 f}{\partial x_p \partial x_1}(x) \\
\frac{\partial^2 f}{\partial x_1 \partial x_2}(x) & \frac{\partial^2 f}{\partial^2 x_2}(x) & \cdots & \frac{\partial^2 f}{\partial x_p \partial x_2}(x) \\
\vdots & \vdots & \cdots & \vdots \\
\frac{\partial^2 f}{\partial x_1 \partial x_p}(x) & \frac{\partial^2 f}{\partial x_2 \partial x_p}(x) & \cdots & \frac{\partial^2 f}{\partial^2 x_p}(x)
\end{bmatrix}
$$

Observe that by Proposition 6.2.3, if we assume the second partial derivatives of $f : G \to \mathbb{R}$ are continuous, the matrix above that represents $D^2 f(x)$ is hermitian because all its mixed partial derivatives are equal. This leads us to introduce the following.

6.6.3. Definition. A hermitian linear transformation is *positive* (or *non-negative*) if all its eigenvalues are positive numbers; it is *negative* or (*non-positive*) if the eigenvalues are all negative numbers. (We will always use the terms "positive" and "negative"; other terms are sometimes seen in the literature.) The hermitian linear transformation is *positive definite* if it is positive and invertible. A *negative definite* linear transformation is defined similarly.

Note that in light of the Spectral Theorem (6.4.29) a hermitian linear transformation is positive definite if all its eigenvalues are strictly positive.

6.6.4. Theorem. *Let G be an open subset of $\mathbb{R}^p$ and let $f : G \to \mathbb{R}$ be a twice continuously differentiable function. If $a \in G$ such that $\nabla f(a) = 0$ and A is the hermitian matrix*

$$
A = \left[\frac{\partial^2 f}{\partial x_i \partial x_j}(a)\right] = [\partial_{ij} f(a)]
$$

the following hold.

(a) *If A is positive definite, then a is a local minimum for f.*
(b) *If A is negative definite, then a is a local maximum for f.*
(c) *If A is invertible and has some eigenvalues that are positive as well as some that are negative, then a is a saddle point.*
(d) *If A is not invertible, the nature of this critical point is undetermined.*

What happens when $p = 1$? In this case $f'(a)$ and $A = f''(a)$ are real numbers. So the statement that A is positive definite is just the condition that $f''(a) > 0$. Therefore when $p = 1$, (a) and (b) are just the conditions we saw in Theorem 2.4.5. Also in this case (c) is meaningless. When $p = 1$ the statement in (d) that A is not invertible is the statement that $f''(a) = 0$; so again this is what we have seen before for critical points of functions of a single variable.

The proof of the general theorem requires a few lemmas.

6.6.5. Lemma. *If A is a hermitian linear transformation on $\mathbb{R}^p$ that is positive definite, then there is a constant $c > 0$ such that $\langle Ax, x \rangle \geq c\|x\|^2$ for all x in $\mathbb{R}^p$.*

Proof. Adopt the notation of The Spectral Theorem 6.4.29, and let $A = \sum_{j=1}^{m} \lambda_j P_j$ be the spectral decomposition of A. Because A is positive definite, $\lambda_j > 0$ for $1 \le j \le m$. Let $c = \min\{\lambda_j : 1 \le j \le m\}$. For any x in $\mathbb{R}^p$, $x = \sum_{j=1}^{m} x_j$ with x_j in $P_j(\mathbb{R}^p)$. Hence $x_j \perp x_i$ for $i \ne j$ and so

$$
\begin{aligned}
\langle Ax, x \rangle &= \left\langle A\left(\sum_{j=1}^{m} x_j\right), \sum_{i=1}^{m} x_i \right\rangle \\
&= \left\langle \sum_{j=1}^{m} \lambda_j x_j, \sum_{i=1}^{m} x_i \right\rangle \\
&= \sum_{j=1}^{m} \lambda_j \|x_j\|^2 \\
&\ge c \sum_{j=1}^{m} \|x_j\|^2 \\
&= c\|x\|^2 \qquad \blacksquare
\end{aligned}
$$

6.6.6. Lemma. *With the hypothesis of* Theorem 6.6.4, *there is a real number $r > 0$ and a function $\eta : \{h \in \mathbb{R}^p : \|h\| < r\} \to [0, \infty)$ such that:*

(a) $\lim_{h \to 0} \eta(h) = 0$; *and*
(b) *for $\|h\| < r$,*

$$
f(a + h) = f(a) + \frac{1}{2}\langle Ah, h \rangle + \eta(h)\|h\|^2
$$

Proof. Choose $r > 0$ such that $B(a; r) \subseteq G$. Define $\eta : B(0; r) \to [0, \infty)$ by $\eta(0) = 0$ and for $0 < \|h\| < r$,

$$
\eta(h) = \frac{f(a + h) - f(a) - \frac{1}{2}\langle Ah, h \rangle}{\|h\|^2}
$$

Clearly (b) holds and we must show that (a) holds as well. Fix a δ with $0 < \delta < r$; in a moment we will specify δ further. Let $\|h\| < \delta$ and define $\phi : (-1, 1) \to \mathbb{R}$ by $\phi(t) = f(a + th)$; we apply Taylor's Theorem (2.5.6) to the function ϕ. By the Chain Rule (6.2.14) we have that $\phi'(t) = \langle \nabla f(a + th), h \rangle$; so $\phi'(0) = 0$. To calculate $\phi''(t)$, apply the Chain Rule (6.5.13) to the composition of ∇f and the function $t \mapsto a + th$. This shows that

$$
\phi''(t) = \frac{d}{dt}\langle \nabla f(a + th), h \rangle = (\nabla f)'(a + th)[h]
$$

But $(\nabla f)' = D^2 f$. Putting these values of the derivative into Taylor's Theorem and using the fact that $\nabla f(a) = 0$ we get that there is a value of t in $[-1, 1]$ such that

$$
f(a + h) = f(a) + \frac{1}{2}\langle D^2 f(a + th)h, h \rangle
$$

Letting $y = a + th$ gives that

$$\left| f(a+h) - f(a) - \frac{1}{2}\langle Ah, h \rangle \right| = \left| \frac{1}{2} \langle \left(A - D^2 f(y) \right) h, h \rangle \right|$$

$$\leq \| A - D^2 f(y) \| \, \| h \|^2$$

So $\eta(h) \leq \| A - D^2 f(y) \|$. Because the second derivatives of f are continuous, if $\epsilon > 0$ we can specify that $0 < \delta < r$ is such that $\| a - D^2 f(x) \| < \epsilon$ whenever $\| a - x \| < \delta$. But y is on the line segment $[a - h, a + h]$ so that it is indeed the case that $\| a - y \| < \delta$. Thus $|\eta(h)| < \epsilon$ when $\| h \| < \delta$. This proves part (a). ∎

Proof of Theorem 6.6.4. (a) Using the preceding two lemmas we have that there is an $r > 0$ such that for $\| h \| < r$

$$f(a+h) - f(a) = \frac{1}{2}\langle Ah, h \rangle + \eta(h)\| h \|^2$$

$$\geq \left(\frac{c}{2} + \eta(h) \right) \| h \|^2$$

Since $\eta(h) \to 0$ as $h \to 0$ and $c > 0$, we can find a $\delta < r$ such that $c/2 + \eta(h) > 0$ for $\| h \| < \delta$. Thus $f(x) > f(a)$ for $0 < \| x - a \| < \delta$, showing that a is a local minimum for f.

The proof of (b) follows from (a) by consideration of $-f$. To prove (c) let λ, μ be two eigenvalues of A with $\lambda > 0$ and $\mu < 0$. Let x and y be eigenvectors for λ and μ, respectively, and assume that $\| x \| = \| y \| = 1$. For $0 < t < \delta$,

$$f(a + tx) - f(a) = \frac{t^2}{2}\langle Ax, x \rangle + \eta(tx)t^2$$

$$= t^2 \left(\frac{1}{2}\lambda + \eta(tx) \right)$$

Again $\eta(tx) \to 0$ as $t \to 0$, so for all sufficiently small t, $f(a + tx) > f(a)$. Similarly, for all sufficiently small s, $f(a + sy) < f(a)$. Hence f has a saddle point at a.

To establish (d) one need only consider various examples. See Exercise 1. ∎

The reader should note that the proof of (c) yields additional information. Using the notation in that proof we have that along the direction determined by the vector x – that is, along the line $a + tx$ – the function has a local minimum at a or when $t = 0$. On the other hand if we go along the direction determined by the vector y the function has a local maximum at a. It is precisely by concentrating on these two directions that we see the saddle nature of the behavior of the function at this critical point.

Now let's reconcile Theorem 6.6.1 with what we have done.

6.6.7. Lemma. *If $\alpha, \beta, \gamma \in \mathbb{R}$ and A is the hermitian matrix*

$$A = \begin{bmatrix} \alpha & \beta \\ \beta & \gamma \end{bmatrix}$$

then A is positive definite if and only if $\alpha > 0$ and $\det A > 0$.

Proof. Assume A is positive definite. Then $0 < \langle Ae_1, e_1 \rangle = \alpha$. Also since $\det A$ is the product of the eigenvalues of A, it must be positive. Now assume that α and $\det A$ are positive. If $z = (x, y) \in \mathbb{R}^2$, then

$$\begin{aligned}
\langle Az, z \rangle &= \alpha x^2 + 2\beta xy + \gamma y^2 \\
&= \alpha \left(x^2 + 2\frac{\beta}{\alpha} xy \right) + \gamma y^2 \\
&= \alpha \left(x + \frac{\beta}{\alpha} y \right)^2 + \left(\gamma - \frac{\beta^2}{\alpha} \right) y^2 \\
&= \alpha \left(x + \frac{\beta}{\alpha} y \right)^2 + \frac{1}{\alpha} (\det A) y^2 \\
&> 0
\end{aligned}$$

∎

We can now see that when $p = 2$, Theorem 6.6.1 is a direct consequence of Theorem 6.6.4 once the criterion for positive definiteness from the preceding lemma is applied to the matrix

$$\begin{bmatrix} f_{11}(a) & f_{12}(a) \\ f_{21}(a) & f_{22}(a) \end{bmatrix}$$

Actually Lemma 6.6.7 can be generalized. For any square real matrix

$$A = \begin{bmatrix} \alpha_{11} & \alpha_{12} & \cdots & \alpha_{1n} \\ \alpha_{21} & \alpha_{22} & \cdots & \alpha_{2n} \\ \vdots & \vdots & \vdots & \vdots \\ \alpha_{n1} & \alpha_{n2} & \cdots & \alpha_{nn} \end{bmatrix}$$

define the *principal minors* of A to be the square matrices

$$A_k = \begin{bmatrix} \alpha_{11} & \alpha_{12} & \cdots & \alpha_{1k} \\ \alpha_{21} & \alpha_{22} & \cdots & \alpha_{2k} \\ \vdots & \vdots & \vdots & \vdots \\ \alpha_{k1} & \alpha_{k2} & \cdots & \alpha_{kk} \end{bmatrix}$$

where $1 \le k \le n$.

6.6.8. Theorem. *A hermitian matrix is positive definite if and only if $\det A_k$ is positive for each of its principal minors.*

For a proof see [5], page 328.

Exercises

(1) Let $f(x, y) = \frac{1}{2}x^2$ and show that f has a local minimum at $(0, 0)$ but $D^2 f(0, 0)$ is not invertible. Give an example of a function g on $\mathbb{R}^2$ that has a local maximum at $(0, 0)$ but $D^2 g(0, 0)$ is not invertible.

(2) Find all local extrema of the function $f(x, y) = x^2 - xy + y^3 - y$ and decide whether each is a local minimum, maximum, or saddle point. If the function has a saddle point at (x_0, y_0), find the eigenvectors corresponding to the two eigenvalues of $D^2 f(x_0, y_0)$.

(3) Find all local extrema of the function $f(x, y) = \sin x + \cos y$ and decide whether each is a local minimum, maximum, or saddle point.

(4) Let $\Omega = \{(x, y, z) \in \mathbb{R}^3 : x^2 + y^2 + z^2 < 1\}$ and define f on Ω by $f(x, y, z) = \sin(\pi x)^2 + \cos(\pi y)^2 - z^2$. Find all the critical points of f in Ω and give the nature of each whenever you can.

(5) Find the critical points of $f(x, y, z, w) = x^2 + (y + w)^2 + \exp[y^2 - \cos(z)]$ on $\mathbb{R}^4$ and, whenever you can, decide whether each is a local minimum, maximum, or saddle point.

(6) Find the critical points of $f(x, y) = x^2 - xy + y^3 - y$ and, whenever you can, decide whether each is a local minimum, maximum, or saddle point.

(7) Using the notation of Theorem 6.6.4, assume that A is positive definite so that f has a local minimum at a. In which direction is the function f increasing the fastest? Prove your assertion.

6.7. Tangent Planes

As we have mentioned and the student no doubt recalls, when $f : (a, b) \to \mathbb{R}$ is differentiable at c in (a, b), $f'(c)$ is the slope of the straight line tangent to the graph of f at the point $(c, f(c))$. Thus the equation of the line tangent to the graph is $y - f(c) + f'(c)(x - c)$. When G is an open subset of $\mathbb{R}^p$, $c \in G$, and $f : G \to \mathbb{R}$ is differentiable, what is the geometric significance of $f'(c)$ or $\nabla f(c)$? First note that the graph of f in this case is the surface $S = \{(x, f(x)) : x \in G\} \subseteq R^{p+1}$. (We are taking the idea of a surface as an intuitive concept and are not precisely defining it. This will be properly defined in Chapter 9 below.) It makes no sense to talk about *the* tangent line to the surface since there are many such lines. Therefore we must increase the level, or dimension, of the discussion. For inspiration we return to the case where $p = 1$ and show that $f'(c)$ has an additional property.

6.7.1. Proposition. *If $f : (a, b) \to \mathbb{R}$ is differentiable and $c \in (a, b)$, then the vector $(f'(c), -1)$ in $\mathbb{R}^2$ is orthogonal to the line tangent to the graph of f at the point $(c, f(c))$.*

Proof. We know that the typical point in the tangent line to the graph at the point $(c, f(c))$ is $(x, f(c) + f'(c)(x - c))$. Thus in $\mathbb{R}^2$

$$\langle (f'(c), -1), (x - c, f'(c)(x - c)) \rangle = 0 \qquad \blacksquare$$

The preceding result rephrases something else the student may recall from calculus. Namely that when $f'(c) \neq 0$, $-[f'(c)]^{-1}$ is the slope of the line perpendicular to the graph of f at the point $(c, f(c))$. The statement in the preceding proposition has an advantage over this last fact since it does not need to exclude the possibility that $f'(c) = 0$.

So we might hope that for f defined on an open subset of $\mathbb{R}^p$ the vector $(\nabla f(c), -1)$ plays a similar role. In fact we'll show that it is perpendicular to all lines tangent to the surface at $(c, f(c))$. But rather than discuss all the tangent lines we will incorporate this into another concept.

6.7.2. Definition. A subset H of $\mathbb{R}^{p+1}$ is called an *affine hyperplane* if there is a vector v in H such that $H - v = \{x - v : x \in H\}$ is a linear subspace of $\mathbb{R}^{p+1}$ having dimension p. If H itself is a linear subspace of $\mathbb{R}^{p+1}$ having dimension p, then H is said to be a *hyperplane*.

For emphasis, a hyperplane has dimension one less than the dimension of the containing space. Let's quickly state the following, whose proof is Exercise 1.

6.7.3. Proposition. *If H is an affine hyperplane, then for all vectors v and w in H, $H - v = H - w$. Consequently, $H - v$ is a linear subspace of $\mathbb{R}^{p+1}$ having dimension p for every vector v in H.*

6.7.4. Example. (a) Any straight line in $\mathbb{R}^2$ is an affine hyperplane.

(b) The affine hyperplanes in $\mathbb{R}^3$ are the translates of planes.

(c) If $a_1, \ldots, a_p \in \mathbb{R}$ and not all of them are 0, then $H = \{x \in \mathbb{R}^{p+1} : \sum_{j=1}^{p} x_j a_j = 1\}$ is an affine hyperplane. In fact if $v \in H$, $H - v = \{y \in \mathbb{R}^{p+1} : \sum_{j=1}^{p} a_j y_j + y_{p+1} = y_{p+1}\} = \{y \in \mathbb{R}^{p+1} : \sum_{j=1}^{p} a_j y_j = 0\}$. (See Exercise 2.) So if $a = (a_1, \ldots, a_p, 0) \in \mathbb{R}^{p+1}$, then $H - v = \{y \in \mathbb{R}^{p+1} : y \perp a\}$ and thus has dimension p.

(d) If H is any affine hyperplane and $x \in \mathbb{R}^{p+1}$, then $H + x$ is an affine hyperplane. In particular this works when H is a linear subspace of $\mathbb{R}^{p+1}$ of dimension p.

(e) If x_0 is any non-zero vector in $\mathbb{R}^{p+1}$ and $u \in \mathbb{R}^{p+1}$, then $\{x + u : x \perp x_0\}$ is an affine hyperplane.

Here is an easy way to manufacture an affine hyperplane. Let $L : \mathbb{R}^{p+1} \to \mathbb{R}$ be a non-zero linear functional. For any real number c, $H_c = \{x \in \mathbb{R}^{p+1} : L(x) = c\}$ is an affine hyperplane. In fact if $v \in H_c$, then $H_c - v = H_0$. It is easy to check that H_0 is a vector subspace of $\mathbb{R}^{p+1}$. Since $L \neq 0$ there is a vector u in $\mathbb{R}^{p+1}$ with $L(u) = 1$. So if $x \in \mathbb{R}^{p+1}$ and $L(x) = b$, $x - bu \in H_0$; that is, $H_0 + \mathbb{R}u = \mathbb{R}^{p+1}$ so that the dimension of H_0 must be p. Note that this is exactly how we got Example 6.7.4(c).

In fact if $L(x) = \sum_{j=1}^{p} a_j x_j$, H_1 is the affine hyperplane in that example. This leads to the following.

6.7.5. Proposition. *If $H \subseteq \mathbb{R}^{p+1}$, the following statements are equivalent.*

(a) *H is an affine hyperplane in $\mathbb{R}^{p+1}$.*
(b) *There is a non-zero linear functional L on $\mathbb{R}^{p+1}$ and a real number c such that $H = \{x \in \mathbb{R}^{p+1} : L(x) = c\}$.*
(c) *There is a non-zero vector a in $\mathbb{R}^{p+1}$ and a real number c such that $H = \{x \in \mathbb{R}^{p+1} : \langle x, a \rangle = c\}$.*

Proof. The equivalence of (b) and (c) is a consequence of the fact that every linear functional L on $\mathbb{R}^{p+1}$ has the form $L(x) = \langle x, a \rangle$ for some a in $\mathbb{R}^{p+1}$ (6.2.8). The fact that (b) implies (a) is in the discussion that preceded the proposition. It remains to prove that (a) implies (b). If H is an affine hyperplane, fix a vector v in H and consider $\mathcal{M} = H - v$. By definition $\mathcal{M}$ is a linear subspace of dimension p. Therefore if $w \in \mathbb{R}^{p+1} \backslash \mathcal{M}$, we have that $\mathbb{R}^{p+1} = \mathcal{M} + \mathbb{R}w$. It follows that for any x in $\mathbb{R}^{p+1}$ there is a unique vector m in $\mathcal{M}$ and a unique scalar α such that $x = m + \alpha w$ (Exercise 3). By the uniqueness of these objects it follows that $L(x) = \alpha$ defines a linear functional on $\mathbb{R}^{p+1}$ and $\mathcal{M} = \{x \in \mathbb{R}^{p+1} : L(x) = 0\}$. (Verify!) If we put $c = L(v)$, then $H = \{x \in \mathbb{R}^{p+1} : L(x) = c\}$. ∎

What about the uniqueness of the vector a in part (c) of the preceding proposition? Here is the answer.

6.7.6. Proposition. *If H is an affine hyperplane, $a_1, a_2 \in \mathbb{R}^{p+1}$, and $c_1, c_2 \in \mathbb{R}$ such that $\{x \in \mathbb{R}^{p+1} : \langle x, a_1 \rangle = c_1\} = H = \{x \in \mathbb{R}^{p+1} : \langle x, a_2 \rangle = c_2\}$, then there is a t in $\mathbb{R}$ such that $a_2 = ta_1$.*

Proof. Let $v \in H$ and consider $\mathcal{M} = H - v$. So both a_1 and a_2 are orthogonal to $\mathcal{M}$. Since dim $\mathcal{M} = p$, it must be that there is a scalar t with $a_2 = ta_1$. ∎

So the vector that is orthogonal to an affine hyperplane is not unique, but it's close to unique. The frequent practice is to take a unit vector η that's orthogonal to H.

Let's establish some notation. As before assume that G is an open subset of $\mathbb{R}^p$, $f : G \to \mathbb{R}$ is a differentiable function, and $c \in G$. Put $S = \{(x, f(x)) : x \in G\} \subseteq \mathbb{R}^{p+1}$; we say that S is the *surface* in $\mathbb{R}^{p+1}$ associated with f. Note that when $\gamma : (-1, 1) \to G$ is a differentiable curve, $\widetilde{\gamma}(t) = (\gamma(t), f(\gamma(t)))$ defines a curve $\widetilde{\gamma} : (-1, 1) \to \mathbb{R}^{p+1}$ and this curve lies in the surface S. We leave it to the reader (Exercise 4) to prove that $\widetilde{\gamma}$ is differentiable and

6.7.7 $$\widetilde{\gamma}'(t) = \big(\gamma'(t), \langle \nabla f(\gamma(t)), \gamma'(t) \rangle\big) \in \mathbb{R}^{p+1}$$

As usual the vector $\widetilde{\gamma}'(t)$ is tangent to the curve defined by $\widetilde{\gamma}$.

6.7.8. Proposition. *Using the notation above, put $\gamma(0) = c$. We have that*

$$(\nabla f(c), -1) \perp \tilde{\gamma}'(0)$$

Proof. Using (6.7.7) and the fact that $\gamma(0) = c$, we have that

$$\tilde{\gamma}'(0) = (\gamma'(0), \langle \nabla f(c), \gamma'(0) \rangle)$$

Hence

$$\begin{aligned}\langle \tilde{\gamma}'(0), (\nabla f(c), -1) \rangle &= \langle (\gamma'(0), \langle \nabla f(c), \gamma'(0) \rangle), (\nabla f(c), -1) \rangle \\ &= \langle \gamma'(0), \nabla f(c) \rangle - \langle \nabla f(c), \gamma'(0) \rangle \\ &= 0 \end{aligned}$$ ∎

6.7.9. Lemma. *With G, f, S as above, if $(c, d) \in S$ and $\tilde{\gamma} : (-1, 1) \to S$ is a smooth curve with $\tilde{\gamma}(0) = (c, f(c))$, then there is a smooth curve $\gamma : (-1, 1) \to G$ with $\gamma(0) = c$ and $\tilde{\gamma}(t) = (\gamma(t), f(\gamma(t))$ for all t in $(-1, 1)$.*

Proof. Consider the projection map of $\mathbb{R}^{p+1}$ onto $\mathbb{R}^p$ and let $\pi : S \to G$ be its restriction to S; so we know π to be continuous and given by $\pi(x, f(x)) = x$. Note that $x \mapsto (x, f(x))$ is a map of G onto S and π is its inverse. In fact, if S has its relative topology, this map is a homeomorphism. (Verify!) Define $\gamma : (-1, 1) \to G$ by $\gamma(t) = \pi(\tilde{\gamma}(t))$. Save for verifying that γ is smooth, it is immediate that γ has all the desired properties. To show smoothness it suffices to show that $\gamma'(t) = \pi(\tilde{\gamma}'(t))$ as in (6.7.7). This is left to the reader (Exercise 5). ∎

Reflect on Proposition 6.7.8. Note that the vector $(\nabla f(c), -1)$ depends only on the function f and the point c in G. In light of the preceding lemma, this proposition says that this vector is orthogonal to every curve lying in S that passes through the point $(c, f(c))$. Thus we make the following definition and say that $(\nabla f(c), -1)$ is orthogonal to the surface defined by f at the point $(c, f(c))$.

6.7.10. Definition. If G be an open subset of $\mathbb{R}^p$, $f : G \to \mathbb{R}$ is a differentiable function, and $c \in G$, call

$$(c, f(c)) + \{x \in \mathbb{R}^{p+1} : x \perp (\nabla f(c), -1)\}$$

the *tangent affine hyperplane* to the graph of f at the point $(c, f(c))$.

Exercises

(1) Prove Proposition 6.7.3.
(2) Verify the statements made in Example 6.7.4(c). Give a basis for $H - v$ when $v \in H$.

(3) Prove that if $\mathcal{M}$ is a subspace of $\mathbb{R}^{p+1}$ of dimension p and $w \in \mathbb{R}^{p+1} \backslash \mathcal{M}$, then for any x in $\mathbb{R}^{p+1}$ there is a unique vector m in $\mathcal{M}$ and a unique scalar α such that $x = m + \alpha w$.

(4) Prove (6.7.7).

(5) In the proof of Lemma 6.7.9 show that γ is smooth.

(6) Let $f : \mathbb{R}^2 \to \mathbb{R}$ be defined by $f(x, y) = 3x^2 - xy$ and find the tangent affine hyperplane to the graph of f when $(x, y) = (1, 2)$.

(7) Let $f(x, y) = \log(3x + y)$ and find the tangent affine hyperplane to the graph of f when $(x, y) = (1, -1)$.

(8) Let $f(x, y, z) = \cos(xy) + \sin(yz)$ and find the tangent affine hyperplane to the graph of f at the point $(x, y, z) = (\pi, -1, \pi/2)$.

(9) Let $G = \{(x, y) \in \mathbb{R}^2 : x^2 + y^2 < 1\}$ and define $f : G \to \mathbb{R}$ by $f(x, y) = \sqrt{1 - x^2 - y^2}$. For any point c in G find the tangent affine hyperplane to the graph of f at the point $(c, f(c))$.

6.8. Inverse Function Theorem

We begin this section with a result from metric spaces. While the only use of this result in this book is the proof of the title result of this section, it has many uses elsewhere in mathematics. There are proofs of the Inverse Function Theorem that do not use this fixed point theorem, but its use lends a simplicity to the argument.

6.8.1. Theorem (Banach[5] Fixed Point Theorem). *Let (X, d) be a complete metric space. If $\phi : X \to X$ is a function such that there is a constant c with $0 < c < 1$ and*

$$d(\phi(x), \phi(y)) \leq cd(x, y)$$

for all x, y in X, then there is a unique point x^* in X with $\phi(x^*) = x^*$.

The point x^* in the theorem is called a *fixed point* of ϕ.

Proof. Take any point x_0 and put $x_{n+1} = \phi(x_n)$ for all $n \geq 0$. Note that for $n \geq 1$, $d(x_{n+1}, x_n) = d(\phi(x_n), \phi(x_{n-1})) \leq cd(x_n, x_{n-1})$. Using mathematical induction we obtain that $d(x_{n+1}, x_n) \leq c^n d(x_1, x_0)$.

---

[5] Stefan Banach was born in 1892 in Krakow, presently in Poland but then part of Austria-Hungary (Polish history is complicated). In 1922 the university in Lvov (a city that during its history has belonged to at least three countries and is presently in Ukraine) awarded Banach his habilitation for a thesis on measure theory. His earlier thesis on Integral Equations is sometimes said to mark the birth of functional analysis. He was one of the stars in a particularly brilliant constellation of Polish mathematicians during this period. Banach and his colleague Hugo Steinhaus in Lvov as well as other mathematicians in Warsaw began publishing a series of mathematical monographs. The first to appear in 1931 was Banach's *Théorie des Opérations Linéaires*, which had an enormous impact on analysis and continues to hold its place as a classic still worth reading. Banach was one of the founders of functional analysis, which had been brewing in academic circles for some time. You will see Banach's name appear often if you ever study the subject. He died in Lvov in 1945 just after the end of World War II and has many results besides this one that bear his name.

Claim. $\{x_n\}$ is a Cauchy sequence.

In fact when $m > n$,

$$d(x_m, x_n) \le d(x_m, x_{m-1}) + \cdots + d(x_{n+1}, x_n)$$
$$\le c^{m-1}d(x_1, x_0) + \cdots + c^n d(x_1, x_0)$$
$$= c^n d(x_1, x_0) \sum_{j=0}^{m-n-1} c^j$$
$$\le c^n d(x_1, x_0) \sum_{j=0}^{\infty} c^j$$
$$= \frac{c^n}{1-c} d(x_1, x_0)$$

Provided we take N sufficiently large this can be made arbitrarily small for $m > n \ge N$.

Since (X, d) is complete there is a point x^* such that $x_n \to x^*$. Note that since $x_n = \phi(x_{n-1})$, we have that $x^* = \lim_n x_n = \lim_n \phi(x_{n-1}) = \phi(x^*)$; thus x^* is a fixed point of ϕ. If y^* is another fixed point of ϕ, then $d(x^*, y^*) = d(\phi(x^*), \phi(y^*)) \le cd(x^*, y^*)$. Since $c < 1$ this is impossible unless $d(x^*, y^*) = 0$. Thus x^* is unique. ∎

There are other proofs of this theorem than the one presented. Finding other proofs is good, but the proof above is Banach's original one and it has always struck me as the "natural" proof. It certainly seems intuitive. Banach's original proof also has the advantage in that it tells you how to find the fixed point.

It is rather easy to manufacture examples illustrating this result. For example, if $\phi : [0, 1] \to [0, 1]$ is defined as the identity function, $\phi(x) = x$ for all x, then ϕ has every point as a fixed point even though it satisfies the inequality but with $c = 1$. If instead we let $\phi(x) = x^2$, there are two fixed points. Also see Exercise 1.

Recall Proposition 2.3.10 where we saw that when $f : (a, b) \to (\alpha, \beta)$ is a differentiable function that is bijective, then the function $f^{-1} : (\alpha, \beta) \to (a, b)$ is differentiable and

$$(f^{-1})'(\zeta) = \frac{1}{f'(f^{-1}(\zeta))}$$

for every ζ in (α, β). Here we wish to extend this result to differentiable functions f from an open subset of $\mathbb{R}^p$ into $\mathbb{R}^p$. Let's underline the fact that this concerns a function from a subset of Euclidean space into the space of the same dimension. Hence the derivative belongs to $\mathcal{L}(\mathbb{R}^p)$ and we can speak of its inverse.

6.8.2. Theorem (Inverse Function Theorem). *Let G be an open subset of $\mathbb{R}^p$ and assume $f : G \to \mathbb{R}^p$ is a continuously differentiable function. If $a \in G$ with $b = f(a)$ and $Df(a)$ is an invertible linear transformation, then there is an open neighborhood U of a such that $\Omega = f(U)$ is open in $\mathbb{R}^p$ and f is a bijection of U onto*

Ω. If $g : \Omega \to U$ is the inverse of $f : U \to \Omega$, then g is continuously differentiable and for every ζ in Ω

$$[Dg](\zeta) = [Df(g(\zeta))]^{-1}$$

Before starting the proof, let's be sure you understand that we already have this result when $p = 1$. In fact the hypothesis in the case that $p = 1$ is that $f'(a) \neq 0$. Thus there is a small open interval $(a - \delta, a + \delta)$ where f is strictly monotonic. Hence by the result referred to before the statement of the theorem, f is bijective and its inverse function is differentiable.

Proof. Let A denote the invertible linear transformation $A = Df(a)$ and let $0 < c < 1/[2\|A^{-1}\|]$. Since $Df : G \to \mathcal{L}(\mathbb{R}^p)$ is continuous there is an $\epsilon > 0$ such that $\|Df(x) - A\| < c$ when $\|x - a\| < \epsilon$. Put $U = B(x; \epsilon)$ and $\Omega = f(U)$. We want to show that f is injective on U and that Ω is open; then we'll tackle showing that the inverse of f on Ω is continuously differentiable. Doing these things is not particularly deep, but it is involved.

For any ζ in Ω define $\phi_\zeta : U \to \mathbb{R}^p$ by

$$\phi_\zeta(x) = x + A^{-1}(\zeta - f(x))$$

Note that a point x is a fixed point of ϕ_ζ if and only if $\zeta = f(x)$. So showing that f is injective on U amounts to showing that ϕ_ζ has a unique fixed point on U. (By the way, we will not use Banach's Fixed Point Theorem here. We will use it later when we show that Ω is open.) Since $\zeta \in \Omega = f(U)$, we know there is at least one fixed point x in U. Now $\phi_\zeta'(x) = 1 - A^{-1}Df(x) = A^{-1}[A - Df(x)]$. Hence when $\|x - a\| < \epsilon$, $\|\phi_\zeta'(x)\| \leq \|A^{-1}\|\|A - f(x)\| \leq c\|A^{-1}\| < \frac{1}{2}$. Using Exercise 2 we have that

6.8.3 $\qquad \|\phi_\zeta(x_1) - \phi_\zeta(x_2)\| \leq \frac{1}{2}\|x_1 - x_2\|$ whenever $\|x_1 - x_2\| < \epsilon$

So if both x_1 and x_2 are fixed points of ϕ_y in U, it must be that $x_1 = x_2$. Thus $f : U \to \Omega$ is injective and therefore bijective.

Now to show that Ω is open. This is done as follows. Fix ζ_0 in Ω, and let x_0 be the unique point in U with $f(x_0) = \zeta_0$. Choose $r > 0$ such that $\overline{B}(x_0; r) \subseteq U$.

Claim. $B(\zeta_0; cr) \subseteq \Omega$ and so Ω is open.

Fix ζ in $B(\zeta_0; cr)$ and define the function ϕ_ζ as above. Thus $\|\phi_\zeta(x_0) - x_0\| = \|A^{-1}(\zeta - \zeta_0)\| < \|A^{-1}\|cr < \frac{1}{2}r$. If $x \in \overline{B}(x_0; r)$, then (6.8.3) implies

$$\|\phi_\zeta(x) - x_0\| \leq \|\phi_\zeta(x) - \phi_\zeta(x_0)\| + \|\phi_\zeta(x_0) - x_0\|$$
$$\leq \frac{1}{2}\|x - x_0\| + \frac{r}{2}$$
$$\leq r$$

In other words, ϕ_ζ maps $\overline{B}(x_0; r)$ into itself. Because $\overline{B}(x_0; r)$ is compact and thus complete, we can apply Banach's Fixed Point Theorem. Therefore ϕ_ζ has a unique fixed point x and this means $f(x) = \zeta$, establishing the claim.

Let $g : \Omega \to U$ be the inverse of f on U and let's show that it is continuously differentiable. Note that the argument above shows that $f : U \to \Omega$ is an open map. That is, if V is an open subset of U, the same reasoning will show that $f(V)$ is open. Equivalently, $f : U \to \Omega$ is a homeomorphism so that $g = f^{-1} : \Omega \to U$ is continuous. This implies we need only show that g is differentiable. Indeed once this is done we can apply the Chain Rule (6.5.13) to $g(f(x)) = x$ for x in U and obtain that $[Dg](f(x)) = [Df(x)]^{-1}$ on U. Thus on Ω we have that $Dg(\zeta) = [Df(f^{-1}(\zeta))]^{-1}$ and, by Proposition 6.4.5, g is continuously differentiable.

To show that g is differentiable we'll use the definition. Fix ζ in Ω and let $k \in \mathbb{R}^p$ with $\|k\|$ sufficiently small that $\zeta + k \in \Omega$. Let $x = f^{-1}(\zeta) = g(\zeta)$ and $x_k = f^{-1}(\zeta + k) = g(\zeta + k)$. Put $S = Df(x)$ and $T = S^{-1}$. If $Dg(\zeta)$ exists, the Chain Rule tells us that it should be that $Dg(\zeta) = T$. By definition to show that g is differentiable we want to show the following:

6.8.4
$$\lim_{k \to 0} \frac{\|g(\zeta + k) - g(\zeta) - T(k)\|}{\|k\|} = 0$$

In what follows it is helpful to note that $k = (\zeta + k) - \zeta = f(x_k) - f(x)$. With this $g(\zeta + k) - g(\zeta) - T(k) = x_k - x - T(k) = -T[k - S(x_k - x)] = -T[f(x_k) - f(x) - S(x_k - x)]$. Hence

$$\frac{\|g(\zeta + k) - g(\zeta) - T(k)\|}{\|k\|} = \frac{\|-T[f(x_k) - f(x) - S(x_k - x)]\|}{\|x_k - x\|}$$
$$\leq \|T\| \frac{\|[f(x_k) - f(x) - S(x_k - x)]\|}{\|x_k - x\|}$$

Since $S = Df(x)$, the right-hand side of this inequality converges to 0 as $\|x - x_k\| \to 0$, but this happens exactly when $\|k\| \to 0$. (Why?) This establishes (6.8.4), showing that g is differentiable. ∎

6.8.5. Corollary. *If G is an open subset of $\mathbb{R}^p$ and $f : G \to \mathbb{R}^p$ is continuously differentiable such that $Df(x)$ is invertible for every x in G, then whenever U is an open subset of G, $f(U)$ is an open subset of $\mathbb{R}^p$.*

The proof of this corollary is left to the reader in Exercise 3. We note that this corollary says that the function f is an open mapping. For this reason it is sometimes called the *Open Mapping Theorem*. We aren't going to use this name as it is more frequently used for another result.

When $f : \mathbb{R}^p \to \mathbb{R}^p$ is differentiable, the $p \times p$ matrix that represents the linear transformation $Df(x)$ in $\mathcal{L}(\mathbb{R}^p)$ is given in (6.5.8). To determine if this linear transformation is invertible, we could take its determinant. This is denoted by

$$\Delta f(x) = \det Df(x)$$

and is called the *Jacobian*[6] of f. We will sometimes refer to this as the *Jacobian determinant*. We do this to make a distinction with the $p \times p$ matrix (6.5.8) which is denoted as

$$Jf(x) = \frac{\partial(f_1, \ldots, f_p)}{\partial(x_1, \ldots, x_p)} = \left[\frac{\partial f_i}{\partial x_j}(x)\right]$$

and is also sometimes referred to as the Jacobian or *Jacobian matrix*. Another definition of the Jacobian matrix is applied to differentiable maps $f : \mathbb{R}^p \to \mathbb{R}^q$ where it is the $q \times p$ rectangular matrix $(\partial f_j / \partial x_i)$. Here the notation is

$$Jf(x) = \frac{\partial(f_1, \ldots, f_q)}{\partial(x_1, \ldots, x_p)}$$

At this stage the reader might be concerned about the possible confusion, but in practice the context will make it clear what is being discussed.

6.8.6. Example. (a) Define $f : \mathbb{R}^2 \to \mathbb{R}^2$ by $f(x, y) = (e^x \cos y, e^x \sin y)$. Computing the Jacobian yields

$$\Delta f(x, y) = \det \begin{bmatrix} e^x \cos y & -e^x \sin y \\ e^x \sin y & e^x \cos y \end{bmatrix} = e^{2x}$$

Thus $Df(a, b)$ is invertible for any (a, b) in the plane so that f is an open mapping on $\mathbb{R}^2$. (See Exercise 4.) Since $f(0, 0) = f(0, 2\pi)$, f does not have an inverse on the entire plane.

(b) Many examples are available in $\mathbb{R}$ to show the existence of local inverses but where the function does not have a global inverse. See Exercise 5.

(c) Here is another example in one variable that illustrates that we must have the continuity of the derivative. Define $f : \mathbb{R} \to \mathbb{R}$ by

$$f(x) = \begin{cases} x + 2x^2 \sin x^{-1} & \text{when } x \neq 0 \\ 0 & \text{when } x = 0 \end{cases}$$

The function f is differentiable everywhere on the real line with $f'(0) = 1$. However f fails to be injective in any neighborhood of 0. The reader is asked to supply the details in Exercise 6.

....

[6] Carl Jacobi was born in Potsdam, Germany in 1804. He was tutored at home by his uncle until he was 12 and entered the gymnasium in Potsdam. After a year he was judged ready for the university, but remained at the gymnasium for another three years because at the time the University of Berlin would not accept students under the age of 18. Four years later he had completed his doctoral dissertation and obtained a post at a prestigious gymnasium in Berlin. He soon became a Christian, permitting him to obtain a position at the university in Berlin (you read that correctly) and then he transferred to Königsberg. His early work on number theory attracted the attention of Gauss. He also obtained results on elliptic functions that attracted Legendre. In 1831 he married and in 1832 was promoted to a full professorship. He conducted fundamental work on partial differential equations, but in 1843 he was diagnosed as a diabetic and this caused him severe problems. He went to Italy to recover. The weather agreed with him and he resumed his publications. It was around this time that he returned to a position in Berlin starting in 1844. There followed a particularly turbulent period in Prussian history and Jacobi fell out of favor because of his political beliefs. In January 1851 he caught the flu, which was a more serious event then than now. In his weakened condition he also contracted smallpox and soon died.

Exercises

(1) (a) Is there a continuous function $\phi : [0, 1] \to [0, 1]$ with no fixed points? (b) Find an example showing that the condition on ϕ in Theorem 6.8.1 is not necessary for there to be a unique fixed point. (c) Find an example of an incomplete metric space (X, d) and a continuous function ϕ of (X, d) into itself such that $d(\phi(x), \phi(y)) \leq cd(x, y)$ for some constant c with $0 < c < 1$ and all x, y in X but where ϕ does not have a fixed point.

(2) If $\epsilon > 0$, $c \geq 0$, $B(x; \epsilon) \subseteq \mathbb{R}^p$, and $\phi : B(x; \epsilon) \to \mathbb{R}^q$ is a continuously differentiable function with $\|D\phi(x)\| \leq c$ for all x in $B(x; \epsilon)$, show that $\|\phi(x_1) - \phi(x_2)\| \leq c\|x_1 - x_2\|$ whenever $\|x_1 - x_2\| < \epsilon$. (Hint: Use Proposition 3.1.10(b).)

(3) Give the details of the proof of Corollary 6.8.5.

(4) In Example 6.8.6(a) note that $f(0, 0) = (1, 0)$ and find the inverse of f defined in a neighborhood of $(1, 0)$.

(5) Give an example of a continuously differentiable function f on $\mathbb{R}$ such that $f'(x) \neq 0$ for every x but such that f does not have a global inverse.

(6) Give the details in Example 6.8.6(c).

(7) Define $f : \mathbb{R}^2 \to \mathbb{R}^2$ by $f(x, y) = (x^2 + 2xy + y^2, 2x + 2y)$. Determine all the points (a, b) where there is a function g defined in a neighborhood of $f(a, b)$ with $f(g(u, v)) = (u, v)$ for all (u, v) in this neighborhood. Justify your answer.

(8) Define $f : \mathbb{R}^3 \to \mathbb{R}^3$ by $f(x, y, z) = (x + xyz, y + xz, z + 2x + 3z^2)$. (a) Compute the Jacobian matrix (not determinant) of f. (b) Does the Inverse Function Theorem imply there is a function g defined in a neighborhood of $f(0, 0, 0)$ such that $f((g(u, v, w)) = (u, v, w)$ in this neighborhood? (c) Does the Inverse Function Theorem imply there a function g defined in a neighborhood of $f(1, 1, 1))$ such that $f((g(u, v, w)) = (u, v, w)$ in this neighborhood?

(9) Define $f : \mathbb{R}^2 \to \mathbb{R}^2$ by

$$f(x, y) = \left(\frac{x^2 - y^2}{x^2 + y^2}, \frac{xy}{x^2 + y^2} \right)$$

(a) Compute the Jacobian matrix (not determinant) of f. (b) Does the Inverse Function Theorem imply there a function g defined in a neighborhood of $f(0, 1)$ such that $f((g(u, v)) = (u, v)$ in this neighborhood?

(10) Recall the Chain Rule (6.5.13) when $p = q = d$: if G and H are open subsets of $\mathbb{R}^p$, $f : G \to \mathbb{R}^p$ is differentiable, $f(G) \subseteq H$, and $g : H \to \mathbb{R}^p$ is differentiable, then $g \circ f : G \to \mathbb{R}^p$ is differentiable and

$$D(f \circ g)(x) = Dg(f(x))[Df(x)]$$

Express the Jacobian determinant $\Delta_{f \circ g}$ in terms of Δ_f and Δ_g.

(11) Define $f : \mathbb{R}^2 \to \mathbb{R}^2$ by $f(x, y) = (xe^y, xy)$ and show that there is a ball B of positive radius and centered at $(1, 0)$ and a continuous function $g : B \to \mathbb{R}^2$ such that $g(1, 0) = (1, 0)$ and $f(g(x, y)) = (x, y)$ for all (x, y) in B. Justify your answer.

(12) Define $f : \mathbb{R}^2 \to \mathbb{R}^2$ by $f(x, y) = (x^2 + y, xy)$. (a) Does the Inverse Function Theorem apply to f at the point $(1, 1)$? If it does not, state why. If it does, state what it implies. (b) Does the Inverse Function Theorem apply to f at the point $(1, 2)$? If it does not, state why. If it does, state what it implies.

6.9. Implicit Function Theorem

I can remember when I was a student and first encountered the Implicit Function Theorem. It perplexed me. There is a certain aspect of the theorem that is mathematically pleasing, but I think at least part of the reason for my confusion was that the proof is complicated and I couldn't understand why we were doing all this work. What was the use of this difficult theorem? Let's see if you have less consternation and irritation than I did on my first encounter. I hope so.

As far as the purpose of the result is concerned, you can Google "implicit function theorem economics" and discover a great interest in this theorem by economists. There are also the uses of this theorem in analysis and especially geometry, some of which you will see later in this book; but I am afraid that to see most of this you must wait until a future course. That's true of much of what you study at this point in your development just as it was for me at the corresponding stage. We are exploring basic properties of functions and these underly most of what mathematics is about. This fundamental understanding has permanent value, but a lot of that value only becomes clear later.

The difficulty in understanding the result cannot be removed by a magic wand, but there are some helpful guide posts. You can regard the Implicit Function Theorem as an extension of the following. If we are given the equation $x^2 + y^2 = 1$, this expresses an interdependence of the real variables x and y. Can I solve for one in terms of the other? Equivalently, can I express one, say y, as a function of x? The answer is clearly no as the graph of this relation in the plane is a circle and that's not the graph of a function of x. On the other hand we can write $y = \sqrt{1 - x^2}$ and this works for some values of x.

Let's recast the above problem by writing $f(x, y) = x^2 + y^2 - 1$. If we set $f(x, y) = 0$ we say that this equation "defines" y implicitly in terms of x (and vice versa, x in terms of y). The word "defines" here is set in quotation marks because for the values of x the corresponding value of y is not unique. So it is impossible to solve for y in terms of x in a global way, if by solving we mean expressing y as a function of x. But as we mentioned $y = \sqrt{1 - x^2}$ works sometimes; that is, it works locally near some values of x. The key here is to consider the partial derivative $\partial f(x, y)/\partial y = 2y$. When this partial derivative is zero, we cannot get a local solution. Indeed, the partial derivative being zero corresponds to the points $(\pm 1, 0)$ on the graph of $f(x, y) = 0$ and an examination of the graph near these points shows that it cannot be the graph of a function $y = h(x)$. As long as we avoid places where this partial derivative is 0 we can indeed find a function h such that where h is defined we have that $f(x, h(x)) = 0$. This condition involving the partial derivative is the key to the general case. Indeed we have the following result that is a special case of the general Theorem 6.9.2 below.

6.9.1. Proposition. *Let G be an open subset of $\mathbb{R}^2$, let $(a, b) \in G$, and let $f : G \to \mathbb{R}$ be a continuously differentiable function such that $f(a, b) = 0$. If $\partial_y f(a, b) \neq 0$, then there exists an open interval I that contains a and a continuously differentiable function $h : I \to \mathbb{R}$ satisfying the following.*

(a) $h(a) = b$.
(b) $f(x, h(x)) = 0$ *for all x in I.*

We aren't going to prove this now; indeed as was mentioned it is a corollary of Theorem 6.9.2. You will be asked later to prove this proposition as a consequence of the theorem.

Some of the difficulty in stating and understanding the general result stems from the complexity of the situation, so let's introduce some notation that might be helpful. We consider the Euclidean space $\mathbb{R}^{p+q}$ and express it as the direct sum $\mathbb{R}^{p+q} = \mathbb{R}^p \oplus \mathbb{R}^q$. See Definition 6.3.7. Thus $\mathbb{R}^{p+q}$ has the elements $x \oplus y$ where $x \in \mathbb{R}^p$ and $y \in \mathbb{R}^q$. This is the true orthogonal direct sum: when $x \in \mathbb{R}^p$ and $y \in \mathbb{R}^q$, then the norm of $x \oplus y$ as an element of $\mathbb{R}^{p+q}$ is precisely $\sqrt{\|x\|^2 + \|y\|^2}$. Note that the order is important. Though mathematically $\mathbb{R}^{p+q}$ and $\mathbb{R}^{q+p}$ are the same, we are making a distinction for the purposes of presenting the Implicit Function Theorem. This theorem gives a condition on a function $f : \mathbb{R}^{p+q} \to \mathbb{R}^q$ at a point a in $\mathbb{R}^p$ such that there is a function h defined in a neighborhood of a and taking values in $\mathbb{R}^q$ for which we have $f(x \oplus h(x)) = 0$. In other words, in the equation $f(x \oplus y) = 0$ the function h locally solves for y in terms of x.

We introduce similar notation to that above but now for linear transformations. If $T \in \mathcal{L}(\mathbb{R}^{p+q}, \mathbb{R}^q)$ we write $T = T_p \oplus T_q$, where $T_p : \mathbb{R}^p \to \mathbb{R}^q$, $T_q : \mathbb{R}^q \to \mathbb{R}^q$, and they are defined as

$$T_p(x) = T(x \oplus 0) \quad \text{and} \quad T_q(y) = T(0 \oplus y)$$

This means that $T(x \oplus y) = T_p(x) + T_q(y)$. (At the risk of seeming pedantic but to make sure all have the right idea about what's going on, we emphasize that the $+$ sign on the right side of this equation is not a mistake; it is not $\oplus$.)

The idea here is that we have an open subset G of $\mathbb{R}^{p+q}$, a continuously differentiable function $f : G \to \mathbb{R}^q$, and we are interested in the set $\{x \oplus y \in \mathbb{R}^{p+q} : f(x \oplus y) = 0\}$. On this set y is implicitly defined as a function of x. We would like to know conditions on f under which we can make this definition explicit. In other words, where can we define y locally as a function of x? The following theorem gives a sufficient condition for this to happen.

6.9.2. Theorem (Implicit Function Theorem). *Let G be an open subset of $\mathbb{R}^{p+q}$, let $a \oplus b \in G$, and let $f : G \to \mathbb{R}^q$ be a continuously differentiable function such that $f(a \oplus b) = 0$. If $T = Df(a \oplus b)$ and $T_q : \mathbb{R}^q \to \mathbb{R}^q$ is invertible, then there exists* *an open subset Ω of $\mathbb{R}^p$ that contains the point a and a continuously differentiable function $h : \Omega \to \mathbb{R}^q$ satisfying the following.*

(a) $h(a) = b$.
(b) $f(x \oplus h(x)) = 0$ *for all x in Ω.*
(c) $Dh(b) = -T_q^{-1}T_p$.

Proof. Begin by defining $F : G \to \mathbb{R}^{p+q}$ by $F(x \oplus y) = x \oplus f(x \oplus y)$. It follows that F is continuously differentiable. (Why?) Also note that $F(a \oplus b) = a \oplus 0$.

Claim. $DF(a \oplus b)$ is an invertible linear transformation in $\mathcal{L}(\mathbb{R}^{p+q})$.

Since $T = Df(a \oplus b)$, we use Proposition 6.5.4 and the fact that $f(a \oplus b) = 0$ to get that for $s \oplus t$ in $\mathbb{R}^{p+q}$ we have $f[(a+s) \oplus (b+t)] = [T(s \oplus t) + R(s \oplus t)]$, where $\|R(s \oplus t)\|/\|(s \oplus t)\| \to 0$ as $\|s \oplus t\| \to 0$. Now

$$F[(a+s) \oplus (b+t)] - F(a \oplus b) = (a+s) \oplus f[(a+s) \oplus (b+t)] - a \oplus 0$$
$$= s \oplus [T(s \oplus t) + R(s \oplus t)]$$
$$= [s \oplus T(s \oplus t)] + [0 \oplus R(s \oplus t)]$$

Since $\|0 \oplus R(s \oplus t)\|/\|s \oplus t\| \to 0$ as $\|s \oplus t\| \to 0$, we have that $DF(a \oplus b)$ is the linear transformation in $\mathcal{L}(\mathbb{R}^{p+q})$ defined by

$$DF(a \oplus b) : s \oplus t \mapsto s \oplus T(s \oplus t)$$

So if $DF(a \oplus b)(s \oplus t) = 0$, then $s = 0$ and $0 = T(s \oplus t) = T_q(t)$. Since the hypothesis is that T_q is invertible, we also get that $t = 0$. Hence $DF(a \oplus b)$ is injective and therefore invertible. (Recall from linear algebra that an injective linear transformation on a finite dimensional space is invertible.) This establishes the claim.

The claim allows us to apply the Inverse Function Theorem to $F : \mathbb{R}^{p+q} \to \mathbb{R}^{p+q}$ and get an open neighborhood U of $a \oplus b$ on which F is a bijection, $V = F(U)$ is open, and the inverse map $H : V \to U \subseteq \mathbb{R}^{p+q}$ is continuously differentiable. Put $\Omega = \{x \in \mathbb{R}^p : x \oplus 0 \in V\}$; since V is open, it follows that Ω is open (Exercise 2). Since $F(a \oplus b) = a \oplus f(a \oplus b) = a \oplus 0 \in V$, $a \in \Omega$. For x in Ω we have that $x \oplus 0 \in V$ and so $H(x \oplus 0) = x \oplus y \in U \subseteq G$ for some y. Hence $x \oplus 0 = F(x \oplus y) = x \oplus f(x \oplus y)$. Because F is injective on U, this $x \oplus y$ is unique. Thus y must be unique. What we have shown is that for every x in Ω there is a unique y in $\mathbb{R}^q$ with $H(x \oplus 0) = x \oplus y$. The uniqueness of this y for each x in Ω means we have defined a function $h : \Omega \to \mathbb{R}^p$ so that for each x in Ω

$$H(x \oplus 0) = x \oplus h(x)$$

Note that h is continuously differentiable since H is (Verify!). Since $F(a \oplus b) = a \oplus 0$, we have that (a) is satisfied. By definition $x \oplus 0 = F(x \oplus h(x)) = x \oplus f(x \oplus h(x))$, so that (b) is satisfied.

It remains to show that (c) holds. For convenience let $g : \Omega \to \mathbb{R}^{p+q}$ be the function defined by $g(x) = x \oplus h(x)$. So for each x in Ω we have that

$Dg(x) \in \mathcal{L}(\mathbb{R}^p, \mathbb{R}^{p+q})$. In fact

$$Dg(x) = 1 \oplus Dh(x)$$

where this 1 stands for the identity linear transformation on $\mathbb{R}^p$ and $Dh(x) \in \mathcal{L}(\mathbb{R}^p, \mathbb{R}^q)$. We know from (b) that $0 = f(g(x))$ so the Chain Rule implies that $0 = [Df](g(x))Dg(x)$. Now let $x = a$. From (a) we have that $h(a) = b$ so we get $0 = Df(a \oplus b)Dg(a) = TDg(a)$. Thus

$$0 = T[1 \oplus Dh(a)] = T_p + T_q Dh(a)$$

Since T_q is invertible, $Dh(a) = -T_q^{-1} T_p$. ∎

At this point you should complete Exercise 1, which asks you to prove Proposition 6.9.1 as a consequence of the theorem.

We will refer to the Implicit Function Theorem as the IPFT. Observe that the IPFT gives a sufficient condition for the existence of the function h. This condition is not necessary. This is like the situation with $f(x) = x^3$ on $\mathbb{R}$, where the function has an inverse even though $f'(0) = 0$. See Exercise 4. Let's also note that in the statement of the IPFT rather than the condition $f(a \oplus b) = 0$ we could have assumed $f(a \oplus b) = c$ for some constant c and then obtained the function h with $f(x \oplus h(x)) = c$ for all x in Ω. In fact we can do this by just substituting the function $f - c$ for f.

The condition in the theorem that T_q is invertible can be interpreted using Jacobians. Since $f : G \to \mathbb{R}^q$, we can write f using its coordinate functions: $f(x \oplus y) = (f_1(x \oplus y), \ldots, f_q(x \oplus y))$. When we write $y = (y_1, \ldots, y_q)$ in $\mathbb{R}^q$, we get that the linear transformation T_q is given by the Jacobian matrix

$$\frac{\partial(f_1, \ldots, f_q)}{\partial(y_1, \ldots, y_q)}(a \oplus b)$$

The invertibility of this matrix is guaranteed by having its determinant non-zero.

6.9.3. Corollary. *Let G be an open subset of $\mathbb{R}^2$, let $a \oplus b \in G$, and let $f : G \to \mathbb{R}$ be a continuously differentiable function such that $f(a \oplus b) = 0$. If $(\partial f/\partial y)(a \oplus b) \neq 0$, then there is an interval (α, β) containing the point a and a continuously differentiable function $h : (\alpha, \beta) \to \mathbb{R}$ satisfying: (a) $h(a) = b$; (b) $f(x \oplus h(x)) = 0$ for all x in (α, β); and (c)*

$$h'(a) = -\frac{\frac{\partial f}{\partial x}(a \oplus b)}{\frac{\partial f}{\partial y}(a \oplus b)}$$

See Exercise 5.

6.9.4. Example. (a) Can the equation $(x^2 + y^2 + 2z^2)^{\frac{1}{2}} = \cos z$ be solved for y in terms of x and z near the point $(0, 1, 0)$? To do this write $f(x, z, y) = (x^2 + y^2 + 2z^2)^{\frac{1}{2}} - \cos z$, where we reversed the alphabetical order of y and z to more easily interpret the IPFT. So $f : \mathbb{R}^3 \to \mathbb{R}$. A calculation shows that

$$\frac{\partial f}{\partial y} = \frac{y}{\sqrt{x^2 + y^2 + 2z^2}}$$

Since this is not 0 at $(0, 1, 0)$ we can apply the IPFT.

(b) Can we solve the above equation for z in terms of x and y near the point $(0, 1, 0)$? Again a calculation shows that

$$\frac{\partial f}{\partial z} = \frac{2z}{\sqrt{x^2 + y^2 + 2z^2}} + \sin z$$

and we have that $(\partial f/\partial z)(0, 1, 0) = 0$. Thus we cannot apply the IPFT.

Exercises

(1) Prove Proposition 6.9.1 as a consequence of the Implicit Function Theorem.
(2) Show that the set Ω in the proof of Theorem 6.9.2 is open. (See Exercise 6.4.17.)
(3) Prove the following version of the Implicit Function Theorem for linear transformations. If $T \in \mathcal{L}(\mathbb{R}^{p+q}, \mathbb{R}^q)$ and T_q is invertible, then the linear transformation $-T_q^{-1}T_p : \mathbb{R}^q \to \mathbb{R}^p$ satisfies

$$T\big(x \oplus \big(-T_q^{-1}T_p(x)\big)\big) = 0$$

for every x in $\mathbb{R}^p$. (It's actually easier to just give a direct proof of this rather than showing how it follows from the IPFT, but seeing how it is a corollary of that result might be instructive.)
(4) Find a smooth function $f : \mathbb{R}^2 \to \mathbb{R}$ with $(\partial f/\partial y)(0 \oplus 0) = 0$ but where there is an open interval (α, β) in $\mathbb{R}$ that contains 0 and a continuous function $h : (\alpha, \beta) \to \mathbb{R}$ such that $h(0) = 0$ and $f(x \oplus h(x)) = 0$ for all x in (α, β). Can you get such an example with h differentiable?
(5) Let G be an open subset of $\mathbb{R}^{p+1}$, let $a \oplus b \in G$, and let $f : G \to \mathbb{R}$ be a continuously differentiable function such that $f(a \oplus b) = 0$. (a) Give a sufficient condition on f that there is an open subset Ω of $\mathbb{R}^p$ that contains a and a continuously differentiable function $h : \Omega \to \mathbb{R}$ such that $h(a) = b$ and $f(x \oplus h(x)) = 0$ for all x in Ω. (b) Can you express this condition in terms of Jacobians? (c) Write out a formula for $h'(a)$.
(6) Consider the two equations $x^2 - y^2 - u^3 + v^2 + 4 = 0$ and $2xy + y^2 - 2u^2 + 3v^4 + 8 = 0$. Show that near the point $(x, y, u, v) = (2, -1, 2, 1)$ we can solve for u and v in terms of x and y.
(7) Let $f(x, y) = e^{xy} + \sin y + y^2 - 1$ and show that there is a continuously differentiable function h defined in a neighborhood U of 2 such that $f(x, h(x)) = 0$ for all x in U. Compute $h'(x)$ for x in U.

(8) Show that the Inverse Function Theorem can be derived as a consequence of the IPFT so that a different proof of the IPFT would yield the Inverse Function Theorem as a corollary. (Such a proof exists.)

6.10. Lagrange Multipliers*

We want to investigate optimization problems for real-valued functions that involve constraints on the possible set where the extremum occurs. For example suppose we are given an open set G in $\mathbb{R}^{p+1}$ and two continuously differentiable functions $f, g : G \to \mathbb{R}$; and the objective is to maximize the value of $f(x)$ subject to the *constraint* that $g(x) = 0$. (We use $\mathbb{R}^{p+1}$ here rather than $\mathbb{R}^p$ in order to be consistent with what comes later.) Such a problem often arises in geometry, physics, and economics, as well as other areas of science. You can consider the set $\{x \in \mathbb{R}^{p+1} : g(x) = 0\}$ as defining a surface in $\mathbb{R}^{p+1}$, and we are seeking to find the maximum value of f on this surface. (We haven't formally defined the concept of a surface in $\mathbb{R}^{p+1}$, but when $p = 2$ this set is a traditional surface in $\mathbb{R}^3$; when $p = 1$ it is a curve in the plane. The general definition will come later and, if you wish, you can just consider it as a subset of $\mathbb{R}^{p+1}$.)

The approach for solving such a problem is to consider the function $F : G \to \mathbb{R}$ defined by $F(x) = f(x) - \lambda g(x)$ for some scalar λ. Why do this? Suppose we can find a λ such that $\nabla F(c) = 0$ and c satisfies the constraint $g(c) = 0$. Since c is a critical point for F there is the possibility that F has a relative extremum at c. But $F(c) = f(c)$ since $g(c) = 0$ and there is the additional possibility that f has a relative extremum at c. Clearly this won't always work. Nevertheless in the result below (see Corollary 6.10.6 and the more general Theorem 6.10.3) we show that under reasonable conditions this is a necessary condition for f to have a relative extremum at c with $g(c) = 0$. That is, we will see that when f does have an extremum at such a c then we can find a scalar λ such that $\nabla(f - \lambda g)(c) = 0$. This discussion may seem abstruse, but after seeing some examples it will become apparent that the method works in many situations.

6.10.1. Definition. Assume $p, q \geq 1$ and G is an open subset of $\mathbb{R}^{p+q}$, $f : G \to \mathbb{R}$ is a continuously differentiable function, $c \in G$, and $g : G \to \mathbb{R}^q$ is a continuously differentiable function with $g(c) = 0$. The function f is said to have a *local maximum* at c subject to the *constraint* $g(c) = 0$ if there is an $r > 0$ such that when $x \in B(c; r)$ and $g(x) = 0$, we have that $f(c) \geq f(x)$. Similarly we can define f to have a *local minimum* at c subject to the constraint $g(c) = 0$. f is said to have a *local extremum* at c subject to the constraint $g(c) = 0$ if f has either a local maximum or minimum at c subject to this constraint.

It is probably needless to point out that the problem of finding the local minimum of f subject to the constraint $g(x) = 0$ is the same as finding the local maximum of $-f$ subject to the same constraint. Also the use of 0 in the constraint is arbitrary; any constant would do. Indeed we could have considered constraints of the form

$g(x) = h(x)$, where h is another continuously differentiable function on G. This reduces to what was described in the preceding definition by simply replacing this constraint by the one $g(x) - h(x) = 0$.

6.10.2. Example. (a) Suppose in the definition above $p = 2$, $q = 1$, and $g : \mathbb{R}^3 \to \mathbb{R}$ is defined by $g(x, y, z) = x^2 + y^2 + z^2 - 1$. We note that the set $\{(x, y, z) : g(x, y, z) = 0\}$ is the surface of the unit ball centered at the origin, the so-called unit sphere. So if we are given a function $f : \mathbb{R}^3 \to \mathbb{R}$, the object would be to maximize or minimize f over this sphere.

(b) Find the maximum and minimum values of $f(x, y, z) = x^2 - y^2$ on the ellipsoid $x^2 + 2y^2 + 3z^2 = 1$. (We'll solve this problem below in Example 6.10.9 after we prove Corollary 6.10.6.) Note that f does have a maximum and a minimum since it is continuous and the ellipsoid is compact.

Adopt the notation in Definition 6.10.1. As in §6.9 we write $\mathbb{R}^{p+q} = \mathbb{R}^p \oplus \mathbb{R}^q$. That is, points in $\mathbb{R}^{p+q}$ are written as $x \oplus y$ with x in $\mathbb{R}^p$ and y in $\mathbb{R}^q$. In particular the distinguished point c in G is written as $c = a \oplus b$. For the set G let

$$G_q = \{y \in \mathbb{R}^q : a \oplus y \in G\}$$

Next define the function $\hat{g} : G_q \to \mathbb{R}^q$ by

$$\hat{g}(y) = g(a \oplus y)$$

Hence $\hat{g}(b) = 0$. The notation g_q seems more appropriate than $\hat{g}$, but for $1 \le i \le q$ we write the coordinate functions $g_i : G \to \mathbb{R}$ of $g : G \to \mathbb{R}^p$ as

$$g_i(x \oplus y) = \langle g(x \oplus y), d_i \rangle$$

where $d_1, \ldots, d_q$ is the standard basis for $\mathbb{R}^q$. So the notation g_q is already taken. With this notation established we can state the theorem, which gives a necessary condition for the constrained extremum to exist.

6.10.3. Theorem (Lagrange[7] Multipliers). *Assuming the notation established in the preceding paragraph, if f has a local extremum at c subject to the constraint $g(c) = g(a \oplus b) = 0$ and*

6.10.4 $\qquad\qquad D\hat{g}(b) : \mathbb{R}^q \to \mathbb{R}^q$ *is invertible*

---

[7] Joseph-Louis Lagrange is claimed by many Italian mathematicians as their countryman, and with justification in spite of his French name. Both his parents were Italian, he was born in Turin, Italy in 1736, and he was baptized Giuseppe Lodovico Lagrangia. He did have French ancestors and as a youth called himself Lagrange. He studied at the college in Turin and came reluctantly to mathematics. In fact he once wrote, "If I had been rich, I probably would not have devoted myself to mathematics," which certainly seems like a non-sequitur to a modern reader. His early work on the calculus of variations attracted the attention of Euler and at the age of 19 he was appointed professor of mathematics at the Royal Artillery School in Turin. He continued to work on analysis and astronomy. Eventually he succeeded Euler as Director of Mathematics at the Berlin Academy in 1766. A year later he married his cousin, Vittoria Conti. She died in 1783 and they had no children. In 1787 he left Berlin to become a member of the Académie des Sciences in Paris, where he remained for the rest of his career. His work covers a wide range of topics from number theory to astronomy, with numerous publications. Napoleon named him to the Legion of Honour and Count of the Empire in 1808. In 1813 he died in Paris.

then there is a vector λ in $\mathbb{R}^q$ such that

6.10.5 $$[Dg(a \oplus b)]^t \lambda = \nabla f(c)$$

We will only prove this result when $q = 1$. The proof of the general theorem is notationally complicated involving matrices and it isn't clear that it's worth the effort. The interested reader can see [15] or [11] for a proof. After stating and proving the corollary we give a more matrix oriented way of viewing the hypothesis and conclusion of the theorem.

6.10.6. Corollary. *Let G be an open subset of $\mathbb{R}^{p+1}$, $c = a \oplus b \in G$, $f : G \to \mathbb{R}$ a continuously differentiable function, and $g : G \to \mathbb{R}$ a continuously differentiable function with $g(c) = 0$. If f has a local extremum at c subject to the constraint $g(c) = 0$ and $G_1 = \{y \in \mathbb{R} : a \oplus y \in G\}$, $\hat{g} : G_1 \to \mathbb{R}$ is defined as $\hat{g}(y) = g(a \oplus y)$, and*

$$[\hat{g}]'(b) = \partial_{p+1} g(c) \neq 0$$

then there is a λ in $\mathbb{R}$ such that

$$\nabla[f - \lambda g](c) = 0$$

Thus $f - \lambda g$ has a critical point at c.

Proof. Writing out the components of the equation $\nabla[f - \lambda g](c) = 0$, we see that we want to find a scalar λ such that

6.10.7 $$\partial_j f(c) - \lambda \partial_j g(c) = 0 \qquad \text{for} \quad 1 \leq j \leq p+1$$

Using the hypothesis that $\partial_{p+1} g(c) \neq 0$ we define

$$\lambda = \frac{\partial_{p+1} f(c)}{\partial_{p+1} g(c)}$$

We recognize that this definition of λ is precisely (6.10.7) when $j = p+1$; it remains to show that we have this equation for the other values of j. To do this we will use the fact that f has a local extremum at c subject to the constraint.

We apply the IPFT. Since $\partial_{p+1} g(c) \neq 0$ there is an open neighborhood Ω of a in $\mathbb{R}^p$ and a continuously differentiable function $h : \Omega \to \mathbb{R}$ such that $h(a) = b$ and $g(x \oplus h(x)) = 0$ for all x in Ω. We will use the observation that because f has a local extremum at c subject to the constraint $g(x) = 0$ this implies that the map $x \mapsto f(x \oplus h(x))$ of Ω into $\mathbb{R}$ has an unconstrained local extremum at a. (Verify!) Therefore by Theorem 6.2.20 all the partial derivatives of the function $\omega : \Omega \to \mathbb{R}$ defined by $\omega(x) = f(x \oplus h(x))$ must vanish at a. We need to compute. Since ω is the composition of f and $x \mapsto x \oplus h(x)$, the Chain Rule implies that for $1 \leq j \leq p$

$$\partial_j \omega(x) = [\partial_j f](x \oplus h(x)) + [\partial f_{p+1}](x \oplus h(x))[\partial_j h(x)]$$

Hence for $1 \leq j \leq p$

6.10.8 $$0 = \partial_j \omega(a) = \partial_j f(c) + \partial_{p+1} f(c) \, \partial_j h(a)$$

Now we need to apply the Chain Rule again. Since $g(x \oplus h(x)) = 0$ on Ω, we have that the partial derivatives of $x \mapsto g(x \oplus h(x))$ are all 0. Therefore for $1 \leq j \leq p$

$$0 = \partial_j[g(x \oplus h(x))]$$
$$= [\partial_j g](x \oplus h(x)) + [\partial g_{p+1}](x \oplus h(x))[\partial_j h(x)]$$

Since $a \oplus h(a) = c$, this last equation implies that

$$0 = \partial_j g(c) + \partial g_{p+1}(c) \partial_j h(a) \qquad \text{for} \quad 1 \leq j \leq p$$

Again using the hypothesis, this implies that

$$\partial_j h(a) = -\frac{\partial_j g(c)}{\partial_{p+1} g(c)} \qquad \text{for} \quad 1 \leq j \leq p$$

Substituting this into (6.10.8) we get that

$$0 = \partial_j f(c) + \partial_{p+1} f(c) - \left[\frac{\partial_j g(c)}{\partial_{p+1} g(c)}\right]$$
$$= \partial_j f(c) - \lambda \partial_j g(c)$$

which is exactly the sought after (6.10.7). ∎

Let's reflect on the preceding proof.

The challenge was to find one number λ that simultaneously solved the $p + 1$ equations (6.10.7). Perhaps it was surprising that we obtained the solution for all the equations by solving one of them, the $(p + 1)$-st. To bring this about we used the existence of the constrained extremum and the Implicit Function Theorem. That will set the pattern for the solution of the general theorem if you consult the references.

6.10.9. Example. Find the maximum and minimum values of $f(x, y, z) = x^2 - y^2$ on the ellipsoid $x^2 + 2y^2 + 3z^2 = 1$. This is the situation in the above corollary where $p = 2$, $q = 1$, and $g(x, y, z) = x^2 + 2y^2 + 3z^2 - 1$. Let $\lambda \in \mathbb{R}$ and note that

$$\nabla(f \quad \lambda g) = \nabla[(x^2 - \lambda x^2) + (-y^2 - 2\lambda y^2) - 3\lambda z^2 - \lambda]$$
$$= \left(2x(1 - \lambda), 2y(-1 - 2\lambda), -6\lambda z\right)$$

Setting these three coordinates equal to 0 we see we must find all possible values of λ such that the following four equations have a solution for x, y, and z:

6.10.10 $\qquad 0 = x(1 - \lambda); \ 0 = y(-1 - 2\lambda); \ 0 = \lambda z; \ 0 = x^2 + 2y^2 + 3z^2 - 1$

If we have $\lambda = 0$, we see that $x = y = 0$ while $0 = 3z^2 - 1$. Thus we have two critical points

$$\left(0, 0, \pm\frac{1}{\sqrt{3}}\right) \qquad \text{when } \lambda = 0$$

Now assume $\lambda \neq 0$. The first observation is that this forces $z = 0$. We also see that we must consider the two cases when $x = 0$ and $x \neq 0$. In the first case we get there are two more critical points

$$\left(0, \pm\frac{1}{\sqrt{2}}, 0\right) \qquad \text{when } \lambda \neq 0 \text{ and } x = 0$$

When $x \neq 0$ it has to be that $\lambda = 1$; consequently $y = 0$. Thus from the last of the equations in (6.10.10) we have that $x = \pm1$. This produces two more critical points

$$(\pm1, 0, 0) \qquad \text{when } \lambda \neq 0 \text{ and } x \neq 0$$

Now it's a matter of calculating the value of f at each of the critical points:

$$f\left(0, 0, \pm\frac{1}{\sqrt{3}}\right) = 0$$
$$f\left(0, \pm\frac{1}{\sqrt{2}}, 0\right) = -\frac{1}{2}$$
$$f(\pm1, 0, 0) = 1$$

Therefore the function f has constrained maximums at $(\pm1, 0, 0)$ and constrained minimums at $(0, \pm1/\sqrt{2}, 0)$.

Let's examine what (6.10.4) and (6.10.5) mean in terms of matrices. When you apply the theorem, you are quite likely to use systems of equations. Let's start by using the Jacobian determinant to interpret the first of these, which is equivalent to

6.10.11 $$\Delta\hat{g}(a) = \frac{\partial(g_1, \ldots, g_q)}{\partial(y_1, \ldots, y_q)}(c) \neq 0$$

Now $Dg(c) \in \mathcal{L}(\mathbb{R}^{p+q}, \mathbb{R}^q)$, so that $[Dg(c)]^t \in \mathcal{L}(\mathbb{R}^q, \mathbb{R}^{p+q})$. Hence when $\lambda \in \mathbb{R}^q$ we have that $[Dg(c)]^t\lambda \in \mathbb{R}^{p+q}$ as is $\nabla f(c)$. Hence the conclusion (6.10.5) can be phrased as saying we want a $\lambda = (\lambda_1, \ldots, \lambda_q)$ in $\mathbb{R}^q$ that satisfies the two sets of equations

6.10.12 $$\sum_{i=1}^{q} \lambda_i \frac{\partial g_i}{\partial y_k}(c) = \frac{\partial f}{\partial y_k}(c) \qquad \text{for } 1 \leq k \leq q$$

and

6.10.13 $$\sum_{i=1}^{q} \lambda_i \frac{\partial g_i}{\partial x_\ell}(c) = -\frac{\partial f}{\partial x_\ell}(c) \qquad \text{for } 1 \leq \ell \leq p$$

We can find a λ in $\mathbb{R}^q$ such that (6.10.12) holds since this set of equations translates to the requirement that

$$D\hat{g}(b)\lambda = \left(\frac{\partial f}{\partial y_1}(c), \ldots, \frac{\partial f}{\partial y_q}(c)\right)$$

and in light of the assumption (6.10.4) we can find such a λ. The objective of the proof of the main theorem is to show that for this same λ the equations (6.10.13) are also satisfied. As we did when we proved Corollary 6.10.6 this is achieved by an application of the Implicit Function Theorem and the fact that when a scalar-valued function has a relative extremum its gradient vanishes.

6.10.14. Example. Minimize $f(x, y, z, w) = x^2 + y^2 + z^2 + w^2$ subject to $x + y + z + w = 10$ and $x - y + z + 3w = 6$. Let's point out that relative to Theorem 6.10.3 $p = 2$, $q = 2$, and $g(x, y, z, w) = (x + y + z + w - 10, x - y + z + 3w - 6)$. Also the two sets in $\mathbb{R}^4$ where $x + y + z + w = 10$ and $x - y + z + 3w = 6$ are affine hyperplanes (6.7.3). So the constraining set where $g = 0$ is the intersection of these two affine hyperplanes. Therefore this example asks to find the (square of the) distance from the origin in $\mathbb{R}^4$ to the intersection of the two affine hyperplanes. This example is from [12] where the reader can find the solution on page 14.

Exercises

(1) Find the extrema of $f(x, y) = x^2 - 4xy + 4y^2$ subject to $x^2 + y^2 = 1$.
(2) A rectangle has perimeter p. Find its largest possible area.
(3) A rectangle has area A. Find its smallest possible perimeter.
(4) (a) Show that $f(x, y, z) = z^2$ has only one critical point on the surface $x^2 + y^2 - z = 0$. (b) Show that at that critical point f has a constrained minimum. (c) Why isn't there a constrained maximum?
(5) Find the extreme values of $f(x, y, z) = x + y^2 + 2z$ subject to $4x^2 + 9y^2 - 36z^2 = 36$.

7 Integration in Higher Dimensions

Here we will extend the development of the Riemann integral of functions defined on the real line, as seen in Chapter 3, to the integral of functions defined on subsets of $\mathbb{R}^p$. The objective of this chapter is not to give the most exhaustive treatment of this integral in $\mathbb{R}^p$, far from it. Those who wish to see such a treatment can consult the books [11] and [15]. In the next level of a course in analysis, a far more general theory of integration is developed. (It's called Lebesgue integration after its discoverer.) What we want to do here is to prepare the reader to tackle the uses of the integral in $\mathbb{R}^p$ that we'll encounter later in this book. In many ways what we seek to do here is to explore and understand situations in $\mathbb{R}^p$ where integration can be reduced to what we did in Chapter 3. In that sense the key is §7.3 below where integrals in $\mathbb{R}^p$ are reduced to a succession of integrals in $\mathbb{R}$.

We'll start with a much less complicated situation where we continue to integrate over an interval in $\mathbb{R}$ but the functions we integrate will be vector valued.

7.1. Integration of Vector-valued Functions

In §6.1 we discussed continuity and differentiability of vector-valued functions $\gamma : [a, b] \to \mathbb{R}^p$ and we saw there were few difficulties in this. Here we discuss the integration of such functions and again we'll see that the theory is almost the same as what we did when we defined the integral of functions $f : [a, b] \to \mathbb{R}$.

We could discuss Riemann sums of functions $\gamma : [a, b] \to \mathbb{R}^p$ in parallel with what we did in Chapter 3, but rather than risk possible boredom we'll just use what we developed there and go forward. Given a function $\gamma : [a, b] \to \mathbb{R}^p$, for $1 \le j \le p$ define a function $\gamma_j : [a, b] \to \mathbb{R}$ by

$$\gamma_j(t) = \langle \gamma(t), e_j \rangle$$

where, as usual, $\{e_1, \ldots, e_p\}$ is the standard basis for $\mathbb{R}^p$. Now define

7.1.1
$$\int_a^b \gamma = \sum_{j=1}^p \left[\int_a^b \gamma_j \right] e_j \in \mathbb{R}^p$$

So $\int_a^b \gamma$ is a vector in $\mathbb{R}^p$. The reader can easily scan the results from Chapter 3 to see that they almost all carry over to this setting. In particular the FTC is valid: If $\gamma : [a, b] \to \mathbb{R}^p$ is continuously differentiable, then $\int_a^b \gamma' = \gamma(b) - \gamma(a)$. One of the few exceptions to this idea that Chapter 3 results carry over to the integration of vector-valued functions is the Mean Value Theorem for integrals (Theorem 3.2.4). See Exercise 1. The interested reader can look as much as (s)he wants at the details of this new integral, but we'll spend the rest of this section looking at the integral from Chapter 3 as applied to a new concept.

Recall (6.1.2) that a curve in $\mathbb{R}^p$ is a continuous function $\gamma : [a, b] \to \mathbb{R}^p$ and $\{\gamma\} = \{\gamma(t) : a \le t \le b\}$ is called its trace.

7.1.2. Definition. If $\gamma : [a, b] \to \mathbb{R}^p$ is a smooth curve, then the *length* or *arc length* of γ is defined as

$$\ell(\gamma) = \int_a^b \|\gamma'(t)\| \, dt$$

If γ is continuous on $[a, b]$ and smooth on the intervals $[t_{i-1}, t_i]$ for $1 \le i \le n$, where $a = t_0 < \cdots < t_n = b$ is a partition on $[a, b]$, then we define the length of γ as

$$\ell(\gamma) = \sum_{i=1}^n \int_{t_{i-1}}^{t_i} \|\gamma'(t)\| \, dt = \sum_{i=1}^n \ell(\gamma([t_{i-1}, t_i]))$$

We should point out that since γ' is continuous as is the function $x \mapsto \|x\|$ from $\mathbb{R}^p$ into $\mathbb{R}$, the function $t \mapsto \|\gamma'(t)\|$ is the composition of two continuous functions; thus it is integrable and the preceding definition is legitimate.

7.1.3. Example. (a) If $x, y \in \mathbb{R}^p$, define $\gamma : [0, 1] \to \mathbb{R}^p$ by $\gamma(t) = ty + (1 - t)x$. So γ traces out the straight line segment in $\mathbb{R}^p$ from x to y. Here $\ell(\gamma) = \int_0^1 \|y - x\| \, dt = \|y - x\|$, as expected.

(b) Recall the definition of a polygon $[x_0, x_1, \ldots, x_n]$ in $\mathbb{R}^p$ (Exercise 5.6.12). If γ is this polygon, then γ is piecewise smooth and $\ell(\gamma) = \sum_{i=1}^n \|x_i - x_{i-1}\|$.

(c) If $f : [a, b] \to \mathbb{R}$ is a continuously differentiable function and $\gamma : [a, b] \to \mathbb{R}^2$ is defined by $\gamma(t) = (t, f(t)) = te_1 + f(t)e_2$, then

$$\ell(g) = \int_a^b \sqrt{1 + |f'(t)|^2} \, dt$$

In fact $\gamma'(t) = e_1 + f'(t)e_2$, so that $\|\gamma'(t)\| = \sqrt{1 + |f'(t)|^2}$.

(d) More generally, if $f : [a, b] \to \mathbb{R}^{p-1}$ is a continuously differentiable function and $\gamma : [a, b] \to \mathbb{R}^p$ is defined by $\gamma(t) = t \oplus f(t) \in \mathbb{R} \oplus \mathbb{R}^{p-1}$, then

$$\ell(g) = \int_a^b \sqrt{1 + \|f'(t)\|^2} \, dt$$

7.1.4. Proposition. *If $\gamma : [a, b] \to \mathbb{R}^p$ is a smooth curve and $\epsilon > 0$, then there is a partition $a = t_0 < t_i < \cdots < t_n = b$ such that*

$$\left| \int_a^b \|\gamma'(t)\| dt - \sum_{i=1}^n \|\gamma(t_i) - \gamma(t_{i-1})\| \right| < \epsilon$$

Proof. The lack of a MVT for vector-valued functions (Exercise 6.1.4) creates a bit of a complication for the proof and we must introduce an auxiliary function F. Let $\gamma(t) = (\gamma_1(t), \ldots, \gamma_p(t))$ for all t in $[a, b]$, let $[a, b]^p$ be the cartesian product of the interval $[a, b]$ with itself p times, and define $F : [a, b]^p \to \mathbb{R}$ by $F(t_1, \ldots, t_p) = \sqrt{\sum_{j=1}^{p} |\gamma_j'(t_j)|^2}$. Observe that F is a continuous function on a compact set, and thus is uniformly continuous. Hence if $\epsilon > 0$ there is a $\delta > 0$ such that for $\tau = (t_1, \ldots, t_p)$, $\sigma = (s_1, \ldots, s_p)$ in $[a, b]^p$ with $\|\tau - \sigma\| < \delta$ we have that $|F(\sigma) - F(\tau)| < \epsilon/2(b-a)$. In particular since $F(t, \ldots, t) = \|\gamma'(t)\|$, we have that if $\{a = t_0 < t_1 < \cdots < t_n = b\}$ satisfies $t_i - t_{i-1} < \delta/\sqrt{p}$ for $1 \le i \le n$, then

7.1.5
$$\left| \int_a^b \|\gamma'(t)\| \, dt - \sum_{i=1}^{n} \|\gamma'(t_i)\|(t_i - t_{i-1}) \right| < \frac{\epsilon}{2}$$

Now we apply the one-dimensional MVT to each γ_j to conclude that for $1 \le j \le p$, $1 \le i \le n$ we can find a point c_{ji} in $[t_{i-1}, t_i]$ with

$$\gamma_j(t_i) - \gamma_j(t_{i-1}) = \gamma_j'(c_{ji})(t_i - t_{i-1})$$

Therefore for each $i = 1, \ldots, n$ we have that

$$F(c_{1i}, \ldots, c_{pi})(t_i - t_{i-1}) = \left(\sum_{j=1}^{p} |\gamma_j'(c_{ji})|^2 \right)^{\frac{1}{2}} (t_i - t_{i-1}) = \|\gamma(t_i) - \gamma(t_{i-1})\|$$

Note that for $0 \le i \le n$ we have that $\|(c_{1i}, \ldots, c_{pi}) - (t_i, \ldots, t_i)\| < \delta$, so that $|F(c_{1i}, \ldots, c_{pi}) - F(t_i, \ldots, t_i)| < \epsilon/2(b-a)$. Hence

$$\sum_{i=1}^{n} \|\gamma(t_i) - \gamma(t_{i-1})\| = \sum_{i=1}^{n} F(c_{1i}, \ldots, c_{pi})(t_i - t_{i-1})$$
$$= \sum_{i=1}^{n} [F(c_{1i}, \ldots, c_{pi}) - F(t_i, \ldots, t_i)](t_i - t_{i-1})$$
$$+ \sum_{i=1}^{n} F(t_i, \ldots, t_i)(t_i - t_{i-1})$$
$$< \frac{\epsilon}{2(b-a)} \sum_{i=1}^{n} (t_i - t_{i-1}) + \sum_{i=1}^{n} \|\gamma'(t_i)\|(t_i - t_{i-1})$$
$$= \frac{\epsilon}{2} + \sum_{i=1}^{n} \|\gamma'(t_i)\|(t_i - t_{i-1})$$

Similarly

$$\sum_{i=1}^{n} \|\gamma(t_i) - \gamma(t_{i-1})\| > -\frac{\epsilon}{2} + \sum_{i=1}^{n} \|\gamma'(t_i)\|(t_i - t_{i-1})$$

Combining these inequalities with (7.1.5) we get that

$$\left| \int_a^b \|\gamma'(t)\| \, dt - \sum_{i=1}^n \|\gamma(t_i) - \gamma(t_{i-1})\| \right| < \epsilon$$

proving the proposition. ∎

So this proposition says that $\ell(\gamma)$ can be approximated as close as desired by the length of an inscribed polygon $[\gamma(t_0), \dots, \gamma(t_n)]$ (Example 7.1.3(b)). This justifies defining the length of the curve γ as we did. Also see Exercise 2.

A problem arises from defining a curve as a function. Is there a difference between the curve $\gamma : [0, \pi] \to \mathbb{R}^2$ defined by $\gamma(t) = (r \cos t, r \sin t)$ and the curve $\rho : [-r, r] \to \mathbb{R}^2$ defined by $\rho(t) = (-t, \sqrt{r^2 - t^2})$? These curves describe the same set of points and they should have the same length. Here is how we address this.

7.1.6. Definition. If $\gamma : [a, b] \to \mathbb{R}^p$ and $\rho : [c, d] \to \mathbb{R}^p$ are two smooth curves, say that ρ is *equivalent* to γ if there is a continuously differentiable, strictly increasing surjection $\tau : [a, b] \to [c, d]$ such that $\gamma(t) = \rho(\tau(t))$ for all t in $[a, b]$. In symbols we write $\gamma \sim \rho$. We also say that ρ is a *reparametrization* of γ.

Note that since the inverse of a continuously differentiable, strictly increasing surjection is a continuously differentiable, strictly increasing surjection, it easily follows that $\sim$ is an equivalence relation on the set of all smooth curves in $\mathbb{R}^p$. (Exercise 7). Clearly equivalent curves trace out the same subset of $\mathbb{R}^p$. Now we show that the definition of the length of a curve is independent of which parametrization of the curve we take.

7.1.7. Proposition. *If γ and ρ are equivalent curves in $\mathbb{R}^p$, then $\ell(\gamma) = \ell(\rho)$.*

Proof. Adopt the notation of the preceding definition. By the COV Theorem (3.2.5)

$$\ell(\rho) = \int_c^d \|\rho'(s)\| ds = \int_a^b \|\rho'(\tau(t))\| \|\tau'(t)\| dt = \int_a^b \|\gamma'(t)\| dt = \ell(\gamma) \quad \blacksquare$$

Strictly speaking we should have defined a curve as an equivalence class of functions using the equivalence relation $\sim$ defined above. Instead we opted for a less stringent approach.

It might have been observed by the reader that the last proposition remains valid if we allow the reparametrization function τ to be strictly decreasing. We insist that $\tau' > 0$ for a reason that will appear in Chapter 8.

Exercises

(1) If $\gamma : [0, 1] \to \mathbb{R}^2$ is defined by $\gamma(t) = (t, t^2)$, then there is no point c in $[0, 1]$ such that $\int_0^1 \gamma = \gamma(c)$.

(2) Show that if $\gamma : [a, b] \to \mathbb{R}^p$ is a smooth curve, then

$$\ell(\gamma) = \sup\left\{ \sum_{i=1}^{n} \|\gamma(t_i) - \gamma(t_{i-1})\| : \{a = t_0 < \cdots < t_n = b\} \right\}$$

(3) Are the curves $\gamma : [0, \pi] \to \mathbb{R}^2$ and $\rho : [-r, r] \to \mathbb{R}^2$ defined by $\gamma(t) = (r\cos t, r\sin t)$ and $\rho(t) = (-t, \sqrt{r^2 - t^2})$ equivalent?

(4) Let $\gamma : [0, 4] \to \mathbb{R}^3$ be the helix defined by $\gamma(t) = (\cos t, \sin t, t^{3/2})$ and compute its length.

(5) Define $\gamma : [0, 2\pi] \to \mathbb{R}^3$ by $\gamma(t) = (e^t \sin t, e^t \cos t, e^t)$ and compute $\ell(\gamma)$.

(6) Compute the length of the graph of the function $f(x) = (x^2)^{\frac{1}{3}}$ with $-1 \le x \le 1$.

(7) Show that $\gamma \sim \rho$ defines an equivalence relation on the set of all smooth curves in $\mathbb{R}^p$.

7.2. The Riemann Integral

In this section X always denotes a bounded subset of $\mathbb{R}^p$ and $f : X \to \mathbb{R}$ is a bounded function.

A *rectangle* in $\mathbb{R}^p$ is a set of the form

$$R = [a_1, b_1] \times \cdots \times [a_p, b_p] = [a, b] \subseteq \mathbb{R}^p$$

where $a = (a_1, \ldots, a_p), b = (b_1, \ldots, b_p)$ are points in $\mathbb{R}^p$, and $a_j < b_j$ for $1 \le j \le p$. (The notation $R = [a, b]$ is a convenient shorthand and the context will usually prevent confusion with a closed interval in $\mathbb{R}$.) We will also refer to the *volume* of R as the number

$$V(R) = (b_1 - a_1) \cdots (b_p - a_p)$$

Two rectangles in $\mathbb{R}^p$ are non-overlapping if their intersection has empty interior. Equivalently, if $R = [a, b]$, $S = [c, d] \subseteq \mathbb{R}^p$, then R and S are non-overlapping if $(a, b) \cap (c, d) = \emptyset$, where I hope the meaning of (a, b) and (c, d) as the open rectangles in $\mathbb{R}^p$ is clear. Equivalently they are non-overlapping if $R \cap S = \partial R \cap \partial S$, where this intersection could be empty. When $X \subseteq \mathbb{R}^p$, say that a finite collection of rectangles $R_1, \ldots, R_m$ is a R-*cover* of X if these rectangles are non-overlapping, $X \subseteq \bigcup_{k=1}^{m} R_k$, and $R_k \cap X \ne \emptyset$ for $1 \le k \le m$. (This last restriction is included to prevent trivialities.) Note that the fact that we are assuming that X is always a bounded set means we can always find a R-cover of X. More on this in the following example.) In $\mathbb{R}^p$ a R-cover will play the role that a partition of an interval did in Chapter 3.

7.2.1. Example. (a) An easy example of a R-cover of a bounded set X is $\mathcal{R} = \{R\}$, where R is a rectangle that contains X.

(b) When $p = 1$, what we call a rectangle is just a closed interval and its volume is the interval's length. A R-cover of $[a, b]$ in this setting is just a partition of the interval, though the way we defined a R-cover $\mathcal{R}$ of $[a, b]$ allows some of the intervals in $\mathcal{R}$ to extend outside of $[a, b]$.

(c) In $\mathbb{R}^2$ a rectangle is what we usually call a closed rectangle and two rectangles in $\mathbb{R}^2$ are non-overlapping if their intersection is contained in the intersection of their sides. If X is a bounded subset of $\mathbb{R}^2$, then we can form a R-cover of X as follows. Draw a grid consisting of vertical and horizontal lines such that these lines do not accumulate anywhere. (You could achieve this by requiring all the vertical and horizontal lines to be separated by at least some minimum distance, though this is not the only way to do this.) This grid defines an infinite collection of non-overlapping closed rectangles. Now discard all the rectangles that are disjoint from X. Because X is bounded, only a finite number of rectangles remain. For future reference note that by making the distance between the vertical and horizontal lines smaller we can make the R-cover finer, a term made precise below.

(d) In $\mathbb{R}^3$ a rectangle is a parallelepiped whose faces are two-dimensional rectangles; that is, it is a box. Two such rectangles are non-overlapping if their intersection is contained in the intersection of their faces.

(e) Extending what was done in (c), if X is a bounded subset of $\mathbb{R}^p$, then we can form a R-cover of X as follows. Fix each coordinate j, $1 \leq j \leq p$, choose a doubly infinite sequence $\{a_j^n : -\infty < n < \infty\}$, where $a_j^n \to \pm\infty$ as $n \to \pm\infty$. Now consider the countable collection of affine hyperplanes $H_j^n = \{x \in \mathbb{R}^p : x_j = a_j^n\}$. The collection of these hyperplanes $\{H_j^n : 1 \leq j \leq p, -\infty < n < \infty\}$ forms a p-dimensional grid in $\mathbb{R}^p$ and the resulting rectangles that meet X forms a R-cover of X. You might follow this when $p = 2$ to see this is what we did in part (c) and then follow it through when $p = 3$.

> *The conceptual difficulties in making the transition in integration from the line to higher dimensional Euclidean spaces already manifest themselves when we consider $\mathbb{R}^2$.*

Maybe that is somewhat contradicted for the reader when (s)he passed from part (c) in the last example to part (e); but that difficulty is more technical than conceptual. As the reader progresses, (s)he can achieve a good deal of understanding by thinking in terms of the plane and seeing what the results say there.

If $\mathcal{Q}$ and $\mathcal{R}$ are two R-covers of X say that $\mathcal{Q}$ is a *refinement* of $\mathcal{R}$ if each rectangle in $\mathcal{Q}$ is contained in a rectangle from $\mathcal{R}$. Let's note that we have that $\bigcup_{Q \in \mathcal{Q}} Q \subseteq \bigcup_{R \in \mathcal{R}} R$, but, unlike when we consider partitions of a compact interval in $\mathbb{R}$, we do not insist that these two unions are the same. Indeed to require that as well as having every rectangle in the finer R-cover $\mathcal{Q}$ meet X is often impossible. Let's note that if $\mathcal{R} = \{R_1, \ldots, R_m\}$ and $\mathcal{S} = \{S_1, \ldots, S_s\}$ are any two R-covers of X, then $\{R_j \cap S_i : 1 \leq j \leq m, 1 \leq i \leq s \text{ and } R_j \cap S_i \neq \emptyset\}$ is a refinement of both $\mathcal{R}$ and $\mathcal{S}$. When we are given a R-cover $\mathcal{R}$ of X and $R_j \in \mathcal{R}$, let

$$M_j^{\mathcal{R}} = M_j = \sup\{f(x) : x \in R_j \cap X\}$$
$$m_j^{\mathcal{R}} = m_j = \inf\{f(x) : x \in R_j \cap X\}$$

and let

$$U(f, \mathcal{R}) = \sum_{j=1}^{m} M_j^{\mathcal{R}} V(R_j) \text{ and } L(f, \mathcal{R}) = \sum_{j=1}^{m} m_j^{\mathcal{R}} V(R_j)$$

(We will use the notation $M_j^{\mathcal{R}}$ and $m_j^{\mathcal{R}}$ when we are discussing more than one R-cover at the same time.) The term $U(f, \mathcal{R})$ is called the *upper sum* of f for the R-cover $\mathcal{R}$ and $L(f, \mathcal{R})$ is called the *lower sum*.

The proof of the next proposition is similar to the proof of Proposition 3.1.2 and is left to the reader as Exercise 2.

7.2.2. Proposition. *Assume R is a rectangle in $\mathbb{R}^p$ that contains X. If $f : X \to \mathbb{R}$ with $m \le f(x) \le M$ for all x in X and $\mathcal{R}$ and $\mathcal{Q}$ are R-covers of X that both refine the R-cover consisting of R alone, then the following hold.*

(a) $mV(R) \le L(f, \mathcal{R}) \le U(f, \mathcal{Q}) \le MV(R)$.
(b) *If $\mathcal{Q}$ is a refinement of $\mathcal{R}$, then*

$$L(f, \mathcal{R}) \le L(f, \mathcal{Q}) \le U(f, \mathcal{Q}) \le U(f, \mathcal{R})$$

(c) *If $\mathcal{Q}$ is a refinement of $\mathcal{R}$, then*

$$0 \le U(f, \mathcal{Q}) - L(f, \mathcal{Q}) \le U(f, \mathcal{R}) - L(f, \mathcal{R})$$

7.2.3. Definition. Using the preceding proposition we have that for any bounded subset X of $\mathbb{R}^p$ and any bounded function $f : X \to \mathbb{R}$

$$\sup\{L(f, \mathcal{R}) : \mathcal{R} \text{ is a R-cover of } X\} \le \inf\{U(f, \mathcal{Q}) : \mathcal{Q} \text{ is a R-cover of } X\}$$

When these two quantities are equal, we set

$$\int_X f = \int_X f(x)dx$$
$$= \sup\{L(f, \mathcal{R}) : \mathcal{R} \text{ is a R-cover of } X\}$$
$$= \inf\{U(f, \mathcal{Q}) : \mathcal{Q} \text{ is a R-cover of } X\}$$

In this situation we say that f is *Riemann integrable* or just *integrable*. The set of all Riemann integrable functions on X is denoted by $\mathcal{R}(X)$.

Note that the above definition only applies to bounded subsets X and bounded functions f. These assumptions may be omitted when we state results but they are always assumed.

In the preceding definitions we see a hint of a difficulty in extending the Riemann integral from the line to higher dimensional Euclidean space. In $\mathbb{R}$ we restrict our attention to integrals over intervals and this allows us to partition them. In $\mathbb{R}^p$ we want to integrate over sets that are not rectangles. (For example, integration over balls and similar sets.) You might say that we should consider covers by sets more general than rectangles, but then how do we define their volume? This lack of simplicity will cause us difficulty as we proceed.

As in §3.1 several results follow. Their statements and proofs follow their analogues in §3.1. The first such result is similar to that of Proposition 3.1.4.

7.2.4. Proposition. *If f is a bounded function, then f is Riemann integrable if and only if for every $\epsilon > 0$ there is a R-cover $\mathcal{R}$ of X such that $U(f, \mathcal{R}) - L(f, \mathcal{R}) < \epsilon$. Moreover $\int_X f$ is the unique number such that for every refinement $\mathcal{Q}$ of $\mathcal{R}$ we have*

$$L(f, \mathcal{Q}) \leq \int_X f \leq U(f, \mathcal{Q})$$

The next is the analogue of Corollary 3.1.5.

7.2.5. Corollary. *If f is a bounded function, then f is Riemann integrable if and only if there is a sequence of R-covers $\{\mathcal{R}_n\}$ of X such that each $\mathcal{R}_{n+1}$ is a refinement of $\mathcal{R}_n$ and $U(f, \mathcal{R}_n) - L(f, \mathcal{R}_n) \to 0$ as $n \to \infty$. When this happens we have that*

$$\int_X f = \lim_{n \to \infty} U(f, \mathcal{R}_n) = \lim_{n \to \infty} L(f, \mathcal{R}_n)$$

The next two results give some elementary properties of the integral that extend their counterparts in §3.1. No proofs will be given.

7.2.6. Proposition. *Assume that X is bounded, R is a rectangle containing X, and $f, g \in \mathcal{R}(X)$.*

(a) *If $f(x) \leq g(x)$ for all x in X, then*

$$\int_X f \leq \int_X g$$

(b) *If $|f(x)| \leq M$ for all x in X, then*

$$\left| \int_X f \right| \leq \int_X |f| \leq MV(R)$$

7.2.7. Proposition. *$\mathcal{R}(X)$ is a vector space over $\mathbb{R}$. Moreover if $f, g \in \mathcal{R}(X)$ and $\alpha, \beta \in \mathbb{R}$, then $\alpha f + \beta g \in \mathcal{R}(X)$ and*

$$\int_X (\alpha f + \beta g) = \alpha \int_X f + \beta \int_X g$$

There are some very complicated subsets of $\mathbb{R}^2$, to say nothing of higher dimensional spaces. The interested reader might play with Exercise 3 for a strange example. The difficulty in developing the Riemann integral for higher dimensions occurs at the boundary of the set X. Here we introduce a collection of sets which are simultaneously abundant and on which there are many integrable functions.

7.2.8. Definition. If $E \subseteq \mathbb{R}^p$, say that E has *volume zero* if for every $\epsilon > 0$ there is a R-cover $\{R_1, \ldots, R_m\}$ of E such that

$$\sum_{j=1}^m V(R_j) < \epsilon$$

A set X in $\mathbb{R}^p$ is called a *Jordan*[1] *set* if it is bounded, cl $X =$ cl [int X], and its topological boundary, ∂X, has volume zero.

We'll see that Jordan sets are the proper places to do Riemann integration. As we progress we'll establish several results implying that a set is a Jordan set. An example of a set that is not a Jordan set follows. To find a closed set that is not a Jordan set is more difficult. But, as we said at the start of this chapter, our aim is not to explore the boundaries of the theory of integration in $\mathbb{R}^p$ but rather to establish places where the theory works and can be applied to the business at hand.

7.2.9. Example. (a) It is easy to see that any finite set in $\mathbb{R}^p$ has volume zero. So in $\mathbb{R}$ the union of a finite number of intervals is a Jordan set.

(b) The set $X = \{(x, y) \in [0, 1] \times [0, 1] : x, y \in \mathbb{Q}\}$ is not a Jordan set since cl $X \neq$ cl [int X] $= \emptyset$. Also, adapting the argument used in Example 5.3.8(a) we see that $\partial X = [0, 1] \times [0, 1]$ and this certainly does not have volume zero.

(c) If $X_1, \ldots, X_n$ are sets of volume zero, then $X = \bigcup_{j=1}^{n} X_j$ has volume zero. Indeed if $\epsilon > 0$ let $\mathcal{R}_j$ be a R-cover of X_j such that the sum of the volumes of the rectangles is less than ϵ/n. We would like to put $\mathcal{R}$ equal to the union of all the rectangles in each of the $\mathcal{R}_j$, $1 \leq j \leq n$. However some of the rectangles in this union may not be overlapping. This, however, can be fixed by considering all the possible intersections of rectangles from the $\mathcal{R}_j$. See Exercise 5. Once this is done we have a R-cover and the sum of the volumes of its rectangles is smaller than ϵ, proving that X has volume zero.

7.2.10. Proposition. *A closed rectangle in $\mathbb{R}^p$ is a Jordan set.*

Proof. Let $R = [a, b] = [a_1, b_1] \times \cdots \times [a_p, b_p]$. If $\epsilon > 0$ there is a $\delta > 0$ such that if we set $S = [a_1 + \delta, b_1 - \delta] \times \cdots \times [a_p + \delta, b_p - \delta]$, then $S \subseteq R$ and $V(R) - V(S) < \epsilon$. The remainder of the proof consists in showing that there is a R-cover $\mathcal{R}$ of ∂R that is contained in $R \backslash S$, thus implying that the sum of the volumes of the rectangles in $\mathcal{R}$ is smaller than ϵ. (You might want to draw pictures of the next argument when $p = 2$. The general argument is a bit cumbersome.) For each $j = 1, \ldots, p$ consider the partition $\{a_j, a_j + \delta, b_j - \delta, b_j\}$ of $[a_j, b_j]$ and form the grid of hyperplanes (as in Example 7.2(e)) determined by these values in each dimension. Let $\mathcal{R} = \{R_1, \ldots, R_m\}$ be the rectangles determined by this grid that

---

[1] Camille Jordan was born in 1838 in Lyon, France. He was educated as an engineer at École Polytechnique and pursued that profession while continuing to explore mathematics. He defended his doctoral thesis in 1861. He continued working as an engineer, though taking time in 1862 to marry. He and his wife had eight children. He made his way back to Paris and became a professor, first at École Polytechnique in 1876 and then at Collège de France in 1883. His mathematical contributions spanned the list of fields under study at the time, from algebra to analysis to the emerging field of topology. He is the author of the Jordan Curve Theorem and Jordan Normal forms of matrices. (To be clear he is not the Jordan in Gauss–Jordan elimination or Jordan algebras.) He also introduced the notion of homotopic paths and worked extensively on finite groups. He retired in 1912. During World War I three of his sons were killed. Another son was a government minister, another a professor of history, and the last son was an engineer. He received many honors during his life, which ended in 1922 in Paris.

intersect ∂R. Thus $\mathcal{R}$ is a R-cover of ∂R, and, moreover, each $R_k \subseteq R \backslash S$. It follows that $\sum_{j=1}^{m} V(R_j) < \epsilon$. ∎

To prove that a set is a Jordan set we need results that tell us when a set has volume zero. The next one will be useful.

7.2.11. Proposition. *If Y is a compact subset of $\mathbb{R}^{p-1}$ and $g : Y \to \mathbb{R}$ is a continuous function, then $W = \{(y, g(y)) : y \in Y\}$ is a compact subset of $\mathbb{R}^p$ having volume zero.*

Proof. The fact that W is compact follows from the observation that $y \mapsto (y, g(y))$ is a continuous function from Y onto W. Let $\epsilon > 0$ and let R be a fixed closed rectangle in $\mathbb{R}^{p-1}$ such that $Y \subseteq R$; put $V = V(R)$. Because Y is compact and g is continuous it follows that g is uniformly continuous. Hence there is a $\delta > 0$ such that $|g(z) - g(y)| < \epsilon/4V$ when $y, z \in Y$ with $\|z - y\| < \delta$. Let $\mathcal{T} = \{T_1, \ldots, T_m\}$ be a R-cover of Y such that each $T_j \subseteq R$ and diam $T_j < \delta$. For $1 \le j \le m$ choose a point y_j in T_j and put

$$R_j = T_j \times [g(y_j) - \epsilon/4V, \; g(y_j) + \epsilon/4V]$$

Observe that $W \subseteq \bigcup_{j=1}^{m} R_j$. In fact, if $(y, g(y)) \in W$, then $y \in Y$ and there is a j such that $y \in T_j$. Thus $\|y - y_j\| < \delta$ and so $|g(y) - g(y_j)| < \epsilon/4V$. Hence $(y, g(y)) \in R_j$. Also $V(R_j) = V(T_j)[\epsilon/2V]$. Thus $\sum_{j=1}^{m} V(R_j) = V[\epsilon/2V] < \epsilon$. Since ϵ was arbitrary this proves that W has volume zero. ∎

Let's point out that if $X \subseteq \mathbb{R}^p$ and we can cover X by a finite number of sets having the form of W in the preceding proposition, then X has volume zero. (See Example 7.2.9(c) and Exercise 4.)

7.2.12. Corollary. *If $a \in \mathbb{R}^p$ and $\delta > 0$, then $X = \text{cl } B(a; r)$ is a Jordan set.*

Proof. Let $a = (a_1, \ldots, a_p)$ and $Y = \{y \in \mathbb{R}^{p-1} : \|y\| \le r\}$. Define $g : Y \to \mathbb{R}$ by $g(y) = \sqrt{r^2 - \|y\|^2}$. By the proposition $W = \{(y, g(y)) : y \in Y\} = \{x \in \mathbb{R}^p : \|x\|^2 \le r^2 \text{ and } x_p \ge 0\}$ has volume zero. It follows that $a + W$ and $a - W$ have volume zero (Why?). Since $\partial X = (a + W) \cup (a - W)$, ∂X has volume zero. ∎

We might mention that the continuous image of a set of volume zero need not have volume zero. Indeed look at the preceding proof and examine the continuous function $(y, g(y)) \mapsto y$ from W onto Y. However when the continuous function has its image in the same Euclidean space the story is sometimes different. See Proposition 7.4.2 below.

The next result is crucial for our progress. It also illustrates how assuming X is a Jordan set can be used to overcome some natural difficulties.

7.2.13. Theorem. *If X is a Jordan set in $\mathbb{R}^p$ and $f : X \to \mathbb{R}$ is uniformly continuous, then f is integrable.*

Proof. As usual X is bounded and R is a rectangle such that $X \subseteq R$; put $V = V(R)$. Let $M > 0$ such that $|f(x)| \le M$ for all x in X. If $\epsilon > 0$, the uniform continuity of f implies there is a $\delta > 0$ such that $|f(x) - f(y)| < \epsilon/8V$ for all x, y in X with $\|x - y\| < \delta$.

Let $\mathcal{S}$ be a R-cover of X such that each rectangle that belongs to $\mathcal{R}$ has diameter smaller than δ. (Do this first when $p = 2$ and then examine the technique in Example 7.2(e).) Consider $\{S \in \mathcal{S} : S \cap \partial X \neq \emptyset\}$. This forms a R-cover of ∂X. Since X is a Jordan set we can find a refinement $\mathcal{B} = \{B_1, \ldots, B_k\}$ of this R-cover of ∂X with $\sum_{i=1}^{k} V(B_i) < \epsilon/4M$. Let $\mathcal{T} = \{R_1, \ldots, R_m\}$ be a collection of non-overlapping rectangles such that $\mathcal{R} = \mathcal{B} \cup \mathcal{T}$ is a R-cover of X that refines $\mathcal{S}$ (Exercise 12). Note that each rectangle in $\mathcal{R}$ has diameter smaller than δ.

For $1 \leq i \leq k$ and $1 \leq j \leq m$ let $M_i^{\mathcal{B}}, m_i^{\mathcal{B}}, M_j^{\mathcal{T}}, m_j^{\mathcal{T}}$ be defined by $M_i^{\mathcal{B}} = \sup\{f(x) \in B_i \cap X\}$, etc. Because of the choice of δ and the fact that each rectangle in $\mathcal{T}$ has diameter smaller than δ, we get that $M_j^{\mathcal{T}} - m_j^{\mathcal{T}} < \epsilon/8V$ for $1 \leq j \leq m$. Hence

$$U(f, \mathcal{R}) - L(f, \mathcal{R}) = [U(f, \mathcal{B}) - L(f, \mathcal{B})] + [U(f, \mathcal{T}) - L(f, \mathcal{T})]$$

$$= \sum_{i=1}^{k} \left[M_i^{\mathcal{B}} - m_i^{\mathcal{B}} \right] V(B_i) + \sum_{j=1}^{m} \left[M_j^{\mathcal{T}} - m_j^{\mathcal{T}} \right] V(R_j)$$

$$< 2M \sum_{i=1}^{k} V(B_i) + \frac{\epsilon}{8V} \sum_{j=1}^{m} V(R_j)$$

$$< 2M \frac{\epsilon}{4M} + \frac{\epsilon}{8V} V$$

$$< \frac{5\epsilon}{8} < \epsilon$$

By Proposition 7.2.4, f is integrable. ∎

Also see Exercise 11.

7.2.14. Corollary. *If X is a compact Jordan set and $f : X \to \mathbb{R}$ is a continuous function, then f is integrable.*

The next result and its corollaries will be used later in this chapter.

7.2.15. Proposition. *If U and X are Jordan sets with U open, X compact, and $U \subseteq X$, then $X \backslash U$ is a Jordan set. If $f : X \to \mathbb{R}$ is a continuous function, then*

$$\int_X f = \int_{X \backslash U} f + \int_U f$$

Proof. Note that $\partial[X \backslash U] = \partial X \cup \partial U$ because $U \subseteq \operatorname{int} X$. So $\partial[X \backslash U]$ is the union of two sets having volume zero and thus has volume zero (Exercise 4). Hence $X \backslash U$ is a Jordan set.

Now let $f : X \to \mathbb{R}$ be a continuous function. Since every continuous function can be written as the difference of two continuous non-negative functions, without loss of generality we may assume that $f \geq 0$. Let $f(y) \leq M$ for all x in

X, $V = V(X)$, and let $\epsilon > 0$. Since X is compact, f is uniformly continuous and there is a $\delta > 0$ such that $|f(x) - f(y)| < \epsilon/V$ when $\|x - y\| < \delta$. Begin by making a R-cover $\mathcal{Z}_1$ of X such that the diameter of every rectangle in in $\mathcal{Z}_1$ is less than $\min\{\delta, \text{dist}\,(\partial X, \partial U)\}$. Because of this restriction no rectangle in $\mathcal{Z}_1$ can meet both ∂X and ∂U. Let $\mathcal{B} = \{B_1, \ldots, B_b\}$ be a R-cover of ∂X that refines $\{Z \in \mathcal{Z}_1 : Z \cap \partial X \neq \emptyset\}$ and such that $\sum_{i=1}^b V(B_i) < \epsilon/2M$. Let $\mathcal{C} = \{C_1, \ldots, C_c\}$ be a R-cover of ∂U that refines $\{Z \in \mathcal{Z}_1 : Z \cap \partial U \neq \emptyset\}$ and such that $\sum_{i=1}^c V(C_i) < \epsilon/2M$. Now let $\mathcal{Z}$ be a R-cover of X that refines $\mathcal{Z}_1$, $\mathcal{B}$, and $\mathcal{C}$. That is, every rectangle in $\mathcal{Z}_1 \cup \mathcal{B} \cup \mathcal{C}$ is the union of rectangles from $\mathcal{Z}$. Note that $\mathcal{Z}$ is a R-cover of X and we still have that each rectangle in $\mathcal{Z}$ has diameter less than $\min\{\delta, \text{dist}\,(\partial X, \partial U)\}$. Let $\mathcal{E} = \{E_1, \ldots E_e\}$ be the rectangles in $\mathcal{Z}$ that meet ∂X and let $\mathcal{R} = \{R_1, \ldots, R_r\}$ those rectangles in $\mathcal{Z}$ that are contained in $\text{int}\,X$. It follows that $\mathcal{Z} = \mathcal{E} \cup \mathcal{R}$.

Now let $\mathcal{Z}_U = \{Z \in \mathcal{Z} : Z \cap U \neq \emptyset\}$, which is a R-cover of U. Since the closed set $\bigcup\{Z \in \mathcal{Z}_U\}$ contains U, it contains $\text{cl}\,U$. Let $\mathcal{D} = \{D_1, \ldots, D_d\}$ be those rectangles in this R-cover that meet ∂U and let $\mathcal{S} = \{S_1, \ldots, S_s\}$ be those contained in U. Hence $\mathcal{Z}_U = \mathcal{S} \cup \mathcal{D}$.

Let $\mathcal{Z}_{X \setminus U} = \mathcal{D} \cup \mathcal{E} \cup \{Z \in \mathcal{Z} : Z \subseteq \text{int}\,(X \setminus U)\}$. Since $\mathcal{D} = \{D_1, \ldots, D_d\}$ are the rectangles in $\mathcal{Z}$ that meet ∂U and $\mathcal{E} = \{E_1, \ldots, E_e\}$ those that meet ∂X, this makes $\mathcal{Z}_{X \setminus U}$ a R-cover of $X \setminus U$. Put $\mathcal{T} = \{Z \in \mathcal{Z} : Z \subseteq \text{int}\,(X \setminus U)\} = \{T_1, \ldots, T_t\}$. Hence $\mathcal{Z}_{X \setminus U} = \mathcal{D} \cup \mathcal{E} \cup \mathcal{T}$.

Now $\mathcal{Z} = \mathcal{E} \cup \mathcal{R} = \mathcal{E} \cup \mathcal{D} \cup \mathcal{S} \cup \mathcal{T}$ and $\mathcal{E}, \mathcal{D}, \mathcal{S}$, and $\mathcal{T}$ are pairwise disjoint collections. Similarly $\mathcal{Z}_U = \mathcal{D} \cup \mathcal{S}$ and $\mathcal{Z}_{X \setminus U} = \mathcal{D} \cup \mathcal{E} \cup \mathcal{T}$. Therefore

$$U(f, \mathcal{Z}) = U(f, \mathcal{Z}_U) + U(f, \mathcal{Z}_{X \setminus U}) - \sum_{i=1}^d M_i^{\mathcal{D}} V(D_i)$$

$$\geq \int_U f + \int_{X \setminus U} f - \sum_{i=1}^d M_i^{\mathcal{D}} V(D_i)$$

$$\geq L(f, \mathcal{Z}_U) + L(f, \mathcal{Z}_{X \setminus U}) - \sum_{i=1}^d [M_i^{\mathcal{D}} + m_j^{\mathcal{D}}] V(D_i)$$

$$= L(f, \mathcal{Z}) - \sum_{i=1}^d [M_i^{\mathcal{D}} + 2m_j^{\mathcal{D}}] V(D_i)$$

Now

$$U(f, \mathcal{Z}) - L(f, \mathcal{Z}) \leq 2M \sum_{i=1}^e V(E_i) + \sum_{j=1}^r [M_j^{\mathcal{R}} - m_j^{\mathcal{R}}] V(R_j)$$

$$< 2M \frac{\epsilon}{2M} + \frac{\epsilon}{V} \sum_{j=1}^r V(R_j)$$

$$< 2\epsilon$$

Also

$$\sum_{i=1}^{d}[M_i^{\mathcal{P}} + m_j^{\mathcal{P}}]V(D_i) < 3M(\epsilon/2M) = \frac{3\epsilon}{2}$$

So $\int_X f$ belongs to the interval $[L(f, \mathcal{Z}), U(f, \mathcal{Z})]$ and $\int_U f + \int_{X\setminus U} f$ belongs to the interval $[L(f, \mathcal{Z}) - 3\epsilon/2, U(f, \mathcal{Z})]$. So the distance between $\int_X f$ and $\int_U f + \int_{X\setminus U} f$ is at most $2\epsilon + 3\epsilon/2$. Since ϵ was arbitrary, this completes the proof. ∎

Also see Exercise 13.

7.2.16. Corollary. *If X is a compact Jordan set and $f : X \to \mathbb{R}$ is a continuous function, then*

$$\int_X f = \int_{\text{int } X} f$$

Proof. Let $U = \text{int } X$ in the preceding proposition and use the fact that because $X\setminus U = \partial X$ has volume zero, $\int_{X\setminus U} f = 0$. ∎

7.2.17. Corollary. *If $U_1, \ldots, U_n$ are pairwise disjoint open Jordan sets, $U = \bigcup_{j=1}^{n} U_j$, and $f : U \to \mathbb{R}$ is uniformly continuous, then U is a Jordan set and*

$$\int_U f = \sum_{j=1}^{n} \int_{U_j} f$$

Proof. It follows that $\partial U \subseteq \bigcup_{j=1}^{n} \partial U_j$ and so U is a Jordan set. If $n = 2$, put $X = \text{cl } U$. Then Proposition 7.2.15 and the preceding corollary imply

$$\int_U f = \int_X f = \int_{U_1} f + \int_{X\setminus U_1} f = \int_{U_1} f + \int_{U_2} f$$

Now use induction. ∎

7.2.18. Corollary. *If $X_1, \ldots, X_n$ are compact Jordan sets such that for $1 \le i < j \le n$, $X_i \cap X_j = \partial X_i \cap \partial X_j$, $X = \bigcup_{j=1}^{n} X_j$, and if $f : X \to \mathbb{R}$ is a continuous function, then X is a compact Jordan set and*

$$\int_X f = \sum_{j=1}^{n} \int_{X_j} f$$

Proof. Since X is the union of a finite number of compact sets, it is compact. Also the hypothesis implies that $\partial X = \bigcup_{j=1}^{n} \partial X_j$; hence X is a Jordan set. Note that the open sets $\{\text{int } X_j : 1 \le j \le n\}$ are pairwise disjoint. Using the preceding results we have that

$$\int_X f = \int_{\text{int } X} f = \sum_{j=1}^{n} \int_{\text{int } X_j} f = \sum_{j=1}^{n} \int_{X_j} f$$

∎

Integration in Higher Dimensions

We conclude this section with a definition.

7.2.19. Definition. If $X \subseteq \mathbb{R}^p$ and the constant function 1 is integrable on X, then define the *volume* of X as

$$V(X) = \int_X 1$$

We note that this means that in light of Theorem 7.2.13 we can define the volume of every Jordan set. Also since each rectangle is a Jordan set (7.2.10) we have just given a second definition of the volume of a rectangle. However in the next section on iterated integrals we'll show that the two definitions are the same (Theorem 7.3.9). The proof of the next result is Exercise 17.

7.2.20. Proposition. (a) *If $U_1, \ldots, U_n$ are pairwise disjoint open Jordan sets and $U = \bigcup_{j=1}^n U_j$, then $V(U) = \sum_{j=1}^n V(U_j)$.*
(b) *If $X_1, \ldots, X_n$ are compact Jordan sets such that for $1 \leq i < j \leq n$, $X_i \cap X_j = \partial X_i \cap \partial X_j$ and $X = \bigcup_{j=1}^n X_j$, then $V(X) = \sum_{j=1}^n V(X_j)$.*

Exercises

(1) If R and S are two rectangles and $S \subseteq R$, show that $V(S) \leq V(R)$.

(2) Prove Proposition 7.2.2 using Proposition 3.1.2 as a guide.

(3) Show that there are open disks $\{\Delta_j\}$ in $\mathbb{R}^2$ of radius r_j having the following properties: (i) cl $\Delta_j \subseteq \mathbb{D}$ and cl $\Delta_j \cap$ cl $\Delta_i = \emptyset$ for $i \neq j$; (ii) $\sum_j r_j < \infty$; (iii) $K = $ cl $\mathbb{D} \backslash \bigcup_j \Delta_j$ has no interior. The set K is called a *Swiss cheese*[2]. Does K have zero area?

(4) Prove that the union of a finite number of sets having volume zero also has volume zero.

(5) If $X_1, \ldots, X_n$ are bounded sets and for $1 \leq j \leq n$ we are given a R-cover $\mathcal{R}_j = \{R_j^1, \ldots, R_j^{m_j}\}$ of X_j, show that there is a R-cover $\mathcal{R}$ of $X = \bigcup_{j=1}^n X_j$ such that if $\mathcal{R} = \{R_1, \ldots, R_q\}$, then $\bigcup_{k=1}^q R_k = \bigcup_{j=1}^n \bigcup_{i=1}^{m_j} R_j^i$. Apply this to verify Example 7.2.9(c).

(6) Can you use Proposition 7.2.11 to give an induction argument that shows that a rectangle is a Jordan set?

(7) If X is a bounded subset of $\mathbb{R}^{p-1}$ and we consider X as a subset of $\mathbb{R}^p$, show that X has volume zero as a subset of $\mathbb{R}^p$.

---

[2] The Swiss cheese was first discussed by the Swiss mathematician, Alice Roth. She was born in 1905 in Bern, where she spent her entire life. The history of mathematics has few women as contributors. Indeed it was almost impossible for a woman to enter higher education until well into the twentieth century. Today this situation has changed significantly and outstanding women mathematicians are numerous. Alice Roth made contributions to rational approximation of analytic functions of a complex variable. She died in Bern in 1977.

(8) Let $Y \subseteq X \subseteq \mathbb{R}^p$, $g : Y \to \mathbb{R}$, $f : X \to \mathbb{R}$. If g is integrable on Y, $f(y) = g(y)$ for all y in Y, and $X \backslash Y$ has volume zero, show that f is integrable on X and $\int_Y g = \int_X f$.

(9) (a) If X and Y are Jordan sets in $\mathbb{R}^p$ and $\mathbb{R}^q$, respectively, then $X \times Y$ is a Jordan set in $\mathbb{R}^{p+q}$. (b) Use part (a) to give another proof of Proposition 7.2.10.

(10) Can you use Exercise 3 to manufacture a compact set that has no interior and does not have volume zero?

(11) If X is a Jordan set and $f : X \to \mathbb{R}$ is a uniformly continuous function, show that for every $\epsilon > 0$ there is a R-cover $\mathcal{R} = \{B_1, \ldots, B_k, R_1, \ldots, R_m\}$ of X such that: (a) $B_1, \ldots, B_k$ are the rectangles in $\mathcal{R}$ that meet ∂X and $\sum_{i=1}^k V(B_i) < \epsilon$; (b) $R_1, \ldots, R_m$ are the rectangles contained in $\mathrm{int}\, X$ and if $x_j \in R_j$ for $1 \le j \le m$, then $\left| \int_X f - \sum_{j=1}^m f(x_j) V(R_j) \right| < \epsilon$. (Hint: See the proof of Theorem 7.2.13.)

(12) Show that the collection of non-overlapping rectangles $\mathcal{T}$ in the proof of Theorem 7.2.13 exists.

(13) Let X_1 and X_2 be disjoint Jordan sets and assume that $f : X_1 \cup X_2 \to \mathbb{R}$ is uniformly continuous. Is it true that $X_1 \cup X_2$ is a Jordan set and $\int_{X_1 \cup X_2} f = \int_{X_1} f + \int_{X_2} f$?

(14) If V is an open set in $\mathbb{R}^p$, $f : V \to \mathbb{R}$ is a bounded continuous function, and $\int_X f = 0$ for every compact Jordan set X contained in V, show that $f(x) = 0$ for every x in V.

(15) Assume X is a Jordan set and for each $n \ge 1$, f_n is an integrable function on X. If $\{f_n\}$ converges uniformly on X to a function f, show that f is integrable on X and $\int_X f_n \to \int_X f$.

(16) Let X be a Jordan set and assume f is an integrable function on X. If there is a compact subset K of $\mathbb{R}$ with $f(X) \subseteq K$ and $g : K \to \mathbb{R}$ is continuous, show that $g \circ f$ is integrable on X.

(17) Prove Proposition 7.2.20.

7.3. Iterated Integration

Here we want to examine certain compact Jordan subsets of $\mathbb{R}^p$ and show that we can calculate the integral of a continuous function as we did in calculus by iterating the integrals; that is by putting together integrals over subsets contained in subspaces of smaller dimension. To get started, however, we establish a result that doesn't involve integration.

7.3.1. Proposition. *If X and Y are compact subsets of $\mathbb{R}^p$ and $\mathbb{R}^q$, respectively, Z is a compact subset of $X \times Y$, and $\mathcal{A}$ is the linear span in $C(Z)$ of the set of restrictions*

$$\{g(x)h(y)|Z : g \in C(X), h \in C(Y)\}$$

then $\mathcal{A}$ is dense in $C(Z)$.

Proof. This proof is an easy consequence of the Stone–Weierstrass Theorem (5.7.11). Observe that $\mathcal{A}$ is a subalgebra of $C(Z)$ that contains the constant functions and separates points in Z (Verify!). Hence the same holds for its closure, so the Stone–Weierstrass Theorem implies $\mathrm{cl}\,\mathcal{A} = C(Z)$. ∎

7.3.2. Corollary. *If X and Y are compact subsets of $\mathbb{R}^p$ and $\mathbb{R}^q$, respectively, and $\mathcal{A}$ is the linear span in $C(X \times Y)$ of*

$$\{g(x)h(y) : g \in C(X), h \in C(Y)\}$$

then $\mathcal{A}$ is dense in $C(X \times Y)$.

7.3.3. Corollary. *Let X be a compact subset of $\mathbb{R}^p$ contained in the rectangle $[a, b]$. If $\mathcal{A}$ is the linear span of*

$$\{f_1(x_1) \cdots f_p(x_p) : f_j \in C[a_j, b_j] \text{ for } 1 \le j \le p\}$$

then $\mathcal{A}$ is dense in $C(X)$.

Proof. This follows from Corollary 7.3.2 by using induction (See Exercise 2) or you can give a direct argument using the Stone–Weierstrass Theorem. ∎

We might also mention that in the definition of $\mathcal{A}$ in Proposition 7.3.1 we could replace $C(X)$ and $C(Y)$ by dense subspaces of these algebras. Consequently in the statement of the last corollary we could assume that each f_j is a polynomial. Now to return to integration.

7.3.4. Proposition. *If X and Y are compact subsets of $\mathbb{R}^p$ and $\mathbb{R}^q$, respectively, Y is a Jordan set, and $f : X \times Y \to \mathbb{R}$ is a continuous function, then the function $F : X \to \mathbb{R}$ defined by*

$$F(x) = \int_Y f(x, y)\, dy$$

is continuous.

Proof. Because Y is a Jordan set it follows from Theorem 7.2.13 that the integral defining the function $F : X \to \mathbb{R}$ makes sense. Let R be a rectangle such that $X \times Y \subseteq R$ and put $V = V(R)$. Since f must be uniformly continuous, for any $\epsilon > 0$ there is a $\delta > 0$ such that $|f(x_1, y_1) - f(x_2, y_2)| < \epsilon/V$ when $\|(x_1, y_1) - (x_2, y_2)\| < \delta$. Thus when $\|x_1 - x_2\| < \delta$

$$|F(x_1) - F(x_2)| \le \int_Y |f(x_1, y) - f(x_2, y)|\, dy \le \frac{\epsilon}{V} V = \epsilon$$

and the proposition follows. ∎

The main reason for proving this last proposition is that, as a consequence, F is a bounded uniformly continuous function on X and hence integrable if we also assume that X is a Jordan set. This is needed for the conclusion in the next theorem to make sense. To do this important theorem the reader should complete Exercise 7.2.9.

7.3.5. Theorem (Fubini's[3] Theorem). *If X and Y are compact Jordan sets of $\mathbb{R}^p$ and $\mathbb{R}^q$, respectively, and $f : X \times Y \to \mathbb{R}$ is a continuous function, then*

$$\int_{X \times Y} f = \int_X \left[\int_Y f(x, y)\, dy \right] dx$$

Proof. Let R and S be rectangles in $\mathbb{R}^p$ and $\mathbb{R}^q$ such that $X \subseteq R$ and $Y \subseteq S$. We first consider the case that $f(x, y) = g(x)h(y)$ where $g \in C(R)$ and $h \in C(S)$. Here we want to show that

7.3.6
$$\int_{X \times Y} f = \left[\int_X g \right] \left[\int_Y h \right]$$

Observe that a R-cover of $X \times Y$ can be written as $\mathcal{R} \times \mathcal{S} = \{R \times S : R \in \mathcal{R}, S \in \mathcal{S}\}$, where $\mathcal{R}$ and $\mathcal{S}$ are R-covers of X and Y. (Verify!) Since f is integrable, Proposition 7.2.4 implies there is a R-cover $\mathcal{R} \times \mathcal{S}$ of $X \times Y$ such that $U(f, \mathcal{R} \times \mathcal{S}) - L(f, \mathcal{R} \times \mathcal{S}) < \epsilon$. Moreover $\int_{X \times Y} f$ is the unique number such that whenever $\mathcal{Q}$ and $\mathcal{T}$ are refinements of $\mathcal{R}$ and $\mathcal{S}$, respectively, then $L(f, \mathcal{Q} \times \mathcal{T}) \le \int_{X \times Y} f \le U(f, \mathcal{Q} \times \mathcal{T})$. Thus to prove (7.3.6) we need only show that for all such $\mathcal{Q}$ and $\mathcal{T}$

7.3.7
$$L(f, \mathcal{Q} \times \mathcal{T}) \le \left[\int_X g \right] \left[\int_Y h \right] \le U(f, \mathcal{Q} \times \mathcal{T})$$

Let's begin. (This argument is notationally cumbersome but conceptually straightforward, so be patient.) Let $\mathcal{Q} = \{Q_1, \ldots, Q_m\}$ and $\mathcal{T} = \{T_1, \ldots, T_t\}$ and define the numbers $M_j^{\mathcal{Q}}(g), m_j^{\mathcal{Q}}(g), M_i^{\mathcal{T}}(h), m_i^{\mathcal{T}}(h)$ in the usual way. Note that $\sup\{f(x, y) : (x, y) \in Q_j \times T_i \cap X \times Y\} = \sup\{g(x)h(y) : x \in Q_j \cap X, y \in T_i \cap Y\} = M_j^{\mathcal{Q}}(g)M_i^{\mathcal{T}}(h)$. Similarly $\inf\{f(x, y) : (x, y) \in Q_j \times T_i \cap X \times Y\} = m_j^{\mathcal{Q}}(g), m_i^{\mathcal{T}}(h)$. Therefore since both g and h are integrable we have that

$$L(f, \mathcal{Q} \times \mathcal{T}) = L(g, \mathcal{Q})L(h, \mathcal{T})$$
$$\le \left[\int_X g \right] \left[\int_Y h \right]$$
$$\le U(g, \mathcal{Q})U(h, \mathcal{T})$$
$$= U(f, \mathcal{Q} \times \mathcal{T})$$

proving (7.3.7) and hence (7.3.6).

Now that we have the theorem when $f(x, y) = g(x)h(y)$ as in (7.3.6), it is a trivial matter to extend the theorem to the linear span $\mathcal{A}$ of all the functions of the form

---

[3] Guido Fubini was born in Venice in 1879. He received his doctorate in 1900 from the University of Pisa, writing a thesis in differential geometry. He took up a position at the University of Catania in Sicily. Shortly after that he was offered a professorship at the University of Genoa and then in 1908 he went to Turin. He began to change the course of his research, gravitating to analysis. During the 1930s he saw the political situation in Italy deteriorating with the rise of Mussolini and anti-semitism. He decided the future was looking bleak for his family, and in 1939, in spite of his deteriorating health, the Fubini family emigrated to New York. His contributions were in several different areas including integral equations, group theory, geometry, continuous groups, and applied mathematics. He was influential in the development of mathematics in Italy. He died in 1943 in New York.

$g(x)h(y)$ where $g \in C(R)$, $h \in C(S)$. If f is any function in $C(X \times Y)$, then Proposition 7.3.1 implies there is a sequence $\{f_n\}$ in $\mathcal{A}$ that converges uniformly on $X \times Y$ to f. For each $n \geq 1$, set $F_n(x) = \int_Y f_n(x, y)\,dy$. So each F_n is in $C(X)$ (7.3.4).

Claim. $F_n(x) \to F(x) = \int_Y f(x, y)\,dy$ uniformly on X.

In fact, if $\epsilon > 0$ there is an N such that $|f_n(x, y) - f(x, y)| < \epsilon/V(S)$ for all $n \geq N$ and all (x, y) in $X \times Y$. Hence for all $n \geq N$ and all x in X we have that $|F_n(x) - F(x)| \leq \int_Y |f_n(x, y) - f(x, y)|\,dy < \epsilon$. This establishes the claim.

By Exercise 7.2.15, $\int_{X \times Y} f_n \to \int_{X \times Y} f$ and $\int_X F_n \to \int_X F$. Therefore

$$\int_{X \times Y} f = \lim_n \int_{X \times Y} f_n$$
$$= \lim_n \int_X \left[\int_Y f_n(x, y)\,dy \right] dx$$
$$= \int_X \left[\int_Y f(x, y)\,dy \right] dx$$

completing the proof. ∎

Reversing the roles of X and Y in Fubini's Theorem we get the next corollary, which is also referred to as Fubini's Theorem.

7.3.8. Corollary. *If X and Y are compact Jordan sets of $\mathbb{R}^p$ and $\mathbb{R}^q$ and $f : X \times Y \to \mathbb{R}$ is a continuous function, then*

$$\int_{X \times Y} f = \int_Y \left[\int_X f(x, y)\,dx \right] dy$$

It has to be mentioned that Fubini's Theorem holds in much greater generality than is stated here. The more general result involves the dramatic extension of integration from what is presented in this book. See [3].

By using induction, Fubini's Theorem can yield other results such as the following one.

7.3.9. Theorem. *If R is the rectangle $\{x \in \mathbb{R}^p : a_j \leq x_j \leq b_j, 1 \leq j \leq p\}$ and $f \in C(X)$, then for any permutation $\{i_1, \ldots, i_p\}$ of $\{1, \ldots, p\}$*

$$\int_R f = \int_{a_{i_1}}^{b_{i_1}} \left[\int_{a_{i_2}}^{b_{i_2}} \cdots \left[\int_{a_{i_p}}^{b_{i_p}} f(x_1, \ldots, x_p) dx_{i_p} \right] \cdots dx_{i_2} \right] dx_{i_1}$$

The preceding results say something about the concept of the volume of a Jordan set (7.2.19). First Fubini's Theorem says that if X is a Jordan set in $\mathbb{R}^p$ and Y is a Jordan set in $\mathbb{R}^q$, then $V(X \times Y) = V(X)V(Y)$. In particular the last theorem says the two definitions of the volume of a rectangle are the same. That is, using the notation of Theorem 7.3.9, $\int_R 1 = (b_1 - a_1) \cdots (b_p - a_p)$.

Now we consider a different type of iterated integration, but first we need a preliminary result.

7.3.10. Proposition. *If X is a compact Jordan set in $\mathbb{R}^{p-1}$ and $\phi, \psi : X \to \mathbb{R}$ are continuous functions such that $\phi(x) \le \psi(x)$ for all x in X, then*

$$Z = \{(x, t) \in \mathbb{R}^{p-1} \times \mathbb{R} : x \in X, t \in \mathbb{R} \text{ and } \phi(x) \le t \le \psi(x)\}$$

is a compact Jordan set in $\mathbb{R}^p$.

Proof. We begin by defining the sets

$$
\begin{aligned}
A &= \{(x, t) : x \in \partial X \text{ and } \phi(x) \le t \le \psi(x)\} \\
B &= \{(x, \phi(x)) : x \in X\} \\
C &= \{(x, \psi(x)) : x \in X\}
\end{aligned}
$$

Claim. $\partial Z = A \cup B \cup C$

(This argument is a bit cumbersome so you might want to follow along by assuming $p = 2$ and drawing a picture.) We start by showing the containment $A \cup B \cup C \subseteq \partial Z$. If $(x, t) \in A$ and $r > 0$, consider the ball $B((x, t); r) \subseteq \mathbb{R}^p$. Since $x \in \partial X$ there is a point c in $\mathbb{R}^{p-1}$ such that $c \in [\mathbb{R}^{p-1} \backslash X] \cap B(x; r)$. It follows that $(c, t) \in [\mathbb{R}^p \backslash Z] \cap B((x, t); r)$. Thus $(x, t) \in \partial Z$ and we have that $A \subseteq \partial Z$. Now assume that $(x, \phi(x)) \in B$ and $r > 0$. If $\phi(x) - r < t < \phi(x)$, then $(x, t) \in (\mathbb{R}^p \backslash Z) \cap B((x, \phi(x)); r)$; hence $B \subseteq \partial Z$. Similarly $C \subseteq \partial Z$.

Now assume $(x, t) \in \partial Z$ and let's show that $(x, t) \in A \cup B \cup C$. Since ∂Z is a part of Z, we have that $x \in X$ and $\phi(x) \le t \le \psi(x)$. Assume that $x \in \operatorname{int} X$ and that $\phi(x) < t < \psi(t)$. Let $r > 0$ and $\delta > 0$ such that $B(x; r) \subseteq \operatorname{int} X$ and $\phi(x) < t - \delta < t + \delta < \psi(x)$. It follows that $U = B(x; r) \times (t - \delta, t + \delta)$ is an open set contained in Z, contradicting the fact that $(x, t) \in \partial Z$. Thus either $x \in \partial X$ or $t = \phi(x)$ or $t = \psi(x)$. In the first case, $(x, t) \in A$; in the second, $(x, t) \in B$; in the third, $(x, t) \in C$, thus establishing the claim.

According to Proposition 7.2.11, B and C have volume zero. If we can show that A has volume zero, it follows from Exercise 7.2.4 that Z is a Jordan set. To show that A has volume zero first observe that since ϕ and ψ are continuous functions on a compact set there are real numbers a and b such that $a \le \phi(x) \le \psi(x) \le b$ for all x in X. Thus $A \subseteq \partial X \times [a, b]$; we show that $\partial X \times [a, b]$ has volume zero. If $\epsilon > 0$ there is a R-cover $\mathcal{R} = \{R_1, \ldots, R_m\}$ of ∂X such that $\sum_{j=1}^m R_j V(R_j) < \epsilon/(b - a)$. It follows that $\{R_1 \times [a, b], \ldots, R_m \times [a, b]\}$ is a R-cover of $\partial X \times [a, b]$ and the sum of the volumes of these rectangles is less than ϵ. ∎

7.3.11. Theorem. *If X is a compact Jordan set in $\mathbb{R}^{p-1}$ and $\phi, \psi : X \to \mathbb{R}$ are continuous functions such that $\phi(x) \le \psi(x)$ for all x in X, and*

$$Z = \{(x, t) \in \mathbb{R}^{p-1} \times \mathbb{R} : x \in X, t \in \mathbb{R} \text{ and } \phi(x) \le t \le \psi(x)\}$$

then for any continuous function $f : Z \to \mathbb{R}$

$$\int_Z f = \int_X \left[\int_{\phi(x)}^{\psi(x)} f(x, t)\, dt \right] dx$$

Proof. First we show that if we define

$$F : X \to \mathbb{R}$$

as $F(x) = \int_{\phi(x)}^{\psi(x)} f(x, t)\,dt$, then F is continuous and so the integral in the statement of the theorem is legitimate. Let $x_n \to x$ in X and observe that

$$
\begin{aligned}
F(x_n) - F(x) &= \int_{\phi(x_n)}^{\psi(x_n)} f(x, t)\,dt - \int_{\phi(x)}^{\psi(x)} f(x, t)\,dt \\
&= \int_{\phi(x_n)}^{\psi(x)} f(x, t)\,dt + \int_{\psi(x)}^{\psi(x_n)} f(x, t)\,dt - \int_{\phi(x)}^{\psi(x)} f(x, t)\,dt \\
&= \int_{\phi(x_n)}^{\phi(x)} f(x, t)\,dt + \int_{\phi(x)}^{\psi(x)} f(x, t)\,dt \\
&\quad + \int_{\psi(x)}^{\psi(x_n)} f(x, t)\,dt - \int_{\phi(x)}^{\psi(x)} f(x, t)\,dt \\
&= \int_{\phi(x_n)}^{\phi(x)} f(x, t)\,dt + \int_{\psi(x)}^{\psi(x_n)} f(x, t)\,dt
\end{aligned}
$$

Let $|f(x, t)| \leq M$ for all (x, t) in Z and let $\epsilon > 0$. We can choose N such that when $n \geq N$, $|\phi(x_n) - \phi(x)| < \epsilon/2M$ and $|\psi(x_n) - \psi(x)| < \epsilon/2M$. From the preceding equation we get that $|F(x_n) - F(x)| < \epsilon$ when $n \geq N$.

Let $[a, b]$ be an interval such that $a \leq \phi(x) \leq \psi(x) \leq b$ for all x in X, let S be a rectangle in $\mathbb{R}^{p-1}$ that contains X, and let $R = S \times [a, b]$. Define $\tilde{f} : R \to \mathbb{R}$ by setting $\tilde{f}(x, t) = f(x, t)$ when $(x, t) \in Z$ and $\tilde{f}(x, t) = 0$ otherwise.

Claim. $\tilde{f}$ is integrable on R and $\int_Z f = \int_R \tilde{f} = \int_S \left[\int_a^b \tilde{f}(x, t)\,dt \right] dx$

To prove the first part of the claim we let $\epsilon > 0$ and invoke Exercise 7.2.11 to find a R-cover $\mathcal{R} = \{B_1, \ldots, B_k, R_1, \ldots, R_m\}$ of Z such that the following two conditions hold: (a) $B_1, \ldots, B_k$ are the rectangles in $\mathcal{R}$ that meet ∂Z and $\sum_{i=1}^k V(B_i) < \epsilon/2M$; (b) $R_1, \ldots, R_m$ are the rectangles contained in $\operatorname{int} X$ and if $(x_j, t_j) \in R_j$ for $1 \leq j \leq m$, then $\left| \int_Z f - \sum_{j=1}^m f(x_j, t_j) V(R_j) \right| < \epsilon/2$. Let $\mathcal{T} = \mathcal{R} \cup \{T_1, \ldots, T_n\}$, where $T_1, \ldots, T_n$ are rectangles contained in R chosen such that $\mathcal{T}$ is a R-cover of R. It follows that $\tilde{f}$ vanishes on each T_k since it is disjoint from Z. Thus $U(\tilde{f}, \mathcal{T}) - L(\tilde{f}, \mathcal{T}) = U(f, \mathcal{R}) - L(f, \mathcal{R}) < \epsilon$. It follows that $\tilde{f}$ is integrable and $\int_Z f = \int_R \tilde{f}$. Though the function $\tilde{f}$ is not continuous on R we can adapt the proof of Fubini's Theorem to complete the proof of the claim. The proof is left to the reader in Exercise 7.

Using the claim we get

$$\int_Z f = \int_R \tilde{f} = \int_S \left[\int_a^b \tilde{f}(x, t)\,dt \right] dx = \int_X \left[\int_{\phi(x)}^{\psi(x)} f(x, t)\,dt \right] dx \qquad \blacksquare$$

7.3.12. Corollary. *Assume* $[c, d] \subseteq \mathbb{R}$ *and* $\phi, \psi : [c, d] \to \mathbb{R}$ *are continuous functions such that* $\phi(x) \leq \psi(x)$ *for all x in* $[c, d]$. *If*

$$X = \{(x, t) \in \mathbb{R}^2 : c \leq x \leq d, \phi(y) \leq t \leq \psi(y)\}$$

and $f : X \to \mathbb{R}$ *is a continuous function, then*

$$\int_X f = \int_c^d \left[\int_{\phi(x)}^{\psi(x)} f(x, t) \, dt \right] dx$$

Exercises

(1) Show that Proposition 7.3.1 can be strengthened a little by replacing $C(X)$ and $C(Y)$ by dense subalgebras of these two spaces.

(2) Give the details needed to prove Corollary 7.3.3.

(3) If $R = [0, 2] \times [-1, 1]$ and $f(x, y) = xy^2$, find $\int_R f$.

(4) If $R = [a_1, b_1] \times [a_2, b_2] \times [a_3, b_3]$, evaluate $\int_R f$ for the following examples of the function $f(x, y, z)$: (a) $f(x, y, z) = x + y + z$; (b) $f(x, y, z) = xyz$; (c) $f(x, y, z) = x$.

(5) Find $\int_X f$ in each of the following cases: (a) $f(x, y) = x + y$, X is the triangle with vertices $(0, 0)$, $(0, 1)$, $(1, 0)$; (b) $f(x, y) = \sin x^2$, $X = \{(x, y) : 0 \leq y \leq 1, y \leq x \leq 1\}$.

(6) Find $\int_X f$ in each of the following cases: (a) $f(x, y, z) = \sqrt{x^3 + z}$, $X = \{(x, y, z) : x^3 \leq z \leq 1, \sqrt{y} \leq x \leq 1, 0 \leq y \leq 1\}$; (b) $f(x, y, z) = x$, $X = \{(x, y, z) : 0 \leq z \leq 1 - x^2, 0 \leq y \leq x^2 + z^2, x \geq 0\}$.

(7) Establish the claim in the proof of Theorem 7.3.11.

7.4. Change of Variables

Recall the Change of Variables Theorem for integration on intervals $[a, b]$ in $\mathbb{R}$ (Theorem 3.2.5). There we are given a continuously differentiable function $\phi : [a, b] \to \mathbb{R}$ and an interval I such that $\phi([a, b]) \subseteq I$. If $f : I \to \mathbb{R}$ is a continuous function, then

$$\int_{\phi(a)}^{\phi(b)} f(x) \, dx = \int_a^b f(\phi(t))\phi'(t) \, dt$$

In this section we'll see an extension of this result to the integral in $\mathbb{R}^p$. The proof of this theorem is very involved and necessitates some preliminary work and several lemmas. We begin by showing that under suitable restrictions the image of a Jordan set is again a Jordan set, but establishing this also requires a few preliminary results.

Let's set some notation. We are given an open subset G of $\mathbb{R}^p$ and a continuously differentiable function $\Phi : G \to \mathbb{R}^p$. For each of the standard basis elements

e_i in $\mathbb{R}^p$ let $\Phi_i : G \to \mathbb{R}$ be defined by $\Phi_i(x) = \langle \Phi(x), e_i \rangle$. The statement of our first lemma is highly technical, but many lemmas are that way.

7.4.1. Lemma. *Let R be a rectangle in $\mathbb{R}^p$ whose shortest side has length s, whose longest side has length ℓ, and assume these lengths satisfy $\ell \le 2s$. If G is an open set containing R, $\Phi : G \to \mathbb{R}^p$ is a continuously differentiable function, and there is a constant M such that $|\partial_{x_j} \Phi_i(x)| \le M$ for all $1 \le i, j \le p$ and all x in G, then there is a rectangle Q in $\mathbb{R}^p$ such that*

$$\Phi(R) \subseteq Q \text{ and } V(Q) \le (2pM)^p V(R)$$

Proof. Let x_0 be the center point of R. If $x \in R$, consider the straight line from x_0 to x: $\gamma(t) = tx + (1-t)x_0$. By the Mean Value Theorem 6.2.15, for $1 \le i \le p$ there is a point t_i in $[0, 1]$ such that

$$\Phi_i(x) - \Phi_i(x_0) = \langle \nabla \Phi_i(\gamma(t_i)), x - x_0 \rangle$$

By hypothesis for each i, $\|\nabla \Phi_i(\gamma(t_i))\| \le \sum_{j=1}^{p} |\partial_{x_j} \Phi_i(\gamma(t_i))| \le pM$. In addition $\|x - x_0\| \le \ell/2$. Thus we have that $|\Phi_i(x) - \Phi_i(x_0)| \le |\langle \nabla \Phi_i(\gamma(t_i)), x - x_0 \rangle| \le \|\nabla \Phi_i(\gamma(t_i))\| \|x - x_0\| \le pM\ell/2$; so the image of the rectangle R under the scalar-valued function Φ_i is contained in the interval with center $\Phi_i(x_0)$ and length $pM\ell$. That is,

$$\Phi_i(R) \subseteq [\Phi_i(x_0) - pM\ell/2, \Phi_i(x_0) + pM\ell/2]$$

Therefore if we let Q be the rectangle in $\mathbb{R}^p$ with center $\Phi(x_0)$ and with each side of length $pM\ell$, we have $\Phi(R) \subseteq Q$ and $V(Q) = (pM\ell)^p$. Now since $s \ge \ell/2$, $V(R) \ge s^p \ge (\ell/2)^p$. Hence

$$V(Q) = (2pM)^p (\ell/2)^p \le (2pM)^p V(R)$$

as we wanted to show. ∎

Let's point out that in the preceding lemma the existence of the constant M is not a formidable restriction. In fact R is a compact subset of G and so we can always find an open set G_1 such that $R \subseteq G_1 \subseteq \text{cl } G_1 \subseteq G$ and $\text{cl } G_1$ is compact (Exercise 5.5.8). Since Φ is continuously differentiable, the constant M exists for the open set G_1.

7.4.2. Proposition. *If X is a compact subset of $\mathbb{R}^p$ having volume zero, G is an open set containing X, and $\Phi : G \to \mathbb{R}^p$ is a continuously differentiable function, then $\Phi(X)$ has volume zero.*

Proof. From the comments preceding the statement of this proposition, we may assume that there is a constant M such that $|\partial_{x_j} \Phi_i(x)| \le M$ for all $1 \le i, j \le p$ and all x in G as in the hypothesis of Lemma 7.4.1. Let $\epsilon > 0$. Since X has volume zero there is a R-cover $\mathcal{R} = \{R_1, \ldots, R_m\}$ of X such that $\sum_{k=1}^{m} V(R_k) < \epsilon/(2pM)^p$. By Exercise 1 we can write each R_k as the union of non-overlapping rectangles having the property that the length of the longest side of each is at most twice the length of its shortest side. But the volume of each R_k is the sum of the volumes of these smaller rectangles. Hence, without loss of generality, we can assume that each R_k in the R-cover $\mathcal{R}$ has the property that the length of its longest side is at most twice the length of its shortest side. By the preceding lemma, for $1 \leq k \leq m$ there is a rectangle Q_k in $\mathbb{R}^p$ such that $\Phi(R_k) \subseteq Q_k$ and $V(Q_k) \leq (2pM)^p V(R_k)$. Hence

$$\sum_{k=1}^{m} V(Q_k) < (2pM)^p \sum_{k=1}^{m} V(R_k) < \epsilon$$

Since $\Phi(X) \subseteq \bigcup_{k=1}^{m} \Phi(R_k) \subseteq \bigcup_{k=1}^{m} Q_k$, we have that $\Phi(X)$ has volume zero. ∎

We want a result that gives conditions on a function $\Phi : G \to \mathbb{R}^p$ such that when X is a Jordan set contained in G, $\Phi(X)$ is a Jordan set. If we knew that $\partial \Phi(X) \subseteq \Phi(\partial X)$, the preceding proposition yields the result. Corollary 6.8.5, a consequence of the Inverse Function Theorem, gives us what we need.

7.4.3. Proposition. *If X is a compact Jordan set in $\mathbb{R}^p$, G is an open set containing X, and $\Phi : G \to \mathbb{R}^p$ is a continuously differentiable function such that $D\Phi(x)$ is invertible for every x in $\mathrm{int}\, X$, then $\Phi(X)$ is a Jordan set.*

Proof. First observe that $\Phi(X)$ is compact. If $x \in \mathrm{int}\, X$, let U be an open set such that $x \in U \subseteq X$. By Corollary 6.8.5, $\Phi(U)$ is open. Since $\Phi(U) \subseteq \Phi(X)$, it must be that $\Phi(U) \subseteq \mathrm{int}\, \Phi(X)$. Hence $\Phi(\mathrm{int}\, X) \subseteq \mathrm{int}\, \Phi(X)$. If $\zeta \in \partial \Phi(X) \subseteq \Phi(X)$, let $x \in X$ such that $\Phi(x) = \zeta$. By what we just proved, it cannot be that $x \in \mathrm{int}\, X$; thus $x \in \partial X$ and we have that $\partial \Phi(X) \subseteq \Phi(\partial X)$. But $\Phi(X)$ has volume zero by Proposition 7.4.2 and the fact that X is a Jordan set. Therefore $\Phi(X)$ is a Jordan set. ∎

7.4.4. Theorem (Change of Variables Theorem). *Let X be a compact Jordan set in $\mathbb{R}^p$. If G is an open subset of $\mathbb{R}^p$ such that $X \subseteq G$ and $\Phi : G \to \mathbb{R}^p$ is a continuously differentiable function such that Φ is injective on $\mathrm{int}\, X$ with $\det[D\Phi(x)] \neq 0$ for all x in $\mathrm{int}\, X$, then for any continuous function $f : \Phi(X) \to \mathbb{R}$*

7.4.5
$$\int_{\Phi(X)} f = \int_X f \circ \Phi \, |\det[D\Phi]|$$

Note that according to Proposition 7.4.3, $\Phi(X)$ is a Jordan set so that the first integral in (7.4.5) makes sense. We might compare this result to the previously quoted result for $p = 1$ and note the differences. The first thing that stands out is the absolute value around the $\det[D\Phi]$ as opposed to not having one around ϕ' in the case when $p = 1$. This is due to the fact that in one-dimensional integration there is a natural direction involved in the integral and this does not happen when $p > 1$. Since $\phi'(x) \neq 0$ on the interval $[a, b]$ we have either $\phi'(x) > 0$ for all x or $\phi'(x) < 0$ for all x. In the latter case the effect is to use $|\phi'(x)|$ in the equation but switch the limits of integration. So we could write the COV Theorem when $p = 1$ as saying that when $\phi'(x) \neq 0$ on $[a, b]$ then

$$\int_{[\phi(a),\phi(b)]} f = \int_{[a,b]} f \circ \phi \, |\phi'|$$

and so the theorem above is indeed a generalization of the COV Theorem we proved in Chapter 3.

We still need some more lemmas. The next one can be phrased as saying that if the Change of Variables Theorem is true for every rectangle, it is true for an arbitrary compact Jordan set. The proof is rather technical so be careful.

7.4.6. Lemma. *Assume X, G, and Φ are as in the statement of* Theorem 7.4.4. *If for every closed rectangle R contained in* $\operatorname{int} X$ *and every continuous function $f : \Phi(X) \to \mathbb{R}$ we have that*

$$\int_{\Phi(R)} f = \int_R f \circ \Phi \, |\det[D\Phi]|$$

then (7.4.5) is valid.

Proof. We begin the proof by making some simplifying assumptions that do nothing to interfere with the generality of the situation. First, since any continuous function can be written as the difference of two positive continuous functions and (7.4.5) remains valid for the difference of two functions, without loss of generality we can assume that $f \geq 0$ on $\Phi(X)$. Define a new function $g(x) = f(\Phi(x))|\det[D\Phi(x)]|$ for all x in X, which we observe is also positive on X. We can also assume that G is a bounded Jordan set with Φ defined and continuously differentiable in an open set that contains $\operatorname{cl} G$. To see this first realize that since X is compact and contained in an open set, there is a $r > 0$ such that $X_r = \{x \in \mathbb{R}^p : \operatorname{dist}(x, X) \leq r\} \subseteq G$ (Exercise 5.5.7). Let $\mathcal{Z}$ be a R-cover of X_r, where each each rectangle in $\mathcal{Z}$ is contained in G. Replace G by $\operatorname{int}[\bigcup\{Z : Z \in \mathcal{Z}\}]$, a bounded open Jordan set. (We needed the bigger set X_r to be sure that X is contained in the replacement open set.) With this replacement we have that Φ is defined in a neighborhood of $\operatorname{cl} G = \bigcup\{Z : Z \in \mathcal{Z}\}$. Since G is a Jordan set we can define $V = V(G)$.

Because $\Phi(X)$ and X are both compact, we have that both f and g are bounded on these domains. Also since Φ and its derivatives are continuous on $\operatorname{cl} G$ we have that there is a constant M such that

$$f(\zeta) \leq M \text{ for } \zeta \in \Phi(X)$$
$$g(x) \leq M \text{ for } x \in X$$
$$|\partial_{x_j}\Phi_i(x)| \leq M \text{ for } x \in \operatorname{cl} G, 1 \leq i, j \leq p$$

Let $\epsilon > 0$.

Claim. There is a R-cover

$$\mathcal{R} = \{B_1, \ldots, B_b, R_1, \ldots, R_r\}$$

of X, where $\{B_1, \ldots, B_b\}$ meet ∂X, $\{R_1, \ldots, R_r\}$ are contained in $\operatorname{int} X$, and the following hold: (a) the longest side of each rectangle in $\mathcal{R}$ is less than twice the length of its shortest side; (b)

$$\sum_{i=1}^b V(B_i) < \epsilon \min\left\{\frac{1}{M(2pM)^p}, \frac{1}{M}\right\}$$

To verify this claim we use methods we have used before and so the details are left to the interested reader. Suffice it to say that (a) follows from Exercise 1 and (b) is a consequence of the fact that ∂X has volume zero.

Let

$$\Omega = \bigcup_{j=1}^{r} \Phi(R_j) \quad \text{and} \quad \Lambda = \Phi(X)\backslash\Omega$$

Note that $\Phi(R_j)$ is a Jordan set by Proposition 7.4.3. Also the Inverse Function Theorem and its corollary imply that $\partial\Phi(R_j) = \Phi(\partial R_j)$. By the corollaries of Theorem 7.2.13, in particular Corollary 7.2.18, we have that

$$\int_{\Phi(X)} f = \int_{\Omega} f + \int_{\Lambda} f$$

$$= \sum_{j=1}^{r} \int_{\Phi(R_j)} f + \int_{\Lambda} f$$

If we set $K = \bigcup_{j=1}^{r} R_j \subseteq \operatorname{int} X$, the hypothesis implies

$$\sum_{j=1}^{r} \int_{\Phi(R_j)} f = \sum_{j=1}^{r} \int_{R_j} f \circ \Phi |\det[D\Phi]| = \int_{K} f \circ \Phi |\det[D\Phi]|$$

Lemma 7.4.1 implies that for each rectangle B_i there is a rectangle Q_i such that $\Phi(B_i) \subseteq Q_i$ and $V(Q_i) \leq (2pM)^p V(B_j)$. Now

$$\Lambda = \Phi(X)\backslash\Omega \subseteq \bigcup_{i=1}^{b} \Phi(B_i) \subseteq \bigcup_{i=1}^{b} Q_i$$

so

$$\int_{\Lambda} f \leq \sum_{i=1}^{b} \int_{Q_i} f$$

$$\leq \sum_{i=1}^{b} MV(Q_i)$$

$$\leq M \sum_{i=1}^{b} (2pM)^p V(B_j)$$

$$< M(2pM)^p \frac{\epsilon}{M(2pM)^p} = \epsilon$$

Therefore

$$\left| \int_{\Phi(X)} f - \int_{K} f \circ \Phi |\det[D\Phi]| \right| < \epsilon$$

On the other hand $\int_X g = \int_K g + \int_{X \setminus K} g$. But $X \setminus K \subseteq \bigcup_{i=1}^b B_i$ and so

$$\int_{X \setminus K} g \leq \sum_{i=1}^b \int_{B_i} g \leq M \sum_{i=1}^b V(B_i) < M \frac{\epsilon}{M} = \epsilon$$

Thus

$$\left| \int_X f \circ \Phi |\det[D\Phi]| - \int_K f \circ \Phi |\det[D\Phi]| \right| < \epsilon$$

Therefore

$$\left| \int_X f \circ \Phi |\det[D\Phi]| - \int_{\Phi(X)} f \right| < 2\epsilon$$

Since ϵ was arbitrary this completes the proof. ∎

Now we parlay the preceding lemma to show that if we prove the COV Theorem locally, then we have a proof of the general theorem.

7.4.7. Lemma. *Assume X, G, and Φ are as in the statement of Theorem 7.4.4. If for every point x in int X there is an open ball $B(x; r)$ such that $\overline{B}(x; r) \subseteq$ int X with the property that for any Jordan region Y contained in $B(x; r)$ and any continuous function $f : \Phi(Y) \to \mathbb{R}$ we have that*

$$\int_{\Phi(Y)} f = \int_Y f \circ \Phi |\det D\Phi|$$

then (7.4.5) is valid.

Proof. We only sketch the proof as it follows lines similar to previous proofs. The reader is asked in Exercise 3 to supply the details. Let $f : \Phi(X) \to \mathbb{R}$ be a continuous function. To show that (7.4.5) is valid, the previous lemma says we need only show that $\int_{\Phi(R)} f = \int_R f \circ \Phi |\det D\Phi|$ for every rectangle $R \subseteq$ int X. Let $B(x; r)$ be an open ball as in the hypothesis, and let R_x be a rectangle that contains x in its interior and such that $R_x \subseteq B(x; r)$. Since the rectangle R is compact we can find a finite number of points $x_1, \ldots, x_m$ such that R is covered by int $R_{x_1}, \ldots,$ int R_{x_m}. Now find non-overlapping rectangles $R_1, \ldots, R_n$ such that $R = \bigcup_{j=1}^n R_j$ and each R_j is contained in one of the rectangles R_{x_i}. Note that for $1 \leq j, k \leq n$, $0 = V(R_j \cap R_k) = V(\Phi(R_j) \cap \Phi(R_k))$. Hence

$$\int_{\Phi(R)} f = \sum_{j=1}^n \int_{\Phi(R_j)} f$$

$$= \sum_{j=1}^n \int_{R_j} f \circ \Phi |\det D\Phi|$$

$$= \int_R f \circ \Phi |\det D\Phi|$$

By the preceding lemma this completes the proof. ∎

The next phase of the proof consists of showing that the theorem is true when Φ is a special type of function. After this we'll show that these special maps can be combined to get any eligible Φ and then we will complete the proof.

7.4.8. Definition. If G is an open subset of $\mathbb{R}^p$ and $H : G \to \mathbb{R}^p$, then H is called a *simple mapping* if there is an integer k with $1 \le k \le p$ and a function $h : G \to \mathbb{R}$ such that for all x in G

$$H(x) = h(x)e_k + \sum_{j \ne k} x_j e_j = x + [h(x) - x_k]e_k$$

Thus a simple function is one that disturbs only one coordinate. Notice that the continuity or differentiability of H is determined by that of h. If H is differentiable at a point a in G then the matrix of $H'(a) = (a_{ij})$ has the following properties. When $i \ne k$, the i-th row of (a_{ij}) has a one in the i-th spot and zeros elsewhere. That is $a_{ij} = 0$ when $j \ne i$ and $a_{ii} = 1$. The k-th row has entries $\partial_{x_1} h(a), \ldots, \partial_{x_k} h(a), \ldots, \partial_{x_p} h(a)$. So the Jacobian determinant takes on the form

$$\det[H'(a)] = \partial_{x_k} h(a)$$

Hence the simple function H has $H'(a)$ invertible if and only if $\partial_{x_k} h(a) \ne 0$.

7.4.9. Lemma. *The Change of Variables Theorem is true if we assume that Φ is a simple function.*

Proof. By Lemma 7.4.6 it suffices to prove this when X is a rectangle $[a, b]$. To simplify the notation we assume that the simple function Φ has the form $\Phi(x) = \phi(x)e_1 + \sum_{j=2}^p x_j e_j$; that is, in the above definition we take $k = 1$. Let $R = [a_2, b_2] \times \cdots \times [a_p, b_p]$, the rectangle in $\mathbb{R}^{p-1}$ such that $X = [a_1, b_1] \times R$. As we pointed out, $\det[\Phi'(x)] = \partial_{x_1} \phi(x)$ and the hypothesis guarantees $\partial_{x_1} \phi(x) \ne 0$ for all x in X. X being a rectangle and hence connected, it must be that $\partial_{x_1} \phi(x)$ is either always positive or always negative; we will assume that $\partial_{x_1} \phi(x) > 0$ for all x in X. (The reader can carry out the similar proof in the other case.) We have that $\Phi(X) = (\phi([a_1, b_1])) \times R$. Hence by Fubini's Theorem

7.4.10 $$\int_{\Phi(X)} f = \int_{(\phi([a_1,b_1])) \times R} f = \int_R \left[\int_{\phi([a_1,b_1])} f(x_1)\, dx_1 \right] dx_2 \cdots dx_p$$

and

7.4.11
$$\int_X f \circ \Phi |\det[\Phi']| = \int_R \left[\int_{a_1}^{b_1} f(\phi(x), x_2, \ldots, x_p) \partial_{x_1} \phi(x)\, dx_1 \right] dx_2 \cdots dx_p$$

For the moment fix $(x_2, \ldots, x_p)$ in R and define $\psi : [a_1, b_1] \to \mathbb{R}$ by $\psi(t) = \phi(t, x_2, \ldots, x_p)$. It follows that $\psi'(t) = \partial_{x_1} \phi(t, x_2, \ldots, x_p)$. So by the COV

Theorem for one variable (3.2.5) we get that

$$\int_{a_1}^{b_1} f(\phi(x), x_2, \ldots, x_p) \partial_{x_1} \phi(x) \, dx_1 = \int_{a_1}^{b_1} f(\psi(t)) \psi'(t) \, dt$$

$$= \int_{\psi(a_1)}^{\psi(b_1)} f$$

$$= \int_{\phi([a_1, b_1])} f(x_1) \, dx_1$$

If we substitute this into (7.4.10) and (7.4.11) we see that the proof is complete. ∎

7.4.12. Definition. A linear transformation in $\mathcal{L}(\mathbb{R}^p)$ is called a *flip* if it interchanges two elements of the standard basis and leaves the others fixed.

So a T in $\mathcal{L}(\mathbb{R}^p)$ is a flip if there are distinct i and j, $1 \le i, j \le p$, such that $T e_i = e_j$, $T e_j = e_i$, and $T e_k = e_k$ when $k \ne i, j$. Notice that if T is a flip, then $T^2 = 1$. We observe that with $\Phi(x) = Tx$, $\Phi'(x) = T$ and so $|\det[\Phi'(x)]| = |\det T| = 1$.

7.4.13. Lemma. *The Change of Variables Theorem is true if we assume that Φ is a flip T.*

Proof. Again we need only prove this for a rectangle. As in the proof of Lemma 7.4.9 let $X = [a, b]$ and, to simplify the notation, we assume that T flips the first two basis vectors. So $T e_1 = e_2$, $T e_2 = e_1$, and $T e_k = e_k$ for $3 \le k \le p$. Let $R = [a_3, b_3] \times \cdots \times [a_p, b_p]$, the rectangle in $\mathbb{R}^{p-2}$ such that $X = [a_1, b_1] \times [a_2, b_2] \times R$. Thus $T(X) = [a_2, b_2] \times [a_1, b_1] \times R$, and

$$\int_{T(X)} f = \int_R \left[\int_{a_1}^{b_1} \left[\int_{a_2}^{b_2} f(s, t, x_3, \ldots, x_p) \, ds \right] dt \right] dx_3 \cdots dx_p$$

and

$$\int_X f \circ T |\det[T]| = \int_R \left[\int_{a_1}^{b_1} \left[\int_{a_2}^{b_2} f(t, s, x_s, \ldots, x_p) \, dt \right] ds \right] dx_3 \cdots dx_p$$

Fubini's Theorem shows these two integrals are the same. ∎

7.4.14. Lemma. *If G is an open subset of $\mathbb{R}^p$ that contains the origin, $\Phi : G \to \mathbb{R}^p$ is continuously differentiable with $\Phi(0) = 0$ and $D\Phi(0)$ invertible, then there is a neighborhood W of 0 such that for all x in W*

7.4.15 $$\Phi(x) = T_1 \cdots T_{p-1} \circ H_p \circ \cdots H_1(x)$$

where: (i) each T_j is either a flip or the identity linear transformation; (ii) each H_j is a simple mapping satisfying $H_j(0) = 0$ and $H_j'(0)$ is invertible.

Proof. We start by introducing the linear projections $P_0, \ldots, P_p$ on $\mathbb{R}^p$, where for all x in $\mathbb{R}^p$: $P_0(x) = 0$ and $P_k(x) = x_1 e_1 + \cdots + x_k e_k$. So P_k projects $\mathbb{R}^p$ onto the subspace defined by the first k members of the standard basis.

Claim. For $1 \leq m \leq p$ there is a neighborhood of 0, W_m, and a continuously differentiable function $\Phi_m : W_m \to \mathbb{R}^p$ such that $\Phi_m(0) = 0$, $D\Phi_m(0)$ is invertible, and for all x in W_m

$$P_{m-1}\Phi_m(x) = P_{m-1}(x)$$

We show this by induction. Take Φ_1 to be the given function Φ with $W_1 = G$. Since $P_0 = 0$, this works. Now assume $1 \leq m < p$ and that we have the function Φ_m and neighborhood W_m as in the statement of the claim. From the equation $P_{m-1}\Phi_m(x) = P_{m-1}(x)$ it follows that for each x in W_m we have

$$\Phi_m(x) = P_{m-1}(x) + \sum_{j=m}^{p} h_j(x)e_j$$

for some scalar-valued functions $h_m, \ldots, h_p$ defined on W_m. Since Φ_m is continuously differentiable, so are each of the functions h_j. Thus

$$D\Phi_m(0)e_m = \sum_{j=m}^{p} [\partial_{x_m} h_j](0)e_j$$

Because $D\Phi_m(0)$ is invertible, there must be a first integer q, $m \leq q \leq p$, with $[\partial_{x_m} h_q](0) \neq 0$. Fix such a q and let T_m be the flip that interchanges this e_q and e_m. (If $q = m$, then $T_m = 1$.) Define $H_m : W_m \to \mathbb{R}^p$ by

$$H_m(x) = x + [h_q(x) - x_m]e_m$$

It follows that H_m is a simple function on W_m that is continuously differentiable and $DH_m(0)$ is invertible. (Verify!) Since $\Phi_m(0) = 0$ it follows that $h_q(0) = 0$ so that $H_q(0) = 0$. We apply the Inverse Function Theorem to H_m and conclude the following: (a) there is an open set U_m with $0 \in U_m \subseteq W_m$; (b) H_m is injective on U_m and $H_m(U_m) = W_{m+1}$, an open set containing 0; (c) $H_m^{-1} : W_{m+1} \to U_m$ is continuously differentiable.

Define $\Phi_{m+1} : W_{m+1} \to \mathbb{R}^p$ for all y in W_{m+1} by

7.4.16 $$\Phi_{m+1}(y) = T_m \Phi_m \circ H_m^{-1}(y)$$

Clearly Φ_{m+1} is continuously differentiable with $\Phi_{m+1}(0) = 0$. By the Chain Rule we have that $D\Phi_{m+1}(0) = T_m[D\Phi_m(H_m^{-1}(0))]D[H_m^{-1}](0) = T_m[D\Phi_m(0)]D[H_m^{-1}](0)$, so that $D\Phi_{m+1}(0)$ is invertible. Finally, if $x \in U_m$,

$$\begin{aligned} P_m \Phi_{m+1}(H_m(x)) &= P_m T_m \Phi_m(x) \\ &= P_m[P_{m-1}x + h_q(x)e_m + \cdots] \\ &= P_{m-1}x + h_q(x)e_m \\ &= P_m H_m(x) \end{aligned}$$

Hence $P_m \Phi_{m+1}(y) = P_m y$ for all y in W_{m+1}, establishing the claim.

Finishing the proof of the proposition now goes quickly. Revisit (7.4.16) with $y = H_m(x)$ for some x in U_m and use the fact that $T_m^2 = 1$ to conclude that

$T_m \Phi_{m+1}(H_m(x)) = T_m^2 \Phi_m(x) = \Phi_m(x)$. Applying this for successive values of $m = 1, \ldots, p$ shows that in some neighborhood of 0 we have that

$$\Phi = \Phi_1 = T_1 \Phi_2 \circ H_1 =, T_1 T_2 \Phi_3 \circ H_2 \circ H_1 = \cdots = T_1 \cdots T_{p-1} \Phi_p \circ H_{p-1} \circ \cdots \circ H_1$$

But according to the claim we have that $P_{p-1} \Phi_p(x) = P_{p-1}(x)$. Now P_{p-1} projects $\mathbb{R}^p$ onto the subspace defined by the first $p-1$ members of the standard basis. This means that there is a function g defined in a neighborhood of 0 such that $\Phi_p(x) = g(x)e_p + \sum_{j<p} x_j e_j$ so that Φ_p is a simple function. ∎

Proof of the Change of Variables Theorem. By Lemma 7.4.7 we need only show that for any point x_0 in int X there is a neighborhood of x_0 on which (7.4.5) is valid. If we replace G by $W = G - x_0$ and Φ by $\Phi_1(x) = \Phi(x + x_0) - \Phi(x_0)$, we note that $0 \in W$, $\Phi_1(0) = 0$, and $D\Phi_1(0)$ is invertible. Hence it suffices to assume that $x_0 = 0 \in G$, $\Phi(0) = 0$, and $D\Phi(0)$ is invertible.

Combine the preceding lemma with the fact that the theorem is valid if Φ is a simple function (Lemma 7.4.9) as well as if Φ is a flip (Lemma 7.4.13), and we see that we need only show that when the theorem is true for functions Λ and Ψ, then it is true for $\Phi = \Lambda \circ \Psi$. In this case we have that

$$\int_{\Phi(X)} f = \int_{\Lambda(\Psi(X))} f$$

$$= \int_{\Psi(X)} f \circ \Lambda |\det[D\Lambda]|$$

$$= \int_X [(f \circ \Lambda) \circ \Psi] |\det[D\Lambda] \circ \Psi||\det[D\Psi]|$$

$$= \int_X f \circ \Phi |\det[D\Phi]|$$

where the last equality is from the Chain Rule and the multiplicativity of the determinant. ∎

7.4.17. Proposition. *If $T \in \mathcal{L}(\mathbb{R}^p)$ and T is invertible, then for any rectangle R in $\mathbb{R}^p$,*

$$V(T(R)) = |\det T| V(R)$$

Proof. This is immediate from the Change of Variables Theorem and the fact that if $\Phi : \mathbb{R}^p \to \mathbb{R}^p$ is defined by $\Phi(x) = T(x)$, then $D\Phi(x) = T$ for all x in $\mathbb{R}^p$. ∎

7.4.18. Example. [Polar Coordinates] (a) Define $\Phi : \mathbb{R}^2 \to \mathbb{R}^2$ by $\Phi(r, \theta) = (r \cos \theta, r \sin \theta)$. A simple computation shows that $\det D\Phi(r, \theta) = r$. Thus if we have any compact Jordan set X in the plane such that int X does not contain the origin and Φ is injective on int X and if $f : X \to \mathbb{R}$ is continuous, then

$$\int_{\Phi(X)} f(x, y) \, d(x, y) = \int_X f(r \cos \theta, r \sin \theta) r \, d(r, \theta)$$

Note that for Φ to be injective on int X it must be that if (r, θ_1), $(r, \theta_2) \in$ int X, then $0 < |\theta_1 - \theta_2| < 2\pi$.

(b) Let $Z = \{(x, y) \in \mathbb{R}^2 : 1 \le x^2 + y^2 \le 4\}$ and evaluate $\int_Z (x^2 + y) dx dy$. If we let $X = \{(r, \theta) : 1 \le r \le 2, 0 \le \theta \le 2\pi\}$ and define Φ as in (a), then $Z = \Phi(X)$ and, on int X, Φ is injective and $\det[D\Phi] \ne 0$. Using Fubini's Theorem we get that

$$\int_Z (x^2 + y) \, d(x, y) = \int_X (r^2 \cos^2 \theta + r \sin \theta) r \, d(r, \theta)$$

$$= \int_1^2 r^2 \left[\int_0^{2\pi} (r \cos^2 \theta + \sin \theta) d\theta \right] dr$$

$$= \frac{15\pi}{4}$$

See Exercise 4.

7.4.19. Example. [Spherical Coordinates] (a) Denote the origin in $\mathbb{R}^3$ by $O = (0, 0, 0)$. For any point P in $\mathbb{R}^3$ let Q be the projection of P onto the xy-plane. We associate the spherical coordinates (r, θ, ϕ) as follows: r is the distance of P to the origin; θ is the angle from the positive x-axis to the line segment OQ; ϕ is the angle from the positive z-axis to the line OP. Using these quantities we get that $(x, y, z) = (r \sin \phi \cos \theta, r \sin \phi \sin \theta, r \cos \phi)$. We define

$$\Phi(r, \theta, \phi) = (r \sin \phi \cos \theta, r \sin \phi \sin \theta, r \cos \phi)$$

A computation shows that

$$D\Phi(r, \theta, \phi) = \frac{\partial(x, y, z)}{\partial(r, \theta, \phi)} = \begin{bmatrix} \sin \phi \cos \theta & -r \sin \phi \sin \theta & r \cos \phi \cos \theta \\ \sin \phi \sin \theta & r \sin \phi \cos \theta & r \cos \phi \sin \theta \\ \cos \phi & 0 & -r \sin \phi \end{bmatrix}$$

and so $\det D\Phi(r, \theta, \phi) = -r^2 \sin \phi$. Hence the Change of Variables Theorem applies to any Jordan set X such that $r^2 \sin \phi \ne 0$ on int X.

(b) Find the volume of a sphere in $\mathbb{R}^3$. The volume of a sphere of radius a is $\frac{1}{8} V(Z)$, where $Z = \{(x, y, z) : x \ge 0, y \ge 0, z \ge 0, \text{ and } x^2 + y^2 + z^2 \le a^2\}$. If we define $X = \{(r, \theta, \phi) : 0 \le r \le a, 0 \le \theta \le \pi/2, 0 \le \phi \le \pi/2\}$, then the function Φ in (a) maps X onto Z. Using the Change of Variables Theorem and Fubini's Theorem we have that

$$V(Z) = \int_Z 1 = \int_X r^2 \sin \phi \, d(r, \theta, \phi)$$

$$= \int_0^a r^2 \left[\int_0^{\pi/2} \left[\int_0^{\pi/2} \sin \phi \, d\phi \right] d\theta \right] dr$$

$$= \frac{\pi a^3}{6}$$

Therefore the volume of the sphere is $4\pi a^3/3$.

Exercises

(1) If R is any rectangle, show that R can be written as the union of a finite number of non-overlapping rectangles $\{R_1, \ldots, R_m\}$ such that each R_k has the property that its longest side is at most twice the length of its shortest side

(2) In the proof of Lemma 7.4.6 show that $\left| \int_{X \setminus Y} f \right| < \epsilon$.

(3) Supply the details of the proof of Lemma 7.4.7.

(4) Give the details of the calculation in Example 7.4.18(b).

(5) Let X be the set in $\mathbb{R}^2$ bounded by the x-axis and the curve $r = 1 - \cos\theta$ where $0 \le \theta \le \pi$; find $\int_X y$.

(6) Find the area of the following subsets X of $\mathbb{R}^2$. (a) $X = \{(x, y) : y \le x \le 4y, 1 \le x + 2y \le 3\}$. (b) $\{(x, y) : 2 \le xy \le 4, 2x \le y \le 5x\}$.

(7) Let X be the top half of the unit ball in $\mathbb{R}^3$ and find $\int_X 16z$.

(8) Let X be the sphere in $\mathbb{R}^3$ centered at the origin with radius 3. When $f(x, y, z) = 10 - x^2 - y^2 - z^2$, show that $\int_X f = 828\pi/5$.

7.5. Differentiation under the Integral Sign

Here we want to explore the idea of the differentiation of functions that are defined by an integral. The proof of the main result (7.5.2) is technically complicated because there are so many things to keep track of. Instead of starting with that we will begin with a special case which is a corollary of the main result. We do this because the proof of this special case is straightforward and it is this result that seems to be used more frequently.

7.5.1. Theorem. *If $[a, b], [c, d] \subseteq \mathbb{R}$, G is an open subset of $\mathbb{R}^2$ that contains $[a, b] \times [c, d]$, and $f : G \to \mathbb{R}$ is bounded and continuously differentiable, then*

$$\frac{d}{dx}\left[\int_c^d f(x, y)dy \right] = \int_c^d \partial_1 f(x, y)dy$$

Proof. We only give the proof for the case that $x \in (a, b)$. Set $F(x) = \int_c^d f(x, y)dy$ so that $F : [a, b] \to \mathbb{R}$. Fix x in (a, b) and let $\delta_1 > 0$ such that $[x - \delta_1, x + \delta_1] \subseteq (a, b)$. Put K equal to the rectangle $[a, b] \times [c, d]$. Note that since $\partial_1 f$ is continuous, it is uniformly continuous on K. Thus we can find $\delta_2 > 0$ such that $|\partial_1 f(w, y) - \partial_1 f(w', y')| < \epsilon/2(d - c)$ when $\|(w, y) - (w', y')\| < \delta_2$. Also the MVT applied to $\partial_1 f$ says that for any y and t with $0 < |t| < \delta_1$ there is a x_1 between x and $x + t$ with

$$\left| \frac{f(x + t, y) - f(x, y)}{t} - \partial_1 f(x_1, y) \right| < \frac{\epsilon}{2(d - c)}$$

We note that x_1 depends on x, y, and t. Let $0 < |t| < \delta = \min\{\delta_1, \delta_2\}$. Note that for any y in $[c, d]$, $\|(x, y) - (x_1, y)\| = |x - x_1| < \delta \leq \delta_2$ so that $|\partial_1 f(x, y) - \partial_1 f(y, x_1)| < \epsilon/2(d - c)$. Hence for $0 < |t| < \delta$ and any y in $[c, d]$ we have that

$$\left| \frac{f(x + t, y) - f(x, y)}{t} - \partial_1 f(x, y) \right|$$

$$\leq \left| \frac{f(x + t, y) - f(x, y)}{t} - \partial_1 f(x_1, y) \right| + |\partial_1 f(x_1, y) - \partial_1 f(x, y)|$$

$$< \frac{\epsilon}{2(d - c)} + \frac{\epsilon}{2(d - c)}$$

$$= \frac{\epsilon}{d - c}$$

Therefore when $0 < |t| < \delta$, we have that

$$\left| \frac{F(x + t) - F(x)}{t} - \int_c^d \partial_1 f(x, y) dy \right|$$

$$\leq \int_c^d \left| \frac{f(x + t, y) - f(x, y)}{t} - \partial_1 f(x, y) \right| dy$$

$$< \epsilon$$

Since ϵ was arbitrary, this completes the proof. ∎

Now we present a more general result.

7.5.2. Theorem. *Consider $[a, b]$ in $\mathbb{R}$ and suppose $\phi, \psi : [a, b] \to \mathbb{R}$ are continuously differentiable with $\phi(x) \leq \psi(x)$ for all x in $[a, b]$. If G is an open subset of $\mathbb{R}^2$ that contains $\{(x, y) : a \leq x \leq b \text{ and } \phi(x) \leq y \leq \psi(x)\}$ and $f : G \to \mathbb{R}$ is continuously differentiable, then $x \mapsto \int_{\phi(x)}^{\psi(x)} f(x, y) dy$ is differentiable and*

$$\frac{d}{dx} \left[\int_{\phi(x)}^{\psi(x)} f(x, y) dy \right] =$$

$$f(x, \psi(x))\psi'(x) - f(x, \phi(x))\phi'(x) + \int_{\phi(x)}^{\psi(x)} \partial_1 f(x, y) dy$$

Proof. Define $F : [a, b] \to \mathbb{R}$ by

$$F(x) = \int_{\phi(x)}^{\psi(x)} f(x, y) dy$$

Fix x in (a, b) and let $d > 0$ such that $[x - d, x + d] \subseteq (a, b)$. (We leave the proof of the theorem when x is an endpoint of $[a, b]$ to the reader as Exercise 1.) When

$0 < |t| < d$, we have that

$$
\begin{aligned}
F(x+t) - F(x) &= \int_{\phi(x+t)}^{\psi(x+t)} f(x+t, y)dy - \int_{\phi(x)}^{\psi(x)} f(x, y)dy \\
&= \int_{\phi(x+t)}^{\phi(x)} f(x+t, y)dy + \int_{\phi(x)}^{\psi(x)} f(x+t, y)dy \\
&\quad + \int_{\psi(x)}^{\psi(x+t)} f(x+t, y)dy - \int_{\phi(x)}^{\psi(x)} f(x, y)dy \\
&= -\int_{\phi(x)}^{\phi(x+t)} f(x+t, y)dy + \int_{\psi(x)}^{\psi(x+t)} f(x+t, y)dy \\
&\quad + \int_{\phi(x)}^{\psi(x)} [f(x+t, y) - f(x, y)]dy
\end{aligned}
$$

We now apply the MVT to the first and second of these three integrals. So for any t with $0 < |t| < d$ we obtain a point y_1 between $\phi(x)$ and $\phi(x+t)$ and y_2 between $\psi(x)$ and $\psi(x+t)$ such that

$$
\begin{aligned}
F(x+t) - F(x) &= -[\phi(x+t) - \phi(x)]f(x+t, y_1) \\
&\quad + [\psi(x+t) - \psi(x)]f(x+t, y_2) \\
&\quad + \int_{\phi(x)}^{\psi(x)} [f(x+t, y) - f(x, y)]dy
\end{aligned}
$$

(The point x is fixed but be conscious of the fact that the points y_1 and y_2 depend on t.) Dividing both sides by t gives that

$$
\begin{aligned}
\frac{F(x+t) - F(x)}{t} &= -\frac{\phi(x+t) - \phi(x)}{t} f(x+t, y_1) \\
&\quad + \frac{\psi(x+t) - \psi(x)}{t} f(x+t, y_2) \\
&\quad + \int_{\phi(x)}^{\psi(x)} \left[\frac{f(x+t, y) - f(x, y)}{t} \right] dy
\end{aligned}
$$

Fix $\epsilon > 0$ and let's make a few observations. First, since both ϕ' and ψ' are continuous, there is a constant C such that $|\phi'(w)| \leq C$ and $|\psi'(w)| \leq C$ for all w in $[a, b]$. Second, the set $K = \{(w, y) : w \in [a, b], \phi(w) \leq y \leq \psi(w)\}$ is compact (Why?). This tells us two things: since f is continuous it is bounded on K and it is uniformly continuous there. So there is a constant M with $|f(w, y)| \leq M$ whenever $(w, y) \in K$ and there is a $\delta' > 0$ such that $|f(w, y) - f(w', y')| < \epsilon/6C$ when $\|(w, y) - (w', y')\| < \delta'$ and these points belong to K.

Consider the first summand in the last equation above; so we are focusing on the function ϕ. There is a δ_{11} with $0 < \delta_{11} < \min\{\delta', d\}$ such that when $0 < |t| < \delta_{11}$,

$$
\left| \phi'(x) - \frac{\phi(x+t) - \phi(x)}{t} \right| < \frac{\epsilon}{6M}
$$

Because ϕ is uniformly continuous on $[a, b]$ there is a $\delta_{12} > 0$ such that $|\phi(w) - \phi(w')| < \frac{1}{2}\delta_{11}$ when $|w - w'| < \delta_{12}$. Put $\delta_1 = \frac{1}{2}\min\{\delta_{11}, \delta_{12}\}$. If $0 < |t| < \delta_1$, then with the choice of y_1 made as above we have that

$$\left| \frac{\phi(x+t) - \phi(x)}{t} f(x+t, y_1) - \phi'(x)f(x, \phi(x)) \right|$$

$$\leq \left| \frac{\phi(x+t) - \phi(x)}{t} - \phi'(x) \right| |f(x+t, y_1)|$$

$$+ |\phi'(x)| |f(x+t, y_1) - f(x, \phi(x))|$$

$$\leq \frac{\epsilon}{6} + C|f(x+t, y_1) - f(x, \phi(x))|$$

Now

$$\|(x+t, y_1) - (x, \phi(x))\|^2 = t^2 + |y_1 - \phi(x)|^2$$

$$< \frac{\delta_{11}^2}{4} + [\phi(x+t) - \phi(x)]^2$$

$$< \frac{\delta_{11}^2}{4} + \frac{\delta_{11}^2}{4} \leq \frac{\delta_{11}^2}{2} < \delta'^2$$

Thus $|f(x+t, y_1) - f(x, \phi(x))| < \epsilon/6C$ and so the above inequality becomes that for $0 < |t| < \delta_1$ we have that

$$\left| \frac{\phi(x+t) - \phi(x)}{t} f(x+t, y_1) - \phi'(x)f(x, \phi(x)) \right| < \frac{\epsilon}{3}$$

Similarly if we focus on the function ψ there is a $\delta_2 > 0$ such that when $0 < |t| < \delta_2$

$$\left| \frac{\psi(x+t) - \psi(x)}{t} f(x+t, y_2) - \psi'(x)f(x, \psi(x)) \right| < \frac{\epsilon}{3}$$

Combining what we have done above we get that when $0 < |t| < \min\{\delta_1, \delta_2\}$

$$\left| \frac{F(x+t) - F(x)}{t} - \int_{\phi(x)}^{\psi(x)} \left[\frac{f(x+t, y) - f(x, y)}{t} \right] dy \right| < \frac{2\epsilon}{3}$$

We're almost done. Note that since $\partial_1 f$ is continuous, it is uniformly continuous on K. Thus we can find $\delta_3 > 0$ such that $\left| \partial_1 f(w, y) - \partial_1 f(w', y') \right| < \epsilon/6C$ when $\|(w, y) - (w', y')\| < \delta_3$ and these points belong to K. Also the MVT applied to $\partial_1 f$ says that for any y and t there is an x_1 between x and $x + t$ with

$$\left| \frac{f(x+t, y) - f(x, y)}{t} - \partial_1 f(x_1, y) \right| < \frac{\epsilon}{6C}$$

Let $0 < \delta < \min\{\delta_1, \delta_2, \delta_3\}$. Fix any y and let $0 < |t| < \delta$. Since $\|(x, y) - (x_1, y)\| = |x - x_1| \le |t| < \delta_3$, we have for the appropriate x_1 that

$$\left| \frac{f(x+t, y) - f(x, y)}{t} - \partial_1 f(x, y) \right| \le \left| \frac{f(x+t, y) - f(x, y)}{t} - \partial_1 f(x_1, y) \right|$$
$$+ |\partial_1 f(x_1, y) - \partial_1 f(x, y)|$$
$$< \frac{\epsilon}{6C} + \frac{\epsilon}{6C} = \frac{\epsilon}{3C}$$

Putting this together with our previous estimates we have that for $0 < |t| < \delta$

$$\left| \frac{F(x+t) - F(x)}{t} - \int_{\phi(x)}^{\psi(x)} \partial_1 f(x, y) dy \right|$$
$$\le \left| \frac{F(x+t) - F(x)}{t} - \int_{\phi(x)}^{\psi(x)} \left[\frac{f(x+t, y) - f(x, y)}{t} \right] dy \right|$$
$$+ \int_{\phi(x)}^{\psi(x)} \left| \frac{f(x+t, y) - f(x, y)}{t} - \partial_1 f(x, y) \right| dy$$
$$< \frac{2\epsilon}{3} + \frac{\epsilon}{3C} C$$
$$= \epsilon$$

Since ϵ was arbitrary, this completes the proof. ∎

One of the important uses of differentiating under the integral sign is that it enables us to compute certain integrals over $\mathbb{R}$ that are very difficult to compute otherwise. Here is an example.

7.5.3. Example. Of importance in probability and Fourier analysis is the improper integral

$$\int_{-\infty}^{\infty} e^{-x^2/2} dx = \sqrt{2\pi}$$

which we want to verify. It isn't clear how to use the above theorems to obtain this, but be patient. Let's first observe that by the symmetry of the integrand we need only prove

$$\int_0^{\infty} e^{-x^2/2} dx = \sqrt{\frac{\pi}{2}}$$

If you play with this integral using the standard tools from calculus, you'll see they all fail you. We take a different approach. Define the function $\alpha : [0, \infty) \to \mathbb{R}$ by

$$\alpha(r) = \left[\int_0^r e^{-x^2/2} dx \right]^2$$

so we want to find $\sqrt{\lim_{r\to\infty} \alpha(r)}$. Note that

$$\alpha'(r) = 2e^{-r^2/2}\left[\int_0^r e^{-x^2/2}dx\right]$$

Now we make a change of variables $x = ry$, where $0 \le y \le 1$, and get that

$$\alpha'(r) = 2e^{-r^2/2}\int_0^1 re^{-r^2y^2/2}dy = \int_0^1 2re^{-\frac{1}{2}(1+y^2)r^2}dy$$

After staring at the integrand and thinking, we see that

$$2re^{-\frac{1}{2}(1+y^2)r^2} = -\frac{\partial}{\partial r}\left[\frac{2\exp\left(-\frac{1}{2}(1+y^2)r^2\right)}{1+y^2}\right]$$

Using Theorem 7.5.1 we therefore have that

$$\alpha'(r) = -2\frac{d}{dr}\int_0^1 \frac{\exp\left(-\frac{1}{2}(1+y^2)r^2\right)}{1+y^2}dy$$

Putting $\beta(r)$ equal to the integral in this last equation, this means that $\alpha'(r) = -2\beta'(r)$ so that $\alpha(r) = -2\beta(r) + C$, for some constant C. We need to evaluate this constant. In fact

$$0 = \lim_{r\to 0}\alpha(r)$$

$$= C - 2\lim_{r\to 0}\int_0^1 \frac{\exp\left(-\frac{1}{2}(1+y^2)r^2\right)}{1+y^2}dy$$

$$= C - 2\int_0^1 \frac{1}{1+y^2}dy \qquad \text{(Why?)}$$

$$= C - \frac{\pi}{2}$$

since this last integral can be evaluated because $(1+y^2)^{-1}$ has arctan y as its primitive. Putting it all together we have that

$$\left[\int_0^r e^{-x^2/2}dx\right]^2 = \frac{\pi}{2} - 2\int_0^1 \frac{\exp\left(-\frac{1}{2}(1+y^2)r^2\right)}{1+y^2}dy$$

Letting $r \to \infty$ shows that $[\int_0^\infty e^{-x^2/2}dx]^2 = \pi/2$, whence the result.

This example was accessed at www.math.uconn.edu/~kconrad/blurbs/analysis/ diffunderint.pdf on 21 Jan 2015. This site contains other examples involving differentiating under the integral sign.

Exercises

(1) Prove Theorem 7.5.2 when x is an endpoint of $[a, b]$.

(2) Can you formulate and prove a version of Corollary 7.5.1 for $[a, b] \times [0, \infty)$?

(3) Show that $\int_0^1 [(x^5 - 1)/\log x] \, dx = \log 6$ by using the function

$$f(t) = \int_0^1 \frac{x^t - 1}{\log x} \, dx$$

and differentiating under the integral sign.

(4) Show that

$$\int_0^\pi \log(1 + t \cos x) = \pi \log \left[\frac{1 + \sqrt{1 - t^2}}{2} \right].$$

8 Curves and Surfaces

The next two chapters are on the same topic, one where geometry and analysis overlap. This present chapter will focus on $\mathbb{R}^2$ and $\mathbb{R}^3$, covering curves and surfaces in those spaces including the theorems of Green, Gauss, and Stokes. Initially, however, we'll encounter the concepts in $\mathbb{R}^p$ since in the basic material there is little difference between what happens in the lower dimensions and the general space. When, however, we discuss in this chapter the concepts leading to the three big theorems, we will focus on $p = 2$ and $p = 3$. Be aware that we will often use heuristic arguments in the present chapter, something that is facilitated by these small dimensions. This is not true in the following chapter where we adopt a different approach that leads to greater rigor. In fact, you could begin reading Chapter 9 after the first section of this chapter and refer back to the present chapter for the heuristics. I don't advise this as I believe seeing the material in this chapter first will better prepare you to understand Chapter 9. Moreover Chapter 9 is not written with this approach in mind so it might be awkward to do this. At the end of the next chapter we will revisit these lower dimensions to supply whatever rigor is absent here.

The historical origins of much of what we do in this chapter and the next started with an effort to use mathematics to study physics. That same statement can be made about most of the mathematics that appears in textbooks. As we progress in this chapter the connection with these historical roots will be evident in some of the language used. For example in the first section we will talk about a particle moving along a curve. You need not know any physics to understand the next two chapters since we deal in abstractions of the physical concepts. Indeed we see the virtue of this abstraction as the mathematics has application to the social and biological sciences as well as the physical.

8.1. Curves

In §6.1 we defined a smooth curve in $\mathbb{R}^p$ as a continuously differentiable function $\gamma : [a, b] \to \mathbb{R}^p$. In §7.1 we introduced the length of a smooth curve, defining it as $\ell(\gamma) = \int_a^b \|\gamma'(t)\| \, dt$. We also said that γ and a second smooth curve $\rho : [c, d] \to \mathbb{R}^p$ are equivalent, $\gamma \sim \rho$, if there is a continuously differentiable, increasing surjection $\tau : [a, b] \to [c, d]$ such that $\gamma(t) = \rho(\tau(t))$ for all t in $[a, b]$. Proposition 7.1.7 showed that equivalent curves have the same length. In this section we continue the study of smooth curves and go more deeply into their analysis. In particular we will explore the orientation or direction of the curve and introduce another integral associated with these curves. Some of the examples and exercises here are from [15].

We begin with some terminology that could have appeared earlier.

8.1.1. Definition. Let $\gamma : [a, b] \to \mathbb{R}^p$ be a curve. We say that $\gamma(a)$ and $\gamma(b)$ are the *starting* and *final* points, respectively. The curve γ is said to be a *simple curve* if γ is injective on the open interval (a, b). Curve γ is a *closed curve* if $\gamma(a) = \gamma(b)$. A simple closed curve is also referred to as a *Jordan curve*. As usual the *trace* of γ is its range and is denoted by $\{\gamma\} = \{\gamma(t) : t \in [a, b]\}$.

8.1.2. Example. (a) Fix $r > 0$ and define $\gamma : [0, 2\pi] \to \mathbb{R}^2$ by $\gamma(\theta) = (r\cos\theta, r\sin\theta)$. So γ is a Jordan curve and it traces out the circle centered at the origin of radius r that moves in the counter clockwise direction.

(b) Let $a, b > 0$ and define $\gamma : [0, 2\pi] \to \mathbb{R}^2$ by $\gamma(t) = (a\cos t, b\sin t)$. The trace of γ is an ellipse centered at the origin. What are its two axes? See Exercise 2. The length of this curve is what is called an elliptic integral and it can seldom be evaluated explicitly; usually it can only be approximated numerically.

(c) If $f : [a, b] \to \mathbb{R}^{p-1}$ is continuously differentiable, define $\gamma : [a, b] \to \mathbb{R}^p$ by $\gamma(t) = t \oplus f(t)$. So γ is the curve in $\mathbb{R}^p$ that traces out the graph of the function f. We note that $\gamma'(t) = 1 \oplus f'(t)$.

8.1.3. Proposition. *Let $\gamma : [a, b] \to \mathbb{R}^p$ be a smooth curve with $\ell = \ell(\gamma)$. If we define $\tau : [a, b] \to \mathbb{R}$ by*

$$\tau(t) = \int_0^t \|\gamma'(u)\| du$$

then τ is an increasing, continuously differentiable function that maps $[a, b]$ onto $[0, \ell]$ and with $\tau'(t) = \|\gamma'(t)\|$.

Proof. We begin by noting that the function $u \mapsto \|\gamma'(u)\|$ is a continuous function. Hence the FTC can be applied to conclude that τ is continuously differentiable and $\tau'(u) = \|\gamma'(u)\|$. Since τ has a positive derivative, it is increasing. Clearly it is a surjection onto $[0, \ell]$. ∎

The preceding proposition, in spite of the brevity of its proof, tells us a lot. At any point t in $[a, b]$, $\tau(t)$ measures the length of the curve γ from the beginning of the curve to the time t. Thus at any t in $[a, b]$, $\tau'(t) = \|\gamma'(t)\|$ is the rate of change of the arc length of the curve with respect to time; that is, $\|\gamma'(t)\|$ is the speed of a particle moving along the curve γ while the vector $\gamma'(t)$ is tangent to the curve and points in the direction the particle is moving.

8.1.4. Definition. If $\gamma : [a, b] \to \mathbb{R}^p$ is a smooth curve and $f : \{\gamma\} \to \mathbb{R}$ is a continuous function, the integral of f along γ is defined as

$$\int_\gamma f = \int_\gamma f \, ds = \int_a^b f(\gamma(t))\|\gamma'(t)\| \, dt$$

We point out that if in this definition we take f to be the constant function 1, then we get the length of the curve γ. This is symbolically captured by using the ds in the integral. The notation is traditional and is related to Proposition 8.1.3 as follows. If τ is as in that proposition and if τ is injective, we could take the inverse $s \mapsto \tau^{-1}(s)$ to give a reparametrization of the curve. This is often called the *natural parametrization* of the curve when it exists.

8.1.5. Definition. Say that a curve $\gamma : [a, b] \to \mathbb{R}^p$ is *regular* if γ is smooth and $\gamma'(t) \neq 0$ for all t in $[a, b]$. The curve γ is *piecewise regular* if there is a partition $\{a = t_0 < t_1 < \cdots < t_n = b\}$ such that γ is regular on $[t_{i-1}, t_i]$ for $1 \leq i \leq n$. We denote this by saying that $\gamma = \gamma_1 + \cdots + \gamma_n$, where for $1 \leq i \leq n$, $\gamma_i : [t_{i-1}, t_i] \to \mathbb{R}^p$ is defined by $\gamma_i(t) = \gamma(t)$.

Observe that the curve in Example 8.1.2(c) is regular. If γ is regular and τ is as in Proposition 8.1.3, then according to that result $\tau'(t) > 0$ for all t. Hence τ is injective and we can form its inverse, thus producing the equivalent curve (Definition 7.1.6) $s \mapsto \gamma(\tau^{-1}(s))$ defined on $[0, \ell]$. This proves the following.

8.1.6. Proposition. *A regular curve can be given its natural parametrization.*

8.1.7. Example. (a) The curve $\gamma : [0, 2\pi] \to \mathbb{R}^2$ defined by $\gamma(\theta) = (\cos \theta, \sin \theta)$ with $\|\gamma'(\theta)\| = \|(-\sin \theta, \cos \theta)\| = 1$. Thus $\ell(\gamma) = 2\pi$. If τ is as in (8.1.3), we have that $\tau(\theta) = \int_0^\theta 1 du = \theta$, so this form of the equation of the unit circle is already in its natural parametrization.

(b) If $f : [a, b] \to \mathbb{R}^{p-1}$ is continuously differentiable and $\gamma(t) = t \oplus f(t)$ as in (8.1.2(c)), then $\|\gamma'(t)\| = \sqrt{1 + \|f'(t)\|^2} > 0$. So γ is regular and has a natural parametrization. If $g : \{\gamma\} \to \mathbb{R}$ is a continuous function, then

$$\int_\gamma g = \int_a^b g(\gamma(t))\sqrt{1 + \|f'(t)\|^2}\, dt$$

8.1.8. Definition. If $\gamma : [a, b] \to \mathbb{R}^p$ is a regular curve, then the *unit tangent vector* for γ at the point $\gamma(t)$ is the vector

$$T(t) = \frac{\gamma'(t)}{\|\gamma'(t)\|}$$

If $F : \{\gamma\} \to \mathbb{R}^p$ is a continuous function, the *line integral* of F along γ is defined as

$$\int_\gamma F \cdot T ds = \int_a^b \langle F(\gamma(t)), \gamma'(t)\rangle\, dt$$

If γ is piecewise regular with $\gamma = \gamma_1 + \cdots + \gamma_n$, then

$$\int_\gamma F \cdot T\, ds = \sum_{i=1}^n \int_{\gamma_i} F \cdot T_i\, ds$$

where for $1 \leq i \leq n$, $T_i(t) = \gamma_i'(t)/\|\gamma_i'(t)\|$ when $t_{i-1} \leq t \leq t_i$.

Strictly speaking we should have used the notation $T_\gamma(t)$ for the unit tangent vector, but the notation $T(t)$ is traditional and assumes we have specified the curve γ. Similarly the notation $\int_\gamma F \cdot T ds$ is also traditional and is sometimes called the *path integral*. The term line integral is frequently also used for the integral in Definition 8.1.4, where we integrate a scalar-valued function. We chose not to use that term to avoid the possibility of ambiguity, but in actuality there is little possibility of this since in one case we integrate a scalar-valued function and in the other a vector-valued function. In fact the earlier definition is used in the preceding one. Indeed

$$\int_\gamma F \cdot T ds = \int_a^b \langle F(\gamma(t)), \gamma'(t) \rangle \, dt$$

$$= \int_a^b \langle F(\gamma(t)), T(t) \rangle \|\gamma'(t)\| \, dt$$

$$= \int_\gamma \langle F \circ \gamma, T \rangle \, ds$$

There is a symbolic representation of $\int_\gamma F \cdot T \, ds$ that we will use and the reader will see in the literature. Namely if $\gamma(t) = \sum_{j=1}^p \gamma_j(t) e_j$ and $F(x) = \sum_{j=1}^p F_j(x) e_j$, then

8.1.9 $$\int_\gamma F \cdot T \, ds = \sum_{j=1}^p \int_\gamma F_j \, dx_j$$

To see where this comes from observe that it means

$$\int_\gamma F \cdot T \, ds = \int_a^b \langle F(\gamma(t)), \gamma'(t) \rangle \, dt = \sum_{j=1}^p \int_a^b F_j(\gamma(t)) \gamma_j'(t) \, dt$$

Symbolically, $x_j = \gamma_j(t)$ and so $dx_j = \gamma_j'(t) dt$. (We'll say more about this in the next chapter when we study differential forms.)

Consider what happens if we have equivalent curves (Definition 7.1.6). That is suppose $\gamma : [a, b] \to \mathbb{R}^p$ and $\rho : [c, d] \to \mathbb{R}^p$ are two smooth curves and there is a continuously differentiable, increasing surjection $\tau : [a, b] \to [c, d]$ such that $\gamma(t) = \rho(\tau(t))$ for all t in $[a, b]$. So $\gamma'(t) = \rho(\tau(t))\tau'(t)$. Note that if ρ is regular, then unless τ is strictly increasing so that $\tau'(t) > 0$ for all t in $[a, b]$, it does not follow that γ is regular. See Exercise 1. On the other hand if γ is regular and $\rho \sim \gamma$, then it must be that ρ is regular and τ is strictly increasing. It also follows that

$$T_\gamma(t) = \frac{\gamma'(t)}{\|\gamma'(t)\|} = \frac{\rho'(\tau(t))\tau'(t)}{\|\rho'(\tau(t))\tau'(t)\|} = T_\rho(\tau(t))$$

With the same notation if both γ and ρ are regular, then

$$\int_\gamma f\,ds = \int_a^b f(\gamma(t))\|\gamma'(t)\|\,dt$$

$$= \int_a^b f(\rho(\tau(t)))\|\rho'(\tau(t))\|\tau'(t)\,dt$$

$$= \int_c^d f(\rho(u))\|\rho'(u)\|\,du$$

Similarly

$$\int_\gamma F \cdot T\,ds = \int_\rho F \cdot T\,ds$$

8.1.10. Example. (a) Let $\gamma : [0, 2\pi] \to \mathbb{R}^2$ be defined by $\gamma(\theta) = (\cos\theta, \sin\theta)$, so that γ represents the unit circle in the plane traversed once in the counterclockwise direction. If $F : \mathbb{R}^2 \to \mathbb{R}^2$ is defined by $F(x, y) = (y, x)$, then

$$\int_\gamma F \cdot T\,ds = \int_0^{2\pi} \langle(\sin\theta, \cos\theta), (-\sin\theta, \cos\theta)\rangle\,d\theta$$

$$= \int_0^{2\pi} [-\sin^2\theta + \cos^2\theta]\,d\theta$$

$$= \int_0^{2\pi} \cos(2\theta)\,d\theta = 0$$

(b) Let γ be as in part (a) and put $G(x, y) = (-y, x)$. So

$$\int_\gamma G \cdot T\,ds = \int_0^{2\pi} \langle(-\sin\theta, \cos\theta), (-\sin\theta, \cos\theta)\rangle\,d\theta = \int_0^{2\pi} \theta\,d\theta = 2\pi$$

(c) Define $\gamma : [0, 1] \to \mathbb{R}^3$ by $\gamma(t) = (t, 2t, -t)$ and let $F : \mathbb{R}^3 \to \mathbb{R}^3$ be defined by $F(x, y, z) = (x + y, x^2, -yz)$. Here

$$\int_\gamma F \cdot T\,ds = \int_0^1 \langle(3t, t^2, 2t^2), (1, 2, -1)\rangle\,dt$$

$$= \int_0^1 3t\,dt = \frac{3}{2}$$

(d) Let γ be the boundary of the square $[0, 1] \times [0, 1]$ in $\mathbb{R}^2$ in the counterclockwise direction and $F(x, y) = (xy, x^2 + y^2)$. Here $\gamma = \gamma_1 + \gamma_2 + \gamma_3 + \gamma_4$, where each of these pieces is a straight line defined on $[0, 1]$ as follows: $\gamma_1(t) = (t, 0)$, $\gamma_2(t) = (1, t)$, $\gamma_3(t) = (1 - t, 1)$, $\gamma_4(t) = (0, 1 - t)$. So the derivatives of these curves are $\gamma_1'(t) = (1, 0)$, $\gamma_2'(t) = (0, 1)$, $\gamma_3'(t) = (-1, 0)$, $\gamma_4'(t) = (0, -1)$.

Hence

$$
\begin{aligned}
\int_\gamma F \cdot T\, ds &= \int_0^1 [\langle (0, t^2), (1, 0)\rangle + \langle (t, 1 + t^2), (0, 1)\rangle \\
&\quad + \langle (1 - t, (1 - t)^2 + 1), (-1, 0)\rangle + \langle (1 - t, (1 - t)^2), (0, -1)\rangle]\, dt \\
&= \int_0^1 [0 + (1 + t^2) - (1 - t) - (1 - t)^2]\, dt \\
&= \int_0^1 [1 + t^2 - 1 + t - (1 - 2t + t^2)]\, dt \\
&= \frac{1}{2}
\end{aligned}
$$

We conclude this section with a brief discussion of curves in $\mathbb{R}^2$. If $\gamma : [a, b] \to \mathbb{R}^2$ is a regular curve and $\gamma(t) = (x(t), y(t))$, then we have functions $t \mapsto x(t)$ and $t \mapsto y(t)$ from $[a, b]$ into $\mathbb{R}$. It is easy to check that each of these functions is smooth and $\gamma'(t) = (x'(t), y'(t))$. If $F : \{\gamma\} \to \mathbb{R}^2$ is a continuous function with $F(x, y) = (P(x, y), Q(x, y))$, then

8.1.11
$$
\int_\gamma F \cdot T\, ds = \int_a^b [P(x(t), y(t))x'(t) + Q(x(t), y(t))y'(t)]\, dt = \int_\gamma P\, dx + Q\, dy
$$

We will frequently encounter expressions such as the right-hand side of the last equation when we study differential forms in the next chapter.

Exercises

(1) Find an example of a regular curve $\rho : [c, d] \to \mathbb{R}^2$ and an increasing function $\tau : [a, b] \to [c, d]$ such that the curve $\tau(t) = \rho(\tau(t))$ is not regular.
(2) Verify the statements in Example 8.1.2(b).
(3) Let $F(x, y) = (x, xy)$ and find $\int_\gamma F \cdot T\, ds$ for the following choices of γ: (a) $\gamma(t) = (t, t)$ with $0 \le t \le 1$; (b) $\gamma(t) = (t, t^2)$ for $0 \le t \le 1$; (c) γ is the square $[0, 1] \times [0, 1]$ in the counterclockwise direction.
(4) Let $F(x, y, z) = (z, x^2, y)$ and find $\int_\gamma F \cdot T\, ds$ for each of the following curves γ: (a) $\gamma(t) = (t, t, t)$ for t in $[0, 1]$; (b) $\gamma(t) = (\cos t, \sin t, t)$ for $0 \le t \le 2\pi$. (c) γ is the closed polygon whose successive vertices are $(0, 0, 0)$, $(2, 0, 0)$, $(2, 3, 0)$, $(0, 0, 1)$, $(0, 0, 0)$.
(5) Find $\int_\gamma F \cdot T\, ds$ for each of the following choices of F and γ. (a) $F(x, y) = (xy, y - x)$ and γ is the parabola $y = x^2$ from $(1, 1)$ to $(3, 0)$. (b) $F(x, y, z) = (\sqrt{x^3 + y^3 + 5}, z, x^2)$ and γ is the intersection of the elliptical cylinder $y^2 + 2z^2 = 1$ with the plane $x = -1$ that is oriented in the clockwise direction when viewed from far out along the positive x-axis. (This last will test your viewing skills.)

(6) If $\gamma : [a, b] \to \mathbb{R}^p$ is a regular curve, let $-\gamma$ be the curve defined on $[-b, -a]$ by $(-\gamma)(t) = \gamma(-t)$. Show that if $F : \{\gamma\} \to \mathbb{R}^p$ is a continuous function, then $\int_{-\gamma} F \cdot T_{-\gamma} \, ds = -\int_{\gamma} F \cdot T_{\gamma} \, ds$. What is the relation between $\{-\gamma\}$ and $\{\gamma\}$?

(7) If f and g are continuous functions of a single variable and F is defined on $\mathbb{R}^2$ by $F(x, y) = (f(x), g(y))$, show that $\int_{\gamma} F \cdot T \, ds = 0$ whenever γ is a regular closed curve in $\mathbb{R}^2$.

8.2. Green's Theorem

In this section we will focus on curves in $\mathbb{R}^2$. They differ quite a bit from curves in $\mathbb{R}^p$ when $p > 2$, and we begin with a result that is unique to the plane. It is a famous result from planar topology that is highly intuitive but difficult to prove. A moment's thought reveals that this theorem is decidedly false in higher dimensional spaces. Recall the definition of Jordan curve as a simple closed curve, which was given at the start of the last section.

8.2.1. Theorem (Jordan[1] Curve Theorem). *If γ is a Jordan curve in $\mathbb{R}^2$, then $\mathbb{R}^2 \backslash \{\gamma\}$ has two components, only one of which is bounded. Moreover $\{\gamma\}$ is the boundary of each of these components.*

We won't prove this theorem. For a proof see [14]. If G is the bounded component of the complement of γ, then G is called the *inside of the curve* γ and γ is the boundary of G. The unbounded component is called the *outside* of γ and its boundary is also γ.

If we assume γ is a regular curve, then it has a natural direction given by its unit tangent vector. Thus $\gamma : [a, b] \to \mathbb{R}^2$ has a direction – as the variable t goes from a to b, $\gamma(t)$ goes from $\gamma(a)$ to $\gamma(b)$. What we want, however, is to discuss the direction of a curve relative to an open subset of $\mathbb{R}^2$ when the curve forms part of the boundary of the set as, for example, is the case with the Jordan Curve Theorem and G is the inside of the curve. Intuitively we want to say that γ is positively oriented relative to G if when we walk along γ in the direction of $T'(t)$, equivalently in the direction of $\gamma'(t)$, then G lies to our left. We saw this for the curve γ in Example 8.1.10(a). We want to make this mathematical. In doing so we generalize the concept to all regular curves, not just those that are Jordan curves.

If γ is regular and $\gamma(t) = (x(t), y(t))$, we have that for each t in $[a, b]$ at least one of $x'(t)$ and $y'(t)$ is not zero. It follows that

$$N(t) = (-y'(t), x'(t))/\|\gamma'(t)\|$$

is a unit vector orthogonal to $T(t)$. In fact, $N(t)$ results from rotating $T(t)$ 90° in the counterclockwise direction. For each t the vector $N(t)$ is called the *unit normal vector* to the curve.

---

[1] A biographical note for Camille Jordan can be found in Definition 7.2.8.

8.2.2. Definition. If a piecewise regular curve $\gamma : [a, b] \to \mathbb{R}^2$ forms part of the boundary ∂G of an open subset G of the plane, say that γ is *positively oriented* relative to G, if for every t in $[a, b]$ where $\gamma'(t)$ exists there is an $\epsilon_0 > 0$ such that when $0 < \epsilon < \epsilon_0$ we have that $\gamma(t) + \epsilon N(t) \in G$.

Intuitively, γ is positively oriented if $N(t)$ is pointing inside G. Just for emphasis, in the preceding definition we are not assuming that γ is a Jordan curve or that it is the entirety of the boundary. Let's look at an example.

8.2.3. Example. Let $0 < r < R$ and put $G = \{(x, y) : r < \sqrt{x^2 + y^2} < R\}$, an annulus. If $\gamma_1(\theta) = (R\cos\theta, R\sin\theta)$ and $\gamma_2(\theta) = (r\cos(2\pi - \theta), r\sin(2\pi - \theta))$ for $0 \le \theta \le 2\pi$, then both γ_1 and γ_2 are positively oriented relative to G. See Exercises 1 and 2.

Now that we have established the idea of orientation of a curve relative to a set, we want to prove one of the fundamental theorems on integrating around a curve in the plane. In fact this is the Fundamental Theorem of Calculus for $\mathbb{R}^2$. First we need to set the stage with a few additional concepts.

8.2.4. Definition. A subset X of $\mathbb{R}^2$ is said to be of *Type I* if there is an interval $[a, b]$ and piecewise regular real-valued functions f_1, f_2 defined on $[a, b]$ such that $f_1(x) \le f_2(x)$ on (a, b) and $X = \{(x, y) \in \mathbb{R}^2 : a \le x \le b, f_1(x) \le y \le f_2(x)\}$. X is said to be of *Type II* if there is an interval $[c, d]$ and piecewise regular real-valued functions g_1, g_2 defined on $[c, d]$ such that $g_1(y) \le g_2(y)$ on (c, d) and $X = \{(x, y) \in \mathbb{R}^2 : c \le y \le d, g_1(y) \le x \le g_2(y)\}$.

The term Type I is traditional though it usually requires only that f_1 and f_2 be continuous functions, as we did when we discussed Fubini's Theorem (7.3.5). We say they are piecewise regular in the definition of Type I because this is the only context in which we will use the term. Similar comments apply to Type II. There are Type I sets that are not also Type II (Why?); but there are sets, for example a disk or a square, that are simultaneously Type I and Type II. Such sets will hold special interest for us.

8.2.5. Proposition. *If X is either Type I or Type II, then X is a compact Jordan set and there is a piecewise regular curve γ whose trace is ∂X and γ is positively oriented relative to* int X.

Proof. Here is the proof for a Type I set; the proof for a Type II set is similar. Use the notation established in the definition above. Proposition 7.3.10 establishes that X is a compact Jordan set. Define the following four curves on $[0, 1]$, where γ_1 and γ_3 trace the graphs of f_1 and f_2, respectively, but in opposite directions; while γ_2

and γ_4 trace vertical straight line segments, but again in opposite directions.

$$\gamma_1(t) = (tb + (1-t)a, f_1(tb + (1-t)a))$$
$$\gamma_2(t) = (b, tf_2(b) + (1-t)f_1(b))$$
$$\gamma_3(t) = (ta + (1-t)b, f_2(ta + (1-t)b))$$
$$\gamma_4(t) = (a, tf_1(a) + (1-t)f_2(a))$$

If we set $\gamma = \gamma_1 + \gamma_2 + \gamma_3 + \gamma_4$, then γ is a piecewise regular curve (Verify!) that traces ∂X. It easily follows that γ is positively oriented relative to int X. ∎

If, in the case of Type I, $f_1(x) < f_2(x)$ for $a < x < b$, then the curve γ in the proof of the preceding proposition is also a Jordan curve and int X is its inside. A similar statement applies to a Type II set. See Exercise 3.

8.2.6. Definition. Let X be a compact Jordan set in $\mathbb{R}^2$ whose boundary is a finite collection of piecewise regular curves. A collection $\{Y_j : 1 \le j \le m\}$ of closed subsets of X is called a *Type I cover* of X if each Y_j is a set that is Type I and the following are satisifed: (a) $X = \bigcup_{j=1}^m Y_j$; (b) when $1 \le i < j \le m$, $Y_i \cap Y_j = \partial Y_i \cap \partial Y_j$; (c) if Y_i, Y_j, and Y_k are three distinct sets, then $\partial Y_i \cap \partial Y_j \cap \partial Y_k$ is either empty or a finite set. The collection $\{Y_j : 1 \le j \le m\}$ of closed subsets of Y is called a *Type II cover* if each set Y_j is a Type II set and conditions (a), (b), and (c) above are also satisfied.

The set X is a *G-set* if it has a Type I cover as well as a Type II cover.

The terminology is not standard. It is given to facilitate the proof of the next theorem. Let's point out that from what we did in Chapter 7 it is automatic that X is a Jordan set if we assume that ∂X is a system of curves as described. We added the assumption that X is a Jordan set for emphasis. To be clear, condition (b) in the definition includes the possibility that $Y_i \cap Y_j = \emptyset$. Note that (b) also implies that int $Y_i \cap$ int $Y_j = \emptyset$ when $i \ne j$. Condition (c) prevents the boundary of one of the sets Y_i from doubling back on itself.

Examples of G-sets abound. Of course any set like an ellipse that is already both Type I and II is an example of a G-set. Here is a description of a general class of G-sets.

Suppose X is a compact subset of $\mathbb{R}^2$ such that $X = $ cl $[$int $X]$ and ∂X consists of a finite number of non-intersecting regular Jordan curves $\gamma_1, \dots, \gamma_m$. Suppose there is a R-cover $\mathcal{R}$ of X that has the following properties. If $R \in \mathcal{R}$ and $R \cap \partial X \ne \emptyset$, then R meets exactly one of the boundary curves γ_k. (See Figure 8.2.1.) Further assume that if R meets γ_k, then $\partial[R \cap X]$ is a Jordan curve consisting of three regular curves: a vertical and a horizontal line segment that meet at some point (a, b) in int X and a part of γ_k that joins the endpoints of these two segments that are different from (a, b). (In Figure 8.2.1, γ_k is γ_3.) The vertical and horizontal line segments form an L and, after suitable rotation, there are four different positions for this L. If we also have that int $[R \cap X]$ has the property that for any two points in this set the line segment joining them is also contained in the set, then $R \cap X$ is simultaneously a

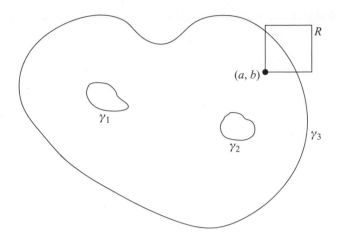

Figure 8.2.1

Type I and a Type II set. Since every rectangle is simultaneously a Type I and Type II set, we have that $\{R \cap X : R \in \mathcal{R}\}$ is both a Type I and a Type II cover of X. Hence X is a G-set.

Frankly it seems to me that we can carry out the procedure described in the preceding paragraph for any set whose boundary consists of a finite number of pairwise disjoint regular Jordan curves. Attempting to write a proof of this with all the details of the argument required, however, doesn't seem worth it.

The following idea will be useful in our discussions for showing a set is a G-set. If X is a compact subset of $\mathbb{R}^2$ with int $X \neq \emptyset$, then a *cut* of X is a simple piecewise regular curve $\lambda : [0, 1] \to X$ such that $\lambda(0), \lambda(1) \in \partial X$ and $\lambda(t) \in \text{int } X$ for $0 < t < 1$. So a cut literally cuts the set X. If int X is connected and λ is a cut, then (int X)\\$\{\lambda\}$ has two components. In practice we will show that a set X is a G-set by making a judicious choice of cuts. See the next example.

8.2.7. Example. Let $0 < a < b < \infty$ and put $X = \{(x, y) : a^2 \leq x^2 + y^2 \leq b^2\}$, an annulus. Let $\lambda_1, \lambda_2, \lambda_3, \lambda_4$ be the cuts defined on $[0, 1]$ as follows; $\lambda_1(t) = (tb + (1 - t)a, 0)$, $\lambda_2(t) = (0, tb + (1 - t)a)$, $\lambda_3(t) = (-tb - (1 - t)a, 0)$, $\lambda_4(t) = (0, -tb - (1 - t)a)$. If X_1, X_2, X_3, X_4 are the sets whose interiors are the components of int $X \setminus \bigcup_{j=1}^4 \{\gamma_j\}$, then these sets show that X is a G-set. Also see Exercise 5.

If X is a G-set and ∂X consists of the traces of the disjoint piecewise regular curves $\gamma_1, \ldots, \gamma_n$, we adopt the notation that for any continuous function $f : \partial X \to \mathbb{R}$

8.2.8
$$\int_{\partial X} f \, ds = \sum_{k=1}^n \int_{\gamma_k} f \, ds$$

Before stating and proving the main theorem, here is a lemma, which is little more than an observation.

8.2.9. Lemma. *Let X be a G-set with $\{X_j\}$ either a Type I or Type II cover.*

(a) *If $F : X \to \mathbb{R}^2$ is a smooth function, then $\int_{\partial X} F \cdot T \, ds = \sum_{j=1}^{m} \int_{\partial X_j} F \cdot T \, ds$.*

(b) *If $g : X \to \mathbb{R}$ is a continuous function, then $\iint_X g = \sum_{j=1}^{m} \iint_{X_j} g$.*

Proof. We only do this for a Type I cover; the proof for a Type II cover is similar. The proof of (b) is immediate since if $\{X_j : 1 \le j \le m\}$ is a Type I cover of X the area of $\bigcup_{j=1}^{m} \partial X_j$ is zero. To prove (a) observe what happens when $1 \le i < j \le m$ and $\partial X_i \cap \partial X_j \ne \emptyset$. If this intersection contains a piecewise regular arc γ, then from Proposition 8.2.5 we know that when γ appears as part of ∂X_i its direction is opposite to that of its direction when it appears as part of ∂X_j. So the contributions of γ to the two integrals $\int_{\partial X_i} g \, ds$ and $\int_{\partial X_j} g \, ds$ are the negative of one another. If $\partial X_i \cap \partial X_j$ has an isolated point, this does not contribute to the corresponding line integrals. This establishes (a). ∎

8.2.10. Theorem (Green's[2] Theorem). *If X is a G-set and $F : X \to \mathbb{R}^2$ is a smooth function with $F(x, y) = (P(x, y), Q(x, y))$, then*

$$\int_{\partial X} F \cdot T \, ds = \iint_X \left(\frac{\partial Q}{\partial x} - \frac{\partial P}{\partial y} \right)$$

Proof. Using (8.1.11) we want to show that

$$\iint_X \left(\frac{\partial Q}{\partial x} - \frac{\partial P}{\partial y} \right) = \int_{\partial X} P \, dx + Q \, dy$$

Separating the sides of the equation we'll show that

8.2.11 $$\iint_X \frac{\partial P}{\partial y} = - \int_{\partial X} P \, dx \quad \text{and} \quad \iint_X \frac{\partial Q}{\partial x} = \int_{\partial X} Q \, dy$$

Let's show the first of these. Since X is a G-set it has a Type I cover. The preceding lemma says it suffices to prove this under the additional assumption that X is a

---

[2] George Green was born in 1793 in Sneinton, England. His father was a baker and George's formal schooling was rather limited. In fact he had only about a year's formal education starting when he was eight. He had to withdraw from school when he was needed back in the bakery. With time his father prospered and purchased several other bakeries and even established his own mill where George worked. Somehow through all this, George continued to study mathematics. How he became familiar with the forefront of mathematics is not known. He never married Jane Smith, the daughter of the manager of his father's mill, but they lived together and eventually had seven children. In 1828 Green published one of his landmark papers on electricity and magnetism. It was offered "on subscription," meaning that to obtain a copy you had to pay a fee. One of the subscribers was Sir Edward Bromhead, an established mathematician, who immediately contacted Green and offered to communicate his future papers to the Royal Society of London. After a delay by Green, he and Bromhead began regular meetings. Green soon produced three new papers, two on electricity and one on hydrodynamics, all of which were published in leading journals. Later Bromhead suggested Green study at Cambridge, where he enrolled in 1833; he became an undergraduate at the age of 40. He graduated four years later and in 1839 he became a Perse fellow. (The fellowship required that the holders be unmarried, which Green was in spite of his having six children at the time. His seventh was born shortly after receiving the fellowship.) Unfortunately his health deteriorated and he died in 1841 in Sneinton.

set of Type I; let $X = \{(x, y) : a \le x \le b \text{ and } f_1(x) \le y \le f_2(x)\}$. Let the positively oriented piecewise regular curve $\partial X = \gamma_1 + \gamma_2 - \gamma_3 + \gamma_4$, where these four curves are defined as in Proposition 8.2.5. Note that since γ_2 and γ_4 are vertical line segments, the value of x is constant (either a or b). Hence $\int_{\gamma_2} P \, dx = \int_{\gamma_4} P \, dx = 0$. As we pointed out in the proof of Proposition 8.2.5, γ_1 traces the graph of f_1 from left to right while γ_3 traces the graph of f_3 but in the opposite direction. Remember this in the next step of this proof.

We now apply Fubini's Theorem to get

$$\iint_X \frac{\partial P}{\partial y} = \int_a^b \left[\int_{f_1(x)}^{f_2(x)} \frac{\partial P}{\partial y}(x, y) \, dy \right] dx$$

$$= \int_a^b [P(x, f_2(x)) - P(x, f_1(x))] \, dx$$

$$= -\int_{\gamma_3} P \, dx - \int_{\gamma_1} P \, dx$$

$$= -\int_{\gamma_3} P \, dx - \int_{\gamma_1} P \, dx - \int_{\gamma_2} P \, dx - \int_{\gamma_4} P \, dx$$

$$= -\int_{\partial X} P \, dx$$

The next step is to show that

8.2.12
$$\int_\gamma Q \, dy = \iint_X \frac{\partial Q}{\partial x}$$

This is left to the reader as Exercise 6. ∎

8.2.13. Example. (a) Find $\int_\gamma F \cdot T \, ds$ when γ is the perimeter of the rectangle $R = [0, 3] \times [1, 4]$ with the counterclockwise direction and $F(x, y) = (x^3, 3x + y^4)$. We could parametrize γ and carry out the process, but it's a bit easier to use Green's Theorem. Here $\partial P/\partial y = 0$, $\partial Q/\partial x = 3$, so $\int_\gamma F \cdot T \, ds = \iint_R (3) = 27$.

(b) Let $F(x, y) = (0, x)$ and let X be any G-set. Green's Theorem says that

$$\int_{\partial X} x \, dy = \text{Area}\,(X)$$

Exercises

(1) Verify Example 8.2.3.
(2) Evaluate $\int_{\partial X} F \cdot T \, ds$ when X is the annulus described in Example 8.2.3 and $F(x, y) = (-y^3, x^2)$.
(3) If $X = \{(x, y) \in \mathbb{R}^2 : a \le x \le b, f_1(x) \le y \le f_2(x)\}$ as in the definition of a Type I set, show that ∂X is a Jordan curve if $f_1(x) < f_2(x)$ when $a < x < b$.

(4) (a) Show that a disk is both a Type I set and a Type II set. (b) Give an example of a Type I set that is not Type II. (c) Give an example of a Type II set that is not Type I.

(5) (a) Show that $X = \{(x, y) : 0 \le x \le 2\pi, |y| \le |\sin x|\}$ is a G-set. (b) Let X be the set in $\mathbb{R}^2$ bounded by the three circles $\{(x, y) : x^2 + y^2 = 1\}$ and $\{(x, y) : \sqrt{(x \pm \frac{1}{8})^2 + y^2} = \frac{1}{8}\}$ and show that it is a G-set.

(6) Prove (8.2.12) as part of the proof of Green's Theorem.

(7) In Green's Theorem what happens if you take $P(x, y) = -y$ and $Q(x, y) = 0$? Suppose $P(x, y) = 0$ and $Q(x, y) = x$?

(8) Find $\int_\gamma F \cdot T\, ds$ when $F(x, y) = (xy, x^2 y^3)$ and γ is the positively oriented perimeter of the triangle with vertices $(0, 0), (1, 0), (1, 2)$.

8.3. Surfaces

We are predisposed to think of a surface as a two-dimensional object inside three-dimensional space and that's what we'll do in this section. Later in §9.1 we'll discuss q-dimensional surfaces that are contained in p-dimensional spaces.

8.3.1. Definition. A *surface domain* in $\mathbb{R}^2$ is a compact Jordan subset R of $\mathbb{R}^2$ such that int R is connected, $R = \mathrm{cl}\,(\mathrm{int}\, R)$, and ∂R is a finite collection of pairwise disjoint piecewise regular curves. A 2-*surface* in $\mathbb{R}^3$, or just a *surface*, is a pair (Φ, R) where R is a surface domain in $\mathbb{R}^2$ and $\Phi : R \to \mathbb{R}^3$ is a continuous function that is smooth on a neighborhood of R. The *trace* of (Φ, R) is the set $\{\Phi\} = \Phi(R) \subseteq \mathbb{R}^3$.

Now that we have stated that a surface is a function, let's mention that we will sometime violate this and refer to a surface as a set S. When we do this, what we mean is that there is a surface (Φ, R) whose trace is the set S. In such a situation (Φ, R) is called a *parametrization* of S. (See Exercise 1.) As you might expect from the treatment of curves, we'll see that such a parametrization is not unique.

To be completely precise, always a virtue, we should have defined an object (with some name) as we defined a surface above, then define an equivalence relation between these objects, and define a surface as an equivalence class of such objects. On the other hand there is another virtue, simplicity of exposition. It seems to me in this case that this latter virtue is more important than the first. We will soon define, however, an equivalence relation between surfaces and show that equivalent surfaces have the same properties we will be interested in exploring.

8.3.2. Example. Let $0 < r < R$ and call the set $S = \{x \in \mathbb{R}^3 : r \le \|x\| \le R\}$ the *corona*. Note that $\partial S = \{x \in \mathbb{R}^3 : \|x\| = r\} \cup \{x \in \mathbb{R}^3 : \|x\| = R\}$. Each component of this boundary is a surface (Exercise 1), but since ∂S is not connected it is not a surface. Each component, however, is.

We'll see more examples of surfaces shortly; we gave the preceding one to underline a facet of the definition. As we noted the definition of a surface (Φ, R) requires that int R, and hence R, be connected and this implies that $\{\Phi\}$ is connected as well as compact. Hence the boundary of a corona is not, strictly speaking, a surface. This is not significant. We want the interior of a surface domain to be connected to apply previous results, but we could have dropped the connectedness assumption and treat one component at a time. Later, in the following section, we'll broaden the concept of a surface where we drop this insistence on connectivity. We'll stick for the moment with the approach taken here, however, as it is more efficacious.

8.3.3. Example. (a) If R is a surface domain in $\mathbb{R}^2$ and $f : R \to \mathbb{R}$, then define $\Phi : R \to \mathbb{R}^3$ by $\Phi(x, y) = (x, y, f(x, y))$. If the function f is smooth in a neighborhood of R, then (Φ, R) is a 2-surface.

(b) If $R = \{u \in \mathbb{R}^2 : \|u\| \leq 1\}$, let $S = \{(u, \sqrt{1 - \|u\|^2}) : u \in R\}$. This set is a hemisphere centered at the origin. If we defined $\Phi : R \to \mathbb{R}^3$ by $\Phi(u) = (u, \sqrt{1 - \|u\|^2})$, then $\{\Phi\} = S$ but Φ fails to be smooth in a neighborhood of R. Exercise 2 gives a parametrization (Ψ, T) that is smooth in a neighborhood of the surface domain T and has this hemisphere as its trace.

(c) A truncated cylinder is the trace of a 2-surface. For example, let $S = \{(u, v, t) \in \mathbb{R}^3 : u^2 + v^2 = 1, 0 \leq t \leq 2\}$, $R = [0, 2\pi] \times [0, 2]$, and define the surface $\Phi : R \to \mathbb{R}^3$ by $\Phi(\theta, t) = (\cos \theta, \sin \theta, t)$.

(d) The torus in $\mathbb{R}^3$ is a 2-surface. In fact let $a > b > 0$, let $R = [-\pi, \pi] \times [-\pi, \pi] \subseteq \mathbb{R}^2$, and define $\Phi : R \to \mathbb{R}^3$ by $\Phi(u, v) = ((a + b \cos v) \cos u, (a + b \cos v) \sin u, b \sin v)$. The reader can verify (Exercise 4) that the image of $\{0\} \times [-\pi, \pi]$ is a circle in the xz-plane centered at $(a, 0, 0)$ and having radius b and that $\Phi(R)$ is a torus.

In analogy with what we did for curves in the last section, we seek a way to orient surfaces. A moment's thought shows this is not as straightforward a task as it was for curves where we oriented by establishing the unit tangent vector $T(t)$ for a regular curve. As an introduction, let's consider a special kind of surface.

Let R be a surface domain in $\mathbb{R}^2$ and suppose we have a function $f : R \to \mathbb{R}$ smooth in a neighborhood of R. We define a 2-surface Φ on R by $\Phi(u, v) = (u, v, f(u, v))$ as in Example 8.3.3(a). In §6.7 we examined a smooth curve $\gamma : (-1, 1) \to R$ and the induced curve $\tilde{\gamma} : (-1, 1) \to \mathbb{R}^3$ defined by $\tilde{\gamma}(t) = (\gamma(t), f(\gamma(t)) = \gamma(t) \oplus f(\gamma(t))$ in $\{\Phi\} \subseteq \mathbb{R}^2 \oplus \mathbb{R}$. We computed

$$\tilde{\gamma}'(0) = (\gamma'(0), \nabla f(\gamma(0)) \cdot \gamma'(0)) = (\gamma'(0), \langle \nabla f(\gamma(0)), \gamma'(0) \rangle)$$

a vector in $\mathbb{R}^3$ that is tangent to the curve $\tilde{\gamma}$ at $t = 0$. In Proposition 6.7.8 we proved that if $\gamma(0) = (u_0, v_0)$, then the vector $(\nabla f(u_0, v_0), -1)$ is orthogonal to $\tilde{\gamma}'(0)$. Note that $(\nabla f(u_0, v_0), -1)$ does not depend on the curve γ and so it is perpendicular to every curve in $\{\Phi\}$ passing through $(u_0, v_0, f(u_0, v_0))$. Thus we have (Definition 6.7.10) that

$$(u_0, v_0, f(u_0, v_0)) + \{x \in \mathbb{R}^3 : x \perp (\nabla f(u_0, v_0), -1)\}$$

is the tangent affine hyperplane to the surface Φ and

$$(\nabla f(u_0, v_0), -1)/\sqrt{\|\nabla f(u_0, v_0)\|^2 + 1}$$

is a unit normal vector to the surface at the point $\Phi(u_0, v_0)$. So we see that when a 2-surface in $\mathbb{R}^3$ is the graph of a smooth function, it has a unit normal vector. We seek to extend this to arbitrary 2-surfaces in $\mathbb{R}^3$.

If (Φ, R) is a 2-surface in $\mathbb{R}^3$, then write $\Phi(u, v) = (\phi_1(u, v), \phi_2(u, v), \phi_3(u, v))$ for (u, v) in R. We want to consider the Jacobians

$$\Delta_{\phi_i, \phi_j}(u_0, v_0) = \frac{\partial(\phi_i, \phi_j)}{\partial(u, v)}(u_0, v_0)$$

The next result hints at the reason for our consideration of these Jacobians. The idea is that under a mild restriction we can use what we did above to show that the surface has a unit normal vector.

8.3.4. Proposition. *If (Φ, R) is a smooth 2-surface with $\Phi = (\phi_1, \phi_2, \phi_3)$ and $(u_0, v_0) \in \text{int } R$ such that $\Delta_{\phi_1, \phi_2}(u_0, v_0) \neq 0$, then there is an open set Ω in $\mathbb{R}^2$ and a smooth function $f : \Omega \to \mathbb{R}$ such that*

$$\Phi(u_0, v_0) \in \{(x, y, f(x, y)) : (x, y) \in \Omega\} \subseteq \Phi(R)$$

Proof. Define $\psi : R \to \mathbb{R}^2$ by $\psi(u, v) = (\phi_1(u, v), \phi_2(u, v))$. Observe that the hypothesis implies that the derivative of ψ at (u_0, v_0) is an invertible linear transformation from $\mathbb{R}^2$ into itself. By the Inverse Function Theorem (6.8.2) there is an open set G with $G \subseteq R \subseteq \mathbb{R}^2$ that contains (u_0, v_0) such that $\Omega = \psi(G)$ is open, ψ is injective on G, and $h = (\psi|G)^{-1} : \Omega \to \mathbb{R}^2$ is continuously differentiable. Define $f : \Omega \to \mathbb{R}$ by

$$f(x, y) = \phi_3(h(x, y))$$

and note that f is a smooth function on Ω. If $(u, v) \in G \subseteq R$ and $\psi(u, v) = (x, y)$, then $h(x, y) = (u, v)$ and

$$\begin{aligned}
\Phi(u, v) &= (\phi_1(u, v), \phi_2(u, v), \phi_3(u, v)) \in \mathbb{R}^3 \\
&= \psi(u, v) \oplus \phi_3(h(x, y)) \in \mathbb{R}^2 \oplus \mathbb{R} \\
&= \psi(u, v) \oplus f(x, y) \in \mathbb{R}^2 \oplus \mathbb{R} \\
&= (x, y, f(x, y)) \in \mathbb{R}^3
\end{aligned}$$

completing the proof. ∎

Observe that the hypothesis in the preceding result that $\Delta_{\phi_1, \phi_2}(u_0, v_0) \neq 0$ can be replaced by $\Delta_{\phi_i, \phi_j}(u_0, v_0) \neq 0$ for any $i \neq j$ and the conclusion will remain that the point $\Phi(u_0, v_0)$ is contained in a small relatively open subset of $\{\Phi\}$ that is a graph. For example if $\Delta_{\phi_2, \phi_3}(u_0, v_0) \neq 0$ we will have an open subset Ω of $\mathbb{R}^2$ and a smooth function $f : \Omega \to \mathbb{R}$ such that

$$\Phi(u_0, v_0) \in \{(f(x, y), x, y) : (x, y) \in \Omega\} \subseteq \Phi(R)$$

A similar conclusion holds if the hypothesis is that $\Delta_{\phi_1, \phi_3}(u_0, v_0) \neq 0$.

Before we can state a condition for there to be a unit normal vector, we need to briefly review the *cross product* of vectors in $\mathbb{R}^3$.

Recall that if $x = (x_1, x_2, x_3)$, $y = (y_1, y_2, y_3) \in \mathbb{R}^3$, then

$$x \times y = (x_2y_3 - x_3y_2, x_3y_1 - x_1y_3, x_1y_2 - x_2y_1)$$

There is a certain cyclic pattern in this definition, but the easiest way to remember it was given in Calculus as the determinant

$$x \times y = \det \begin{bmatrix} i & j & k \\ x_1 & x_2 & x_3 \\ y_1 & y_2 & y_3 \end{bmatrix} = \det \begin{bmatrix} e_1 & e_2 & e_3 \\ x_1 & x_2 & x_3 \\ y_1 & y_2 & y_3 \end{bmatrix}$$

where the unit vectors i, j, k are now being labeled e_1, e_2, e_3. You can see that these equations are the same as the definition above by expanding the determinant by minors using the first row. Here are the basic properties of the cross product.

8.3.5. Proposition. *If $x, y, z \in \mathbb{R}^3$ and $\alpha \in \mathbb{R}$, the following hold.*

(a) $x \times x = 0$ *and* $x \times y = -y \times x$.
(b) $(\alpha x) \times y = x \times (\alpha y) = \alpha(x \times y)$.
(c) $x \times (y + z) = (x \times y) + (x \times z)$.
(d) $\langle x \times y, z \rangle = \langle x, y \times z \rangle = \det \begin{bmatrix} x_1 & x_2 & x_3 \\ y_1 & y_2 & y_3 \\ z_1 & z_2 & z_3 \end{bmatrix}$.
(e) $x \times (y \times z) = \langle x, z \rangle y - \langle x, y \rangle z$.
(f) $\|x \times y\|^2 = \|x\|^2 \|y\|^2 - \langle x, y \rangle^2$.
(g) *If $x \times y \neq 0$, then $x \times y$ is orthogonal to both x and y.*

Proof. The proof of much of this proposition is routine and in Exercise 6 the reader is asked to supply the details. (a) Regard the determinant definition of the cross product and realize that part (a) is a consequence of two facts about determinants. $x \times x = 0$ is a consequence of the fact that a determinant is 0 when two rows are identical. $x \times y = -y \times x$ is a consequence of the fact that when two rows in a determinant are interchanged, the new determinant is the negative of the other.

(b) and (c). Just use the definition of the cross product,

(d), (e), and (f). By the definition of the cross product, $x \times y = (x_2y_3 - x_3y_2, x_3y_1 - x_1y_3, x_1y_2 - x_2y_1)$. Take the inner product with z. Now write out the cross product of $y \times z$ and take its inner product with x. Compare the two calculations to see that they are equal. Expand the determinant in (d) by minors using the first row to see that this equals the formula for $\langle x, y \times z \rangle$. (e) Compute both sides and compare. (Ugh!) Follow the same instructions to show (f).

(g) By (d) and (a), $\langle x \times y, y \rangle = \langle x, y \times y \rangle = 0$. Similarly, $\langle y \times x, x \rangle = \langle y, x \times x \rangle = 0$. ∎

Now to fix some notation that will help us explore what we mean by an orientation of a 2-surface.

8.3.6. Definition. If (Φ, R) is a 2-surface in $\mathbb{R}^3$ with $\Phi = (\phi_1, \phi_2, \phi_3)$, then

$$\Phi_u = \frac{\partial \Phi}{\partial u} = \left(\frac{\partial \phi_1}{\partial u}, \frac{\partial \phi_2}{\partial u}, \frac{\partial \phi_3}{\partial u} \right) \in \mathbb{R}^3$$

$$\Phi_v = \frac{\partial \Phi}{\partial v} = \left(\frac{\partial \phi_1}{\partial v}, \frac{\partial \phi_2}{\partial v}, \frac{\partial \phi_3}{\partial v} \right) \in \mathbb{R}^3$$

Note that Φ_u and Φ_v are precisely the first and second columns in the matrix of $D\Phi(u, v)$ in $\mathcal{L}(\mathbb{R}^2, \mathbb{R}^3)$. So Φ_u, Φ_v are continuous functions from R into $\mathbb{R}^3$. Thus we obtain another function $\Phi_u \times \Phi_v : R \to \mathbb{R}^3$; by a calculation

$$\Phi_u \times \Phi_v = \det \begin{bmatrix} i & j & k \\ (\phi_1)_u & (\phi_2)_u & (\phi_3)_u \\ (\phi_1)_v & (\phi_2)_v & (\phi_3)_v \end{bmatrix}$$

$$= \left(\Delta_{\phi_2, \phi_3}, \Delta_{\phi_3, \phi_1}, \Delta_{\phi_1, \phi_2} \right)$$

For any point (u, v) in R, define

8.3.7 $\quad N_\Phi(u, v) = (\Phi_u \times \Phi_v)(u, v) = \left(\frac{\partial(\phi_2, \phi_3)}{\partial(u, v)}, \frac{\partial(\phi_3, \phi_1)}{\partial(u, v)}, \frac{\partial(\phi_1, \phi_2)}{\partial(u, v)} \right)$

8.3.8. Example. Let Ω be an open subset of $\mathbb{R}^2$, let R be a surface domain contained in Ω, and assume $f : \Omega \to \mathbb{R}$ is a smooth function. If Φ is given by $\Phi(u, v) = (u, v, f(x, y))$, then $\Phi_u = (1, 0, f_u)$ and $\Phi_v = (0, 1, f_v)$. It follows that $N_\Phi(u, v) = (-\nabla f(x, y), 1)$. See Exercise 7.

8.3.9. Proposition. *If (Φ, R) is a 2-surface in $\mathbb{R}^3$ and there is a point (u_0, v_0) in int R such that $N_\Phi(u_0, v_0) \neq 0$, then $N_\Phi(u_0, v_0)$ is orthogonal to every smooth curve that lies in $\{\Phi\}$ and passes through the point $\Phi(u_0, v_0)$.*

Proof. Let $w_0 = (u_0, v_0) \in \text{int } R$ and let $\zeta_0 = \Phi(u_0, v_0) \in \{\Phi\}$. First note that if $N_\Phi(w_0) \neq 0$, then at least one of its coordinates is not 0; without loss of generality we may assume that $\Delta_{\phi_1, \phi_2}(w_0) \neq 0$. By Proposition 8.3.4 there is an open set Ω in $\mathbb{R}^2$ and a smooth function $f : \Omega \to \mathbb{R}$ such that

$$\zeta_0 \in S = \{(u, v, f(u, v)) : (u, v) \in \Omega\} \subseteq \Phi(R)$$

Thus to prove the proposition it suffices to assume that Φ is given by $\Phi((u, v) = (u, v, f(u, v))$ for a smooth function $f : \Omega \to \mathbb{R}$ as in Example 8.3.8. Using that example and Proposition 6.7.8, this finishes the proof. ∎

In analogy with Definition 6.7.10 we present the following.

8.3.10. Definition. If (Φ, R) is a 2-surface in $\mathbb{R}^3$ and $(u_0, v_0) \in \text{int } R$ such that $N_\Phi(u_0, v_0) \neq 0$, then the *tangent plane* to $\{\Phi\}$ at the point $\Phi(u_0, v_0)$ is defined as

$$\Phi(u_0, v_0) + \{\zeta \in \mathbb{R}^3 : \zeta \perp N_\Phi(u_0, v_0)\}$$

Here is another definition.

8.3.11. Definition. A 2-surface (Φ, R) in $\mathbb{R}^3$ is said to be *regular* if $N_\Phi(u, v) \neq 0$ for all (u, v) in int R.

So for a regular surface we have a well-defined normal vector at every point of the interior of its domain. Now we want to examine what happens when we change variables.

8.3.12. Proposition. *If (Φ, R) and (Ψ, Q) are 2-surfaces in $\mathbb{R}^3$ and $\tau : Q \to R$ is a mapping that is smooth in some neighborhood of Q such that $\Psi = \Phi \circ \tau$, then for all (u, v) in int Q we have that*

$$N_\Psi(u, v) = \Delta_\tau(u, v) N_\Phi(\tau(u, v)) = \frac{\partial(\tau_1, \tau_2)}{\partial(u, v)} N_\Phi(\tau(u, v))$$

Proof. This is basically an application of the Chain Rule. As usual we set $\Phi = (\phi_1, \phi_2, \phi_3)$ and $\Psi = (\psi_1, \psi_2, \psi_3)$. For $1 \leq i < j \leq 3$ and the fact that $(\psi_i, \psi_j) = (\phi_i, \phi_j) \circ \tau$, a computation shows that $\Delta_{(\psi_i, \psi_j)}(u, v) = \Delta_\tau(u, v) \Delta_{(\phi_i, \phi_j)}(u, v)$ for all (u, v) in Q. The result now follows directly by computation. ∎

8.3.13. Definition. Two regular 2-surfaces (Φ, R) and (Ψ, Q) in $\mathbb{R}^3$ are *equivalent* if there is a neighborhood Ω of Q and a bijective regular mapping τ from Ω onto a neighborhood of R such that $\tau(Q) = R$ and $\Psi = \Phi \circ \tau$.

It follows that this is indeed an equivalence relation between 2-surfaces (Exercise 8). Technically we should define a 2-surface as an equivalence class of what we have defined as a 2-surface. Indeed in many books this is the approach taken. Our approach of defining a surface as a function was taken for simplicity and ease of exposition. Nevertheless we will still have to show that various properties of surfaces remain the same for equivalent surfaces.

Note that in practice when we want to define a map τ to demonstrate that two surfaces are equivalent, we won't specify the neighborhood Ω on which τ is defined but only define τ on the set Q. The definition of τ on a neighborhood will be clear.

Exercises

(1) Let $R > 0$ and put $S = \{x \in \mathbb{R}^3 : \|x\| = R\}$. Find a surface (Φ, R) such that $\{\Phi\} = S$.

(2) Let $T = [0, 2\pi] \times [0, \pi]$ and define $\Psi : T \to \mathbb{R}^3$ by

$$\Psi(u, v) = (\cos u \cos v, \sin u \cos v, \sin v)$$

Show that (Ψ, T) is a surface whose trace is the hemisphere as in Example 8.3.3.

(3) Verify the statement in Example 8.3.3(c).

(4) In Example 8.3.3(d), what is the image of the segment $[-\pi, \pi] \times \{v\}$? What are the images of the segments $[-\pi, \pi] \times \{\pm\pi\}$? Verify that $\Phi([-\pi, \pi] \times [-\pi, \pi])$ is a torus.

(5) If $R = [0, 2\pi] \times [0, b]$ and $\Phi : R \to \mathbb{R}^3$ is defined by $\Phi(u, v) = (v \cos u,$ $v \sin u, v)$, show that Φ is a 2-surface whose trace is the truncated cone $\{(x, y, z) : z = \sqrt{x^2 + y^2}, 0 \le z \le b\}$.

(6) Supply the missing details in the proof of Proposition 8.3.5.

(7) In Exercise 2 above calculate N_Ψ.

(8) Show that the concept of equivalence defined in Definition 8.3.13 is an equivalence relation on the collection of all 2-surfaces in $\mathbb{R}^3$.

8.4. Integration on Surfaces

In this rather long section we'll define an integral on a regular 2-surface using the normal vector and explore its properties. We'll also introduce orientable surfaces and the concept of the positive orientation of a surface relative to a solid it surrounds.

8.4.1. Definition. If (Φ, R) is a regular 2-surface in $\mathbb{R}^3$ and $f : \{\Phi\} \to \mathbb{R}$ is a continuous function, define the *surface integral* of f over Φ as

$$\iint_\Phi f \, d\sigma = \int_R f \circ \Phi \, \|N_\Phi\| = \int_R f(\Phi(u, v)) \, \|N_\Phi(u, v)\| \, dudv$$

and define the *surface area* of $\{\Phi\}$ as

$$\sigma(\Phi) = \int_R \|N_\Phi((u, v))\| \, dudv$$

Why define the integral this way?

It is possible to look at this formula in a way that makes it seem natural once we justify the the formula for $\sigma(\Phi)$; let's undertake this. Imagine approximating the area of $\{\Phi\}$ by the sum of small rectangles contained in the tangent planes to the surface. The area of each of these small rectangles is close to the length of the normal vector $\|N_\Phi\|$ times the area of the underlying rectangle in the surface domain R. Once you accept this we see that this leads to the integral defining $\sigma(\Phi)$ in the definition. We'll avoid the details but a web search may shed some light on such an undertaking. Once this is established we see that the definition of $\iint_\Phi f \, d\sigma$ is obtained by approximating by the sum of the values of the function f times a small amount of area of the underlying surface.

The reasons for looking at this integral are rooted in physics, but we won't go into this; the interested reader can do a web search for "vector fields" and "surface integral."

We'll look at some examples below that show that the above formula gives the value of the surface area we expect. First let's point out that the surface area and surface integrals for equivalent regular surfaces are equal (Exercise 6).

8.4.2. Example. (a) Let's start with a rectangle in the xy-plane considered as a subset of $\mathbb{R}^3$. Let $R = [a, b] \times [c, d] \subseteq \mathbb{R}^2$ and define $\Phi(u, v) = (u, v, 0)$.

Here $\Phi_u(u, v) = (1, 0, 0)$ and $\Phi_v(u, v) = (0, 1, 0)$ for all (u, v). It follows that $N_\Phi(u, v) = (0, 0, 1)$. Thus

$$\sigma(\Phi) = \int_R \|(0, 0, 1)\| \, dudv = (b - a)(d - c)$$

as expected. We might give a preview here of something we will soon do, namely discuss the surface area of a box in $\mathbb{R}^3$. This is an example of a non-smooth surface, but we can call it a piecewise smooth surface. We can approach the calculation of this surface area by partitioning the surface of the box into six smooth surfaces, calculating each of their areas and adding them together.

(b) How about a circular disk in the xy-plane? The approach is similar to what we did in part (a) and this is left for the reader as Exercise 1.

(c) Calculate the surface area of a truncated cylinder of radius 1. Recall from Example 8.3.3(c) that we let $R = [0, 2\pi] \times [0, 2]$ and define the surface Φ : $R \to \mathbb{R}^3$ by $\Phi(\theta, t) = (\cos\theta, \sin\theta, t)$; and $\{\Phi\} = \{(x, y, t) \in \mathbb{R}^3 : x^2 + y^2 = 1, 0 \le t \le 2\}$. So $\Phi_\theta(\theta, t) = (-\sin\theta, \cos\theta, 0)$ and $\Phi_t(\theta, t) = (0, 0, 1)$. Hence $N_\Phi(\theta, t) = (\cos\theta, \sin\theta, 0)$ for all (θ, t) in R and $\|N_\Phi\| = 1$. Thus $\sigma(\Phi) = \int_0^{2\pi} \int_0^2 dtd\theta = 4\pi$. If you slit the cylinder vertically and lay it flat, you get a rectangle with sides of length 2π and 2. So again the formula from Definition 8.4.1 yields the correct answer.

(d) The area of a graph of $f : R \to \mathbb{R}$. Here $\Phi(u, v) = (u, v, f(u, v))$, so that $N_\Phi(u, v) = (-f_u(u, v), -f_v(u, v), 1)$. Thus

$$\sigma(\Phi) = \int_R \sqrt{1 + f_u(u, v)^2 + f_v(u, v)^2} \, dudv$$

We need to introduce additional concepts.

I ask the reader to be patient when (s)he reads the next definition as it may seem strange. We'll have a discussion after the statement that I think will help. Remember that an open disk in $\mathbb{R}^2$ is a set of the form $B((a, b); r)$.

8.4.3. Definition. If (Φ, R) is a smooth surface in $\mathbb{R}^3$ and (x, y, z) is a point in $\{\Phi\}$, say that (x, y, z) is an *interior point* of Φ if there is a homeomorphism of an open disk in $\mathbb{R}^2$ onto a relatively open subset of $\{\Phi\}$ that takes the center of the disk onto (x, y, z). Let int Φ denote the set of interior points of Φ. A point of $\{\Phi\}$ is a *boundary point* if it is not an interior point. We denote the set of boundary points by $\partial\Phi$. The surface is said to be *closed* if it has no boundary points.

The first comment is that this terminology is standard even though it introduces an ambiguity with the topological terms interior, boundary, and closed as well as the notation for the interior and boundary. On the other hand the definition of an interior point and boundary point of a surface is consistent with the topological notions if we think of the relative topology. The definition of a closed surface is completely at odds, however, with its topological cousin. This all seems unfortunate but in practice it should not lead to confusion. We'll see some examples shortly.

Realize that the trace of a two-surface, $\{\Phi\}$, is the continuous image of the compact set R so topologically it is always closed in $\mathbb{R}^3$. Also it is a two-dimensional object sitting inside $\mathbb{R}^3$ and so it cannot have any interior points in the topological sense. Thus in the topological sense, every point is a boundary point. With such observations we see that to a large extent the topological versions of the words are irrelevant for a discussion of a surface and won't arise very often. Nevertheless to aid in making this distinction we wiil use the symbols int Φ and $\partial\Phi$ for the interior and boundary points of the surface as defined above and always use $\{\Phi\}$ to discuss topological properties of the trace of the surface.

It is useful to be aware of an intuitive or visual way of defining the interior points of a surface. Imagine (and do not take this literally) you are standing at some point of the surface and you decide to move a small amount. If it is the case that you can move in any direction while remaining on the surface, then you are standing at an interior point. If there is some direction such that no matter how small a step you take in that direction you must fall off the surface, then you are standing at a boundary point.

Now let's consider a few examples. Consider the surface Φ whose trace is the rectangle sitting inside the xy-plane in $\mathbb{R}^3$ as introduced in Example 8.4.2(a) above. Here $\partial\Phi$ is the set of four edges while the other points are in int Φ. The sphere in $\mathbb{R}^3$ is a closed surface since every point is an interior point. The same is true of the torus. The hemisphere, however, is not a closed surface.

The proof of the next result is an application of the Inverse Function Theorem and is left to the reader as Exercise 7.

8.4.4. Proposition. *If (Φ, R) is a regular 2-surface, then the image of the topological interior of R is contained in* int Φ.

8.4.5. Definition. Say that a regular surface (Φ, R) is *oriented* if Φ is a C'' function and for (u_1, v_1), (u_2, v_2) in R with $\Phi(u_1, v_1) = \Phi(u_2, v_2)$ in int Φ, it follows that

$$\frac{N_\Phi(u_1, v_1)}{\|N_\Phi(u_1, v_1)\|} = \frac{N_\Phi(u_2, v_2)}{\|N_\Phi(u_2, v_2)\|}$$

If (Φ, R) is oriented and $(x_0, y_0, z_0) = \Phi(u_0, v_0) \in$ int Φ, we define the *unit normal* to be the vector-valued function $\mathbf{n} :$ int $\Phi \to \mathbb{R}^3$ by

$$\mathbf{n}(x_0, y_0, z_0) = \frac{N_\Phi(u_0, v_0)}{\|N_\Phi(u_0, v_0)\|}$$

When (Φ, R) is oriented with unit normal $\mathbf{n}$ and $F : \{\Phi\} \to \mathbb{R}^3$ is a continuous function, then the *oriented surface integral* of F on Φ is defined as

$$\iint_\Phi F \cdot \mathbf{n}\, d\sigma = \iint_R \langle (F \circ \Phi)(u, v), N_\Phi(u, v) \rangle\, du dv$$

See Exercise 8.

The condition that $\mathbf{n} = N_\Phi / \|N_\Phi\|$ is the same unit vector at points where Φ takes on the same value means that the vector N_Φ points in the same direction at such points. Hence the definition of $\mathbf{n}$ is unambiguous.

The reader may be perplexed when $\iint_\Phi F \cdot \mathbf{n}\, d\sigma$ is defined since N_Φ is not normalized in the integral on the right-hand side of the formula. But this happens because we defined $d\sigma = \|N_\Phi(u,v)\|\, dudv$ in (8.4.1) so that this occurrence of $\|N_\Phi(u,v)\|$ cancels the one in the denominator of the definition of $\mathbf{n}$. Finally we have used the boldfaced notation $\mathbf{n}$ for unit normal after having said previously that we won't use boldface to denote vectors. Alas, sometimes clarity must trump doctrine. The letter n is used too often as a subscript or as the number of elements in a set to let it be used in its naked state as an important vector-valued function. In addition this is the traditional notation.

8.4.6. Example. When (Φ, R) is a regular surface in $\mathbb{R}^3$, suppose the continuous function $F : \{\Phi\} \to \mathbb{R}^3$ is given by

$$F(u,v) = (P(u,v), Q(u,v), R(u,v))$$

Using (8.3.7) we have that $\iint_\Phi F \cdot \mathbf{n}\, d\sigma$ equals

$$\iint_\Phi F \cdot \mathbf{n}\, d\sigma = \iint_R \langle (P \circ \Phi, Q \circ \Phi, R \circ \Phi), (\Delta_{\phi_2,\phi_3}, \Delta_{\phi_3,\phi_1}, \Delta_{\phi_1,\phi_2}) \rangle \, dudv$$

$$= \iint_R \left[(P \circ \Phi)\Delta_{\phi_2,\phi_3} + (Q \circ \Phi)\Delta_{\phi_3,\phi_1} + (R \circ \Phi)\Delta_{\phi_1,\phi_2} \right] dudv$$

Symbolically we write this as

8.4.7 $$\iint_\Phi F \cdot \mathbf{n}\, d\sigma = \iint_\Phi P\, dydz + \iint_\Phi Q\, dzdx + \iint_\Phi R\, dxdy$$

We'll have more to say about this when we encounter differential forms in the next chapter.

8.4.8. Example. The *Möbius*[3] *Strip* is the most famous of all non-oriented surfaces. I suspect that every reader has come across this surface in some form, and many of you may have constructed one by giving a strip of paper a single twist. If you do this you see that you obtain a surface with "one side." This is

---

[3] August Ferdinand Möbius was born in 1790 in Schulpforta, Saxony; his father was a dancing teacher and died when Möbius was three years old. He was educated at home until he was 13 and then was enrolled in school. In 1809 he entered the University of Leipzig. In spite of his interest in mathematics, he began by studying law at the urging of his family. He soon decided to follow his own inclinations and switched to mathematics, astronomy, and physics. In 1813 he went to Göttingen to study astronomy with Gauss. He then went to Halle to study mathematics with Pfaff, Gauss's teacher. He soon completed a doctoral thesis in mathematics and astronomy and in 1816 he was appointed to the faculty at the University of Leipzig. In spite of not being able to secure a post at Leipzig that was more prestigious than the one he had and in spite of offers from other universities, he remained at Leipzig. (Apparently he was not a good teacher.) Finally in 1844 he received an offer from the University of Jena; to counter this, Leipzig offered him a full professorship in astronomy. He made contributions to projective geometry and other areas of mathematics and astronomy. His discovery of the strip described here was part of a larger work on one sided surfaces. It must be mentioned that the strip was independently discovered at about the same time by Johann Listing. Möbius died in 1868 in Leipzig.

the trait of not being oriented, but let's examine this strip from a mathematical point of view and match it with the definition of an oriented surface given above. Let $R = [-\pi, \pi] \times [-1, 1]$ and define $\Phi : R \to \mathbb{R}^3$ by $\Phi(u, v) = ((2 + v \sin(u/2)) \cos u, (2 + v \sin(u/2)) \sin u, v \cos(u/2))$. Sketching any surface in $\mathbb{R}^3$ is a challenge, but doing this one is particularly difficult and will tax your artistic abilities; but it does give the Möbius Strip. Note that if u is a constant, then as v varies from -1 to 1 a straight line segment L_u is transcribed. As u progresses from $-\pi$ to π this segment L_u rotates. It is this rotation that creates a lack of orientation and leads us to say it is a surface with only one side. In Exercise 9 the reader is asked to show the (Φ, R) is not an oriented surface.

8.4.9. Definition. If two oriented surfaces (Φ, R) and (Ψ, Q) are equivalent with a smooth, bijective mapping $\tau : Q \to R$ such that $\Psi = \Phi \circ \tau$ as in Definition 8.3.13, then they are *orientation equivalent* if the Jacobian of τ satisfies $\Delta_\tau(u, v) > 0$ for all (u, v) in Q.

As we pointed out when we defined equivalent surfaces earlier, since the surfaces are regular it follows that $\Delta_\tau(u, v) \neq 0$ for all (u, v) in Q. But because Q is connected it follows that the Jacobian must be always positive or always negative.

8.4.10. Proposition. *Let (Φ, R) and (Ψ, Q) be equivalent oriented surfaces with a smooth, bijective mapping $\tau : Q \to R$ such that $\Psi = \Phi \circ \tau$ and let F be a continuous function from $\{\Phi\} = \{\Psi\}$ into $\mathbb{R}^3$.*

(a) *If the surfaces are orientation equivalent, then $\iint_\Psi F \cdot \mathbf{n} \, d\sigma = \iint_\Phi F \circ \tau \cdot \mathbf{n} \, d\sigma$.*

(b) *If the surfaces are not orientation equivalent, then $\iint_\Psi F \cdot \mathbf{n} \, d\sigma = -\iint_\Phi F \circ \tau \cdot \mathbf{n} \, d\sigma$.*

Proof. (a) Assume $\Delta_\tau(u, v)$ is always positive. By Proposition 8.3.12, $N_\Psi(s, t) = \Delta_\tau(s, t) N_\Phi(\tau(s, t))$. Hence

$$\iint_\Psi F \cdot \mathbf{n} \, d\sigma = \iint_Q \langle (F \circ \Psi)(s, t), N_\Psi(s, t) \rangle \, ds dt$$

$$= \iint_Q \langle (F \circ \Phi \circ \tau)(s, t), N_\Phi(\tau(s, t)) \rangle \Delta_\tau(s, t) \, ds dt$$

$$= \iint_Q H \circ \tau(s, t) |\Delta_\tau(s, t)| \, ds dt$$

where $H : R \to \mathbb{R}$ is defined by $H(u, v) = \langle F(\Phi(u, v)), N_\Phi(u, v) \rangle$. Now using the COV Theorem (7.4.4) we get that

$$\iint_\Psi F \cdot \mathbf{n} \, d\sigma = \iint_R H(u, v) \, du dv = \iint_\Phi F \cdot \mathbf{n} \, d\sigma$$

The proof of (b) is Exercise 10. ∎

8.4.11. Example. (a) Let $R = [a, b] \times [c, d]$ and $Q = [0, 1] \times [0, 1]$. Define $\tau : Q \to R$ by

$$\tau(s, t) = ((1 - s)a + sb, (1 - t)c + td)$$

A calculation shows that $\Delta_\tau(s, t) = (b - a)(d - c) > 0$ everywhere on the unit square. Thus every regular surface with surface domain a rectangle is orientation equivalent to a surface with domain the unit square.

(b) Let (Φ, R) be a regular surface and put $Q = \{(v, u) : (u, v) \in R\}$. Define $\tau : Q \to R$ by $\tau(v, u) = (u, v)$. A computation shows that $D_\tau = \begin{bmatrix} 0 & 1 \\ 1 & 0 \end{bmatrix}$ so that $\Delta_\tau(v, u) = -1$ everywhere on Q. If (Ψ, Q) is the regular surface defined by $\Psi = \Phi \circ \tau$, then this surface is equivalent to (Φ, R) but not orientation equivalent. In fact $\mathbf{n}_\Psi(v, u) = -\mathbf{n}_\Phi(u, v)$. In other words, if we are given any regular surface (Φ, R) there is an equivalent surface (Ψ, Q) such that at every point of $\{\Phi\} = \{\Psi\}$, $\mathbf{n}_\Psi = -\mathbf{n}_\Phi$.

In (8.2.2) we defined what it means for a piecewise smooth curve γ in $\mathbb{R}^2$ that forms part of the boundary of an open set G in $\mathbb{R}^2$ to be positively oriented relative to G. We'd like to extend this to curves that form part of the boundary of an oriented surface (Φ, R). In fact for an oriented surface (Φ, R) it is often the case that $\partial\Phi$ consists of a finite number of piecewise smooth curves. (For example, if Φ is a hemisphere or a rectangle.) We would like to induce an orientation on the curves in $\partial\Phi$ relative to the orientation of (Φ, R). This can be done in a precise manner, but that requires more concepts and will be postponed until the next chapter when we discuss differential forms in higher dimensional Euclidean space. Here we give a heuristic description of this induced orientation.

Suppose we are given a point $(x, y, z) = \Phi(u, v)$ on the surface. Near this point the surface has two sides, including one where the normal vector $\mathbf{n}$ is pointing. That side where $\mathbf{n}$ is pointing is called the *positive side* of Φ near this point. Assume γ is one of the curves that make up $\partial\Phi$, and imagine you are walking along γ with your head pointing in the direction of $\mathbf{n}$; that is, your head is in the positive side of Φ. We say you are going in the *positive direction* of γ relative to Φ provided $\{\Phi\}$ is on your left. This is also called the *right-hand orientation* of $\partial\Phi$. Notice that this is consistent with our definition of the positive direction of a curve in $\mathbb{R}^2$ given in (8.2.2). In fact let (Φ, R) be a surface with $\{\Phi\}$ a subset of the xy-plane in $\mathbb{R}^3$ and $\partial\Phi$ is a piecewise regular curve γ. Assume the unit normal vector $\mathbf{n}$ for Φ points upward into the half-space $\{(x, y, z) : z \geq 0\}$. Then the positive direction of γ as discussed above is the counterclockwise direction as discussed in §8.2.

It must be emphasized that the positive direction of γ depends on the orientation of the surface whose boundary includes γ. We'll see that in the following example, especially part (b).

8.4.12. Example. (a) Consider the hemisphere centered at the origin of radius r that lies above the xy-plane; so $R = \{(u, v) \in \mathbb{R}^2 : u^2 + v^2 \leq r^2\}$ and $\Phi(u, v) = (u, v, \sqrt{r^2 - u^2 - v^2})$. (This Φ is not regular in a neighborhood of R as required;

a correct parametrization can be found in Exercise 8.3.2.) Here $\partial\{\Phi\}$ is the circle $\{(u, v, 0) : u^2 + v^2 = r^2\}$. It follows that

$$N_\Phi = \left(\frac{u}{\sqrt{r^2 - u^2 - v^2}}, \frac{v}{\sqrt{r^2 - u^2 - v^2}}, 1 \right)$$

(See Exercise 2.) So the positive direction for this surface is upwards and the positive direction for $\partial\{\Phi\}$ is to go along this circle in the counterclockwise direction.

(b) Suppose $0 < a < b < \infty$, $R = \{(u, v) \in \mathbb{R}^2 : a^2 \le u^2 + v^2 \le b^2\}$, and $\Phi(u, v) = (u, v, 0)$. So $\{\Phi\} = R \times \{0\}$ and $\partial\{\Phi\}$ consists of two circles γ_1 and γ_2 with $\{\gamma_1\} = \{(u, v, 0) : u^2 + v^2 = b^2\}$, $\{\gamma_2\} = \{(u, v, 0) : u^2 + v^2 = a^2\}$. The unit normal vector for $\{\Phi\}$ is e_3, the positive direction for γ_1 is counterclockwise, and the positive direction for γ_2 is clockwise.

Now we extend the ideas of regular surfaces and oriented surfaces to piecewise versions to accommodate examples such as the boundary of a cube in $\mathbb{R}^3$.

8.4.13. Definition. If (Φ_i, R_i) is a surface for $1 \le i \le n$ and $\Sigma = \bigcup_{i=1}^n \{\Phi_i\}$, then Σ is called a *piecewise regular surface* if the following conditions are satisfied: (i) each (Φ_i, R_i) is a regular surface; (ii) for $i \ne j$ either $\{\Phi_i\} \cap \{\Phi_j\} = \emptyset$ or $\{\Phi_i\} \cap \{\Phi_j\} \subseteq \partial\Phi_i \cap \partial\Phi_j$; (iii) for three distinct indices in $\{1, \ldots, n\}$, $\{\Phi_i\} \cap \{\Phi_j\} \cap \{\Phi_k\}$ is either empty or finite. The *boundary of* Σ, $\partial\Sigma$, is defined as the closure of the set of points x in $\mathbb{R}^3$ such that for some index i, $x \in \partial\Phi_i$ but $x \notin \partial\Phi_j$ for all $j \ne i$.

The surface area of Σ is defined by

$$\sigma(\Sigma) = \sum_{i=1}^n \sigma(\Phi_i)$$

If $f : \Sigma \to \mathbb{R}$ is a continuous function, then

$$\iint_\Sigma f \, d\sigma = \sum_{i=1}^n \iint_{\Phi_i} f \, d\sigma$$

There is a certain inconsistency in the above definition in that the surface Σ is described as a set of points rather than a function. There are some things we could do to make it a function, but they could easily be considered contrived and it doesn't seem worth the effort. It's better to live with this inconsistency. Besides the surface of a cube, there are other examples of sets Σ that we want to consider that are not surfaces as in Definition 8.3.1 since they are not connected. The first such example is ∂S where S is the corona of Example 8.3.2. Since ∂S is not connected it is not a surface but it does satisfy the definition of a piecewise regular surface. Let's point out a similarity to the definition above and that of a G-set (8.2.6).

The preceding definition seems plain enough, but if Σ is described geometrically as a set of points in $\mathbb{R}^3$, it may be difficult to find the regular surfaces (Φ_i, R_i) needed to make the piecewise regular surface fit the definition. Let's look at an example taken from [15].

8.4.14. Example. If Σ is the surface of the tetrahedron in $\mathbb{R}^3$ resulting from taking the topological boundary of the solid bounded by the four planes $x = 0$, $y = 0$, $z = 0$, and $x + y + z = 1$, compute $\iint_\Sigma f \, d\sigma$ when $f(x, y, z) = x + y^2 + z^3$. Σ has four faces $\{(\Phi_i, R) : 1 \leq i \leq 4\}$, where each face has the same surface domain $R = \{(u, v) : 0 \leq u \leq 1, 0 \leq v \leq 1, u + v \leq 1\}$ (the triangle in $\mathbb{R}^2$ with vertices $(0, 0), (0, 1), (1, 0)$); and for each (u, v) in R we define $\Phi_1(u, v) = (u, v, 0)$, $\Phi_2(u, v) = (0, u, v)$, $\Phi_3(u, v) = (u, 0, v)$, $\Phi_4(u, v) = (u, v, 1 - u - v)$. A computation shows that $\|N_{\Phi_i}\| = 1$ for $i = 1, 2, 3$ and $\|N_{\Phi_4}\| = \sqrt{3}$. Setting up the integrals and performing the computations (Exercise 15) shows that

$$\iint_\Sigma f \, d\sigma = \frac{3}{10}(2 + \sqrt{3})$$

Defining what is meant by a piecewise regular surface to be orientable is tricky. For $1 \leq i \leq n$ let (Φ_i, R_i) be a surface such that $\Sigma = \bigcup_{i=1}^n \{\Phi_i\}$ is a piecewise regular surface. If we want to have Σ orientable, we clearly must have each Φ_i orientable; this is not sufficient as the reader can see by using the Möbius Strip (Exercise 19). We need to be sure the direction of the various normals N_{Φ_i} are consistent. This is easy with something like the surface of a box where for each of the six surfaces that correspond to the surfaces of the box we can choose the outward (or inward) pointing normal vector.

The next definition is again heuristic, but it suffices for the examples we will see. In the next chapter we will use a different approach and make it all precise.

8.4.15. Definition. For $1 \leq i \leq n$ let (Φ_i, R_i) be a surface such that $\Sigma = \bigcup_{i=1}^n \{\Phi_i\}$ is a piecewise regular surface and each $\partial\Phi_i$ is a piecewise regular curve that is positively oriented relative to Φ_i. Σ is a *a piecewise orientable surface* if the following hold: (a) each (Φ_i, R_i) is oriented; (b) if $1 \leq i < j \leq n$ and $\partial\Phi_i \cap \partial\Phi_j \neq \emptyset$, then the positive direction of any curve in this intersection relative to Φ_i is the opposite of the positive direction of this curve relative to Φ_j; (c) the surfaces (Φ_i, R_i) are so oriented that the directions of the normal vectors N_{Φ_i} are consistent on each component of Σ.

When $\Sigma = \bigcup_{i=1}^n \{\Phi_i\}$ is an oriented piecewise regular surface and $F : \Sigma \to \mathbb{R}^3$ is a continuous function, define the surface integral by

$$\iint_\Sigma F \cdot \mathbf{n} \, d\sigma = \sum_{i=1}^n \iint_{\Phi_i} F \cdot \mathbf{n}_{\Phi_i} \, d\sigma$$

The meaning of condition (c) is imprecise, but for most examples it becomes clear. For example with a box it must be that the normal vectors for each face must point outward or they must all point inwards. Or consider the boundary of the corona, Σ, a piecewise regular surface. Here we could make the normal vector on the outer boundary point away from the origin and the normal vector on the inner boundary point toward the origin. We can similarly orient the boundary of a cylinder.

Exercises

(1) Give the details to show that the definition of surface area in (8.4.1), when applied to $\{\Phi\} = \{(x, y, 0) : x^2 + y^2 \leq r^2\}$, yields $\sigma(\{\Phi\}) = \pi r^2$.

(2) Let (Φ, R) be the hemisphere of radius r as in Example 8.4.12(a). (a) Verify the formula for N_Φ given there. (b) Find $\sigma(\Phi)$. (c) Find $\iint_\Phi f \, d\sigma$ when $\{\Phi\}$ is the hemisphere of radius r centered at the origin and lying above the xy-plane and $f(x, y, z) = \sqrt{z}$.

(3) Find the surface area of the torus (Example 8.3.3(d)).

(4) Find $\iint_\Phi z \, d\sigma$, where $\{\Phi\}$ is the upper half of the sphere centered at the origin in $\mathbb{R}^3$ and having radius 2.

(5) Find $\iint_\Phi \sqrt{x^2 + y^2} \, d\sigma$ if $\{\Phi\}$ is the surface of the circular cone sitting on the xy-plane with base $\{(x, y) : x^2 + y^2 \leq 2\}$ and height 3.

(6) If (Φ, R) and (Ψ, Q) are equivalent regular surfaces, show that $\sigma(\{\Phi\}) = \sigma(\{\Psi\})$; and if f is a continuous real-valued function on $\{\Phi\} = \{\Psi\}$, then show that $\iint_\Phi f \, d\sigma = \iint_\Psi f \, d\sigma$.

(7) Prove Proposition 8.4.4.

(8) If (Φ, R) is given as the graph of a smooth function $f : R \to \mathbb{R}$ (Example 8.4.2(d)), is it oriented?

(9) In the Möbius Strip (Example 8.4.8) find distinct points (u_1, v_1) and (u_2, v_2) in R such that $\Phi(u_1, v_1) = \Phi(u_2, v_2)$ but

$$N_\Phi(u_1, v_1)/\|N_\Phi(u_1, v_1)\| \neq N_\Phi(u_2, v_2)/\|N_\Phi(u_2, v_2)\|$$

(10) Prove part (b) of Proposition 8.4.10.

(11) Show that there is an oriented surface (Φ, R) such that $\{\Phi\} = \{(x, y, z) : z = x^2 + y^2 \text{ and } 0 \leq z \leq 4\}$ and determine the positive direction of $\partial\{\Phi\}$.

(12) Consider the surface (Φ, R) where $R = [0, 2\pi] \times [0, 2]$ and $\Phi(u, v) = (\cos u, \sin u, v)$. Determine the positive direction of each curve that makes up $\partial\{\Phi\}$.

(13) Let $R = [0, 2] \times [0, 2]$ and define $\Phi : R \to \mathbb{R}^3$ by $\Phi(u, v) = (u, v, 4u \quad v)$. If $F(x, y, z) = (3x^2, -2xy, 8)$, find $\iint_\Phi F \cdot \mathbf{n} \, d\sigma$.

(14) (a) Find a surface (Φ, R) such that $\{\Phi\} = \{(x, y, z) : x^2 + z^2 \leq 1, y = x^2 + z^2\}$, show that it is oriented, and determine the positive direction of the curves comprising $\partial\{\Phi\}$. (b) Find $\iint_\Phi F \cdot \mathbf{n} \, d\sigma$ when $F(x, y, z) = (0, y, -z)$.

(15) Carry out the computations in Example 8.4.14.

(16) Find $\iint_\Sigma f \, d\sigma$ when $f(x, y, z) = x^2 + yz$ and Σ is the topological boundary of the box $[0, 1] \times [0, 1] \times [0, 1]$.

(17) Find $\iint_\Sigma (y + z) d\sigma$ where Σ is the surface in $\mathbb{R}^3$ bounded by the cylinder $x^2 + y^2 = 3$, the plane $y + z = 4$, and the disk $x^2 + y^2 \leq 3$.

(18) If $\quad \Sigma = \{(x, y, z) : z \geq 0, x^2 + y^2 + z^2 = 1\} \cup \{xyz : z = 0, x^2 + y^2 \leq 1\} \quad$ and $F(x, y, z) = (2x, 2y, 2z)$, find $\iint_\Sigma F \cdot \mathbf{n} \, d\sigma$.

(19)   Consider the Möbius Strip X (8.4.8) and for $k = 1, 2$ let $R_k = [\pi(k-2), \pi(k-1)] \times [-1, 1]$. Show that for the map Φ we can consider X as the piecewise regular surface $(\Phi, R_1) \cup (\Phi, R_2)$ and each (Φ, R_k) is orientable.

8.5. The Theorems of Gauss and Stokes

This last section of the chapter contains two important theorems, but first we need two pieces of notation that the reader may recall from Calculus.

8.5.1. Definition. Let X be an open subset of $\mathbb{R}^3$ and suppose $F : X \to \mathbb{R}^3$ is a smooth function with $F = (P, Q, R)$. The *curl* and *divergence* of F are the functions curl $F : X \to \mathbb{R}^3$ and div $F : X \to \mathbb{R}$ defined by

$$\text{curl } F = \left(\frac{\partial R}{\partial y} - \frac{\partial Q}{\partial z}, \frac{\partial P}{\partial z} - \frac{\partial R}{\partial x}, \frac{\partial Q}{\partial x} - \frac{\partial P}{\partial y} \right)$$

and

$$\text{div } F = \frac{\partial P}{\partial x} + \frac{\partial Q}{\partial y} + \frac{\partial R}{\partial z}$$

The readers who have not seen these before can be forgiven their bewilderment, but realize that to remember the formulas we can do the following. Recall the definition of the gradient of a differentiable function $f : G \to \mathbb{R}$, where G is an open subset of $\mathbb{R}^3$, and define the symbol

$$\nabla = \left(\frac{\partial}{\partial x}, \frac{\partial}{\partial y}, \frac{\partial}{\partial z} \right)$$

Treating ∇ as a vector in $\mathbb{R}^3$ we have that symboically

$$\text{curl } F = \nabla \times F \text{ and div } F = \nabla \cdot F$$

The concepts of the curl and divergence of a function originated with the study of fluid flow.

Just as we did when we proved Green's Theorem, we want to discuss particular types of Jordan sets in $\mathbb{R}^3$.

8.5.2. Definition. Here we say that a Jordan subset X of $\mathbb{R}^3$ is *Type I* if there is a compact rectangle S in $\mathbb{R}^2$ and regular functions $f_1, f_2 : S \to \mathbb{R}$ such that $f_1(x, y) \leq f_2(x, y)$ for all (x, y) in S and $X = \{(x, y, z) : (x, y) \in S, f_1(x, y) \leq z \leq f_2(x, y)\}$; X is *Type II* if there is a compact rectangle S in $\mathbb{R}^2$ and regular functions $g_1, g_2 : S \to \mathbb{R}$ such that $g_1(x, z) \leq g_2(x, z)$ for all (x, z) in S and $X = \{(x, y, z) : (x, z) \in S, g_1(x, z) \leq y \leq g_2(x, z)\}$; X is *Type III* if there is a compact rectangle S in $\mathbb{R}^2$ and regular functions $h_1, h_2 : S \to \mathbb{R}$ such that $h_1(y, z) \leq h_2(y, z)$ for all (y, z) in S and $X = \{(x, y, z) : (y, z) \in S, h_1(y, z) \leq x \leq h_2(y, z)\}$.

Once again, as we said in the preface to this chapter, at a few points in the remainder of this section we will rely more on the reader's intuition and, perhaps, call on him/her to fill in more of the details. Most of this will occur when we deal with orienting piecewise regular surfaces. In the next chapter these things will be rectified. The next proof is a dramatic example of this approach.

8.5.3. Proposition. *If X is either a Type I, Type II, or Type III set as defined above, then ∂X is a piecewise orientable surface.*

Proof. We only prove this for Type I sets, so let X be such a set with the notation as in the definition and assume the rectangle $S = [a, b] \times [c, d]$. Put

$$\partial X_a = \{(a, y, z) : c \leq y \leq d, f_1(a, y) \leq z \leq f_2(a, y)\}$$
$$\partial X_b = \{(b, y, z) : c \leq y \leq d, f_1(b, y) \leq z \leq f_2(b, y)\}$$
$$\partial X_c = \{(x, c, z) : a \leq x \leq b, f_1(x, c) \leq z \leq f_2(x, c)\}$$
$$\partial X_b = \{(x, d, z) : c \leq x \leq b, f_1(x, d) \leq z \leq f_2(x, d))\}$$
$$\partial X_1 = \{(x, y, f_1(x, y)) : (x, y) \in S\}$$
$$\partial X_2 = \{(x, y, f_2(x, y)) : (x, y) \in S\}$$

It follows that ∂X is the union of these six surfaces, now called *faces*, and is a piecewise regular surface. Moreover we leave it to the reader to show that each face is orientable and if we take the normal vector to each face pointing outward from X (that is away from int X), ∂X is positively oriented. ∎

In analogy with Definition 8.2.6 we make the following definition.

8.5.4. Definition. Let X be a compact Jordan subset of $\mathbb{R}^3$ whose topological boundary is a piecewise regular surface that is positively oriented. Let $\{X_j : 1 \leq j \leq m\}$ be a collection of closed subsets of X. We say that $\{X_j : 1 \leq j \leq m\}$ is a *Type I cover* of X if each set X_j is Type I and the following are satisfied: (a) $X = \bigcup_{j=1}^m X_j$; (b) when $1 \leq i < j \leq m, X_i \cap X_j = \partial X_i \cap \partial X_j$; (c) if X_i, X_j, and X_k are three distinct sets, then $\partial X_i \cap \partial X_j \cap \partial X_k$ is either empty or a finite set; (d) each ∂X_j is oriented and if $1 \leq i < j \leq m$ and $X_i \cap X_j$ is not finite, then $\mathbf{n}_{\Phi_i}(x, y, z) = -\mathbf{n}_{\Phi_j}(x, y, z)$ for each (x, y, z) in $X_i \cap X_j$. The collection $\{X_j : 1 \leq j \leq m\}$ is called a *Type II cover* (respectively, *Type III cover*) if each set X_j is a Type II (respectively, Type III) set and the analogues of conditions (a), (b), (c), and (d) above are also satisfied.

A *G-set X* in $\mathbb{R}^3$ is a compact Jordan subset of $\mathbb{R}^3$ whose topological boundary is a piecewise regular surface with positive orientation and such that it has a Type I cover as well as a Type II cover and a Type III cover.

Note that as a consequence of condition (d) in the above definition it follows that for any smooth function $F : X \to \mathbb{R}^3$ we have that

8.5.5
$$\iint_{\partial X} F \cdot \mathbf{n}\, d\sigma = \sum_{j=1}^m \iint_{\partial X_j} F \cdot \mathbf{n}\, d\sigma$$

8.5.6. Theorem (Gauss's[4] Theorem). *If X is a G-set in $\mathbb{R}^3$ and $F : X \to \mathbb{R}^3$ is a smooth function, then*

$$\iint_{\partial X} F \cdot \mathbf{n}\, d\sigma = \iiint_X \operatorname{div} F$$

(Gauss's Theorem is also called the *Divergence Theorem*.)

Proof. If $F = (P, Q, R)$, then we have that

$$\iiint_X \operatorname{div} F = \iiint_X \frac{\partial P}{\partial x} + \iiint_X \frac{\partial Q}{\partial y} + \iiint_X \frac{\partial R}{\partial z}$$

and from (8.4.7)

$$\iint_{\partial X} F \cdot \mathbf{n}\, d\sigma = \iint_{\partial X} P\, dydz + \iint_{\partial X} Q\, dzdx + \iint_{\partial X} R\, dxdy$$

This reduces the proof to establishing three separate equalities that only involve the scalar-valued functions P, Q, or R one at a time:

$$\iiint_X \frac{\partial P}{\partial x} = \iint_{\partial X} P\, dydz$$

$$\iiint_X \frac{\partial Q}{\partial y} = \iint_{\partial X} Q\, dzdx$$

$$\iiint_X \frac{\partial R}{\partial z} = \iint_{\partial X} R\, dxdy$$

Let's show that

$$\iint_{\partial X} R\, dxdy = \iiint_X \frac{\partial R}{\partial z}$$

---

[4] Johann Carl Friedrich Gauss was born in Brunswick, Germany in 1777. At the age of seven he began his schooling and from the start exhibited extraordinary talent in mathematics, making small discoveries of known results. As a teenager he discovered the law of quadratic reciprocity and the prime number theorem. In 1795 he enrolled in Göttingen University, but left in 1798 without a diploma in spite of great success that included the construction of a regular 17-gon with ruler and compass, the first progress in such questions since the time of the ancient Greeks. He returned to the university and received the degree the next year. In 1801 he published his book *Disquisitiones Arithmeticae*, which was devoted to number theory except for the last section that contained his work on the 17-gon. He also began extensive work in astronomy and made many significant contributions, leading to his appointment as director of the Göttingen observatory in 1807. Two years earlier he married. Sadly his wife died four years later after giving birth to their second son, who also died shortly after his mother. A year later he married his first wife's best friend and they had three children. Gauss's professional life was long and filled with discoveries. His motto was "Few but Ripe." This meant he never published anything unless he had fully developed the subject. He kept a notebook in which he wrote his ideas, many of which were eventually rediscovered and were significant for the progress of mathematics. He is thought by some to have been the greatest mathematician ever. While there is no doubt that he was one of the giants of the profession, I've always thought the comparison of achievement of people from different epochs to be unwise. Suffice it to say that he did work in different areas of mathematics and whatever he turned his attention to resulted in extraordinary advancement. He died in Göttingen in 1855.

Since X is a G-set there is a Type I cover $\{X_j : 1 \leq j \leq m\}$. Since the volume of each ∂X_j is zero, we have that

$$\iiint_X \frac{\partial R}{\partial z} = \sum_{j=1}^m \iiint_{X_j} \frac{\partial R}{\partial z}$$

Therefore by (8.5.5) we need only prove that for $1 \leq j \leq m$

$$\iiint_{X_j} \frac{\partial R}{\partial z} = \iint_{\partial X_j} R \, dxdy$$

To simplify the notation we can assume that X is Type I and use the notation of Definition 8.5.2. From Fubini's Theorem and the Fundamental Theorem of Calculus we have that

$$\iiint_X \frac{\partial R}{\partial z} = \iint_S \left[\int_{f_1(x,y)}^{f_2(x,y)} \frac{\partial R}{\partial z}(x, y, z) \, dz \right] dxdy$$

$$= \iint_S [R(x, y, f_2(x, y)) - R(x, y, f_1(x, y))] dxdy$$

Now let's examine $\iint_{\partial X} R \, dxdy$. As in Proposition 8.5.3, ∂X is composed of six faces: $X_a, X_b, X_c, X_d, X_1, X_2$. Note that on the first four of these faces either the variable x or variable y is held constant. Thus the integral of R over these faces is 0. Therefore we have (Exercise 1) that

8.5.7 $$\iint_{\partial X} R \, dxdy = \iint_S [R(x, y, f_2(x, y)) - R(x, y, f_1(x, y))] dxdy$$

(Why the minus sign?) We note this is the same as the value of $\iiint_X \frac{\partial R}{\partial z}$.

Now X also has a Type II and a Type III cover and we can use these to give proofs that $\iint_{\partial X} P \, dydz = \iiint_X \frac{\partial P}{\partial x}$ and $\iint_{\partial X} Q \, dzdx = \iiint_X \frac{\partial Q}{\partial y}$. ∎

8.5.8. Example. This example shows how Gauss's Theorem can be used to simplify the evaluation of an integral.

(a) Define $F : \mathbb{R}^2 \to \mathbb{R}^2$ by $F(x, y, z) = (5x, 3y, 0)$ and let X be the unit sphere in $\mathbb{R}^3$ centered at the origin and having radius 2. So X is a G-set. (Why?) Find $\iint_{\partial X} F \cdot \mathbf{n} \, d\sigma$. Note that $\operatorname{div} F = 5 + 3 + 0 = 8$. So $\iint_{\partial X} F \cdot \mathbf{n} \, d\sigma = \iiint_X 8 = 8\operatorname{Vol}X$. From Example 7.4.19(b) we know the formula for the volume of this sphere is $4\pi(2)^3/3 = 32\pi/3$, so $\iint_{\partial X} F \cdot \mathbf{n} \, d\sigma = 256\pi/3$. This was easier than parametrizing the boundary of the sphere.

(b) Let $F(x, y, z) = (2x - z, x^2y, -xz^2)$ and $X = [0, 1] \times [0, 1] \times [0, 1]$ and evaluate $\iint_{\partial X} F \cdot \mathbf{n} \, d\sigma$. Here $\operatorname{div} F = 2 + x^2 - 2xz$ so

$$\iint_{\partial X} F \cdot \mathbf{n} \, d\sigma = \int_0^1 \int_0^1 \int_0^1 (2 + x^2 - 2xz) \, dxdydz = \frac{11}{6}$$

This is easier than integrating $F \cdot \mathbf{n}$ over the six different sides of ∂X.

8.5.9. Example. Find $\iiint_X z^2$ when X is the unit sphere in $\mathbb{R}^3$ centered at the origin. Here we want to use Gauss's Theorem to express this as an integral over ∂X. To start we need a function F on $\mathbb{R}^3$ such that div $F = z^2$. We're free to choose any so let's pick the easiest: $F(x, y, z) = (0, 0, \frac{1}{3}z^3)$. Now we need to parametrize ∂X. Let $S = [0, 2\pi] \times [-\pi/2, \pi/2]$ and define $\Phi : S \to \partial X \subseteq \mathbb{R}^3$ by

$$\Phi(\theta, v) = (\cos\theta \cos v, \sin\theta \cos v, \sin v)$$

It follows that

$$\Phi_\theta \times \Phi_v = \det \begin{vmatrix} i & j & k \\ -\sin\theta\cos v & \cos\theta\cos v & 0 \\ -\cos\theta\sin v & -\sin\theta\sin v & \cos v \end{vmatrix}$$

$$= (\cos\theta\cos^2 v, \sin\theta\cos^2 v, \sin v\cos v)$$

Hence

$$\iiint_X z^2 = \iint_{\partial X} \left(0, 0, \frac{1}{3}z^3\right) \cdot \mathbf{n}\, d\sigma$$

$$= \iint_S \left(0, 0, \frac{1}{3}(\sin v)^3\right) \cdot (\cos\theta\cos^2 v, \sin\theta\cos^2 v, \sin v\cos v)\, d\theta dv$$

$$= \frac{1}{3} \int_0^{2\pi} \int_{-\pi/2}^{\pi/2} (\sin v)^3 \sin v \cos v\, dv dt$$

$$= \frac{2\pi}{3} \int_{-\pi/2}^{\pi/2} (\sin v)^4 \cos v\, dv$$

$$= \frac{2\pi}{3}\frac{2}{5} = \frac{4\pi}{15}$$

The next celebrated theorem examines the case of a surface like a hemisphere in $\mathbb{R}^3$ whose boundary is a curve. We only prove this theorem for a special case where the surface is the graph of a smooth function. Later in Theorem 9.5.1 we'll prove a generalization of this theorem and so it doesn't seem justified to do the extra work to prove the result here in $\mathbb{R}^3$.

8.5.10. Theorem (Stokes's[5] Theorem). *If R is a G-set in $\mathbb{R}^2$ and (Φ, R) is a regular surface in $\mathbb{R}^3$ such that $\partial\Phi$ is a piecewise regular curve with positive orientation*

[5] George Gabriel Stokes was born in County Sligo, Ireland in 1819. His father was a church minister and he remained deeply religious his entire life. His early education was near home, but at the age of 13 he went to school in Dublin where he stayed with his uncle. At 16, in 1835, he enrolled in Bristol College in England where his mathematical talent was recognized. Two years later he entered Cambridge University where he remained as a student and then a faculty member for the rest of his life. In 1842 he published papers on hydrodynamics, and in particular incompressible fluids, a subject he focused on for much of his life and from which the present theorem emerged. He also made significant contributions to the study of light. In 1857 he began a courtship with Mary Susanna Robinson, the daughter of an astronomer at Armagh Observatory in Ireland. The courtship followed a path different from romances today in that it was accompanied by an extensive exchange of letters. They married that year and this meant he had to resign his fellowship at Cambridge. (Five years later the university did away with the rule that prohibited fellows from being married.) He and his wife had five children. As his career progressed he turned his attention to administration and from 1887 to 1892 he was one of three members of Parliament from the university. He died in Cambridge in 1903.

relative to Φ, *then for any smooth function* $F : \{\Phi\} \to \mathbb{R}^3$,

$$\int_{\partial\Phi} F \cdot T \, ds = \iint_{\Phi} \text{curl } F \cdot \mathbf{n} \, d\sigma$$

Proof of a Special Case. Assume $\Phi(x, y) = (x, y, f(x, y))$, where f is a twice continuously differentiable function $f : S \to \mathbb{R}$ defined on the G-set S. Thus for Φ the unit normal vector is $\mathbf{n} = N/\|N\|$ where $N = (-f_x, -f_y, 1)$. Putting $F = (P, Q, R)$ and using (8.1.9) we have that

$$\int_{\partial\Phi} F \cdot T \, ds = \int_{\partial\Phi} P \, dx + Q \, dy + R \, dz$$

To simplify matters we'll assume that the topological boundary of the G-set S is parametrized by a single piecewise regular curve γ defined on $[a, b]$ with $\gamma(t) = (x(t), y(t))$. (If ∂S consists of a finite number of such curves, the argument that follows is the same but notationally more cumbersome.) Thus $\partial\Phi$ is parametrized by the function on $[a, b]$ defined by $t \mapsto (x(t), y(t), f(x(t), y(t)))$.

Now in the integral above we have

$$dz = \frac{\partial f}{\partial x} \, dx + \frac{\partial f}{\partial y} \, dy$$

so that symbolically the integral above becomes

$$\int_{\partial\{\Phi\}} F \cdot T \, ds = \int_{\gamma} \left(P + R\frac{\partial f}{\partial x}\right) dx + \left(Q + R\frac{\partial f}{\partial y}\right) dy$$

To evaluate this integral we will use Green's Theorem. To do this we need to calculate the partial derivatives of the functions in parentheses, and to do this we use the Chain Rule to get

$$\frac{\partial}{\partial x}\left(Q + R\frac{\partial f}{\partial y}\right) = \frac{\partial Q}{\partial x} + \frac{\partial Q}{\partial z}\frac{\partial f}{\partial x} + \frac{\partial R}{\partial x}\frac{\partial f}{\partial y} + \frac{\partial R}{\partial z}\frac{\partial f}{\partial x}\frac{\partial f}{\partial y} + R\frac{\partial^2 f}{\partial x \partial y}$$

and

$$\frac{\partial}{\partial y}\left(P + R\frac{\partial f}{\partial x}\right) = \frac{\partial P}{\partial y} + \frac{\partial P}{\partial z}\frac{\partial f}{\partial y} + \frac{\partial R}{\partial y}\frac{\partial f}{\partial x} + \frac{\partial R}{\partial z}\frac{\partial f}{\partial y}\frac{\partial f}{\partial x} + R\frac{\partial^2 f}{\partial y \partial x}$$

Now because f has continuous second partial derivatives, $f_{xy} = f_{yx}$. Therefore

$$\frac{\partial}{\partial x}\left(Q + R\frac{\partial f}{\partial y}\right) - \frac{\partial}{\partial y}\left(P + R\frac{\partial f}{\partial x}\right)$$

$$= \left(\frac{\partial R}{\partial y} - \frac{\partial Q}{\partial z}\right)\left(-\frac{\partial f}{\partial x}\right) + \left(\frac{\partial P}{\partial z} - \frac{\partial R}{\partial x}\right)\left(-\frac{\partial f}{\partial y}\right) + \left(\frac{\partial Q}{\partial x} - \frac{\partial P}{\partial y}\right)$$

$$= \text{curl } F \cdot N$$

Putting all this together with Green's Theorem, we have that

$$\int_{\partial\Phi} F \cdot T \, ds = \iint_{S} \text{curl } F \cdot N = \iint_{\Phi} \text{curl } F \cdot \mathbf{n} \, d\sigma \qquad \blacksquare$$

8.5.11. Example. Green's Theorem is a special case of Stokes's Theorem, at least when the G-set X has its boundary as a single piecewise regular curve. In that case using the notation of both Green's Theorem, let $S = X$ and $F(x, y) = (P(x, y), Q(x, y), 0)$. Define $\Phi : S \to \mathbb{R}^3$ by $\Phi(x, y) = (x, y, 0)$. In this case $\mathbf{n} = e_3$ and (curl F) $\times e_3 = \frac{\partial Q}{\partial x} - \frac{\partial P}{\partial y}$, the term that appears in Green's Theorem. The formula in Stokes's Theorem now yields Green's Theorem.

8.5.12. Example. Let Σ be the upper hemisphere $x^2 + y^2 + z^2 = 1$ with upper pointing normal vector and evaluate $\iint_\Sigma (x^3 e^y, -3x^2 e^y, 0) \cdot \mathbf{n} \, d\sigma$. If we want to do this using Stokes's Theorem we must find a surface (Φ, S) such that $\Sigma = \{\Phi\}$ and a function F such that curl $F = (x^3 e^y, -3x^2 e^y, 0)$. The surface is the easiest: let $S = \{(x, y) \in \mathbb{R}^2 : x^2 + y^2 \le 1\}$ and define $\Phi(x, y) = (x, y, \sqrt{1 - x^2 + y^2})$. To find $F = (P, Q, R)$ we want the partial derivatives to satisfy

$$x^3 e^y = R_y - Q_z, \quad -3x^2 e^y = P_z - R_x, \quad 0 = Q_x - P_y$$

The process here is trial and error, so using the last of these three equations let's start by guessing that $P = Q = 0$. If we do this the first two equations become $R_y = x^3 e^y$, $R_x = 3x^2 e^y$. Aha! $R = x^3 e^y$ works. The appropriate curve for $\partial\{\Phi\} = \partial\Sigma$ is the unit circle in the xy-plane in the counterclockwise direction; that is, $\gamma : [0, 2\pi] \to \mathbb{R}^3$ given by $\gamma(\theta) = (\cos\theta, \sin\theta, 0)$. By Stokes's Theorem we have

$$\iint_\Sigma (x^3 e^y, -3x^2 e^y, 0) \cdot \mathbf{n} \, d\sigma = \int_\gamma F \cdot T \, ds$$
$$= \int_\gamma R \, dz$$
$$= 0$$

since z is constant along γ.

Exercises

Some of the exercises below were taken from websites where I could find no authors or other attribution.

(1) Verify (8.5.7).
(2) Let S be the sphere $x^2 + y^2 + z^2 = 9$ in $\mathbb{R}^3$ and let $F : \mathbb{R}^3 \to \mathbb{R}^3$ be defined by $F(x, y, z) = (3x, 2y, 0)$. (a) Find a G-set X in $\mathbb{R}^3$ with $\partial X = S$. (b) Use Gauss's Theorem to evaluate $\iint_S F \cdot \mathbf{n} \, d\sigma$.
(3) Let S be the boundary of the unit ball in $\mathbb{R}^3$ with positive orientation. If $F(x, y, z) = (x^3, y^3, z^3)$, evaluate $\iint_S F \cdot \mathbf{n} \, d\sigma$.
(4) Let $X = \{(x, y, z) \in \mathbb{R}^3 : -1 \le x \le 1, -1 \le y \le 1, 0 \le z \le 2\}$ and define $F : \mathbb{R}^3 \to \mathbb{R}^3$ by $F(x, y, z) = (y^2 z, y^3, xz)$. Find $\iint_{\partial X} F \cdot \mathbf{n} \, d\sigma$.
(5) If X is the unit sphere in $\mathbb{R}^3$ of radius 1, use Gauss's Theorem to evaluate $\iiint_X z^2$.

(6) Let S be the boundary of the unit ball in $\mathbb{R}^3$ with positive orientation. If $F(x, y, z) = (x^3, y^3, z^3)$, evaluate $\iint_S F \cdot \mathbf{n} \, d\sigma$.

(7) Compute $\int_\gamma \langle F, \mathbf{T} \rangle \, ds$ where γ is the circle in $\mathbb{R}^3$ $\{(x, y, z) : x^2 + z^2 = 1, y = 0\}$ with the counterclockwise direction and $F(x, y, z) = (x^2 z + \sqrt{x^3 + x^2 + 2}, xy, xy + \sqrt{z^3 + z^2 + 2})$.

(8) Define $\gamma : [0, 2\pi] \to \mathbb{R}^3$ by $\gamma(\theta) = (0, 2 + 2\cos\theta, 2 + 2\sin\theta)$ and define $F : \mathbb{R}^3 \to \mathbb{R}^3$ by $F(x, y, z) = (x^2 e^{5z}, x\cos y, 3y)$. Evaluate $\int_\gamma F \cdot T \, ds$ by using Stokes's Theorem.

(9) Let S be the hemisphere $z = \sqrt{1 - x^2 - y^2}$. If $\mathbf{n}$ is the outward pointing normal, and $F(x, y, z) = (x, x, x^2 y^3 \log(z + 1))$, find $\iint_{\partial S} \mathrm{curl} F \cdot \mathbf{n} \, d\sigma$.

(10) Let $S = \{(x, y, z) : x^2 + y^2 + z^2 = 1$ and $z \geq 0\}$. (a) Give a parametrization (Φ, R) for S. (b) Use Stokes's Theorem to evaluate $\iint_\Phi (x^3 e^y, -3x^2 e^y, 0) \cdot \mathbf{n} \, d\sigma$.

9 Differential Forms

Differential forms constitutes an approach to multivariable calculus that simplifies the study of integration over surfaces of any dimension in $\mathbb{R}^p$. This topic introduces algebraic techniques into the study of higher dimensional geometry and allows us to recapture with rigor the results obtained in the preceding chapter.

There are many approaches to defining and exploring differential forms, all leading to the same objective. One is to just define a form as a symbol with certain properties. That was the approach taken when I first encountered the subject. Though this appeals to many and ultimately leads to the same objective, it strikes me as inconsistent with the approach we have taken. So in this chapter I have decided to follow the approach in [7] where a form is defined in terms that are more in line with what I think is the background of readers of this book. The reader might also want to look at [1], [9], and [10] where there are different approaches.

9.1. Introduction

Here we will define differential forms and explore their algebraic properties and the process of differentiating them. The first definition extends that of a surface as given in the preceding chapter.

9.1.1. Definition. Let $q \geq 1$. A *q-surface domain* or just a *surface domain* is a compact Jordan subset R of $\mathbb{R}^q$ such that int R is connected and $R = \text{cl}(\text{int } R)$. A *q-surface* in $\mathbb{R}^p$ is a pair (Φ, R) where R is a q-surface domain and Φ is a function from R into $\mathbb{R}^p$ that is smooth on some neighborhood of R. The *trace* of (Φ, R) is the set $\{\Phi\} = \Phi(R) \subseteq \mathbb{R}^p$. If G is an open subset of $\mathbb{R}^p$ and $\{\Phi\} \subseteq G$, we say (Φ, R) is a q-surface in G; let $\mathcal{S}_q(G)$ be the collection of all q-surfaces contained in G. We define a 0-surface in G as a constant function with value in G.

A comparison of the above definition with the definition of a 2-surface in $\mathbb{R}^3$ (8.3.1) reveals a difference. In (8.3.1) we required the boundary of the surface domain to consist of curves; here we do not. Indeed we cannot. If $R \subseteq \mathbb{R}^4$, for example, this would prohibit all but a few possible choices of R. As in the last chapter we will sometimes describe a set S in $\mathbb{R}^p$ and say it is a q-surface. What we mean by this is that there is a q-surface (Φ, R) as defined above with $\{\Phi\} = S$; again (Φ, R) will be called a *parametrization* of S.

9.1.2. Example. Of course all the examples from the previous chapter are available, but here are two more. (a) Let $R = \{u \in \mathbb{R}^{p-1} : \|u\| \leq 1\}$ and let S be the unit hemisphere in $\mathbb{R}^p$: $S = \{(u, \zeta) \in \mathbb{R}^{p-1} \oplus \mathbb{R} = \mathbb{R}^p : u \in R \text{ and } \zeta = \sqrt{1 - \|u\|^2}\}$. This is a $p-1$ surface in $\mathbb{R}^p$, though as we mentioned in Example 8.3.3(b) the function $\Phi : R \to \mathbb{R}^p$ defined by $\Phi(u) = (u, \sqrt{1 - \|u\|^2})$ is not a parametrization since it fails to be smooth in a neighborhood of R. In Exercise 8.3.2 a parametrization when $p = 3$ was given, here is one when $p = 4$. Let $T = [0, 2\pi] \times [0, 2\pi] \times [0, \pi] \subseteq \mathbb{R}^3$ and define $\Psi : T \to \mathbb{R}^4$ by

$$\Psi(u, v, w) = (\cos u \cos v \cos w, \sin u \cos v \cos w, \sin v \cos w, \sin w)$$

The reader can check that (Ψ, T) is a parametrization of S when $p = 4$. Also see Exercise 1.

(b) Let $1 \leq q < p$, let R be any surface domain in $\mathbb{R}^q$, fix indices $i_1, \ldots, i_q$ in $\{1, \ldots, p\}$, and let x_0 be a fixed vector in $\mathbb{R}^p$. If $\Phi : R \to \mathbb{R}^p$ is defined by $\Phi(u) = x_0 + \sum_{k=1}^{q} u_k e_{i_k}$, where $e_1, \ldots, e_p$ is the standard basis for $\mathbb{R}^p$, then (Φ, R) is a q-surface in $\mathbb{R}^p$. (Exercise 2)

9.1.3. Definition. Let G be an open subset of $\mathbb{R}^p$. A *0-form* on G is a continuous function $f : G \to \mathbb{R}$. When $q \geq 1$, a *differential form of order q* or simply a *q-form* on G is a function $\omega : \mathcal{S}_q(G) \to \mathbb{R}$ defined as follows. Let $\mathcal{I}$ be a finite collection of subsets $I = \{i_1, \ldots, i_q\}$ of $\{1, \ldots, p\}$, each of which has q elements, and suppose that for each I in $\mathcal{I}$ we are given a continuous function $f_I = f_{i_1, \ldots, i_q} : G \to \mathbb{R}$. For any $\Phi : R \to G$ in $\mathcal{S}_q(G)$ we define

9.1.4
$$\omega(\Phi) = \int_\Phi \omega = \int_R \sum_{I \in \mathcal{I}} f_{i_1, \ldots, i_q}(\Phi(u)) \frac{\partial(x_{i_1}, \ldots, x_{i_q})}{\partial(u_1, \ldots, u_q)} du$$

This is symbolically written as

$$\omega = \sum_{I \in \mathcal{I}} f_{i_1, \ldots, i_q} \, dx_{i_1} \wedge \cdots \wedge dx_{i_q}$$

The q-form ω is *smooth* if the coefficient functions $f_{i_1, \ldots, i_q}$ are continuously differentiable. If each coefficient function is continuously differentiable up to and including the derivatives of order $n \geq 1$, then ω is said to be a *q-form of order n*.

Whew! That's a complicated definition.

Let's take it apart and examine it and then we'll look at some examples. First note that a q-form is a function defined on the set of functions $\mathcal{S}_q(G)$. This is not the only example you'll ever see of a function whose domain is another collection of functions. In fact you can interpret an integral as being just such a function. As we said $\mathcal{S}_0(G) = G$, so a 0-form f on G acts on a point in G by evaluating the function at the point. Next, for any $q \geq 1$ the term $dx_{i_1} \wedge \cdots \wedge dx_{i_q}$ in the definition can be treated at this stage solely as notation. (The symbol $\wedge$ is read "wedge.") Later we will give it more content and develop $\wedge$ as a form of multiplication for differential forms, but now it's just notation. The first equality in (9.1.4), $\omega(\Phi) = \int_\Phi \omega$, is just an alternate way of writing the value of the function ω at an element of its domain. It is, however, a useful way of interpreting the value of ω at a point in its domain and this interpretation will be amplified later. As stated, the indices $i_1, \ldots, i_q$ are contained in $\{1, \ldots, p\}$, and there is no guarantee that they are distinct; again, we'll see more on this in the next paragraph. There are, however, only a finite number of the continuous functions $f_{i_1,\ldots,i_q}$. Yes, the sum inside the integral in (9.1.4) could have been put outside the integral sign; that's just the way we chose to do it.

The Jacobian that appears in (9.1.4) arises as follows. If $\Phi : R \to G \subseteq \mathbb{R}^p$ has coordinate functions $\Phi(u) = (\phi_1(u), \ldots, \phi_p(u))$, then the Jacobian is the determinant of the mapping from R into $\mathbb{R}^q$ defined by $u = (u_1, \ldots, u_q) \mapsto (\phi_{i_1}(u), \ldots, \phi_{i_q}(u)) \in \mathbb{R}^q$. That is,

9.1.5
$$\frac{\partial(x_{i_1}, \ldots, x_{i_q})}{\partial(u_1, \ldots, u_q)} = \frac{\partial(\phi_{i_1}, \ldots, \phi_{i_q})}{\partial(u_1, \ldots, u_q)}$$

Remember that Φ is assumed smooth in a neighborhood of R so the derivatives present no problem and the Jacobian is a continuous function. Also note that if the indices $i_1, \ldots, i_q$ are not distinct, the Jacobian is the determinant of a matrix with identical columns and so it must be 0. Finally, if the formula in (9.1.4) reminds you of the Change of Variable Theorem (7.4.4), it should. Suppose $q = p$ and $\Phi \in \mathcal{S}_p(G)$. So here we have a surface domain R contained in $\mathbb{R}^p$ and a function $\Phi : R \to G \subseteq \mathbb{R}^p$ that almost satisfies the hypothesis of Change of Variable Theorem (7.4.4). If Φ does indeed satisfy this hypothesis, then for $f \in C(G)$,

$$\int_{\Phi(R)} f = \int_R f \circ \Phi \, |\det[D\Phi]|$$

Except for the absolute value sign around the Jacobian of Φ, this is precisely the formula for $\omega(\Phi)$ when ω is the p-form $f dx_1 \wedge \cdots \wedge dx_p$. For a q-form ω with $q < p$, as the notation $\omega(\Phi) = \int_\Phi \omega$ suggests, we want to think of integrating ω over the surface Φ where we define such an integral by transferring to the integral over the surface domain in the way stipulated in formula (9.1.4).

Now for some examples. Study them carefully. You will notice that we have seen differential forms many times before without ever calling them by this name.

9.1.6. Example. (a) Suppose $q = p = 1$, $R = [a, b] \subseteq \mathbb{R}^q = \mathbb{R}$, and G is an open subset of $\mathbb{R}^p = \mathbb{R}$. Here the only 1-form possible is $\omega = f dx$. A 1-surface in G is a continuously differentiable function $\Phi : [a, b] \to G$ and, using the Change of Variables Theorem,

$$\omega(\Phi) = \int_R f(\Phi(u))\Phi'(u)du = \int_{\Phi([a,b])} f(x)dx$$

We therefore see again the idea mentioned in the paragraph preceding this example that differential forms are a way of defining an integration of functions over a surface. This will be reinforced as we develop the theory.

(b) Let $\gamma : [a, b] \to \mathbb{R}^p$ be a regular curve. As we pointed out, γ is a 1-surface if we assume γ is smooth in a neighborhood of $[a, b]$. Assume the trace of γ is contained in the open set G, and let $F : G \to \mathbb{R}^p$ be a continuous function. If $e_1, \ldots, e_p$ is the usual basis for $\mathbb{R}^p$, then we set $\gamma_j(t) = \langle \gamma(t), e_j \rangle$ and $F_j(x) = \langle F(x), e_j \rangle$. So if we define the one form $\omega = \sum_{j=1}^{p} F_j(x) dx_j$ on G, then by definition

$$\omega(\gamma) = \int_\gamma \omega = \int_{[a,b]} \sum_{j=1}^{p} F_j(\gamma(t)) \gamma_j'(t) dt$$

$$= \int_a^b \langle F(\gamma(t)), \gamma'(t) \rangle \, dt$$

$$= \int_\gamma F \cdot T \, ds$$

So we see that the line integral defined in (8.1.8) is an example of a 1-form. In fact the general 1-form on $\mathbb{R}^p$ is $\omega = \sum_{j=1}^{p} f_j(x) dx_j$ for continuous functions $f_1, \ldots, f_p$.

(c) Let (Φ, R) be a 2-surface in $\mathbb{R}^3$ and recall the definition of $N_\Phi = \Phi_u \times \Phi_v$ (8.3.7). Consider the 2-form $\omega = \|N_\Phi(u, v)\| \, du \wedge dv$. So

$$\omega(\Phi) = \iint_R f(\Phi(u, v)) \|N_\Phi(u, v)\| \, du dv = \iint_\Phi f \, d\sigma$$

the surface integral defined in (8.4.1).

(d) Again let (Φ, R) be a regular 2-surface in $\mathbb{R}^3$ with $\{\Phi\} \subseteq G$, an open subset of $\mathbb{R}^3$. If $F : G \to \mathbb{R}^3$ is a continuous function denoted by $F = (P, Q, R)$, let ω be the 2-form on G defined by

$$\omega = P \, dy \wedge dz + Q \, dz \wedge dx + R \, dx \wedge dy$$

It follows that $\omega(\Phi) = \iint_\Phi F \cdot \mathbf{n} \, d\sigma$, the oriented surface integral as in (8.4.7).

(e) Let $\gamma : [a, b] \to \mathbb{R}^3$ be a 1-surface in $\mathbb{R}^3$ and let ω be the 1-form $\omega = x dy + y dx$. It follows that

$$\int_\gamma \omega = \int_a^b [\gamma_1(t) \gamma_2'(t) + \gamma_2(t) \gamma_1'(t)] dt = \gamma_1(b) \gamma_2(b) - \gamma_1(a) \gamma_2(a)$$

(f) A general $(p - 1)$-form on $\mathbb{R}^p$ is

$$\omega = \sum_{j=1}^{p} f_j(x) \, dx_1 \wedge \cdots \wedge \widehat{dx_j} \wedge \cdots \wedge dx_p$$

where the notation $\widehat{dx_j}$ means that the term dx_j is missing.

We now proceed to develop some basic operations on q-forms, first an arithmetic and then a differential calculus for forms. To set the scene fix an open subset G in $\mathbb{R}^p$ and define $\mathcal{F}_q(G)$ to be the set of q-forms on G. If $\omega, \omega_1, \omega_2 \in \mathcal{F}_q(G)$, then, since they are all functions on $\mathcal{S}_q(G)$, for $c \in \mathbb{R}$ we define

$$(c\,\omega)(\Phi) = c\,\omega(\Phi) \text{ and } (\omega_1 + \omega_2)(\Phi) = \omega_1(\Phi) + \omega_2(\Phi)$$

for every Φ in $\mathcal{S}_q(G)$. With these operations we see that $\mathcal{F}_q(G)$ is a vector space.

We want to define a multiplication of forms, which is more complex. To do this we first examine the representation of differential forms.

9.1.7. Proposition. *Let* $\omega = f(x)dx_{i_1} \wedge \cdots \wedge dx_{i_q} \in \mathcal{F}_q(G)$ *and let k and m be distinct integers in* $\{1, \ldots, q\}$.

(a) *If* $i_k = i_m$, *then* $\omega = 0$.
(b) *If k and m are distinct integers in* $\{1, \ldots, q\}$ *and*

$$\{i_1, \ldots, i_q\} = \{j_1, \ldots, j_q\}$$

with $i_k = j_m$, $i_m = j_k$, *and* $i_\ell = j_\ell$ *when* $\ell \neq k, m$ *then* $\omega = -f(x)dx_{j_1} \wedge \cdots \wedge dx_{j_q}$.

In words, part (a) of this proposition says that if any two indices in the definition of ω are equal, then $\omega = 0$; part (b) says that interchanging exactly two indices produces the q-form $-\omega$. In particular, when $q = 2$, $dx_i \wedge dx_j = -dx_j \wedge dx_i$.

Proof. (b) This follows from Corollary 6.4.22: if A and B are two $q \times q$ matrices such that B is the same as A except that two rows are interchanged, then $\det B = -\det A$. Therefore if $\eta = f(x)dx_{j_1} \wedge \cdots \wedge dx_{j_q}$ and if $\Phi \in \mathcal{S}_q(G)$, then

$$\eta(\Phi) = \int_R f(\Phi(x)) \frac{\partial(x_{j_1}, \ldots, x_{j_q})}{\partial(u_1, \ldots, u_q)} du$$

$$= - \int_R f(\Phi(x)) \frac{\partial(x_{i_1}, \ldots, x_{i_q})}{\partial(u_1, \ldots, u_q)} du$$

$$= -\omega(\Phi)$$

(a) We again use Corollary 6.4.22 to get that when a $q \times q$ matrix has two identical rows, its determinant is 0. If $\Phi \in \mathcal{S}_q(G)$, then in the definition of $\omega(\Phi)$ the Jacobian in that definition has two identical rows; so that $\omega(\Phi) = 0$. ∎

9.1.8. Corollary. *If* $\omega = f(x)dx_{i_1} \wedge \cdots \wedge dx_{i_q} \in \mathcal{F}_q(G)$ *and* $\omega \neq 0$, *then*

$$\omega = \pm f(x)dx_{j_1} \wedge \cdots \wedge dx_{j_q} \in \mathcal{F}_q(G)$$

where $j_1 < \cdots < j_q$ *and* $\{j_1, \ldots, j_q\} = \{i_1, \ldots, i_q\}$.

The preceding proposition and its corollary allow us to give a canonical representation of q-forms. First observe that by (9.1.7(a)), when $q > p$ we have that $\mathcal{F}_q(G) = (0)$ for any open subset G of $\mathbb{R}^p$. Next if $1 \leq i_1 < \cdots < i_q \leq p$, we put $I = \{i_1, \ldots, i_q\}$ and set

$$dx_I = dx_{i_1} \wedge \cdots \wedge dx_{i_q}$$

This is called a *basic q-form*. A small counting argument shows there are $p!/q!$ $(p-q)!$ different basic q-forms on $\mathcal{S}_q(G)$.

9.1.9. Corollary. *If ω is a q-form on G, then*

$$\omega = \sum_I f_I dx_I$$

where I ranges over all strictly ordered q-tuples of the integers $1, \ldots, p$ and each $f_I \in C(G)$.

Proof. If we are given any ω in $\mathcal{F}_q(G)$, by Corollary 9.1.8 we can replace the coefficient functions in the representation of ω as it appears in the definition as follows. (i) Put $f_I = 0$ if no permutation of the index set I appears. (ii) If some permutation of the strictly increasing q-tuple I does appear, then let f_I be sum of all the functions $\pm f_{i_1,\ldots,i_q}$ where $\{i_1, \ldots, i_q\}$ is a permutation of I and the $\pm$ sign is chosen according to how $\{i_1, \ldots, i_q\}$ is permuted to obtain I. If this is done, we obtain the representation as it appears in the corollary. ∎

Such a representation of a q-form is called its *standard representation*.

9.1.10. Proposition. *If $\omega = \sum_I f_I(x)dx_I$ is a q-form in its standard representation, then $\omega = 0$ in $\mathcal{F}_q(G)$ if and only if $f_I = 0$ for every I.*

Proof. One way is clear: if each $f_I = 0$, then ω is the zero q-form. Conversely assume that $\omega(\Phi) = 0$ for every Φ in $\mathcal{S}_q(G)$ and assume there is a strictly ordered q-tuple $K = \{k_1 < \cdots < k_q\}$ and a point x in G with $f_K(x) > \epsilon > 0$. By the continuity of f_K there is a positive number $a > 0$ such that $f_K(w) > \epsilon$ when $|x_j - w_j| \leq a$ for $1 \leq j \leq p$. Let $R = \{u \in \mathbb{R}^q : |u_i| \leq a$ when $1 \leq i \leq q\}$, note that the rectangle R is a surface domain, and define $\Phi : R \to \mathbb{R}^p$ by $\Phi(u) = x + \sum_{i=1}^q u_i e_{k_i}$. As in Example 9.1.2(g), $\Phi \in \mathcal{S}_q(G)$.

Now a computation shows that

$$\frac{\partial(x_{k_1}, \ldots, x_{k_q})}{\partial(u_1, \ldots, u_q)} = 1$$

while for $I \neq K$,

$$\frac{\partial(x_{i_1}, \ldots, x_{i_q})}{\partial(u_1, \ldots, u_q)} = 0$$

since it is the determinant of a matrix having at least one row of zeros. Thus

$$\omega(\Phi) = \int_R f_K(\Phi(u))du \geq \epsilon V(R) > 0$$

a contradiction. ∎

If you know the language of modules, the preceding proposition says that the basic q-forms constitute a basis for $\mathcal{F}_q(G)$ as a module over the ring $C(G)$. If you don't know this language, don't worry about it. The comment was made for the edification of those who know the language and nothing later in this book depends on what was just said. What is important about the preceding proposition is that when a differential form is expressed in its standard representation that representation is unique.

Now we want to multiply forms. Since each ω in $\mathcal{F}_q(G)$ is a function taking values in $\mathbb{R}$, we might be tempted to define this as we would the product of any such functions: $(\omega_1 \omega_2)(\Phi) = \omega_1(\Phi)\omega_2(\Phi)$. That's legitimate as a definition, but there is a big problem with this. Namely this product is not always a differential form as a little experimentation will show. We need a definition where the product of two forms is again a form. What we will define is the product of forms of different dimensions. We start with the case of a 0-form times a q-form.

9.1.11. Definition. If $\omega = \sum_I f_I(x)dx_I \in \mathcal{F}_q(G)$ and $f \in C(G)$, define

$$f\omega = \sum_J f(x)g_J(x)dx_J$$

That was natural enough. To define the product of other forms we need to start with basic forms of different dimensions and, when we multiply them, produce a form in the sum of the dimensions.

9.1.12. Definition. Let $1 \leq q, r \leq p$ and suppose $I = \{i_1, \ldots, i_q\}$ and $J = \{j_1, \ldots, j_r\}$, where $i_1 < \cdots < i_q$ and $j_1 < \cdots < j_r$. The *product* or *wedge product* of the q-form dx_I and the r-form dx_J is the (q+r)-form denoted by $dx_I \wedge dx_J$ and defined by

$$dx_I \wedge dx_J = dx_{i_1} \wedge \cdots \wedge dx_{i_q} \wedge dx_{j_1} \wedge \cdots \wedge dx_{j_r}$$

In analogy with this definition and notation we could denote the product of a 0-form f and a q-form ω by $f \wedge \omega$. Also note that we have that $0 \wedge \omega = 0$. Maintaining the notation of this definition, we observe that if $I \cap J \neq \emptyset$, then $dx_I \wedge dx_J = 0$ since at least one index is repeated (9.1.7a). In particular if $q + r > p$, $dx_I \wedge dx_J = 0$. If $I \cap J = \emptyset$, let $[I, J]$ be the reordering of $I \cup J$ into an increasing sequence. It follows that

$$dx_{[I,J]} = \pm dx_I \wedge dx_J$$

(See Exercise 3.)

9.1.13. Proposition. *If I, J, K are, respectively, an ordered q-index, an ordered r-index, and an ordered s-index; then*

$$(dx_I \wedge dx_J) \wedge dx_K = dx_I \wedge (dx_J \wedge dx_K)$$

Proof. The proof of this proposition is not difficult, but it is somewhat tedious. The first step is, however, easy: if any pair of the sets I, J, K have an element in common, then both sides of the above equation are 0. Assuming that I, J, K are pairwise disjoint, the idea is to let $dx_{[I,J]} = (-1)^\alpha dx_I \wedge dx_J$ as in Exercise 3, and analogously $dx_{[J,K]} = (-1)^\beta dx_J \wedge dx_K$ and $dx_{[I,K]} = (-1)^\gamma dx_I \wedge dx_K$. If $[I, J, K]$

is the ordering of $I \cup J \cup K$, we must show that

$$(dx_I \wedge dx_J) \wedge dx_K = (-1)^{\alpha}(-1)^{\beta+\gamma} dx_{[I,J,K]}$$
$$dx_I \wedge (dx_J \wedge dx_K) = (-1)^{\beta}(-1)^{\alpha+\gamma} dx_{[I,J,K]}$$

Thus completing the proof. Carrying out the details is left to the reader (Exercise 4). ∎

9.1.14. Definition. If $\omega = \sum_I f_I(x)dx_I \in \mathcal{F}_q(G)$ and $\eta = \sum_J g_J(x)dx_J \in \mathcal{F}_r(G)$, each in its standard representation, then their product, denoted by $\omega \wedge \eta$, is defined as the (q+r)-form on G

$$\omega \wedge \eta = \sum_{I,J} f_I(x)g_J(x)dx_I \wedge dx_J$$

Let's underline that in the preceding definition we took two forms in their standard representation and multiplied them, but the form that is the product is not in its standard representation.

9.1.15. Proposition. *If* $\omega, \omega_1, \omega_2 \in \mathcal{F}_q(G)$, $\eta, \eta_1, \eta_2 \in \mathcal{F}_r(G)$, *and* $\sigma \in \mathcal{F}_s(G)$, *then:*

(a) $(\omega_1 + \omega_2) \wedge \eta = \omega_1 \wedge \eta + \omega_2 \wedge \eta$;
(b) $\omega \wedge (\eta_1 + \eta_2) = \omega \wedge \eta_1 + \omega \wedge \eta_2$; *and*
(c) $\omega \wedge (\eta \wedge \sigma) = (\omega \wedge \eta) \wedge \sigma$.

The proof of this proposition establishing the distributive and associative laws for the multiplication of differential forms is left to the reader (Exercise 5). As we have already seen, the multiplication is not commutative.

Now we discuss differentiating smooth q-forms.

9.1.16. Definition. If f is a smooth 0-form on G, then its derivative is the 1-form

$$df = \sum_{i=1}^{p}(D_i f)(x)dx_i = \sum_{i=1}^{p}\frac{\partial f}{\partial x_i}(x)dx_i$$

If $\omega = \sum_I f_I(x)dx_i$ is the standard representation of a smooth q-form on G, then its *derivative* is the (q+1)-form

$$d\omega = \sum_I (df_I) \wedge dx_I$$

9.1.17. Example. (a) Return to Example 9.1.6(a) where $p = q = 1$. If ω is the 1-form $\omega = f dx$ where f is a smooth function on $[a, b]$, then $d\omega = df \wedge dx = 0$ since it is a 2-form on the one-dimensional space. If we consider the 0-form f, then $df = f'(x)dx$. Thus we can interpret the Fundamental Theorem of Calculus as saying that the integral of the 1-form df over the 1-surface equals the integral of the 0-form f over its boundary, though we would have to give $\partial[a, b]$ an orientation to obtain the minus sign in $f(b) - f(a)$.

(b) Let $p = q = 3$, suppose $X \subseteq \mathbb{R}^3$, and assume $F : X \to \mathbb{R}^3$ is a smooth function given by $F(x, y, z) = (P(x, y, z), Q(x, y, z), R(x, y, z))$. Put $\eta = Pdy \wedge dz + Qdz \wedge dx + Rdx \wedge dy$ as in Example 9.1.6(d). So using the fact that $dy \wedge dz \wedge dx = -dy \wedge dx \wedge dz = dx \wedge dy \wedge dz$ and $dz \wedge dx \wedge dy = -dx \wedge dz \wedge dy = dx \wedge dy \wedge dz$

$$
\begin{aligned}
d\eta &= dP \wedge (dy \wedge dz) + dQ \wedge (dz \wedge dx) + dR \wedge (dx \wedge dy) \\
&= \frac{\partial P}{\partial x} dx \wedge dy \wedge dz + \frac{\partial Q}{\partial y} dy \wedge dz \wedge dx + \frac{\partial R}{\partial z} dz \wedge dx \wedge dy \\
&= \left(\frac{\partial P}{\partial x} + \frac{\partial Q}{\partial y} + \frac{\partial R}{\partial z} \right) dx \wedge dy \wedge dz \\
&= (\operatorname{div} F) \, dx \wedge dy \wedge dz
\end{aligned}
$$

Recall Gauss's Theorem where for an appropriate solid X with ∂X a 2-surface we have $\iint_{\partial X} F \cdot \mathbf{n} \, d\sigma = \iiint_X \operatorname{div} F$. Rephrasing this using the form η and its derivative we have that Gauss's Theorem says that

$$
\int_{\partial X} \eta = \int_X d\eta
$$

That is, the integral of the derivative $d\eta$ over the inside of X equals the integral of the η over the boundary of X. With this interpretation we can rightfully call Gauss's Theorem an extension of the FTC to $\mathbb{R}^3$.

(c) As in part (b) let $p = q = 3$, suppose $S \subseteq \mathbb{R}^3$, and assume $F : S \to \mathbb{R}^3$ is a smooth function given by $F(x, y, z) = (P(x, y, z), Q(x, y, z), R(x, y, z))$. This time put $\omega = Pdx + Qdy + Rdz$, and compute $d\omega$. Using the rules for differentiating and multiplying forms we obtain

$$
\begin{aligned}
d\omega &= dP \wedge dx + dQ \wedge dy + dR \wedge dz \\
&= \left(\frac{\partial P}{\partial y} dy + \frac{\partial P}{\partial z} dz \right) \wedge dx \\
&\quad + \left(\frac{\partial Q}{\partial x} dx + \frac{\partial Q}{\partial z} dz \right) \wedge dy + \left(\frac{\partial R}{\partial x} dx + \frac{\partial R}{\partial y} dy \right) \wedge dz \\
&= \left(\frac{\partial R}{\partial y} - \frac{\partial Q}{\partial z} \right) dy \wedge dz + \left(\frac{\partial P}{\partial z} - \frac{\partial R}{\partial x} \right) dz \wedge dx + \left(\frac{\partial Q}{\partial x} - \frac{\partial P}{\partial y} \right) dx \wedge dy
\end{aligned}
$$

You may recognize the coefficients of the basic forms $dy \wedge dz$, $dz \wedge dx$, $dx \wedge dy$, which have been arranged here in a certain rotational pattern, as forming the components of curl F. We will return to this later in this chapter when we prove a generalized version of Stokes's Theorem.

(d) Suppose ω is a basic q-form: $\omega = dx_I = dx_{i_1} \wedge \cdots \wedge dx_{i_q}$. It follows that $d\omega = 0$. In fact using the definition of differentiation of q-forms shows $d\omega = d(1 \cdot dx_I) = d(1) \wedge dx_I = 0$.

(e) Consider $\omega = f dx_1$, the 1-form on $G \subseteq \mathbb{R}^p$ where $p \geq 2$ and f is smooth. So

$$d\omega = df \wedge dx_1 = \left(\sum_{i=1}^{p} \frac{\partial f}{\partial x_i}(x) dx_i \right) \wedge dx_1$$

$$= \sum_{i=2}^{p} \frac{\partial f}{\partial x_i}(x) \, dx_i \wedge dx_1$$

We now examine how differentiation of forms interacts with the algebraic operations as well as what happens when we take multiple derivatives. Regarding this latter statement, if $\omega = \sum_I f_I dx_I \in \mathcal{F}_q(G)$, say that ω is a form of class C'' if each f_I has two continuous derivatives. That is, for each I, $Df_I : G \to \mathbb{R}^p$ exists and also $D^2 f_I = D(Df_I) : G \to \mathcal{L}(\mathbb{R}^p)$ exists and is continuous. Equivalently, each f_I has continuous second partial derivatives.

9.1.18. Theorem. *If G is an open subset of $\mathbb{R}^p$, ω a smooth element of $\mathcal{F}_q(G)$, and η a smooth element of $\mathcal{F}_r(G)$, then the following hold.*

(a) *When $r = q$, $d(\omega + \eta) = d\omega + d\eta$.*
(b) *$d(\omega \wedge \eta) = (d\omega) \wedge \eta + (-1)^q \omega \wedge (d\eta)$.*
(c) *If ω is of class C'', then*

$$d(d\omega) = 0$$

Proof. The proof of (a) is routine. To prove (b) it suffices to show this under the assumption that $\omega = f dx_I$ and $\eta = g dx_J$. (Why?) First if $q = r = 0$ so that $\omega = f$ and $\eta = g$, then the product rule for differentiating functions shows that (b) holds here. When $q, r \geq 1$, we have $\omega \wedge \eta = fg dx_I \wedge dx_J$. If I and J have a term in common, then both sides of the equation in (b) are 0. So assume $I \cap J = \emptyset$. Thus

$$d(\omega \wedge \eta) = d(fg dx_I \wedge dx_J)$$
$$= \pm d(fg dx_{[I,J]})$$
$$= \pm(gdf + fdg) \wedge dx_{[I,J]}$$
$$= (gdf + fdg) \wedge dx_I \wedge dx_j$$

where at each stage of these equalities the choice of a plus or minus sign is the same. Now using the fact that dg is a 1-form and applying Proposition 9.1.7(b) q times we have that $dg \wedge dx_I = (-1)^q dx_I \wedge dg$. Therefore the above equalities become

$$d(\omega \wedge \eta) = (gdf) \wedge dx_I \wedge dx_J + (fdg) \wedge dx_I \wedge dx_J$$
$$= (df \wedge dx_I) \wedge (gdx_J) + (-1)^q(fdx_I) \wedge (dg \wedge dx_J)$$
$$= (d\omega) \wedge \eta + (-1)^q \omega \wedge (d\eta)$$

proving (b).

(c) First we prove this when $\omega = f$, a 0-form. In this case, remembering that $D_{ij}f = D_{ji}f$ since f has continuous second partial derivatives and that $dx_i \wedge dx_j = -dx_j \wedge dx_i$,

$$d(df) = d\left(\sum_{i=1}^{p}(D_i f)(x)dx_i\right)$$

$$= \sum_{i=1}^{p} d(D_i f) \wedge dx_i$$

$$= \sum_{i=1}^{p}\sum_{j=1}^{p}(D_{ij}f)(x)dx_j \wedge dx_i$$

$$= -\sum_{j=1}^{p}\sum_{i=1}^{p}(D_{ji}f)(x)dx_i \wedge dx_j$$

$$= -d(df)$$

so that it must be that $d(df) = 0$.

Now let $q \geq 1$ and assume that $\omega = f dx_I$ so that, by definition, $d\omega = (df) \wedge dx_I$. Using part (b) this implies that $d(d(\omega)) = (d(df)) \wedge dx_I - (df) \wedge d(dx_I) = -(df) \wedge d(dx_I)$. But from Example 9.1.17(d) we know that $d(dx_I) = d(1) \wedge dx_I = 0$. So that $d(d(\omega)) = 0$ when $\omega = f dx_I$. The case of a general q-form now follows from part (a). ∎

9.1.19. Example. (a) Let G be an open subset of $\mathbb{R}^p$ and suppose f is a continuously differentiable function on G. If $\gamma : [a, b] \to G$ is any regular curve and $\omega = df$, then Example 9.1.6(b) and the Chain Rule applied to $f \circ \gamma$ shows that

$$\omega(\gamma) = \int_\gamma df = \int_a^b \langle (Df)(\gamma(t)), \gamma'(t)\rangle\, dt$$

$$= \int_a^b (f \circ \gamma)'(t)\, dt$$

$$= f(\gamma(b)) - f(\gamma(a))$$

So the value of $df(\gamma)$ does not depend on the precise curve γ but only on its starting and final points, $\gamma(a)$ and $\gamma(b)$. Note that we can interpret this as an analogue of the Fundamental Theorem of Calculus.

(b) Consider the 1-form $\omega = x dy$ in $\mathbb{R}^2$. If $\gamma : [0, 2\pi] \to \mathbb{R}^2$ is defined by $\gamma(\theta) = (\cos\theta, \sin\theta)$, then γ traces the unit circle in the counterclockwise direction. We have that

$$\omega(\gamma) = \int_0^{2\pi} \cos\theta\, d(\sin\theta) = \int_0^{2\pi} \cos^2\theta\, d\theta = \pi$$

From part (a) of this example we see that there is no continuously differentiable function f on $\mathbb{R}^2$ such that $x dy = df$. More on such a phenomenon in §9.6.

Exercises

(1) (a) Give the details for Example 9.1.2(a). (b) In Example 9.1.2(a) give a parametrization of the hemisphere S when $p = 5$. Can you give a parametrization for arbitrary p?

(2) Verify Example 9.1.2(b).

(3) In the definition of two basic forms $dx_I \wedge dx_J$ show that $dx_{[I,J]} = (-1)^\alpha dx_I \wedge dx_J$, where α is the number of differences $\{ j_\ell - i_k : 1 \le k \le q, 1 \le \ell \le r \}$ that are negative.

(4) Supply the details in the proof of Proposition 9.1.13.

(5) Prove Proposition 9.1.15. (I know this is tedious, but this is one of those instances where if you just read the proof it does no good and does not have any meaning. If, on the other hand, you work it out on your own you will gain extra familiarity with the concepts.)

The next two exercises are from [1].

(6) Let ω be the 2-form on $\mathbb{R}^4$ defined by $\omega = x_2 \, dx_1 \wedge dx_3 - x_4 \, dx_3 \wedge dx_4$; let (Φ, R) be the 2-surface in $\mathbb{R}^4$ where $R = \{ (x, y) : x^2 + y^2 \le 1 \}$ and $\Phi(x, y) = (x, x - y, 3 - x + xy, -3y)$. Compute $\int_\Phi \omega$.

(7) Let ω be the 3-form in $\mathbb{R}^4$ defined by $\omega = x_3 \, dx_1 \wedge dx_2 \wedge dx_4$; let (Φ, R) be the 3-surface where $R = \{ (x, y, z) : x^2 + y^2 + z^2 \le 1 \}$ and $\Phi(x, y, z) = (yz, x^2, 1 - 3y + z, xy)$. Compute $\int_\Phi \omega$.

(8) For each of the following choices of forms ω and η, compute $\omega \wedge \eta$ and write the product in its standard representation. (a) $\omega = y \, dx + x \, dy$ and $\eta = x \, dx \wedge dz + y \, dy \wedge dz$. (b) $\omega = (x + y) \, dx + z \, dy + z \, dz$ and $\eta = (y - z) \, dx + (x - z) \, dy + (y - z) \, dz$.

(9) For each of the following choices of the form ω compute $d\omega$ and write it in its standard representation. (a) $\omega = f \, dx \wedge dy + g \, dx \wedge dz + h \, dy \wedge dz$. (b) $\omega = (x + y^2) \, dx \wedge dy + x^3 \, dy \wedge dz$.

9.2. Change of Variables for Forms

Suppose G is an open subset of $\mathbb{R}^p$ and H is an open subset of $\mathbb{R}^m$. If $T : G \to H$ is a continuously differentiable mapping and $\Phi : R \to G$ is in $\mathcal{S}_q(G)$, then $T \circ \Phi : R \to H$ is in $\mathcal{S}_q(H)$. If $\omega \in \mathcal{F}_q(H)$, we want to show that this change of variables leads naturally to a q-form ω_T in $\mathcal{F}_q(G)$ and to explore the relation between these forms. This is a fundamental step in exploring forms and also in the orientation of q-surfaces.

Note that we automatically have a function $\Phi \mapsto \omega(T \circ \Phi)$ with domain $\mathcal{S}_q(G)$. This does not mean, however, that this is a q-form on G. To be a q-form it must satisfy additional properties as stated in the definition. The process of finding what we will define as the change of variables for ω is rather straightforward, nevertheless you must be careful because the notation gets a bit complicated.

Let $\omega = \sum_I g_I(y)dy_I \in \mathcal{F}_q(H)$, written in its standard representation. To get the sought for q-form ω_T on G, two things seem clear: the index sets should remain the same and the coefficient functions should be $f_I(x) = g_I(T(x))$ for each x in G and each index set I. The more complicated step will be to get the corresponding basic q-forms on G. To do this examine the components of $T : G \to H$,

$$y = (y_1, \ldots, y_m) = T(x) = (t_1(x), \ldots, t_m(x))$$

where it follows that for $1 \le i \le m, t_i : G \to \mathbb{R}$ is continuously differentiable. Thus we can obtain the 1-forms

$$dt_i = \sum_{j=1}^{p} (D_j t_i)(x) \, dx_j$$

for $1 \le i \le m$. This leads us to the following.

9.2.1. Definition. If G and H are open sets in $\mathbb{R}^p$ and $\mathbb{R}^m$, respectively, and $\omega = \sum_I g_I(y)dy_I \in \mathcal{F}_q(H)$ expressed in its standard representation, then for any continuously differentiable mapping $T = (t_1, \ldots, t_m) : G \to H$, define the q-form ω_T on G as

$$\omega_T = \sum_I g_I(T(x))dt_{i_1} \wedge \cdots \wedge dt_{i_q}$$

where $I = \{i_1 < \cdots < i_q\}$.

Observe that since each dt_{i_j} is a 1-form and the indices i_j are distinct, $dt_{i_1} \wedge \cdots \wedge dt_{i_q}$ is a q-form so that ω_T is indeed a q-form on G. Also each part of the definition of ω_T involves the mapping T as well as the given q-form ω.

To simplify the statements and notation below, maintain the notation we have used so far: G always denotes an open subset of $\mathbb{R}^p$, H an open subset of $\mathbb{R}^m$, and $T : G \to H$ is a continuously differentiable function.

9.2.2. Proposition. *If $\omega \in \mathcal{F}_q(H)$ and $\eta \in \mathcal{F}_r(H)$, the following hold.*

(a) *When $r = q$, $(\omega + \eta)_T = \omega_T + \eta_T$.*
(b) *$(\omega \wedge \eta)_T = \omega_T \wedge \eta_T$.*
(c) *If ω is continuously differentiable and T is a twice continuously differentiable function, then $d(\omega_T) = (d\omega)_T$.*

Proof. (a) This is a matter of examining the definition and is left to the reader (Exercise 1(a)).

(b) This proof is also easy, though it is cumbersome. Let's first examine what happens to basic q and r-forms. If $I = \{i_1 < \cdots < i_q\}, J = \{j_1 < \cdots < j_r\}, \omega = dx_I$, and $\eta = dx_J$, then $\omega_T = dt_{i_1} \wedge \cdots \wedge dt_{i_q}, \eta_T = dt_{j_1} \wedge \cdots \wedge dt_{j_r}$. So

$$\omega \wedge \eta = dx_{i_1} \wedge \cdots \wedge dx_{i_q} \wedge dx_{j_1} \wedge \cdots \wedge dt_{x_r}$$
$$(\omega \wedge \eta)_T = dt_{i_1} \wedge \cdots \wedge dt_{i_q} \wedge dt_{j_1} \wedge \cdots \wedge dt_{y_r}$$

So clearly (b) is satisfied for basic forms. To show that (b) holds for all q-forms is now routine (Exercise 1(b)).

(c) First assume ω is a 0-form: $\omega = g$ for a continuously differentiable function g. So $\omega_T = g \circ T$ and $d\omega = \sum_{i=1}^{m}(D_i g)(y)dy_i \in \mathcal{F}_1(H)$. Hence. using the Chain Rule, we get

$$d(\omega_T) = \sum_{j=1}^{p}[D_j(g \circ T)](x)dx_j$$

$$= \sum_{j=1}^{p}\left[\sum_{i=1}^{m}(D_i g)(T(x))(D_j t_i)(x)\right]dx_j$$

$$= \sum_{i=1}^{m}\left[\sum_{j=1}^{p}(D_j t_i)(x)dx_j\right](D_i g)(T(x))$$

$$= \sum_{i=1}^{m}(D_i g)(T(x))dt_i$$

$$= (d\omega)_T$$

Now assume ω is a basic q-form on H: $\omega = dy_I = dy_{i_1} \wedge \cdots \wedge dy_{i_q}$ so that $\omega_T = dt_{i_1} \wedge \cdots \wedge dt_{i_q}$. As we saw in Example 9.1.17(d), $d\omega = 0$ and also $d(\omega_T) = 0$.

Finally assume $\omega = gdy_I$ so that $\omega_T = (g \circ T)dt_{i_1} \wedge \cdots \wedge dt_{i_q}$. Therefore

$$d(\omega_T) = d(g \circ T) \wedge dt_{i_1} \wedge \cdots \wedge dt_{i_q}$$

$$= (dg)_T \wedge dt_{i_1} \wedge \cdots \wedge dt_{i_q}$$

$$= [(dg) \wedge dy_I]_T$$

$$= (d\omega)_T$$

The general case now follows by using part (a). ∎

See Exercise 2.

9.2.3. Proposition. *Let G, H, and W be open sets with $G \subseteq \mathbb{R}^p$, $H \subseteq \mathbb{R}^m$, and $W \subseteq \mathbb{R}^d$; let $T : G \to H$ and $S : H \to W$ be continuously differentiable mappings. If $\omega \in \mathcal{F}_q(W)$ and $ST : G \to W$ is the composition of S and T, then $(\omega_S)_T = \omega_{ST}$.*

Proof. Just to be clear, $\omega_S \in \mathcal{F}_q(H)$ and both $(\omega_S)_T$ and $\omega_{ST} \in \mathcal{F}_q(G)$. To set the notation let $T(x) = (t_1(x), \ldots, t_m(x))$ for each x in G, $S(y) = (s_1(y), \ldots, s_d(y))$ for each y in H, and $ST(x) = (r_1(x), \ldots, r_d(x))$ for each x in G. Let's note that for $1 \leq k \leq d$,

$$r_k(x) = s_k(T(x)) = s_k(t_1(x), \ldots, t_m(x))$$

so that by the Chain Rule

$$(D_i r_k)(x) = \sum_{j=1}^{p}(D_j s_k)(T(x))(D_i t_j)(x)$$

For ω a 0-form, the result is the usual chain rule for functions. Now let's prove the proposition when ω is the 1-form $\omega = dz_k$ on W, where $1 \le k \le d$. So $\omega_{ST} = dr_k$ and $\omega_S = ds_k = \sum_{j=1}^{m}(D_j s_k)(y)dy_j$. Using the Chain Rule we have that

$$(\omega_S)_T = (ds_k)(T(x))$$

$$= \sum_{j=1}^{m}(D_j s_k)(T(x))dt_j$$

$$= \sum_{j=1}^{m}(D_j s_k)(T(x))\left[\sum_{i=1}^{p}(D_i t_j)(x)dx_i\right]$$

$$= \sum_{i=1}^{p}\left[\sum_{j=1}^{m}(D_j s_k)(T(x))(D_i t_j)(x)\right]dx_i$$

$$= \sum_{i=1}^{p}(D_i r_k)(x)dx_i$$

$$= dr_k = \omega_{ST}$$

Now if ω and η are forms on W, Proposition 9.2.2 shows that $((\omega \wedge \eta)_S)_T = (\omega_S \wedge \eta_S)_T = (\omega_S)_T \wedge (\eta_S)_T$. Since the result holds when ω is a 0-form, this observation combined with what we did above completes the proof of the proposition. ∎

Assume $\omega \in \mathcal{F}_q(G)$ and $\Phi : R \to G$ is in $\mathcal{S}_q(G)$. So we can consider Φ as a change of variables to obtain a q-form on R, the surface domain for Φ that is contained in $\mathbb{R}^q$. That is, using the notation in Definition 9.2.1 we have $\omega_\Phi \in \mathcal{F}_q(R)$. So for any q-surface Ψ in $\mathcal{S}_q(R)$ we can form $\omega_\Phi(\Psi) = \int_\Psi \omega_\Phi$. After thinking about the hypothesis of the next result, you might predict the conclusion though the proof takes some effort.

9.2.4. Proposition. *If $\omega \in \mathcal{F}_q(G)$, $\Phi : R \to G$ is in $\mathcal{S}_q(G)$ with $R \subseteq \mathbb{R}^q$, and $\Psi : R \to \mathbb{R}^q$ is defined by $\Psi(u) = u$ for all u in R, then*

$$\int_\Phi \omega = \int_\Psi \omega_\Phi.$$

Proof. It suffices to prove the proposition under the assumption that

$$\omega = f(x)dx_{i_1} \wedge \cdots \wedge dx_{i_q}$$

(Why?) Assume $\Phi(u) = (\phi_1(u), \ldots, \phi_p(u))$ with $\phi_i : R \to \mathbb{R}$ for $1 \le i \le p$. Recall from the definition of a q-form and (9.1.5) that

9.2.5
$$\int_\Phi \omega = \int_R f(\Phi(u)) \frac{\partial(\phi_{i_1}, \ldots, \phi_{i_q})}{\partial(u_1, \ldots, u_q)} du$$

Let $J(u)$ be the Jacobian in the preceding equation; recall that this is the determinant of the matrix $A(u) = [a_{mn}(u)]$, where for $1 \le m, n \le q$

$$a_{mn}(u) = (D_n \phi_{i_m})(u)$$

Thus $d\phi_{i_m} = \sum_{n=1}^q a_{mn}(u) du_n$. By a herculean manipulation of symbols we get that

$$d\phi_{i_1} \wedge \cdots \wedge d\phi_{i_q} = \sum [a_{1,n_1}(u) \cdots a_{q,n_q}(u)] du_{n_1} \wedge \cdots \wedge du_{n_q}$$

where in this sum $n_1, \ldots, n_q$ ranges independently over $1, \ldots, q$. Now $\sigma(m) = n_m$ for $1 \le m \le q$ defines a permutation in S_q. From Theorem 9.1.18 we know that $du_m \wedge du_k = -du_k \wedge du_m$. Thus when $\sigma(m) = n_m$ for $1 \le m \le q$,

$$du_{n_1} \wedge \cdots \wedge du_{n_q} = \text{sign}(\sigma) du_1 \wedge \cdots \wedge du_q$$

Remembering that $J(u) = \det A(u)$ and using Definition 6.4.18, we get

$$d\phi_{i_1} \wedge \cdots \wedge d\phi_{i_q} = \sum_{\sigma \in S_q} [a_{1,\sigma(1)}(u) \cdots a_{q,\sigma(q)}(u)] du_{\sigma(1)} \wedge \cdots \wedge du_{\sigma(q)}$$

$$= \sum_{\sigma \in S_q} \text{sign}(\sigma) [a_{1,\sigma(1)}(u) \cdots a_{q,\sigma(q)}(u)] du_1 \wedge \cdots \wedge du_q$$

$$= J(u) du_1 \wedge \cdots \wedge du_q$$

Applying this equation to (9.2.5) we get

$$\int_\Phi \omega = \int_R f(\Phi(u)) J(u) du$$

$$= \int_\Psi f(\Phi(u)) J(u) du_1 \wedge \cdots \wedge du_q$$

$$= \int_\Psi \omega_\Phi \qquad \blacksquare$$

Combining Proposition 9.2.3 and Proposition 9.2.4 will prove the following, which is the Change of Variable Formula for forms.

9.2.6. Theorem. *Let G be an open subset of $\mathbb{R}^p$ and assume $T : G \to \mathbb{R}^d$ is a continuously differentiable mapping with H an open subset of $\mathbb{R}^d$ that contains $T(G)$. If $\Phi \in \mathcal{S}_q(G)$ and $\omega \in \mathcal{F}_q(H)$, then*

$$\int_{T \circ \Phi} \omega = \int_\Phi \omega_T$$

Proof. Suppose $\Phi : R \to G$ with $R \subseteq \mathbb{R}^q$, and let $\Psi : R \to \mathbb{R}^q$ be defined by $\Psi(u) = u$ as in Proposition 9.2.4. By that proposition $\int_{T \circ \Phi} \omega = \int_\Psi \omega_{T \circ \Phi}$. But

Proposition 9.2.3 says this last integral becomes $\int_\Psi \omega_{T\circ\Phi} = \int_\Psi (\omega_T)_\Phi$. Applying (9.2.4) once again we have that the last integral becomes $\int_\Psi (\omega_T)_\Phi = \int_\Phi \omega_T$, whence the theorem. ∎

See Exercise 3.

Exercises

(1) (a) Give the details of the proof of Proposition 9.2.2(a). (b) Supply the missing details in the proof of Proposition 9.2.2(b).

(2) Where in the proof of Proposition 9.2.2 was use made of the assumption that T is twice continuously differentiable?

(3) Show that both Proposition 9.2.3 and Proposition 9.2.4 are special cases of Theorem 7.4.19.

9.3. Simplexes and Chains
. . . .

Before getting into the heart of this section, we introduce an elementary concept that will play a role.

9.3.1. Definition. A subset X of $\mathbb{R}^p$ is *convex* provided that when $x, y \in X$, the line segment $[x, y] = \{(1 - t)x + ty : 0 \le t \le 1\} \subseteq X$.

9.3.2. Proposition.

(a) *Every ball in $\mathbb{R}^p$ is a convex set.*
(b) *A convex set is connected.*
(c) *The intersection of any collection of convex sets is convex.*

The proof of the preceding proposition is left to the reader in Exercise 1, where a few additional facts about convex sets also appear. Part (c) of the proposition leads us to an additional concept. For any non-empty subset E of $\mathbb{R}^p$ we define the *convex hull* of E as the intersection of all convex subsets containing E. This is denoted by $\text{con}(E)$. Note that the collection of all convex subsets containing E is non-empty since $\mathbb{R}^p$ is one such set and $\text{con}(E)$ is convex by (c).

9.3.3. Lemma. *A subset X of $\mathbb{R}^p$ is convex if and only if when $x_1, \ldots, x_n \in X$ and $t_1, \ldots, t_n \ge 0$ with $\sum_{i=1}^n t_i = 1$ we have that $\sum_{i=1}^n t_i x_i \in X$.*

Proof. If X satisfies the stated condition of the lemma when $n = 2$, then this is the definition of convexity. Conversely, assume X is convex and $x_1, \ldots, x_n \in X$ and $t_1, \ldots, t_n \ge 0$ with $\sum_{i=1}^n t_i = 1$. The proof that $\sum_{i=1}^n t_i x_i \in X$ is by induction.

The statement $n = 1$ is trivial and $n = 2$ is the definition. Assume $n \geq 3$ and the statement for $n - 1$ is true. Note that

$$\sum_{i=1}^{n} t_i x_i = (1 - t_n) \sum_{i=1}^{n-1} \frac{t_i}{1 - t_n} x_i + t_n x_n$$

But by the induction hypothesis, $\sum_{i=1}^{n-1} \frac{t_i}{1-t_n} x_i \in X$. The case n now follows. ∎

9.3.4. Proposition. *If $E \subseteq \mathbb{R}^p$, then*

$$\mathrm{con}(E) = \left\{ \sum_{i=1}^{n} t_i x_i : x_1, \ldots, x_n \in E, \text{ each } t_i \geq 0 \text{ and } \sum_{i=1}^{n} t_i = 1 \right\}$$

Proof. Denote the set on the right of the above displayed equation as S. It is left as Exercise 2 to show that S is convex. Since $E \subseteq S$ it follows that $\mathrm{con}(E) \subseteq S$. Now assume that X is a convex set that contains E. By the preceding lemma, $S \subseteq X$ and the proposition follows. ∎

9.3.5. Corollary. *If E is a finite subset of $\mathbb{R}^p$, then $\mathrm{con}(E)$ is compact.*

Proof. If $E = \{x_1, \ldots, x_n\}$, let

$$T = \left\{ (t_1, \ldots, t_n) \in \mathbb{R}^n : t_i \geq 0 \text{ and } \sum_{i=1}^{n} t_i = 1 \right\}$$

If we define $f : T \to \mathbb{R}^p$ by $f(t_1, \ldots, t_n) = \sum_{i=1}^{n} t_i x_i$, then f is continuous and $f(T) = \mathrm{con}(E)$ by the proposition. On the other hand it is immediate that T is a compact subset of $\mathbb{R}^n$ since it is closed and bounded, so $f(T)$ is compact. ∎

Now we define a convex subset of $\mathbb{R}^q$ that will be very important for our discussion, in fact a key building block in this development.

9.3.6. Definition. If $\{e_1, \ldots, e_q\}$ is the standard basis in $\mathbb{R}^q$, the *unit simplex* in $\mathbb{R}^q$ is the set

$$C^q = \left\{ \sum_{j=1}^{q} a_j e_j : 0 \leq a_j \leq 1 \text{ for } 1 \leq j \leq q \text{ and } \sum_{j=1}^{q} a_j \leq 1 \right\}$$

$$= \{(a_1, \ldots, a_q) \in \mathbb{R}^q : a_j \geq 0 \text{ for } 1 \leq j \leq q \text{ and } a_1 + \cdots + a_q \leq 1\}$$

So $C^1 = [0, 1]$ and $C^2 = \{(x, y) : x \geq 0, y \geq 0, \text{ and } x + y \leq 1\}$, the triangle in the plane with vertices $(0, 0)$, $(0, 1)$, $(1, 0)$. C^3 is the tetrahedron in $\mathbb{R}^3$ with the four vertices $\{(0, 0, 0), (1, 0, 0), (0, 1, 0), (0, 0, 1)\}$. We easily see that C^q is a convex set. In fact, C^q is the convex hull of $\{0, e_1, \ldots, e_q\}$.

Now for the next building block. Say that a mapping $\sigma : \mathbb{R}^q \to \mathbb{R}^p$ is *affine* if there is a linear transformation $A : \mathbb{R}^q \to \mathbb{R}^p$ and a vector v_0 in $\mathbb{R}^p$ such that $\sigma(u) = v_0 + A(u)$ for all u in $\mathbb{R}^q$. Equivalently $\sigma : \mathbb{R}^q \to \mathbb{R}^p$ is an affine mapping if $A : \mathbb{R}^q \to \mathbb{R}^p$, defined by $A(u) = \sigma(u) - \sigma(0)$, belongs to $\mathcal{L}(\mathbb{R}^q, \mathbb{R}^p)$. (Exercise 3.) We note in passing the connection between affine maps and the notion of an affine hyperplane defined in (6.7.2): if $p = 1$ and $\sigma : \mathbb{R}^q \to \mathbb{R}$ is an affine mapping, then $\sigma^{-1}(0)$ is an affine hyperplane. The proof of the next proposition is Exercise 4.

9.3.7. Proposition. *If $\sigma : \mathbb{R}^q \to \mathbb{R}^p$ is an affine map and X is a convex subset of $\mathbb{R}^q$, then $\sigma(X)$ is a convex subset of $\mathbb{R}^p$.*

9.3.8. Definition. An *affine q-simplex* is a q-surface (σ, C^q) such that σ is an affine map.

Of course an affine map is defined on all of $\mathbb{R}^q$, but in the preceding definition we are restricting σ to the unit simplex. Also there is a double usage of the term "simplex" here: once when we discuss the unit simplex (a set) and then when we define an affine q-simplex (a function). This shouldn't cause any confusion as long as we consider the context in which the terms are used.

So the trace of an affine q-surface is a convex set, but there are convex subsets that are not the trace of an affine q-simplex; for example the closed disk in $\mathbb{R}^2$ (Why?). Note that every line segment in $\mathbb{R}^p$ is the trace of an affine 1-simplex. For the ease of notation in the remainder of this chapter, $e_1, \ldots, e_q$ will always denote the standard basis elements in $\mathbb{R}^q$ and we set $e_0 = 0$.

9.3.9. Proposition. *Let σ be an affine q-simplex and for $0 \le j \le q$ put $v_j = \sigma(e_j)$.*

(a) *σ is uniquely determined by the $q + 1$ points $v_0, \ldots, v_q$. That is, if τ is also an affine q-simplex and $\tau(e_j) = v_j$ for $0 \le j \le q$, then $\tau = \sigma$.*
(b) *If $(a_1, \ldots, a_q) \in C^q$, then*

$$\sigma(a_1, \ldots, a_q) = v_0 + \sum_{j=1}^{q} a_j(v_j - v_0).$$

Proof. Clearly part (a) follows from part (b). To prove (b) let $A(u) = \sigma(u) - \sigma(0) = \sigma(u) - v_0$. Since σ is affine, $A \in \mathcal{L}(\mathbb{R}^q, \mathbb{R}^p)$. If $1 \le j \le q$, then $A(e_j) = \sigma(e_j) - v_0$, from which (b) follows. ∎

In light of the last proposition we can also define σ as the ordered $(q+1)$-tuple of points in $\mathbb{R}^p$:

$$\sigma = [v_0, v_1, \ldots, v_q]$$

Our use of the word "ordered" in the last sentence is a precursor of our next undertaking: giving an affine q-simplex σ an *orientation*, determined by the order of the points $v_0, \ldots, v_q$ when we write $\sigma = [v_0, v_1, \ldots, v_q]$.

Suppose κ is a permutation of $0, 1, \ldots, q$ and $i_j = \kappa(j)$ for $0 \le j \le q$. Let $\sigma_\kappa = [v_{i_0}, \ldots, v_{i_q}] = [v_{\kappa(0)}, \ldots, v_{\kappa(q)}]$. So σ_κ is a new affine q-simplex with the same trace as σ but a different orientation. If σ is an oriented affine 1-simplex in $\mathbb{R}^p$ and κ is the permutation $\kappa(0) = 1, \kappa(1) = 0$, then σ_κ is the same line segment traced in the opposite direction. As another example consider $C^2 \subseteq \mathbb{R}^2$. As mentioned previously this is the triangle with vertices $v_0 = (0, 0)$, $v_1 = (1, 0)$, and

$v_2 = (0, 1)$. So if we take $\sigma = [v_0, v_1, v_2]$ this orients ∂C^2 in the counterclockwise or positive direction. Considering $\sigma_\kappa = [v_0, v_2, v_1]$ orients the boundary in the clockwise direction. More complicated interpretations apply to higher dimensional affine q-simplexes.

Recall (6.4.16) that for any permutation κ of $0, \ldots, q$ we have $\text{sign}(\kappa) = \pm 1$. If $\text{sign}(\kappa) = +1$, we say that σ and σ_κ have the *same orientation* and they are *equivalent affine q-simplexes*. If $\text{sign}(\kappa) = -1$, the two q-simplexes are said to have the *opposite orientation*.

The next result establishes the effect that changing the orientation of an affine q-simplex has on the value of a differential form when it is evaluated at the simplex. To start, let's examine the interaction between a q-form and an affine q-simplex. As above let $\sigma(u) = v_0 + A(u)$ for some A in $\mathcal{L}(\mathbb{R}^q, \mathbb{R}^p)$ and all u in $\mathbb{R}^q$, and let G be an open subset of $\mathbb{R}^p$ that contains the trace of σ, $\{\sigma\} = \sigma(C^q)$. Recall that $q \le p$ and we let $\mathcal{I}$ be the subsets $I = \{i_1 < \cdots < i_q\}$ of $\{1, \ldots, p\}$. The typical q-form on G in its standard representation is given by $\omega = \sum_{I \in \mathcal{I}} f_I dx_I \in \mathcal{F}_q(G)$. We seek to give specificity to the formula (9.1.4) for $\int_\sigma \omega$ when σ is an affine q-simplex. In this the important thing is to calculate the Jacobians involved.

The matrix of the linear transformation A has size $p \times q$. Using the explication and notation employed to obtain (9.1.5) we have that for every u in $\mathbb{R}^q$, $\sigma(u) = (\phi_1(u), \ldots, \phi_p(u)) \in \mathbb{R}^p$. Thus for $1 \le i \le p$, ϕ_i corresponds to the i-th row of the matrix of A. Since σ is an affine map, each $\phi_i : \mathbb{R}^q \to \mathbb{R}$ is an affine map. If for $I = \{i_1 < \cdots < i_q\}$ in $\mathcal{I}$ we let A_I be the $q \times q$ matrix with the same columns as A and the $i_1, \ldots, i_q$ rows we get, using the notation of (9.1.4) and (9.1.5),

$$\frac{\partial(x_{i_1}, \ldots, x_{i_q})}{\partial(u_1, \ldots, u_q)} = \det A_I$$

Thus (9.1.4) becomes

9.3.10
$$\int_\sigma \omega = \int_{C^q} \sum_{I \in \mathcal{I}} f_I(\sigma(u)) \det A_I \, du$$

9.3.11. Proposition. *If σ is an affine q-simplex, κ is a permutation of $0, 1, \ldots, q$, and σ_κ is as above, then for any q-form on an open set G containing $\sigma(C^q)$ we have that $\int_{\sigma_\kappa} \omega = \text{sign}(\kappa) \int_\sigma \omega$.*

Proof. Let $\sigma = [v_0, \ldots, v_q]$ so that for all u in $\mathbb{R}^q$, $\sigma(u) = v_0 + A(u)$ for some A in $\mathcal{L}(\mathbb{R}^p, \mathbb{R}^q)$. First we assume that κ is the following transposition; fix j with $1 \le j \le q$ and let $\kappa(0) = j, \kappa(j) = 0$, and $\kappa(i) = i$ when $i \neq 0, j$. So κ is a transposition and $\text{sign}(\kappa) = -1$. Hence $\sigma_\kappa(x) = v_j + B(x)$ where $B \in \mathcal{L}(\mathbb{R}^q, \mathbb{R}^p)$ and is defined by $B(e_j) = v_0 - v_j$ and $B(e_i) = v_i - v_j$ when $i \neq j$. (Pay close attention as the following argument requires it.) Note that in arriving at the formula (9.3.10) above and the definition of the $q \times q$ matrix A_I, the columns of A were not disturbed and the choice of I in $\mathcal{I}$ determined the q rows. We therefore concentrate on relating the columns of B to those of A so we can establish that $\det B_I = -\det A_I$ for each I in $\mathcal{I}$.

The columns of A are the vectors $x_i = A(e_i) = v_i - v_0$ for $1 \leq i \leq q$. The columns of B are $v_1 - v_j, \ldots, v_{j-1} - v_j, -v_j, v_{j+1} - v_j, \ldots, v_q - v_j = x_1 - x_j, \ldots, x_{j-1} - x_j, -x_j, x_{j+1} - x_j, \ldots, x_q - x_j$. Now form the matrix $\overline{B}$ by subtracting the j-th column of B from each of the others. Thus the columns of $\overline{B}$ are $x_1, \ldots, x_{j-1}, -x_j, x_{j+1}, \ldots, x_q$. Note that for each I in $\mathcal{I}$, $\det \overline{B}_I = \det B_I$. But the columns of $\overline{B}$ are identical to those of A except for the j-th, which is the negative of that column in A. Therefore we have that $\det \overline{B}_I = -\det A_I$ and we therefore have that $\int_{\sigma_\sigma} \omega = -\int_\sigma \omega$.

The case where σ is any transposition is similar, though the notation is more cumbersome. Since every permutation σ of $\{1, \ldots, p\}$ is the composition of transpositions, the proposition follows by combining this result for transpositions with Proposition 9.2.3. ∎

The plot thickens.

In fact I'm afraid the plot gets quite thick with an exceptionally high concentration of definitions before we get to the main result of this section.

9.3.12. Definition. Let G be an open subset of $\mathbb{R}^p$. An *affine q-chain* Γ in G is a finite collection $\Gamma = \{\sigma_1, \ldots, \sigma_r\}$ of oriented affine q-simplexes in G. An affine q-simplex may appear more that once in Γ and the number of times it is repeated is called its *multiplicity*. If $\omega \in \mathcal{S}_q(G)$, then we write

$$\int_\Gamma \omega = \sum_{i=1}^r \int_{\sigma_i} \omega$$

The *trace* of Γ is defined as

$$\{\Gamma\} = \bigcup_{i=1}^r \{\sigma_i\}$$

We define

$$\Gamma = \sigma_1 + \cdots + \sigma_r = \sum_{i=1}^r \sigma_i$$

This is the definition of addition on the set of affine q-simplexes in an open subset G of $\mathbb{R}^p$. In fact just as we defined q-forms as a certain type of function on the set of q-surfaces, each affine q-simplex σ in G can be thought of as a function defined on the set $\mathcal{F}_q(G)$, where $\sigma(\omega) = \int_\sigma \omega$ for all ω in $\mathcal{F}_q(G)$. If σ and τ are two such affine q-simplexes, then $\Gamma = \{\sigma, \tau\}$ is an affine q-chain in G and for each ω in $\mathcal{F}_q(G)$ we define

9.3.13 $$(\sigma + \tau)(\omega) = \int_{\{\sigma,\tau\}} \omega = \int_\Gamma \omega = \int_\sigma \omega + \int_\tau \omega = \sigma(\omega) + \tau(\omega)$$

Notice that Proposition 9.3.11 says that when σ is an oriented affine q-simplex, we have a negative for this addition. Indeed if κ is an odd permutation of $0, 1, \ldots, q$, then in this definition of addition $\sigma_\kappa = -\sigma$. That is $\sigma + \sigma_\kappa = 0$.

I think the following examples hint at the usefulness of introducing affine q-chains.

9.3.14. Example. (a) The unit square $I^2 = [0, 1] \times [0, 1]$ in $\mathbb{R}^2$ is the trace of a 2-chain. In fact the four vertices of I^2 are $0, e_1, e_1 + e_2$, and e_2. Define σ, τ on C^2 by $\sigma(u) = u$ and $\tau(u) = e_1 + e_2 - u$ for all u in C^2. So these are the affine simplexes $\sigma = [0, e_1, e_2]$ and $\tau = [e_1 + e_2, e_2, e_1]$. We have that $I^2 = \{\sigma\} \cup \{\tau\}$. We might point out that the directions given the line segment joining the vertices e_1 and e_2 by σ and τ are the opposite of one another.

(b) Generalizing part (a), the unit cube in $\mathbb{R}^q$, I^q, is the trace of a q-chain in $\mathbb{R}^q$. This follows as in part (a). Note that if $X_1 = C_q$ and $X_2 = \{(a_1, \ldots, a_q) \in I^q : a_1 + \cdots + a_q \geq q - 1\}$, then $X_1 \cup X_2 = I^q$ and $X_1 \cap X_2 = \{a \in I_q : a_1 + \cdots + a_q = 1\}$ (Exercise 6). Define σ and τ on C^q by $\sigma(u) = u$ and $\tau(u) = e_1 + \cdots + e_q - u$. So $\sigma = [0, e_1, \ldots, e_q]$ and a calculation reveals that $\tau(e_j) = e_1 + \cdots + e_{j-1} + e_{j+1} + \cdots + e_q$. Hence

$$\tau = [(e_1 + \cdots + e_q), (e_2 + \cdots + e_q), \ldots, (e_1 + \cdots + e_{q-1})]$$

Thus $I^q = \{\sigma\} \cup \{\tau\}$, the trace of a q-chain.

9.3.15. Definition. If $\sigma = [v_0, \ldots, v_q]$ is an oriented affine q-simplex, its *boundary* is defined to be the affine (q − 1)-chain

$$\partial\sigma = \sum_{j=0}^{q}(-1)^j[v_0, \ldots, v_{j-1}, v_{j+1}, \ldots, v_q]$$

So once again we are using the addition of affine simplexes with emphasis on their orientation. Let's call attention to the ambiguity in the notation: the use of the symbol ∂ in $\partial\sigma$. The trace of σ, $\{\sigma\}$, has no interior if $q < p$. Hence the topological boundary of $\{\sigma\}$ is the trace itself. However the trace of $\partial\sigma$ consists of the image under σ of the topological boundary of C^q.

9.3.16. Example. (a) If σ is the 1-simplex $\sigma = [v_0, v_1]$ in $\mathbb{R}^p$, then $\partial\sigma = v_1 - v_0$, where this is the difference of two 0-simplexes.

(b) If $\sigma = [v_0, v_1, v_2]$, an affine 2-simplex, then $\partial\sigma = [v_1, v_2] - [v_0, v_2] + [v_0, v_1]$. Note that $\{\sigma\}$ is a triangle in $\mathbb{R}^p$ and this formulation of its boundary gives its customary counterclockwise (positive) orientation.

(c) Using the notation in the preceding definition, let $\sigma_0 = [v_1, \ldots, v_q]$. So σ_0 is an affine (q − 1)-simplex defined on the unit simplex $C^{q-1} \subseteq \mathbb{R}^{q-1}$ by $\sigma_0(u) = v_1 + A_0(u)$, where A_0 is the linear transformation from $\mathbb{R}^{q-1}$ into $\mathbb{R}^p$ defined by $A_0(e_i) = v_{i+1} - v_1$ for $1 \leq i \leq q-1$. Similarly for $1 \leq j \leq q$ if $\sigma_j = [v_0, \ldots, v_{j-1}, v_{j+1}, \ldots, v_q]$ as in Definition 9.3.15, then σ_j is an affine (q − 1)-simplex defined on the unit simplex $C^{q-1} \subseteq \mathbb{R}^{q-1}$ by $\sigma_j(u) = v_0 + A_j(u)$, where

$A_j : \mathbb{R}^{q-1} \to \mathbb{R}^p$ is the linear transformation defined by $A_j(e_i) = v_i - v_0$ when $1 \le i \le j - 1$ and $A_j(e_i) = v_{i+1} - v_0$ when $j + 1 \le i \le q - 1$.

The exploration of the idea of a q-simplex and affine q-chains was preliminary to the next concept and that in Definition 9.3.18 that follows. The idea is to first take a q-simplex in $\mathbb{R}^p$ and then apply a C'' map to it.

9.3.17. Definition. If H is an open subset of $\mathbb{R}^d$, an *oriented q-simplex of class C''* in H is a q-surface (Φ, C^q) in H that arises as follows. There is an open subset G in $\mathbb{R}^p$, a C'' mapping $T : G \to H$, and there is an oriented affine q-simplex $\sigma : C^q \to G$ such that

$$\Phi = T \circ \sigma = T\sigma$$

The *boundary* of Φ is defined by

$$\partial\Phi = T \circ \partial\sigma = T(\partial\sigma)$$

Let's note a few things. First, $T\sigma$ is just a shorthand notation for $T \circ \sigma$; similarly for $T(\partial\sigma)$. Second, the surface domain of Φ is the unit q-simplex C^q. Next Φ is usually not an affine q-simplex, hence the word "affine" does not appear in its name. What we have done is to introduce a collection of q-surfaces in H that we can orient by using the orientation of an affine q-simplex.

9.3.18. Definition. If H is an open subset of $\mathbb{R}^d$, an *oriented q-chain of class C''* in H is a finite set $\Psi = \{\Phi_1, \ldots, \Phi_r\}$ of oriented q-simplexes of class C'' on H. If $\omega \in \mathcal{F}_q(H)$, then

$$\int_\Psi \omega = \sum_{i=1}^r \int_{\Phi_i} \omega$$

and we use the notation $\Psi = \sum_{i=1}^r \Phi_i$. The *boundary* of Ψ is defined as

$$\partial\Psi = \sum_{i=1}^r \partial\Phi_i$$

Note that in the notation $\Psi = \sum_{i=1}^r \Phi_i$ we are introducing an addition for oriented q-simplexes of class C''. In fact we have begun to consider a duality between q-simplexes of class C'' and differential forms. We defined a q-form as a certain type of function from surfaces into $\mathbb{R}$. We also want to think of surfaces and oriented q-chains of class C'' as functions from q-forms into $\mathbb{R}$. So if $\Psi = \{\Phi_1, \ldots, \Phi_r\}$ as in the preceding definition, we can consider the function $\Psi : \mathcal{F}_q(H) \to \mathbb{R}$ defined by

$$\Psi(\omega) = \int_\Psi \omega$$

So when we say that

$$\Psi = 0$$

we mean the value of every q-form at Ψ is 0. Let's also point out that in some ways an oriented q-chain is a generalization of a piecewise regular surface (8.4.13) to surfaces in $\mathbb{R}^p$. (In which ways is it a generalization and in which ways is it not?) If we look at Example 9.3.14 above we see a hint of this.

9.3.19. Example. Again let G be an open subset of $\mathbb{R}^p$, H an open subset of $\mathbb{R}^d$, and $T : G \to H$ a C'' mapping. If $\Gamma = \sum_{i=1}^{r} \sigma_i$ is an affine q-chain in G, and $\Phi_i = T\sigma_i$ for $1 \le i \le r$, we will write

$$\Psi = \sum_{i=1}^{r} \Phi_i = T\left(\sum_{i=1}^{r} \sigma_i\right) = \sum_{i=1}^{r} T(\sigma_i)$$

and

$$\partial\Psi = \sum_{i=1}^{r} T(\partial\sigma_i)$$

9.3.20. Example. Here we want to show how some of the examples seen in the preceding chapter fit into the current landscape. Since many of those examples had surface domains that were rectangles (see Exercise 8.3.2), Example 9.3.14 will be useful. For example let's see how a 2-surface whose surface domain is the square I^2 can be handled. We start by assuming the surface is (T, I^2), T is C'', and this surface lies in an open subset H of $\mathbb{R}^d$. So there is an open subset G of $\mathbb{R}^2$ that contains I^2 and on which T is C''. Adopt the notation in Example 9.3.14(a) and let $X_1 = \sigma(C^2)$ and $X_2 = \tau(C^2)$ so that $X_1 \cup X_2 = I^2$. Write $G = G_1 \cup G_2$, where G_k is open and contains X_k for $k = 1, 2$. If $\Phi_1 = T\sigma : C^2 \to H$ and $\Phi_2 = T\tau : C^2 \to H$, then Φ_1 and Φ_2 are oriented 2-simplexes and $\Psi = \{\Phi_1, \Phi_2\}$ is an oriented q-chain of class C''. Observe that $\partial\Phi_1 = T(\partial\sigma)$ and $\partial\Phi_2 = T(\partial\tau)$.

Now suppose we are given a rectangle $R = [a, b] \times [c, d]$ and a surface (S, R) of class C'' that lies in the open subset $H \subseteq \mathbb{R}^d$. If we define $A : \mathbb{R}^2 \to \mathbb{R}^2$ by $A(x, y) = ((1 - x)a + xb, (1 - y)c + yd)$, then A is a linear transformation and $A(I^2) = R$. Thus $T = S \circ A : I^2 \to H$ is of class C'' and the discussion in the preceding paragraph applies. For use in the next section when we discuss orientation, note that $\det A = (b - a)(d - c) > 0$.

Exercises

(1) (a) Prove Proposition 9.3.2. (b) Give three examples in $\mathbb{R}^3$ of compact convex sets that are not balls. (c) Show that the union of two convex sets may not be convex.
(2) Show that the set S defined in the proof of Proposition 9.3.4 is convex.
(3) Show that a function $\sigma : \mathbb{R}^q \to \mathbb{R}^p$ is an affine mapping if and only if $A : \mathbb{R}^q \to \mathbb{R}^p$, defined by $A(u) = \sigma(u) - \sigma(0)$ is linear.
(4) Prove Proposition 9.3.7.

(5) Let σ and τ be two affine q-simplexes. If we define $(\sigma \dotplus \tau)(u) = \sigma(u) + \tau(u)$ for all u in C^q, is $(\sigma \dotplus \tau)(u)$ the same as $\sigma + \tau$ as defined in 9.3.12?

(6) Supply the details needed for Example 9.3.14(b).

(7) Write out the details of Example 9.3.20 for the hemisphere in $\mathbb{R}^3$ as parametrized in Exercise 8.3.2. What can be said about the common parts of $\partial\Phi_1$ and $\partial\Phi_2$?

(8) Can you represent the 3-cube in $\mathbb{R}^3$ as the trace of a 3-chain?

9.4. Oriented Boundaries

In the preceding section we introduced the idea of an oriented q-chain of class C'', which was a finite set of functions. Now we want to use this and the natural orientation of an affine simplex to orient the boundary of certain smooth surfaces. This is similar to what we did in Definition 8.2.2 when we oriented a closed curve in $\mathbb{R}^2$ relative to the open set it surrounded and Definition 8.4.15 when we introduced a piecewise orientable surface. The first step is to orient the boundary of the unit simplex C^p for $p \geq 2$.

9.4.1. Definition. If $p \geq 2$ and σ is the identity map on C^p, then σ is an oriented p-simplex. Indeed, using the notation of the preceding section, $\sigma = [0, e_1, \ldots, e_p]$. Thus the $(p-1)$-chain $\partial\sigma$ orients the boundary of C^p. We refer to this as the *positive orientation* of ∂C^p. Whenever we use the notation ∂C^p we assume it has this positive orientation.

9.4.2. Example. (a) Recall Example 9.3.16(b) where we saw that the positive orientation orientation of the boundary of the 2-simplex C^2 is $\partial\sigma = \partial[0, e_1, e_2] = [0, e_1] + [e_1, e_2] - [0, e_2] = [0, e_1] + [e_1, e_2] + [e_2, 0]$. This gives the boundary of the simplex C^2 the direction of going from 0 to e_1 to e_2 and back to 0, the counterclockwise direction.

(b) The positively oriented boundary of $C^3 = [0, e_1, e_2, e_3]$ is $[e_1, e_2, e_3] - [0, e_2, e_3] + [0, e_1, e_3] - [0, e_1, e_2]$.

Now suppose that G is an open subset of $\mathbb{R}^p$ that contains C^p and that $\Phi : G \to \mathbb{R}^p$ is a C'' mapping that is injective on C^p and whose Jacobian is positive on int C^p. It follows from the Inverse Function Theorem that $X = \Phi(C^p) = \mathrm{cl}\,\Phi(\mathrm{int}\,C^p)$ and that $\Phi(\mathrm{int}\,C^p)$ is an open set.

9.4.3. Definition. Let G be an open subset of $\mathbb{R}^p$ that contains C^p and let $\Phi : G \to \mathbb{R}^p$ be a C'' mapping that is injective on int C^p and whose Jacobian is strictly positive there. If $X = \Phi(C^p)$, then the *positively oriented boundary* of X is the $(p-1)$-chain

$$\partial X = \partial\Phi = \Phi(\partial\sigma)$$

where σ is the identity map on C^p.

If $\Phi_i : G \to \mathbb{R}^p$ is such a map for $1 \le i \le r$, $X_i = \Phi_i(C^p)$, and int $X_i \cap$ int $X_j = \emptyset$ for $i \ne j$, put $\mathcal{X} = \bigcup_{i=1}^r X_i$. Then

$$\partial \mathcal{X} = \partial X_1 + \cdots + \partial X_r = \partial \Phi_1 + \cdots + \partial \Phi_r$$

is called the *positively oriented boundary* of $\mathcal{X}$.

Before going further we must point out that when $p = 2$, we now have two defini-tions of the positive orientation of ∂X when X is as above. Indeed if $p = 2$, $\Phi(\partial C^2)$ is a curve. Since Φ is smooth on G with a strictly positive Jacobian, this Jacobian is strictly positive in a neighborhood of C^2. Thus this curve is piecewise regular and so Definition 8.2.2 gives a concept of ∂X being positively oriented. Are these two concepts the same? Yes! Using linear algebra and some effort we can prove this. We choose, however, not to do this proof. Why? The use of these concepts of orienta-tion when $p = 2$ is, in both cases, for the statement and proof of Green's Theorem. This was done in (8.2.10) when $p = 2$ and will be extended below in Theorem 9.7.1 as a consequence of a generalization of Stokes's Theorem. In each case the value of the integrals over the boundary of the set is the same, whereas if the orienta-tions differed they would be the negative of one another. (This is established in this more general setting for $p \ge 2$ in Proposition 9.3.11; in two-space it is immediate from the definition of the integral over a curve (8.1.8).) Thus the two concepts of positively oriented curves must be the same.

> *Why have we introduced oriented q-chains and*
> *the concepts in the preceding definition?*
> *Where is all this leading?*

As we said in the preceding paragraph we are headed towards obtaining a Gener-alized Stokes's Theorem (9.5.1), for which we now have all the terminology needed to read the statement but not to begin the proof. What we will see is that this result also yields Green's Theorem (8.2.10) and Gauss's Theorem (8.5.6) as consequences. How? What is the relationship? You may recall that when we gave the versions of these results in $\mathbb{R}^2$ and $\mathbb{R}^3$ we needed to express surfaces as the union of non-overlapping surfaces of a special type. (Especially recall the introduction of a G-set in connection with Green's Theorem.) In defining sets of the type $\mathcal{X}$ in the last def-inition we have extended and made precise the concept of a G-set in higher dimen-sions. Realize that C^2 is both a Type I and a Type II set, while C^3 is simultaneously Type I, II, and III. We saw that each of the expressions in Green's, Gauss's, and Stokes's Theorems are examples of differential forms. Now we must explore the interaction between these extended notions and arbitrary differential forms.

9.4.4. Proposition. *Let G be an open subset of $\mathbb{R}^p$ that contains C^p and for $i = 1, 2$ let $\Phi_i : G \to \mathbb{R}^p$ be a C'' mapping that is injective on a neighborhood of C^p and whose Jacobian is strictly positive on int C^p. If $\Phi_1(C^p) = X = \Phi_2(C^p)$ and if ω is a $(p-1)$-form defined on some open set containing X, then $\int_{\partial \Phi_1} \omega = \int_{\partial \Phi_2} \omega$.*

The complicated proof of this proposition, which won't be used in the rest of this chapter, is left to the interested reader. Here is an outline of the proof. Let H be an open subset of $\mathbb{R}^p$ that contains $\Phi_1(C^p) = \Phi_2(C^p)$ and on which Φ_2^{-1} is defined. Let $T : H \to G$ be defined by $T = \Phi_1 \circ \Phi_2^{-1}$. Note that $T(\partial\Phi_2) = \partial\Phi_1$. By Theorem 9.2.6 we have

$$\int_{\partial\Phi_1} \omega = \int_{T\circ\partial\Phi_2} \omega = \int_{\partial\Phi_2} \omega_T$$

Now we need to use the properties of T and the definition of ω_T to show that $\int_{\partial\Phi_2} \omega_T = \int_{\partial\Phi_2} \omega$, and here is where the complications arise.

9.4.5. Example. (a) Recall Example 9.3.14 on the unit square $I^2 = [0, 1] \times [0, 1]$ in $\mathbb{R}^2$, where we defined $\sigma_1 = [0, e_1, e_2]$ and $\sigma_2 = [e_1 + e_2, e_2, e_1]$ and observed that $I^2 = \{\sigma_1\} \cup \{\sigma_2\}$. Therefore

$$\partial\sigma_1 = [e_1, e_2] - [0, e_2] + [0, e_1]$$
$$\partial\sigma_2 = [e_2, e_1] - [e_1 + e_2, e_1] - [e_1 + e_2, e_2]$$

When we add these boundaries together we get $[e_1, e_2] + [e_2, e_1] = 0$ so that

$$\partial I^2 = [0, e_1] + [e_1, e_1 + e_2] + [e_1 + e_2, e_2] + [e_2, 0]$$

and this orients the boundary of I^2 in the counterclockwise direction.

(b) This is actually a continuation of the preceding part. Assume (Φ, I^2) is a 2-surface with surface domain I^2. As a function on 2-forms we have $\Phi = \Phi \circ \sigma_1 + \Phi \circ \sigma_2$. Thus $\partial\Phi = \partial(\Phi \circ \sigma_1) + \partial(\Phi \circ \sigma_2) = \Phi(\partial\sigma_1) + \Phi(\partial\sigma_2) = \Phi(\partial I^2)$.

(c) We continue with the preceding parts. Recall from §8.2 where we defined a subset $\mathcal{X}$ to be Type I. Let $f_1, f_2 : [a, b] \to \mathbb{R}$ be C'' functions such that $f_1(x) < f_2(x)$ on (a, b) and set $\mathcal{X} = \{(x, y) \in \mathbb{R}^2 : a \le x \le b, f_1(x) \le y \le f_2(x)\}$. Assume that $\Phi : I^2 \to \mathcal{X}$ is defined by

$$\Phi(u, v) = ((1 - u)a + ub, (1 - v)f_1((1 - u)a + ub) + vf_2((1 - u)a + ub))$$
$$= (\phi(u, v), \psi(u, v))$$

It is easy to check that Φ is injective on int I^2.

We have that

$$\phi_u(u, v) = b - a$$
$$\phi_v(u, v) = 0$$
$$\psi_u(u, v) = (b - a)[(1 - v)f_1'((1 - u)a + ub) + vf_2'((1 - u)a + ub)]$$
$$\psi_v(u, v) = f_2((1 - u)a + ub) - f_1((1 - u)a + ub)$$

A further computation shows that

$$\frac{\partial(\phi, \psi)}{\partial(u, v)} = (b - a)[f_2((1 - u)a + ub) - f_1((1 - u)a + ub)] > 0$$

on int I^2. So if we let $T_i : C^2 \to \mathbb{R}^2$ be defined by $T_i = \Phi \circ \sigma_i$, we have the situation of Definition 9.4.3.

9.4.6. Example. Let $\mathcal{X} = [0, \pi] \times [0, 2\pi]$ and define $\Phi : \mathcal{X} \to \mathbb{R}^3$ by

$$\Phi(u, v) = (\sin u \cos v, \sin u \sin v, \cos u)$$

The reader might note that $\{\Phi\}$ is the unit sphere in $\mathbb{R}^3$. Here $\partial \Phi = \Phi(\partial \mathcal{X}) = \gamma_1 + \gamma_2 + \gamma_3 + \gamma_4$, where these are the four curves in $\mathbb{R}^3$ defined on $[0, \pi]$, $[0, 2\pi]$, $[0, \pi]$, $[0, 2\pi]$, respectively, by the formulas

$$\gamma_1(u) = \Phi(u, 0) = (\sin u, 0, \cos u)$$
$$\gamma_2(v) = \Phi(\pi, v) = (0, 0, -1)$$
$$\gamma_3(u) = \Phi(\pi - u, 2\pi) = (\sin u, 0, -\cos u)$$
$$\gamma_4(v) = \Phi(0, 2\pi - v) = (0, 0, 1)$$

Let ω be any 1-form defined in a neighborhood of $\{\Phi\}$. Since γ_2 and γ_4 are constant we have that $\int_{\gamma_2} \omega = \int_{\gamma_4} \omega = 0$. Observing that $\gamma_3(u) = \gamma_1(\pi - u)$, it follows by an application of Example 9.1.6(a) that $\int_{\gamma_3} \omega = -\int_{\gamma_1} \omega$. Therefore we have that

$$\int_{\partial \Phi} \omega = 0$$

for any 1-form. Equivalently, $\partial \Phi = 0$ (as a function on forms).

Exercises

(1) Carry out an analysis similar to that in Examples 9.4.5(a) and (b) for C^3.
(2) Repeat Example 9.4.6 but this time for a parametrization of the unit hemisphere above the xy-plane.

9.5. Stokes's Theorem

Here is the main result of this chapter, whose proof will occupy us for the rest of the section.

9.5.1. Theorem (Generalized Stokes's Theorem). *If H is an open subset of $\mathbb{R}^p$ and Ψ is a q-chain of class C'' in H, then for every smooth $(q-1)$-form ω in H,*

$$\int_{\Psi} d\omega = \int_{\partial \Psi} \omega$$

In §9.7 we'll see how the theorems of Green, Gauss, and Stokes from Chapter 8 follow from this theorem, but now let's point out that the above result extends the Fundamental Theorem of Calculus. (You might want to look at Example 9.1.6(a).) In Theorem 9.5.1 take $q = 1$ and let ω be the 0-form defined by the smooth function $f : [a, b] \to \mathbb{R}$. For H take any open set that contains $[a, b]$ and for the 1-chain let $\Psi : [a, b] \to H \subseteq \mathbb{R}$ be the identity map, $\Psi(x) = x$. Consequently $\int_\Psi d\omega = \int_a^b f'(x)\, dx$ and $\int_{\partial\Psi} \omega = f(b) - f(a)$; Theorem 9.5.1 says they are equal, thus establishing the FTC. (Of course the number of times the FTC was used to get to this point is probably large, so this is not an independent proof. In fact the FTC is used directly in the proof of Theorem 9.5.1 below.)

Proof of The Generalized Stokes's Theorem. To begin we show that what has to be proved can be reduced to a simpler environment. To begin note that it suffices to show that $\int_\Phi d\omega = \int_{\partial\Phi} \omega$ when Φ is a single d-simplex of class C''. By definition $\Phi = T \circ \sigma$ where $\sigma = [0, v_1, \ldots, v_q]$ defined on C^q and $T : G \to H \subseteq \mathbb{R}^p$ is a mapping of class C'' on an open subset G of $\mathbb{R}^q$ that contains $\sigma(C^q)$. By the rules for differentiating forms (9.2.2(c)) and the Change of Variables Theorem for forms (9.2.6) we have that

$$\int_\Phi d\omega = \int_{T\sigma} d\omega = \int_\sigma (d\omega)_T = \int_\sigma d(\omega_T)$$

Since $\partial\Phi = \partial(T\sigma) = T(\partial\sigma)$, we also have that

$$\int_{\partial\Phi} \omega = \int_{T(\partial\sigma)} \omega = \int_{\partial\sigma} \omega_T$$

This means that to prove the theorem we need only prove

9.5.2
$$\int_\sigma d\zeta = \int_{\partial\sigma} \zeta$$

for every smooth (q − 1)-form ζ.

As we saw prior to the proof, if $q = 1$ this amounts to the Fundamental Theorem of Calculus. So we can assume that $q \geq 2$. Suppose $1 \leq r \leq q$ and f is a smooth function on G. It suffices to establish (9.5.2) when

$$\zeta = f\, dx_1 \wedge \cdots \wedge dx_{r-1} \wedge dx_{r+1} \wedge \cdots \wedge dx_q$$

since the arbitrary (q − 1)-form is the sum of such forms as this. Now

$$\partial\sigma = [v_1, \ldots, v_q] + \sum_{i=1}^{q} (-1)^i \tau_i$$

where $\tau_i = [0, v_1, \ldots, v_{i-1}, v_{i+1}, \ldots, v_q]$ for $1 \leq i \leq q$.

We now need to do some tedious calculations. Put

$$\tau_0 = [v_r, v_1, \ldots, v_{r-1}, v_{r+1}, \ldots, v_q]$$

and observe that τ_0 can be obtained from $[v_1, \ldots, v_q]$ by $r - 1$ successive interchanges with its left-hand neighbor. Hence

$$\partial\sigma = (-1)^{r-1}\tau_0 + \sum_{i=1}^{q} (-1)^i \tau_i$$

Recall that for $0 \le i \le q$ each τ_i is defined on C^{q-1}. Consider $i = 0$ and fix $u = (u_1, \ldots, u_{q-1})$ in C^{q-1}. We have from (9.3.9) that

$$\tau_0(u) = v_r + \sum_{i=1}^{r-1} u_i(v_i - v_r) + u_r(v_{r+1} - v_r) + \cdots + u_{q-1}(v_{q-1} - v_r)$$

Thus if $u \in C^{q-1}$ and we put $x = \tau_0(u) \in \mathbb{R}^q$, then

9.5.3
$$x_j = \begin{cases} u_j & (1 \le j < r) \\ 1 - (u_1 + \cdots + u_{q-1}) & (j = r) \\ u_{j-1} & (r < j \le q) \end{cases}$$

Similarly if $1 \le i \le q$ and $x = \tau_i(u)$ for some u in C^{q-1}, then

9.5.4
$$x_j = \begin{cases} u_j & (1 \le j < i) \\ 0 & (j = i) \\ u_{j-1} & (i < j \le q) \end{cases}$$

For $0 \le i \le q$ let J_i be the Jacobian of the map

$$(u_1, \ldots, u_{q-1}) \mapsto (x_1, \ldots, x_{r-1}, x_{r+1}, \ldots, x_q)$$

as defined above using the simplexes τ_i. Observe that when $i = 0$ or $i = r$, this is the identity map. (Verify!) Hence $J_0 = 1 = J_r$. When $i \ne 0, r$, then the fact that the above formula for x has $x_i = 0$ implies that when we compute the determinant J_i, its associated square matrix has a row of zeros. Thus $J_i = 0$ when $i \ne 0, r$. From the definition of the action of ζ this implies that $\int_{\tau_i} \zeta = 0$ when $i \ne 0, r$. Therefore

$$\int_{\partial\sigma} \zeta = (-1)^{r-1} \int_{\tau_0} \zeta + \sum_{i=1}^{q} (-1)^i \int_{\tau_i} \zeta$$
$$= (-1)^{r-1} \int_{\tau_0} \zeta + (-1)^r \int_{\tau_r} \zeta$$
$$= (-1)^{r-1} \int_{C^{q-1}} [f(\tau_0(u)) - f(\tau_r(u))]\, du$$

But we also have that

$$d\zeta = (D_r f)(x)\, dx_r \wedge dx_1 \wedge \cdots \wedge dx_{r-1} \wedge dx_{r+1} \wedge \cdots \wedge dx_q$$
$$= (-1)^{r-1}(D_r f)(x)\, dx_1 \wedge \cdots \wedge dx_q$$

so that

$$\int_\sigma d\zeta = (-1)^{r-1} \int_{C^q} (D_r f)(x)\, dx$$

To evaluate this last integral we use Fubini's Theorem and first integrate with respect to x_r over the interval $[0, 1 - (x_1 + \cdots + x_{r-1} + x_{r+1} + \cdots + x_q)] = [0, 1 - y_r]$.

Doing this we get

$$\int_0^{1-y_r} (D_r f)(x_1, \ldots, x_q)\, dx_r = f(x_1, \ldots, x_{r-1}, 1 - y_r, x_{r+1}, \ldots, x_q)$$
$$- f(x_1, \ldots, x_{r-1}, 0, x_{r+1}, \ldots, x_q)$$

Using (9.5.3) we see that

$$y_r = x_1 + \cdots + x_{r-1} + x_{r+1} + \cdots + x_q = u_1 + \cdots + u_{q-1}$$

Hence

$$f(x_1, \ldots, x_{r-1}, 1 - y_r, x_{r+1}, \ldots, x_q) = f(\tau_0(u))$$

Now using (9.5.4) we have that

$$f(x_1, \ldots, x_{r-1}, 0, x_{r+1}, \ldots, x_q) = f(\tau_r(u))$$

Therefore

$$\int_\sigma d\zeta = (-1)^{r-1} = \int_{C^{q-1}} [f(\tau_0(u)) - f(\tau_r(u))]\, du = \int_{\partial\sigma} \zeta$$

completing the proof. ∎

9.6. Closed and Exact Forms*

In $\mathbb{R}$ every continuous function has a primitive by the FTC. Namely, if f is a continuous function on some interval (a, b) that contains the point c, then $F(x) = \int_c^x f$ has the property that $F'(x) = f(x)$ for all x in (a, b). In this section we will be concerned with an analogous problem of deciding if a form in $\mathbb{R}^p$ has a primitive. This is not always the case, as we will soon see. Here is the basic language.

9.6.1. Definition. Let G be an open subset of $\mathbb{R}^p$ and let ω be a q-form on G. The form ω is *closed* if it is smooth and $d\omega = 0$; ω is *exact* if there is a smooth $(q-1)$-form η on G such that $\omega = d\eta$.

We know from Theorem 9.1.18 that if ω is smooth and exact, then it is closed. When $p > 1$ there are closed forms that are not exact, and the analogous problem referred to in the introduction to this section is which closed forms are exact. Before giving an example let's establish a few facts.

9.6.2. Proposition. *Let G be an open subset of $\mathbb{R}^p$ and let ω be a C'' q-form on G.*

(a) *If $q = 1$ and $\omega = \sum_{j=1}^p f_j dx_j$, then ω is closed if and only if $D_i f_j = D_j f_i$ for all i, j.*
(b) *If ω is closed and Φ is a $(q+1)$-chain of class C'', then $\int_{\partial\Phi} \omega = 0$.*
(c) *If ω is exact and Φ and Ψ are two q-chains in G with $\partial\Phi = \partial\Psi$, then $\int_\Phi \omega = \int_\Psi \omega$. Consequently, if $\partial\Phi = 0$, then $\int_\Phi \omega = 0$ for every exact form ω.*

Proof. (a) This is straightforward. Note that

$$d\omega = \sum_{j=1}^{p}(df_j) \wedge dx_j$$

$$= \sum_{j=1}^{p}\left(\sum_{i=1}^{p}(D_i f_j) \wedge dx_i\right) \wedge dx_j$$

$$= \sum_{i<j}(D_i f_j - D_j f_i)dx_i \wedge dx_j$$

(b) This is a direct application of Stokes's Theorem (9.5.1). (How is Example 9.1.6(e) related to this part?)

(c) Let $\omega = d\eta$. Again we apply Stokes's Theorem to get that $\int_\Phi \omega = \int_\Phi d\eta = \int_{\partial\Phi} \eta = \int_{\partial\Psi} \eta = \int_\Psi \omega$. ∎

Note that showing that a form ω is exact means solving several partial differential equations simultaneously. For example if we have a 1-form $\omega = \sum_{j=1}^{p} f_j dx_j$ on $G \subseteq \mathbb{R}^p$, then to show that ω is exact we must find a smooth function $g : G \to \mathbb{R}$ such that $\omega = dg$; this means that $D_j g = f_j$ for $1 \leq j \leq p$. When ω is a q-form and $q \geq 2$, the differential equations that are to be solved are more complex. See Exercise 1.

9.6.3. Example. If G is an open subset of $\mathbb{R}^2$, note that when γ is a smooth curve in G we have that γ is a 1-chain in G. If γ is a closed curve, then $\partial\gamma = 0$. Thus if ω is an exact 1-form on G it must be that $\int_\gamma \omega = 0$ by part (a) of the preceding proposition. Let G be the punctured plane $\mathbb{R}^2\backslash\{(0, 0)\}$. Define ω on G by

$$\omega = \frac{x\,dy - y\,dx}{x^2 + y^2}$$

It is left to the reader to verify that ω is a closed form. Now define $\gamma : [0, 2\pi] \to G$ by $\gamma(\theta) = (\cos\theta, \sin\theta)$. So γ is the unit circle in the positive direction. The reader can verify that $\int_\gamma \omega = 2\pi$ (Exercise 2). By what we said at the start of this example, ω cannot be exact. The impediment to exactness of closed forms is that puncture at the origin.

Proposition 9.6.2(c) gives a necessary condition for a form to be exact. We want to discover sufficient conditions on an open subset G of $\mathbb{R}^p$ such that every closed form on G is exact.

At the beginning of §9.3 we discussed convex subsets of $\mathbb{R}^p$. We revisit this topic here. The proof of the next proposition is left to the reader (Exercise 4).

9.6.4. Proposition. (a) *Any ball in $\mathbb{R}^p$ is convex as is any affine hyperplane or half-space.*
(b) *If X is a convex subset of $\mathbb{R}^p$ and $T \in \mathcal{L}(\mathbb{R}^p, \mathbb{R}^q)$, then $T(X)$ is a convex subset of $\mathbb{R}^q$.*

(c) *If Y is a convex subset of $\mathbb{R}^q$ and $T \in \mathcal{L}(\mathbb{R}^p, \mathbb{R}^q)$, then $T^{-1}(Y)$ is a convex subset of $\mathbb{R}^p$.*

Note that the punctured plane is not convex.

It holds that whenever G is an open convex subset of $\mathbb{R}^p$, every closed form on G is exact. To establish this would involve more background than we deem appropriate, because the argument is very involved and after all the effort convexity is not the most general hypothesis we can impose on G to have the same conclusion. Instead we will focus on $\mathbb{R}^2$, where the proof is straightforward but not trivial. A proof of the convex case in $\mathbb{R}^p$ can be found in [7], page 278. A statement and proof of the more general result can be found in [9], page 457.

If α and β are two points in $\mathbb{R}^2$, let $[\alpha, \beta]$ denote the straight line segment from α to β. We begin the proof for $\mathbb{R}^2$ with a lemma.

9.6.5. Lemma. *Let G be an open disk in $\mathbb{R}^2$, let $(a, b) \in G$, and let $\eta = P\,dx + Q\,dy$ be a smooth closed form on G. If $(x, y) \in G$, then*

$$\int_{[(a,b),(x,y)]} P\,dx + Q\,dy = \int_a^x P(t, b)\,dt + \int_b^y Q(x, t)\,dt$$

Proof. First note that

$$\int_a^x P(t, b)\,dt = \int_{[(a,b),(x,b)]} (P\,dx + Q\,dy)$$

and $\int_b^y Q(x, t)\,dt = \int_{[(x,b),(x,y)]} (P\,dx + Q\,dy)$. Now note that the line segments $[(a, b), (x, b)]$, $[(x, b), (x, y)]$, and $[(x, y), (a, b)]$ form the sides of a triangle; denote by X this triangle together with its inside. By Green's Theorem we have that

$$\int_{\partial X} P\,dx + Q\,dy = \iint_X (D_1 Q - D_2 P)$$

But Proposition 9.6.2(a) says the condition that ω is closed is equivalent to the condition that $D_2 P = D_1 Q$. So we have that $\int_{\partial X} P\,dx + Q\,dy = 0$ and the lemma follows. ∎

9.6.6. Theorem. *If G is an open convex subset of $\mathbb{R}^2$, then every smooth closed form on G is exact.*

Proof. Assume that ω is a smooth form on G that is closed. Thus $\omega = P\,dx + Q\,dy + f\,dx \wedge dy$. So

$$0 = d\omega$$
$$= dP \wedge dx + dQ \wedge dy + df \wedge dx \wedge dy$$
$$= (D_1 Q - D_2 P)dx \wedge dy$$

Thus we see that the 1-form $\eta = P\,dx + Q\,dy$ is also closed.

9.6.7. Claim. The form η is exact.

First assume that G is an open disk with center (a, b). Define $F :\to \mathbb{R}$ by

$$F(x, y) = \int_{[(a,b),(x,y)]} P\,dx + Q\,dy$$

We claim that $dF = \eta$. To show this we must show that $D_1 F = P$ and $D_2 F = Q$. By the lemma and the FTC we have that $D_2 F(x, y) = D_2 \int_b^y Q(x, t)\,dt = Q(x, y)$. On the other hand first employing Theorem 7.5.2 about differentiating under the integral sign and then the assumption that ω is closed, we get

$$D_1 F(x, y) = D_1 \int_a^x P(t, b)\,dt + D_1 \int_b^y Q(x, t)\,dt$$

$$= P(x, b) + \int_b^y \frac{\partial Q(x, t)}{\partial x}\,dt$$

$$= P(x, b) + \int_b^y \frac{\partial P(x, t)}{\partial t}\,dt$$

$$= P(x, b) + P(x, y) - P(x, b)$$

$$= P(x, y)$$

Now assume G is any open convex set and fix a point (c, d) in G. By convexity $[(c, d), (x, y)] \subseteq G$ for every (x, y) in G. Define $F(x, y) = \int_{[(c,d),(x,y)]} P\,dx + Q\,dy$. Note that the line segment $[(c, d), (x, y)]$ can be covered by a finite number of open disks, each contained in G. By what we did before we have that $D_1 F = P$ and $D_2 F = Q$ in each of these disks (Why?). It follows that $dF = \eta$ in all of G, establishing the claim.

It remains to show that $f\,dx \wedge dy$ is exact. Let $U = \{x \in \mathbb{R} : (x, y) \in G$ for some $y\}$. Since U is the projection of G onto the first coordinate, it is open and connected. Hence it is an open interval. Fix a point c in U and define $H : G \to \mathbb{R}$ by

$$H(x, y) = \int_c^x f(t, y)\,dt$$

The FTC implies $D_1 H = f$. Hence $d(H\,dy) = dH \wedge dy = (D_1 H\,dx + D_2 H\,dy) \wedge dy = f\,dx \wedge dy$.

Letting $\lambda = H + \eta$, we get that $d\lambda = \omega$ and so ω is exact. ∎

9.6.8. Proposition. *Let G be an open subset of $\mathbb{R}^2$ such that every closed form on G is exact. If U is an open subset of $\mathbb{R}^2$ such that there is a C'' bijection $T : G \to U$, then every closed form on U is exact.*

Proof. Let ω be a closed form on U. Thus ω_T is a form on G and, by Proposition 9.2.2(c), $d(\omega_T) = (d\omega)_T = 0$. By hypothesis there is a form η on G such that $\omega_T = d\eta$. If $S = T^{-1} : U \to G$, let $\zeta = \eta_S$, which is a form on U. Using Proposition 9.2.3 we have that $d\zeta = d(\eta_S) = (d\eta)_S = (\omega_T)_S = (\omega)_{TS} = \omega$, so that ω is exact. ∎

See Exercise 5.

Exercises

(1) In $G \subseteq \mathbb{R}^3$ if ω is the 2-form $\omega = f\,dxdy + g\,dxdz + h\,dxdy$, phrase the problem of showing that ω is exact in terms of solving differential equations.

(2) Show that $\int_\gamma \omega = 2\pi$ in Example 9.6.3.

(3) Let $G = \mathbb{R}^3 \backslash \{0\}$, punctured three-space, and put

$$\omega = \frac{x\,dy \wedge dz + y\,dz \wedge dx + z\,dx \wedge dy}{(x^2 + y^2 + z^2)^{3/2}}$$

(a) Show that ω is closed. (b) Let Φ be as in Example 9.4.6 and show that $\int_\Phi \omega = 4\pi$. (c) Conclude that ω is not exact.

(4) Verify the statements in Example 9.6.4.

(5) Let U be the open half-annulus $\{(x,y) \in \mathbb{R}^2 : -2 < x < -1 \text{ or } 1 < t < 2, 0 < y < 2\}$ and show that every closed form on U is exact.

9.7. Denouement

In this section we want to derive the theorems of Green, Gauss, and Stokes in $\mathbb{R}^2$ and $\mathbb{R}^3$ as a consequence of the generalized Stokes's Theorem. We begin with Green's Theorem 8.2.10. The setup for Green's Theorem is as follows. Let G be an open subset of $\mathbb{R}^2$ that contains C^2 and for $1 \le j \le r$ let $\Phi_j : G \to \mathbb{R}^2$ be a C'' mapping that is injective on $\operatorname{int} C^2$ and whose Jacobian is positive there. Put $X_j = \Phi_j(C^2)$, and assume $\operatorname{int} X_i \cap \operatorname{int} X_j = \emptyset$ for $i \ne j$; put $\mathcal{X} = \bigcup_{j=1}^r X_j$. Give $\partial \mathcal{X}$ the positive orientation as in Definition 9.4.3. We note that each ∂X_j is a smooth closed curve in $\mathbb{R}^2$. The reader can mull over the connection between this setup and the assumption that $\mathcal{X}$ is a G-set given in (8.2.10). Establishing this connection is left to the inclined reader. (Exercise 1.) Be sure to first go over Example 9.4.5(c).

9.7.1. Theorem (Green's Theorem). *Let G be an open subset of $\mathbb{R}^2$ that contains C^2, let $\mathcal{X}$ be as described above, and let Ω be an open subset of $\mathbb{R}^2$ containing $\mathcal{X}$. If $F : \Omega \to \mathbb{R}^2$ is a smooth function with $F(x,y) = (P(x,y), Q(x,y))$, then*

$$\int_{\partial \mathcal{X}} F \cdot T\,ds = \iint_{\mathcal{X}} \left(\frac{\partial Q}{\partial x} - \frac{\partial P}{\partial y} \right)$$

Proof. Put $\omega = P\,dx + Q\,dy$. By the Generalized Stokes's Theorem (9.5.1)

$$\int_{\partial \mathcal{X}} \omega = \int_{\mathcal{X}} d\omega$$

Now

$$d\omega = (\partial_y P) \wedge dy + (\partial_x Q)) \wedge dx = \left(\frac{\partial Q}{\partial x} - \frac{\partial P}{\partial y} \right) dx \wedge dy$$

so

$$\int_{\mathcal{X}} d\omega = \iint_{\mathcal{X}} \left(\frac{\partial Q}{\partial x} - \frac{\partial P}{\partial y} \right)$$

From Example 9.1.6(b) we know that $\omega(\gamma) = \int_{\gamma} F \cdot T \, ds$ for any smooth curve γ. Given that $\partial \mathcal{X}$ consists of smooth curves, this completes the proof. ∎

Again we adopt the setup from Definition 9.4.3, this time in $\mathbb{R}^3$. Let G be an open subset of $\mathbb{R}^3$ that contains C^3 and for $1 \le j \le r$ let $\Phi_j : G \to \mathbb{R}^3$ be a C'' mapping that is injective on int C^3 and whose Jacobian is positive there. Also assume that with $X_j = \Phi_j(C^3)$ we have that int $X_i \cap$ int $X_j = \emptyset$ for $i \ne j$; put $\mathcal{X} = \bigcup_{j=1}^{r} X_j$. Give $\partial \mathcal{X}$ its positive orientation as in (9.4.3). Again there is a connection between this setup and the assumption in Theorem 8.5.6 that $\mathcal{X}$ is a G-set; it is left to the reader to explore this connection (Exercise 2).

9.7.2. Theorem (Gauss's Theorem). *Let G be an open set in $\mathbb{R}^3$ containing C^3, let $\mathcal{X}$ be as described above, and let Ω be an open subset of $\mathbb{R}^3$ containing $\mathcal{X}$. If $F : \Omega \to \mathbb{R}^3$ is a smooth function, then*

$$\iint_{\partial \mathcal{X}} F \cdot \mathbf{n} \, d\sigma = \iiint_{\mathcal{X}} \text{div } F$$

Proof. Let

$$F(x, y, z) = (P(x, y, z), Q(x, y, z), R(x, y, z))$$

In Example 9.1.17(b) we saw that for the 2-form $\omega = P dy \wedge dz + Q dz \wedge dx + R dx \wedge dy$,

$$d\omega = (\text{div } F) \, dx \wedge dy \wedge dz$$

So using (8.4.7) and the generalized Stokes's Theorem we have that

$$\iint_{\partial \mathcal{X}} F \cdot \mathbf{n} \, d\sigma = \iint_{\partial \mathcal{X}} P \, dy dz + \iint_{\partial X} Q \, dz dx + \iint_{\partial X} R \, dx dy$$

$$= \int_{\partial \mathcal{X}} \omega$$

$$= \int_{\mathcal{X}} d\omega$$

$$= \iiint_{\mathcal{X}} \text{div } F$$ ∎

Coordinating the statement in the theorem below with Theorem 8.5.10 is up to the reader.

9.7.3. Theorem (Stokes's Theorem). *Let Ω be an open subset of $\mathbb{R}^3$ and let Φ be a regular 2-surface contained in Ω. If $F : \Omega \to \mathbb{R}^3$ is a smooth function, then*

$$\int_{\partial \Phi} F \cdot T \, ds = \iint_{\Phi} \text{curl } F \cdot \mathbf{n} \, d\sigma$$

Proof. Put $F(x, y, z) = (P(x, y, z), Q(x, y, z), R(x, y, z))$. For efficiency in expression we write curl $F = C_1(F)e_1 + C_2(F)e_2 + C_3(F)e_3$. Let $\omega = Pdx + Qdy + Rdz$ and $\eta = C_1(F)dy \wedge dz + C_2(F)dz \wedge dx + C_3(F)dx \wedge dy$. As in Example 9.1.17(c), we have that $\eta = d\omega$. From Example 9.1.6(b) applied to a chain of curves rather than one, we know that

$$\int_{\partial \Phi} \omega = \int_{\partial \Phi} F \cdot T \, ds$$

From (8.4.7) we know that

$$\int_{\Phi} d\omega = \int_{\Phi} \eta = \iint_{\Phi} \text{curl } F \cdot \mathbf{n} \, d\sigma$$

Therefore Stokes's Theorem (8.5.10) follows from the generalized version. ■

Exercises

(1) This is a rather involved exercise. Can you show that a G-set (8.2.6) in $\mathbb{R}^2$ is an oriented q-chain for an appropriate value of q?

(2) This is also a rather involved exercise. Can you show that a G-set (8.5.4) in $\mathbb{R}^3$ is an oriented q-chain for an appropriate value of q?

Bibliography

[1] K. Bryan, "Differential Forms," www.rose-hulman.edu/~bryan/lottamath/difform.pdf

[2] P. Cameron, "Ten Chapters of the Algebraic Art," www.maths.qmul.ac.uk/~pjc/notes/intalg.pdf

[3] J. B. Conway, *A Course in Abstract Analysis*, AMS (2012).

[4] J. B. Conway, *A Course in Point Set Topology*, Springer–Verlag (2013).

[5] K. Hoffman and R. Kunze, *Linear Algebra*, Prentice Hall Publishing (1971).

[6] E. Landau, *Foundations of Analysis*, AMS Chelsea Publishing (2001).

[7] W. Rudin, *Principles of Mathematical Analysis*, McGraw-Hill Education; 3rd edition (1976).

[8] R. Schwartz, "Dedekind cuts," www.math.brown.edu/~res/INF/handout3.pdf

[9] J. Shurman, *Multivariable Calculus*, http://people.reed.edu/~jerry/211/vcalc.pdf

[10] R. Sjamaar, "Manifolds and differential forms," www.math.cornell.edu/~sjamaar/manifolds/manifold.pdf

[11] William F. Trench, *Introduction to Real Analysis* (2013). Books and Monographs. Book 7. http://digitalcommons.trinity.edu/mono/7

[12] William F. Trench, "The Method of Lagrange Multipliers" 2013 Available at: http://works.bepress.com/william_trench/130

[13] H. A. Thurston, *The Number System*, Dover (2012).

[14] H. Tverberg, "A proof of the Jordan Curve Theorem," *Bull. London Math. Soc.* **12** 1980, 34–38.

[15] W. R. Wade, *An Introduction to Analysis*, Pearson (2009).

Index of Terms

Index of Symbols